ADVANCES IN THE MECHANICS AND THE FLOW OF GRANULAR MATERIALS

VOLUME II

ADVANCES IN THE MECHANICS AND THE FLOW OF

GRANULAR MATERIALS

VOLUME II

Mohsen Shahinpoor, Editor
Granular Materials Research Laboratory
Clarkson College of Technology
Potsdam, New York 13676, USA

FIRST EDITION 1983

TRANS TECH PUBLICATIONS

Advances in the Mechanics and the Flow of Granular Materials
Volume II

Distributed in North America by

KARL DISTRIBUTORS
16 Bearskin Neck, Rockport, MA 01966, USA

and worldwide by

TRANS TECH PUBLICATIONS
P.O. Box 266, D-3392 Clausthal-Zellerfeld
Federal Republic of Germany

and

TRANS TECH S.A.
Trans Tech House, CH-4711 Aedermannsdorf
Switzerland

ISBN 087849-050-7 Pf/cancelled
Printed in the United States of America

PRINTED AND BOUND IN THE UNITED STATES OF AMERICA

CONTENTS

XIII – POROSITY AND PORE SIZE DURING CERAMIC PROCESSING
O. J. Whittemore

GENERAL STATISTICAL/MECHANICAL CONSIDERATION IN GRANULAR MATERIALS

XIV – STATISTICAL MECHANICS OF GRANULAR MATERIALS
B. F. Backman, C. B. Brown, P. W. Jowitt, and J. Munro

XV – MACROSCOPIC AND MICROSCOPIC DESCRIPTIONS OF THE MECHANICS OF GRANULAR MATERIALS
Ken-Ichi Kanatani

XVI – APPLICATION OF MARKOV PROCESS TO MECHANICS OF SOIL AND GRANULAR MATERIAL
R. Kitamura

XVII – SOME RECENT RESULTS ON THE BEHAVIOR OF GRANULAR MATERIALS
M. Shahinpoor

ELECTROMAGNETIC, THERMAL, AND GENERAL TRANSPORT PROPERTIES OF GRANULAR MATERIALS

XVIII – MECHANICS OF SEPARATING PARTICULATE SOLIDS IN LIQUID DIELECTRICS AND MAGNETICS
U. Ts. Andres

XIX – DYNAMICS OF AN ELECTRIC (PARTICULATE) SUSPENSION
G. M. Colver

XX – RHEOLOGICAL AND MAGNETIC PROPERTIES IN RECORDING MATERIALS
A. R. Corradi

GRANULAR MATERIALS RAPID FLOWS, GENERAL DYNAMICS, AND FLUIDIZATION

XXXI – A KINETIC THEORY FOR THE RAPID FLOW OF ROUGH IDENTICAL SPHERICAL PARTICLES AND THE EVOLUTION OF FLUCTUATIONS
M. Shahinpoor and G. Ahmadi

XXXII – GEOMETRICAL FOUNDATION OF THE ANALYSIS OF GRANULAR MATERIALS FLOWS
M. Shimbo

XXXIII – DYNAMIC INTERACTIONS OF GRANULAR MATERIALS
S. L. Soo

XXXIV – DEBRIS FLOW AND DEBRIS FLOW DEPOSITION
T. Takahashi

XXXV – RHEOLOGY OF POWDER SYSTEMS
K. Umeya

MECHANICS OF GRANULAR MATERIALS AND SOILS

XXXVI – AN ELASTIC-PLASTIC ISOTROPIC CONSTITUTIVE MODEL FOR SAND
G. Y. Baladi

XXXVII – MECHANICAL BEHAVIOR OF SOILS DESCRIBED BY HYPOELASTICITY EQUATIONS
L. Dragusin

XXXVIII – A DISCRETE LIMIT ANALYSIS OF FOUNDATION STRUCTURES
T. Kawai and N. Takeuchi

PREFACE

This two-volume book on the mechanics and the flow of granular materials is a collection of 46 papers contributed by some of the world's leading experts on the subject. It is meant to serve as a survey of recent works on the mechanics, flow and general transport properties of granular materials, i.e., assemblies (aggregates) of distinct (discrete) unit entities (grains, particles, etc.). However, the fundamental purpose of this book is to provide a foundation for bridging the gaps that have existed among seemingly distinct areas of research that nevertheless have been directed towards the same goal, i.e., the understanding of the behavior of granular materials. Examples of diverse fields dealing with all aspects of granular materials are pharmacology, agriculture, ecology, geotechnics, geophysics, stellar physics, nuclear engineering, defense, manufacturing, mathematics-statistics, molecular physics and biomedicine-biometrics. In all of these fields of activities the behavior of granular assemblies is of fundamental and critical importance and poses interesting problems. Problems of immediate interest include the granular jams in capsule and other granular transport, the phenomena of vibratory densification, fluidization, and liquefaction, random packing and critical phenomena, determination of the velocity distribution function for colliding rough inelastic particles, dynamics of planetary granular rings and electromagneto-granular anomalies. Many fundamental problems remain to be solved. For example, can it be shown that the critical density of random close packing of equal sphere is $(2/\pi) \approx 0.6366197$?

The mechanics and flow of granular materials and bulk solids will undoubtedly become much more important in the very near future.

The first volume of this book concentrates on the following areas:

1. Random Packing of Granular Materials
2. General Morphological Characterization and Measurements
3. Statistical Mechanical Considerations
4. Electrical, Thermal and General Transport Properties.

The second volume of this book concentrates on the following main areas:

1. Granular Materials Rapid Flow, General Dynamics and Fluidization
2. Granular Materials and Soil Mechanics.

I hope that the publication of this book will materially contribute to the growth of our understanding of the behavior and the characteristics of granular materials. I would like to thank many friends and colleagues who have generously helped me throughout the preparation of these volumes. Among these I would specifically like to thank Dr. Alexander A. Hayday, Chairman of the Mechanical and Industrial Engineering Department at Clarkson (where the Granular Materials Research Laboratory is housed), Dr. N. L. Ackermann, Chairman of the Civil and Environmental Engineering Department at Clarkson, and Professors G. Ahmadi and S. P. Lin for their continuous support and encouragement throughout the preparation of these volumes. I wish to thank all of the authors for their cooperation. I would also like to thank a number of research experts who have been most kind in helping me prepare these two volumes, who, however, due to pressure from other commitments, could not make a contribution in time to be included in these two volumes. Among these are Professors N. L. Ackermann, G. K. Batchelor, A. Blinowski, C. Brennen, M. M. Carroll, H. Colijn, S. C. Cowin, P. Cundall, J. Curran, R. Davies, R. A. Davis, K. J. Euler, J. Ghaboussi, R. L. Glicksman, A. S. Goldberg, D. A. Grivas, M. E. Harr, K. Hashiguchi, G. M. Homsy, I. Ishibashi, J. T. Jenkins, D. Jeffrey, G. De Josselin de Jong, J. R. Johanson, B. H. Kaye, R. P. King, J. Konishi, M. M. Mehrabadi, T. P. Meloy, N. Moroto, Z. Mróz, S. Nemat-Nasser, S. Niisek, J. W. Nunziato, M. Oda, R. Oliver, G. D. Parfitt, S. O. Passman, H. B. Poorooshasb, C. H. Rochester, B. Rohani, P. W. Rowe, R. H. Sabersky, M. Satake, S. B. Savage, P. J. Sherrington, A. Shokooh, A. J. M. Spencer, O. L. Strack, F. Tatsuoka, I. G. Vardoulakis, O. T. Walton, K. Yamaguchi and O. C. Zienkiewicz.

I hope to be able to incorporate the contributions from the above experts in future volumes on the mechanics and flow of granular materials.

Special thanks go to my wife Jamileh for her patience during the preparation of this work and to Ms. Rhonda L. Price for typing and preparation of the manuscripts.

Finally I would like to thank Dr.-Ing. Reinhard H. Wöhlbier, Publisher of Trans Tech Publications, and his staff both in the USA and the Federal Republic of Germany for the generous help, encouragement and cooperation extended to me throughout the preparation of this work.

January 1983

Mohsen Shahinpoor
Potsdam, NY 13676

Whatsoever things are true, whatsoever things are honest, whatsoever things are just, whatsoever things are pure, whatsoever things are lovely, whatsoever things are of good report; If there be any virtue, and if there be any praise, think on these things.

Philippians 4:8

LIST OF CONTRIBUTORS

Numbers in parentheses indicate the pages on which the authors' contributions begin.

G. Ahmadi, Department of Mechanical and Industrial Engineering, Clarkson College of Technology, Potsdam, N.Y., USA (497), (641)

U. Ts. Andres, Department of Mineral Resources Engineering, Royal School of Mines, Imperial College of Science and Technology, London, England (331)

B. F. Backman, Boeing Commercial Airline Company, Seattle, WA, USA (259)

G. Y. Baladi, Geomechanics Division, Structures Laboratory, U.S. Army Engineer Waterways Experiment Station, Vicksburg, Mississippi, USA (745)

A. Bar-Cohen, Department of Mechanical Engineering, Ben Gurion University, Beer Sheva, Israel (529)

J. Bodziony, Strata Mechanics Research Institute, Polish Academy of Sciences, Krakow, Poland (129)

J. K. Beddow, Division of Chemical and Materials Engineering, The University of Iowa, Iowa City, USA (107)

J. G. Berryman, Lawrence Livermore National Laboratory, P.O. Box 808, L-200, Livermore, CA, USA (1)

C. B. Brown, Department of Civil Engineering, University of Washington, Seattle, WA, USA (259)

J. L. Chermant, Equipe Materiaux-Microstructure, Laboratoire de Crystallographie, Physique et Chimie du Solide, L.A. 251 Ismra-Universite, Caen, France (157)

A. R. Corradi, Sonorex Tape S.P.A. — Cazzogo, S.M., Italy (375)

G. M. Colver, Department of Mechanical Engineering, Iowa State University, Ames, Iowa, USA (355)

M. Coster, Equipe Materiaux-Microstructure, Laboratoire de Crystallographie, Physique et Chimie du Solide, L.A. 251 Ismra-Universite, Caen, France (157)

L. W. De Backer, Agricultural Engineering Department, University of Louvian, Louvian-La-Neuve, Belgium (339)

L. Dragusin, Catedra Mathematici II, Institutul Politechnic Bucuresti, Bucarent, Romania (761)

J. L. Finney, Crystallography Department, Birkbeck College, London, England (19)

D. Geldart, School of Chemical Engineering and Powder Technology, University of Bradford, Bradford, West Yorkshire, England (553)

K. Gotoh, Department of Energy Engineering, Toyohashi University of Technology, Tenpaku-Cho, Toyohashi, Japan (41)

H. Hatano, Research Laboratory of Resources Utilization, Tokyo Institute of Technology, Yokohama, Japan (565)

J. Hidaka, Department of Chemical Engineering, Doshisha University, Kyoto City, Japan (53)

K. Hutter, Laboratory of Hydraulics, Hydrology and Glaciology, ETH Zurich, Switzerland (613)

K. Iinoya, Department of Industrial Engineering Aichi Institute of Technology, Toyota, Japan (589)

M. Ishida, Research Laboratory of Resource Utilization, Tokyo Institute of Technology, Yokohama, Japan (565)

K. Ishihara, Department of Civil Engineering, University of Tokyo, Japan (903), (929)

J. P. Jernot, Equipe Materiaux-Microstructure, Laboratoire de Cristallographie, Physique et Chimie du Solide, L.A. 251, Ismra-Universite, Caen, France (157)

W. S. Jodrey, Rural Route 4, Amherst, N.S., Canada (75)

P. W. Jowitt, Department of Civil Engineering, Imperial College of Science and Technology, London, England (259)

K. I. Kanatani, Department of Computer Science, Gunma University, Kiryu, Gunma, Japan (273)

T. Kawai, Institute of Industrial Science, University of Tokyo, Tokyo, Japan (779)

R. Kitamura, Department of Ocean Civil Engineering, Kagoshima University, Kagoshima, Japan (285)

J. Kloubek, The J. Heyrovský Institute of Physical Chemistry and Electrochemistry, Czechoslovak Academy of Sciences, Prague, Czechoslovakia (179)

W. Kraj, Strata Mechanics Research Institute, Polish Academy of Sciences, Krakow, Poland (129)

S. Lowell, Quantachrome Corporation, Syosett, New York, USA (193)

K. Makino, Department of Chemical Engineering, Kyoto University, Kyoto City, Japan (53)

E. Margarita-Eglit, Department of Hydromechanics, Moscow State University, Moscow, USSR (577)

H. Masuda, Department of Chemical Engineering, University of Hiroshima, Higashi-Hiroshima, Japan (589)

H. Matsuoka, Department of Civil Engineering, Nogoya Institute of Technology, Nagoya, Japan (813), (837)

D. R. McKenzie, School of Physics, The University of Sydney, Sydney, Australia (415)

R. C. McPhedran, School of Physics, The University of Sydney, Sydney, Australia (415)

J. Munro, Department of Civil Engineering, Imperial College of Science and Technology, London, England (259)

T. Nakai, Department of Civil Engineering, Nagoya Institute of Technology, Nagoya, Japan (837)

V. N. Nikolaevski, Institute of Physics of the Earth, USSR Academy of Sciences, Moscow, USSR (855)

H. Ochiai, Department of Civil Engineering, Nagasaki University, Nagasaki, Japan (871)

S. Ogawa, Computer Center, National Aerospace Laboratory, Chofu, Tokyo, Japan (601)

C. Orr, Micromeritics Instrument Corporation, Norcross, GA, USA (207)

N. Phan-Thien, California Institute of Technology, Pasadena, CA, USA and Department of Mechanical Engineering, The University of Sydney, Sydney, Australia (415)

T. Scheiwiller, Laboratory of Hydraulics, Hydrology and Glaciology, ETH, Zurich, Switzerland (613)

M. Shahinpoor, Granular Materials Research Laboratory and Department of Mechanical and Industrial Engineering, Clarkson College of Technology, Potsdam, N.Y., USA (297), (641)

M. Shimbo, Division of Information Engineering, Hokkaido University, Sapporo, Japan (669)

J. E. Shields, Quantachrome Corporation, Syosset, New York, USA (193)

S. L. Soo, Department of Mechanical and Industrial Engineering, University of Illinois at Urbana-Champaign, Urbana, IL, USA (675)

N. G. Stanley-Wood, Postgraduate School of Studies in Powder Technology, University of Bradford, Bradford, West Yorkshire, England (221)

T. Takahashi, Disaster Prevention Research Institute, Kyoto University, Kyoto, Japan (699)

N. Takeuchi, Systems Research Laboratory, Kyowa Engineering Consultants Co., LTD, Tokyo, Japan (779)

Y. Tanabashi, Department of Civil Engineering, Nagasaki University, Nagasaki, Japan (871)

Y. Tobita, Department of Civil Engineering, Tohoku University, Sendai, Japan (885)

E. M. Tory, Department of Mathematics and Computer Science, Mount Allison University, Sackville, Canada (75)

I. Towhata, Department of Civil Engineering, University of Tokyo, Tokyo, Japan (903)

K. Umeya, Applied Chemistry Department, Faculty of Engineering, Tohoku University, Sendai, Japan (719)

O. J. WHITTEMORE, Department of Mining, Metallurgical and Ceramic Engineering, University of Washington, Seattle, Washington, USA (249)

Y. YAMADA, Department of Civil Engineering, University of Tokyo, Tokyo, Japan (929)

T. YAMANOUCHI, Department of Civil Engineering (Suiko), Kyushu University, Fukuoka, Japan (871)

E. YANAGISAWA, Department of Civil Engineering, Tohoku University, Sendai, Japan (947)

H. YOSHIDA, Department of Chemical Engineering, University of Hiroshima, Higashi-Hiroshima, Japan (589)

SECTION 5

GRANULAR MATERIALS RAPID FLOWS, GENERAL DYNAMICS, AND FLUIDIZATION

A GENERALIZED CONTINUUM THEORY FOR FLOW OF GRANULAR MATERIALS

Goodarz Ahmadi

Department of Mechanical Engineering
Clarkson College of Technology
Potsdam, NY 13676, USA

ABSTRACT

Generalized nonlinear constitutive equations for the flow of granular materials are formulated by consideration of the effects of rotation of granules. Basic equations of motions of granular materials with compressible as well as incompressible grains are derived. Examples of Couette and gravity flows are presented and discussed.

1. INTRODUCTION

The mechanics of granular materials is of fundamental interest to the fields of soil mechanics, powder technology as well as bulk solid transport.

The early investigation of the mechanical behavior of soil and granular materials was carried out by Coulomb [1]. The Mohr-Coulomb theory of limiting equilibrium was formulated by Kötter [2]. Further generalization has been carried out by Sokoloviskii [3]. Drucker and Prager [4], and Shield [5] have applied the theory of elastic-plastic solid for describing the mechanics of granular materials.

A theory of fluid saturated porous media was developed by Biot [6,7] and further studied by Deresiewicz [8]. The continuum theory of mixtures was studied by Truesdell and Toupin [9], Truesdell [10], Eringen and Ingram [11], Green and Naghdi [12] and Bowen [13]. Recently, a continuum theory of granular materials was formulated by Goodman and Cowin [14,15], and later extended by Jenkins [16], Passman [17], Ahmadi and Sohrabpour [18], Shahinpoor and Ahmadi [19] and Ahmadi [20,21]. Goodman-Cowin theory is based on the theory of structured media and the Ericksen-Leslie [22-25] theory of liquid crystals.

The theory of media with nonvanishing couple stresses is first studied by D. and F. Cosserat [26]. Modern development of the theory of media with couple stresses are due to Truesdell and Toupin [9], Ericksen [22,23], Toupin [27], Mindlin and Tiersten [28], Eringen and Suhubi [29], Mindlin [30] and Eringen [31]. Recently, developments have been carried out by Eringen [32-34], Ahmadi, Koh and Goldschmidt [35,36], Shahinpoor and Ahmadi [37], Ahmadi and Firoozbakhsh [38] and Ahmadi [39].

Rapid flows of granular materials were studied by Bagnold [40,41], Savage [42], Jenkins and Cowin [43], McTigue [44], Blinowski [45], Ogawa [46], Kanatani [47,48], Shahinpoor and Lin [49], Shahinpoor and Siah [50], Ogawa, Umemura and Oshima [51], Savage and Jeffrey [52], Jenkins and Savage [53] and Ahmadi and Shahinpoor [54,55] among others. Extensive reviews of the subject were provided in [42,43] and [56,57].

In [58] a continuum theory for granular materials is presented where the effects of variations of the solid volume fraction and the rotation of the granules are simultaneously included in the formulation. The equilibrium and slow motion of granular media were discussed at length in [58]. In the present study, the theory of [58] is generalized by the developments of nonlinear constitutive equations appropriate for the rapid flow of granular media. Explicit equations of motion of granular materials with compressible as well as incompressible granules are derived. It is shown that the present theory contains the models of Bagnold [41,42], Shahinpoor and Lin [50], McTigue [45,59], Goodman and Cowin [14,15] and Savage [42], as special cases. Our derivation allows for variation of the solid volume fraction in a self contained fashion and there is no need for the apriori assumption about its space-dependent as is done in [59]. Several examples of shear flows of granular materials are also presented and discussed.

2. BASIC EQUATIONS

Let $\nu(x,t)$ be the solid volume fraction (or solid volume distribution function) which is defined as the local ratio of the volume occupied by the solid particles to the sum of the volumes of solids and voids. The macroscopic behavior of a granular materials is then represented by a distributed (continuous) solid body whose mass density is given by

$$\rho = \nu\rho^{s} \quad , \tag{1}$$

where ρ^{s} is the mass density of the solid granular particles.

The distributed solid body must satisfy the basic laws of motion of continuum mechanics. Accordingly, the following field equations must be satisfied.

Conservation of mass

$$\frac{\partial\rho}{\partial t} + \nabla \cdot (\rho \underset{\sim}{v}) = 0. \tag{2}$$

Balance of linear momentum

$$\rho \frac{dv_i}{dt} = t_{ji,j} + \rho\, b_i \; . \tag{3}$$

Balance of angular momentum

$$\rho j \frac{d\nu_i}{dt} = m_{ji,j} + \varepsilon_{ijk} t_{jk} + \rho l_i \quad . \tag{4}$$

Balance of equilibrated force

$$\rho k \frac{d\dot{\nu}}{dt} = h_{i,i} + \rho \ell + \hat{g} \; . \tag{5}$$

Conservation of equilibrated inertia

$$\frac{dk}{dt} - 2\dot{\nu} k = 0 \; . \tag{6}$$

Conservation of energy

$$\rho\dot{e} = t_{kl} d_{lk} + m_{kl} \nu_{l,k} + h_k \dot{\nu}_{,k} - \hat{g}\dot{\nu} + q_{k,k} + \rho r. \tag{7}$$

Entropy Inequality

$$\rho\dot{\eta} - (q_k/\theta)_{,k} - \rho r/\theta \geq 0 \; . \tag{8}$$

The microdeformation rate densor D_{lk} is defined by

$$D_{lk} = v_{l,k} - \varepsilon_{klm} \nu_m \tag{9}$$

In these equations the following notations have been employed:

v_i = velocity vector,
ν_i = gyration vector,
t_{ij} = stress tensor,
b_i = body force per unit mass,
m_{ij} = couple stress tensor,
j = microinertia,
l_i = body couple per unit mass,
ν = solid volume distribution,
k = equilibrated inertia,
ℓ = equilibrated force per unit mass,

$\hat{g}$ = internal equilibrated force,
h_i = equilibrated stress vector,
e = internal energy density per unit mass,
r = internal heat source per unit mass,
q_k = heat flux vector pointing outward,
η = entropy density per unit mass,
θ = temperature.

Throughout this paper the regular cartesian tensor notation is employed with indicies after a comma denoting partial derivative and dot on the top of a letter standing for the total time derivative. The free energy of the distributed solid body is defined by

$$\psi = e - \eta\theta \quad . \tag{10}$$

Eliminating the heat source distributions r between equations (7) and (8) and using equation (10) we find an alternative form of the entropy inequality

$$-\rho\,(\dot{\psi} + \eta\,\dot{\theta}) + q_k\,\theta_{,k}/\theta + t_{kl}\,D_{lk} + m_{kl}\,\nu_{l,k}$$
$$+ h_k\,\dot{\nu}_{,k} - \hat{g}\,\dot{\nu} \geq 0 \ . \tag{11}$$

Let the stresses be decomposed into elastic and dissipative parts, i.e.

$$t_{kl} = {}_E t_{kl} + {}_D t_{kl} \ , \tag{12}$$

$$m_{kl} = {}_E m_{kl} + {}_D m_{kl} \ , \tag{13}$$

$$h_k = {}_E h_k + {}_D h_k \quad . \tag{14}$$

Based on thermodynamics arguments a set of constitutive equations for the elastic parts of the stresses were derived in [58]. These are

$${}_E t_{kl} = -\pi\delta_{kl} - \rho\,\frac{\partial\psi}{\partial\nu_{,k}}\,\nu_{,l} - \frac{\partial\psi}{\partial\phi_{m,k}}\,\phi_{m,l} \quad , \tag{15}$$

$${}_E m_{kl} = \rho\,\frac{\partial\psi}{\partial\phi_{l,k}} \quad , \tag{16}$$

$$E^{h}{}_{k} = \rho \frac{\partial \psi}{\partial \nu,_{k}} \quad , \tag{17}$$

and

$$\hat{g} = \hat{\pi} - \rho \frac{\partial \psi}{\partial \nu} \quad , \tag{18}$$

where ϕ_i is microrotation and is related to gyration vector by

$$\dot{\phi}_i = \nu_i \quad . \tag{19}$$

The thermodynamic pressure and micropressure are defined respectively by

$$\pi = - \frac{\partial \psi}{\partial \rho^{-1}} \quad , \tag{20}$$

and

$$\hat{\pi} = - 2 \rho \frac{\partial \psi}{\partial k} k. \tag{21}$$

Equations (15)-(21) are for a general granular material with no constraint. However, if the material of the granular grains is incompressible (i.e. ρ^{s} = const), then according to equation (1) the bulk mass density becomes directly proportional to the solid volume fraction. The equation of conservation of mass (2) then simplifies to

$$\frac{\partial \nu}{\partial t} + \nabla \cdot (\nu \underset{\sim}{v}) = 0 \ . \tag{22}$$

In other words the kinematic variables ν and $\underset{\sim}{v}$, can not vary independently. As a consequence of the constraint of equation (22), the expressions for the elastic part of the stress tensor and the thermodynamic pressure become [58]

$$E^{t}{}_{kl} = (-\pi + \nu\hat{\pi} - \nu\hat{g})\delta_{kl} - \rho \frac{\partial \psi}{\partial \nu,_{k}} \nu,_{l} - \rho \frac{\partial \psi}{\partial \phi_{m,k}} \phi_{m,l} \quad , \tag{23}$$

$$\pi = \rho \nu \frac{\partial \psi}{\partial \nu} \quad , \tag{24}$$

and $\hat{g}$ remain unrestricted.

Equations (23) and (24) together with (16)-(17) form the basic constitutive equations for the elastic parts of the stresses of a granular material with incompressible grains.

The nature of the free energy of granular materials has been the subject of much discussions [14-16, 18-20]. In [58] the following generalized form was assumed

$$\rho\psi = \rho_o \psi_o (\rho, \theta, k) + \alpha(\nu) \nu,_k \nu,_k + \frac{1}{2}\gamma(\nu)\phi_{m,k} \phi_{m,k} , \tag{25}$$

where α and γ are scalar functions of solid volume fraction ν. The requirement of positive definiteness of the free energy function implies that

$$\alpha(\nu) \geq 0 , \quad \gamma(\nu) \geq 0 . \tag{26}$$

Different functional form for $\alpha(\nu)$ were suggested in the past [16,19,60]. In [58], it was shown that at the equilibrium, for a constant porosity* granular sample with incompressible grains to satisfy the Mohr-Columb condition of limiting equilibrium, γ must be given by

$$\gamma = \gamma_o \nu^{-\text{Cotan}\ \phi} , \quad \text{(Equilibrium)} \tag{27}$$

where ϕ is the angle of internal friction.

With the free energy function as given by equation (25), the explicit expression, for the constitutive equations of the elastic parts of the stresses become

$$E^t{}_{kl} = - \pi \delta_{kl} - 2\alpha(\nu) \nu,_k \nu,_l - \gamma(\nu) \phi_{m,k} \phi_{m,l} , \tag{28}$$

$$E^t{}_{kl} = \gamma(\nu) \phi_{l,k} , \tag{29}$$

$$E^h{}_k = 2\alpha(\nu) \nu,_k . \tag{30}$$

---

*porosity is one minus ν.

From equation (18), the expression for $\hat{g}$ is found to be

$$\hat{g} = \hat{\pi} - \nu \frac{d}{d\nu} (\frac{\alpha}{\nu}) \nu,_k \nu,_k - \frac{\nu}{2} \frac{d}{d\nu} (\frac{\gamma}{\nu}) \phi_{m,k} \phi_{m,k} \quad . \tag{31}$$

When the grains are incompressible, the elastic part of the stress tensor is given by equation (23). Therefore

$$_E t_{kl} = [\hat{\pi}\nu - \hat{g}\nu - \nu^2 \frac{d}{d\nu} (\frac{\alpha}{\nu}) \nu,_i \nu,_i - \frac{1}{2}\nu^2 \frac{d}{d\nu} (\frac{\gamma}{\nu}) \phi_{m,i} \phi_{m,i}] \delta_{kl}$$

$$- 2\alpha \nu,_k \nu,_l - \gamma \phi_{m,k} \phi_{m,l} , \tag{32}$$

with the thermodynamic pressure being given by

$$\pi = \nu^2 [\frac{d}{d\nu} (\frac{\alpha}{\nu}) \nu,_i \nu,_i + \frac{1}{2} \frac{d}{d\nu} (\frac{\gamma}{\nu}) \phi_{m,i} \phi_{m,i}] \quad . \tag{33}$$

Assuming that ψ_o is zero in the expression for the free energy function as given by equation (25), it follows that

$$\hat{\pi} = 0 \ . \tag{34}$$

Furthermore, if it is assumed that the equilibrated inertia k, equilibrated force ℓ as well as the dissapative part of the equilibrated stress vector are zero, from equation of balance of equilibrated forces (5), it follows that

$$\hat{g} = - h_{i,i} = - 2(\alpha \nu,_i),_i \quad . \tag{35}$$

The elastic part of the stress tensor then becomes

$$_E t_{kl} = [2\alpha\nu \nu,_{ii} + \frac{d}{d\nu}(\alpha\nu)\nu,_i \nu,_i - \frac{\nu^2}{2} \frac{d}{d\nu}(\frac{\gamma}{\nu}) \phi_{m,i} \phi_{m,i}] \delta_{kl}$$

$$- 2\alpha \nu,_k \nu,_l - \gamma \phi_{m,k} \phi_{m,l} \ . \tag{36}$$

Equation (36) gives the final form of the constitutive equation for the elastic part of the stress tensor for granular materials with incompressible grains.

Using the decompositions of (12) - (14) and the constitutive relations for the elastic parts of the stresses in entropy inequality (11) it follows that the dissipative parts of the stresses must satisfy the entropy inequality given by

$$ {}_Dt_{kl}\, D_{lk} + {}_Dm_{kl}\, \nu_{l,k} + {}_Dh_k\, \dot{\nu}_{,k} + q_k\, (\ell n\, \theta)_{,k} \geq 0 \ . \tag{37} $$

3. CONSTITUTIVE EQUATIONS FOR RAPID FLOWS

The basic laws of motion (2) - (8) together with the constitutive equations for the elastic parts of stresses (25) - (28) and (36) are sufficient for analysis of the equilibrium state of granular materials. When the granular media is set in motion, the dissipative parts of the stresses become quite large and control the motion. Specially, in the rapid flow regime of granular materials, the dissipative stresses dominate the dynamics of the flow with the effects of the elastic parts of the stresses being negligibly small. More specifically, for the flow of cohesionless granular material the material parameter, $\gamma \simeq o$ since we do not expect that the finite continuous rotation of granules to produce strain energy. That would simplify the expressions for the elastic parts of stresses to some extent.

We propose the following functional constitutive equations for the dissipative parts of stresses:

$$ {}_Dt_{kl} = {}_Dt_{kl}\, (D_{ij}) \ , \tag{38} $$

$$ {}_Dm_{kl} = {}_Dm_{kl}\, (\nu_{i,j}) \ , \tag{39} $$

$$ {}_Dh_k = {}_Dh_k\, (\nu_{,i}) \ . \tag{40} $$

Note that it is possible to develop more general constitutive equations by the use of the "principle of equipresence" of continuum mechanics. However, with the lack of experimental motivations and for sake of simplicity of analysis we restrict ourselves to the simple constitutive relationships given by equations (38) - (40). Linear constitutive equations of the form given by (38) - (40) were derived in [58] which are suitable for slow motion of granular media. Here the intent is to develop more general non-linear constitutive equations appropriate to rapid flows of granular materials.

For an isotropic media, the following set of constitutive equations for the dissipative parts of the stresses are proposed

$$_D t_{ij} = \lambda D_{kk} \delta_{ij} + (2\mu + \zeta_o |II_D|^{\frac{1}{2}}) D_{(ij)} + \kappa D_{ji}$$

$$+ \zeta_1 D_{(jm)} D_{(mj)} , \tag{41}$$

$$_D m_{ij} = \alpha_v \nu_{k,k} \delta_{ij} + \beta_v \nu_{i,j} + \gamma_v \nu_{j,i} , \tag{42}$$

$$_D h_i = \eta_v \dot{\nu}_{,i} . \tag{43}$$

Where II_D is the second principle invariant of the symmetric part of the microdeformation rate tensor $D_{(ij)}$ and is given by

$$II_D = \frac{1}{2} (D_{ii} D_{jj} - D_{(ij)} D_{(ji)}) . \tag{44}$$

The coefficients of viscosity λ, μ, ζ_o, κ, ζ_1, α_v, β_v, γ_v and η_v are functions of the solid volume distribution function ν. The entropy inequality (37) imposes the following restrictions on the coefficients of viscosity

$$3\lambda + 2\mu + \kappa \geq 0 , \quad 2\mu + \kappa \geq 0 , \quad \kappa \geq 0 ,$$

$$3\alpha_v + \beta_v + \gamma_v \geq 0 , \quad |\beta_v| \leq \gamma_v , \quad \gamma_v \geq 0 ,$$

$$\zeta_0 \geq \quad , \quad \zeta_1 \leq \zeta_o/\sqrt{3}* , \quad \eta_v \geq 0 \tag{45}$$

It is possible to formulate more general nonlinear constitutive equations by the use of the theory of invariants [61-63]. However, in the absence of any experimental evidence we avoid the introduction of further complexities.

When the coefficients of the nonlinar terms ζ_0 and ζ_1 are set equal to zero, equations (41) - (43) reduce to the linear constitutive equations of [58].

---

*This restruction was based on demanding that entropy inequality (24) be satisfied for isochoric flows.

In the limiting case of negligible microrotational effects and constant solid volume fraction, equations (42) and (43) are omitted and equation (41) with κ being dropped reduces to

$$D^t_{ij} = \lambda\, d_{kk}\, \delta_{ij} + (2\mu + \zeta_0 |II_d|^{\frac{1}{2}})\, d_{ij} + \zeta_1\, d_{ik} d_{kj} \,, \tag{46}$$

where†

$$d_{ij} = \frac{1}{2}\,(v_{i,j} + v_{j,i}) \,, \tag{47}$$

and

$$II_d = \frac{1}{2}\,(d_{ii} d_{jj} - d_{ij} d_{ji}) \,. \tag{48}$$

When the linear terms in (46) are neglected the constitutive equation of Shahinpoor and Lin [49] is recovered.

For a simple shear flow the component of stresses become

$$D^t_{12} = \mu \frac{\partial u}{\partial y} + \frac{\zeta_0}{4\sqrt{2}} \left|\frac{\partial u}{\partial y}\right| \frac{\partial u}{\partial y} \,, \tag{49}$$

$$D^t_{11} = \frac{\zeta_1}{4} \left(\frac{\partial u}{\partial y}\right)^2 . \tag{50}$$

For high shear rate when the nonlinear term in (49) becomes dominant, equations (49) and (50) become identical to those obtained by Bagnold [40] for the rapid shear flows. Similar constitutive equations were obtained in [54,55] based on statistical arguments. Thus, equations (41) - (43) appear to be the simplest constitutive equations which are applicable to slow as well as rapid flows of granular materials. These equations do include the effects of rotations of granules with in the limits of linear terms.

The results of Bagnold [40] also suggests that ζ_0 and ζ_1 are proportional to $(\nu_m-\nu)^{-2}$ where ν_m is the maximum ν for flow. However, further experimental data are required to truely justify such variations and also to find the ν-dependence of other coefficients of viscosity.

---

†Note that d_{ij} is equal to D_{ij} with ν_i being set equal angular velocity.

4. EQUATIONS OF MOTION

Explicit equations of motion of a granular medium could be obtained by direct substitution of the derived constitutive equations into the basic equations of balance (3) - (5). We consider the cases of general unrestricted granular materials and the materials with incompressible granules separately.

Compressible Granules

The constitutive relations for the granular materials with compressible grains are given by equations (28) - (31) and (41) - (43). The equations of balance of linear momentum, angular momentum and equilibrated force then become,

$$\rho\, \dot{v}_i = -\,\pi,_i - 2(\alpha\, \nu,_i\, \nu,_j),_j - (\gamma\, \phi_{m,i}\, \phi_{m,j}),_j + (\lambda\, v_{j,j}),_i$$

$$+ [(\mu + \frac{1}{2}\zeta_0 |II_D|^{\frac{1}{2}})(v_{i,j} + v_{j,i})],_j + [\kappa(v_{i,j} - \varepsilon_{jik}\nu_k)],_j$$

$$+ \frac{1}{4}[\zeta_1(v_{i,k} + v_{k,i})(v_{k,j} + v_{j,k})],_j + \rho\, b_i\ , \qquad (51)$$

$$\rho j \dot{\nu}_i = (\gamma\, \phi_{i,j}),_j + (\alpha_v\, \nu_{j,j}),_i + (\beta_v\, \nu_{j,i}),_j + (\gamma_v\, \nu_{i,j}),_j$$

$$+ \kappa\, \varepsilon_{ijk}\, v_{k,j} - 2\kappa\, \nu_i + \rho\, \ell_i\ , \qquad (52)$$

$$\rho k \ddot{\nu} = 2(\alpha\, \nu,_i),_i + (\eta_v\, \dot{\nu},_i),_i + \hat{\pi} - \nu \frac{d}{d\nu}(\frac{\alpha}{\nu})\, \nu,_i\, \nu,_i$$

$$- \frac{\nu}{2}\frac{d}{d\nu}(\frac{\gamma}{\nu})\, \phi_{i,j}\, \phi_{i,j} + \rho\ell\ , \qquad (53)$$

where

$$II_D = \frac{1}{2}[v_{i,i}\, v_{j,j} - \frac{1}{2}(v_{i,j}\, v_{i,j} + v_{i,j}\, v_{j,i})]\ . \qquad (54)$$

Equations (51) - (53) and the equations of conservation of mass and equilibrated inertia (2) and (6) together with two appropriate equations of state for π and $\hat{\pi}$ form a complete set of eleven equations for eleven unknowns v_i, ν_i (or ϕ_i), ν, ρ, k, π and $\hat{\pi}$. These equations cover the full range of equilibrium, slow and rapid flows of granular materials with compressible grains in the most general case.

Incompressible Granules

When the material of granular grains are incompressible, the equation of mass conservation (2) reduces to equation (22) and the constitutive equation for the elastic part of the stress tensor is given by the modified form of equations (32) or (36). The equations of state for pressures are also given explictly by equations (33) and (34). The equation of balance of linear momentum now becomes

$$\rho^s\nu\,\dot{v}_i = 2\alpha\,\nu\,\nu,_{jji} + 2\nu\,\frac{d\alpha}{d\nu}(\nu,_i\,\nu,_{jj} + \nu,_j\nu,_{ji}) + \nu\frac{d^2\alpha}{d\nu^2}\nu,_i\nu,_j\nu,_j$$

$$- \gamma\,\phi_{m,kk}\,\phi_{m,i} - \frac{d\gamma}{d\nu}\,(\nu\,\phi_{m,ji}\,\phi_{m,j} + \nu,_j\,\phi_{m,j}\,\phi_{m,i})$$

$$- \frac{\nu}{2}\frac{d^2\gamma}{d\nu^2}\,\nu,_i\,\phi_{m,j}\,\phi_{m,j} + (\lambda\,v_{j,j}),_i + [(\mu + \frac{1}{2}\zeta_0|II_D|^{\frac{1}{2}})$$

$$(v_{i,j} + v_{j,i})],_j$$

$$+ [\kappa(v_{i,j} - \varepsilon_{jik}\nu_k)],_j + \frac{1}{4}[\zeta_1(v_{i,k} + v_{k,i})$$

$$(v_{k,j} + v_{j,k})],_j + \rho^s\nu b_i \quad . \tag{55}$$

Equation of balance of angular momentum for the granular media with incompressible grains becomes identical to equation (53) with ρ being replaced by $\rho^s\nu$.

Equations (55), (53) together with equation (22) are now forming seven equations governing the variations of seven unknowns, v_i, ν_i (or ϕ_i) and ν. For a given set of appropriate boundary and initial conditions, the solutions to this set of equations would give the velocity, microrotation (or microgyration) as well as the solid volume fraction as functions of space and time.

The field equations derived in this section governs the general motion or equilibrium of a granular medium. They are also applicable to the fields where part of the granular material is at equilibrium and the other part is in motion. However, if only the flow or equilibrium are of interest considerable simplifications of the field equations are possible. For instance for analysis of equilibrium of granular material the dissipative part of stresses drop from the equations completely. On the other hand, for flow of granular materials, the coefficient γ is zero which leads to some simplifications of the equations of motion. Furthermore, if ν does not vary appreciably, the coefficients of viscosity may be treated as constants.

5. PLANE SHEAR FLOWS

To compare various approximation, we consider the example of unsteady plane shear flows of a body of granular material with incompressible grains, we assume that

$$\underset{\sim}{v} = [v_x(y,t),\ 0,\ 0], \quad \nu = \nu(y,t) ,$$
$$\underset{\sim}{\nu} = [0,\ 0,\ \nu_z(y,t)]. \tag{56}$$

With γ being zero, equation of balance of linear momentum (55) leads to

$$\rho^s\nu \frac{\partial v_x}{\partial t} = \frac{\partial}{\partial y}\,[(\mu + \frac{\zeta_0}{8}|\frac{\partial v_x}{\partial y}|)\frac{\partial v_x}{\partial y}] + \frac{\partial}{\partial y}\,[\kappa(\frac{\partial v_x}{\partial y} + \nu_z)$$
$$+ \rho^s\nu b_x , \tag{57}$$

$$0 = 2\alpha\nu \frac{\partial^3\nu}{\partial y^3} + 4\nu \frac{d\alpha}{d\nu}\frac{\partial\nu}{\partial y}\frac{\partial^2\nu}{\partial y^2} + \nu \frac{d^2\alpha}{d\nu^2}(\frac{\partial\nu}{\partial y})^3$$
$$+ \frac{1}{4}\frac{\partial}{\partial y}\,[\zeta_1(\frac{\partial v_x}{\partial y})^2] + \rho^s\nu b_y . \tag{58}$$

Equation (22) is identically satisfied and equation of balance of angular momentum (52) becomes

$$\rho^s j\nu \frac{\partial\nu_z}{\partial t} = \frac{\partial}{\partial y}\,(\gamma_v \frac{\partial\nu_z}{\partial y}) - \kappa\,(\frac{\partial v_x}{\partial y} + 2\nu_z) , \tag{59}$$

where the body couple is assumed to be zero. The expression for thermodynamic pressure is given by equation (33) reduces to

$$\pi = \nu^2 \frac{d}{d\nu} (\frac{\alpha}{\nu}) (\frac{\partial \nu}{\partial y})^2 \quad . \tag{60}$$

Equations (57) - (60) are four equations for finding the four unknowns v_x, ν_z, ν and π. However, before attempting to solve these equations, the functional dependence of the material parameter α and the coefficients of viscosity must be prescribed. For the sake of simplicity of analysis we treat these parameter as being constants. Such an assumption is justified if these coefficients are smoothly varying functions of solid volume fraction (that is expected to be so) and ν does not vary appreciably across the flow field. For steady shear flows equations (57) - (59) reduce to

$$\frac{\partial}{\partial y} [(\mu + \kappa + \frac{\zeta_0}{8} |\frac{\partial v_x}{\partial y}|) \frac{\partial v_x}{\partial y}] + \kappa \frac{\partial \nu_z}{\partial y} + \rho^s \nu \, b_x = 0 \ , \tag{61}$$

$$2\alpha\nu \frac{\partial^3 \nu}{\partial y^3} + \frac{\zeta_1}{2} \frac{\partial v_x}{\partial y} \frac{\partial^2 v_x}{\partial y^2} + \rho^2 \nu \, b_y = 0 \ , \tag{62}$$

$$\gamma_v \frac{\partial^2 \nu_z}{\partial y^2} - \kappa \, (2 \, \nu_z + \frac{\partial v_x}{\partial y}) = 0 \ . \tag{63}$$

If the rotational effects of the granules be neglected (i.e. κ=0), equation (61) becomes

$$\frac{\partial}{\partial y} [(\mu + \frac{\zeta_0}{8} |\frac{\partial v_x}{\partial y}|) \frac{\partial v_x}{\partial y}] + \rho^s \nu \, b_x = 0 \ . \tag{64}$$

When the second term in the parenthesis is much larger than the coefficient μ, equation (64) reduces to the momentum equation with Bagnold's [40] constitutive relations as given by (49).

We now consider two special examples of plane shear flows, namely a Couette flow and a gravity flow.

Couette Flow

For a simple Couette flow, the body force $\underset{\sim}{b}$ is zero and the boundary conditions for rough granular particle are

$$v_s = 0 \ , \quad \nu_z = 0 \ , \quad \frac{\partial \nu}{\partial y} = 0 \quad \text{at } y = 0 \ , \tag{65}$$

$$v_x = U_0 \ , \quad \nu_z = 0 \ , \quad \frac{\partial \nu}{\partial y} = 0 \quad \text{at } y = h \ . \tag{66}$$

When the granules and the walls are smooth, the particles are free to rotate and the boundary condition $\nu_z = 0$ must be replaced by $\nu_z = \frac{1}{2}\frac{\partial v_x}{\partial y}$. Conservation of mass implies that

$$\int_0^h \rho^s \nu \, dy = \rho \, \nu_0 \, h = \text{mass flow rate} \ , \tag{67}$$

where ν_0 is a constant uniform solid volume fraction.

In the absence of rotational effect and when Bagnold's [40] equation for shear stress is employed, the equation of motion becomes

$$\frac{\partial}{\partial y} \left[\frac{\zeta_0}{8} \left|\frac{\partial v_x}{\partial y}\right| \frac{\partial v_x}{\partial y}\right] = 0 \ . \tag{68}$$

Equation (68) implies that the velocity profile is linear, i.e.

$$\bar{v}_x = \eta \ , \tag{69}$$

where dimensionless quantities are defined by

$$\bar{v}_x = v_x/U_0 \ , \quad \eta = y/h \ . \tag{70}$$

Even when the Bagnold's constitutive equation is supplemented by an extra linear term as given by equation (49), the velocity profile remains linear. Thus, Bagnold's expression for the shear stress leads to a velocity profile similar to that of a Newtonian fluid.

Let us now consider another limit when the coefficient ζ_0 is negligible. Equation (61) then reduces to

$$(\mu + \kappa) \frac{\partial^2 v_x}{\partial y^2} + \frac{\partial \nu_z}{\partial y} = 0 \, . \tag{71}$$

The solutions of equations (63) and (71) satisfying boundary conditions (65) and (66) are given by

$$\bar{v}_z = \frac{\eta - (K/2Q \sinh Q)[\cosh Q\eta + \cosh Q - \cosh Q(1-\eta) - 1]}{1 - K(\cosh Q - 1)/Q \sinh Q} \, , \tag{72}$$

$$\bar{\nu}_z = \frac{\sinh Q - \sinh Q\eta - \sinh Q(1-\eta)}{2\, K(\cosh Q - 1)/Q - 2 \sinh Q} \, ,$$

where the nondimensional quantities are defined by

$$\bar{\nu}_z = \nu_z h/U_0, \quad K = \kappa/(\mu+\kappa) \, ,$$

$$Q^2 = \frac{(2\mu+\kappa)\kappa h^2}{(\mu+\kappa)\gamma} \, . \tag{74}$$

The expression for the shear stress becomes

$$\bar{\tau}_0 = \frac{(2-K)/2(1-K)}{1-K(\cosh Q-1)/Q \sinh Q} \, , \tag{75}$$

where

$$\bar{\tau}_0 = \tau_0 h/\mu U_0 \, , \tag{76}$$

and the shear stress is constant across the flow field. When parameter Q is small, that is

$$\gamma/\kappa h^2 >> 1 , \tag{77}$$

employing Taylor series expansions, the expressions for the non-dimensional velocity, gyration and shear stress become, respectively,

$$\bar{v}_x = \eta \left[1 - \frac{KQ^2}{12(2-K)} (1-\eta)(1-2\eta)\right] , \tag{78}$$

$$\bar{\nu}_z = - \frac{Q^2}{2(2-K)} (\eta-\eta^2) , \tag{79}$$

$$\bar{\tau}_0 = \frac{1}{1-K} \left[1 - \frac{KQ^2}{12(2-K)}\right] . \tag{80}$$

Equation (78) clearly shows the deviation of the velocity profile from that of a Newtonian fluid. Examples of the nondimensional velocity and granular gyration distributions are shown in figures 1 and 2, respectively. The solid curves correspond to the exact equations (72) and (73) and the dashed curves are from the approximate expressions (78) and (79). Numerical evaluations show that expression (78) gives a remarkably accurate velocity profiles even for relatively large values of $Q \leq 2$. The approximate expression (79) for gyration remains accurate only for $Q \leq 0.5$.

To find the variations of solid volume fraction across the flow, equation (62) must be solved. The linearized form of equation (62) for Couette flow becomes

$$\frac{\partial^3 \nu}{\partial y^3} + \frac{\zeta_1}{4\alpha\nu_0} \frac{\partial v_x}{\partial y} \frac{\partial^2 v_x}{\partial y^2} = 0 . \tag{81}$$

Employing approximate expression (78) for the velocity distribution, the solution to equation (81) which satisfies conditions (65) - (67) becomes

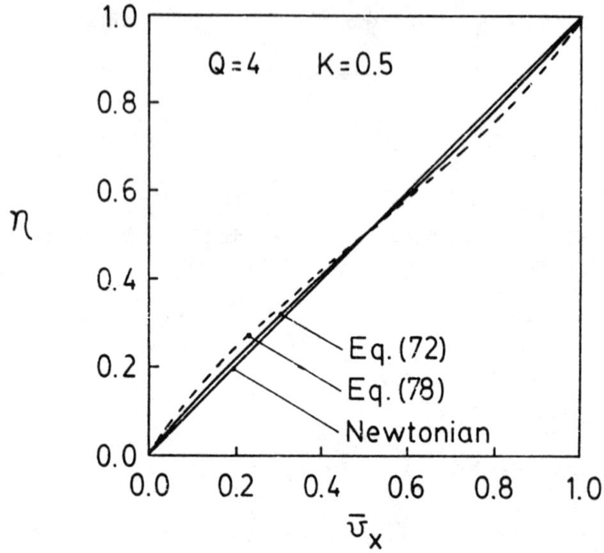

Fig. 1. Velocity distributions for the Couette flow.

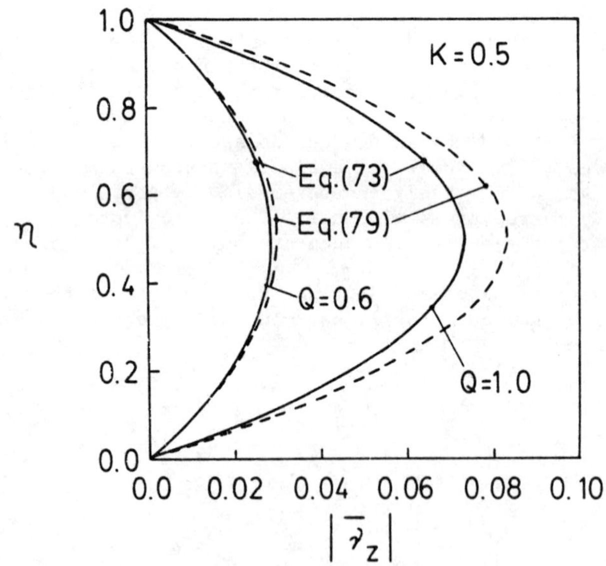

Fig. 2. Variations of dimensionless gyration for the Couette flow.

$$\bar{\nu} = \eta^2(1-\eta)^2 - 1/30 \ , \qquad (82)$$

where

$$\bar{\nu} = \left(\frac{96(2-K)\alpha\nu_0}{KQ^2\zeta_1 U_0^2}\right)(\nu-\nu_0) \ . \qquad (83)$$

The deviation of solid volume fraction form its mean value ν_0 is shown in figure 3. The linearization approximation about ν_0 is valid as long as parameter $Q\zeta_1 U_0^2/\alpha$ is small.

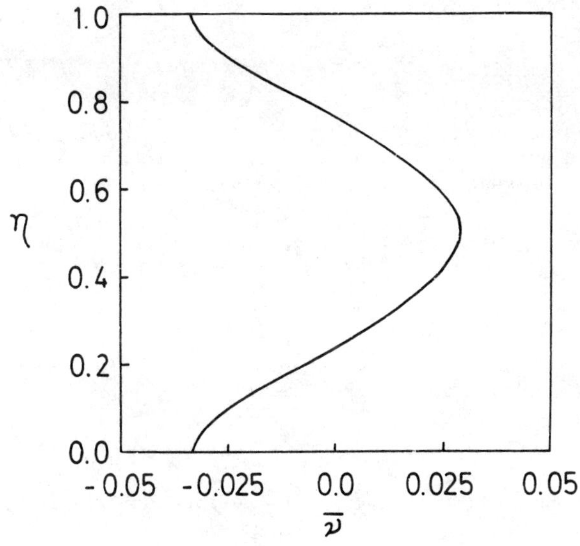

Fig. 3. Solid volume fraction profile in the Couette flow.

Gravity Flow

For a gravity flow of granular materials along an incline, the component of the body force are given by

$$b_x = g \sin \theta , \quad b_y = -g \cos \theta \tag{84}$$

where θ is the angle of inclination and g is the acceleration of gravity.

The boundary conditions (65) and (67) of Couette flow remain valid for the present gravity flow. The boundary conditions at the free surface becomes

$$t_{21} = 0 , \quad m_{23} = 0 , \quad t_{22} = 0 , \quad \text{at } y = h . \tag{85}$$

That is, the shear and normal stresses as well as the couple stress must vanish. For Bagnold's constitutive equations in the absence of rotational effect, the velocity profile is given by

$$\bar{v}_x = 1 - (1-\eta)^{3/2} , \tag{86}$$

where

$$\bar{v}_x = v_x/v_M , \tag{87}$$

with $v_M = v_x(h)$ is the maximum velocity at the free surface.

In the limiting case when ζ_0 is negligibly small, the first two boundary conditions of (85) become

$$(\mu+\kappa) \frac{\partial v_x}{\partial y} + \kappa \, \nu_z = 0 , \quad \frac{\partial \nu_z}{\partial y} = 0 \quad \text{at } y = h . \tag{88}$$

With $\tilde{b}$ given by equation (84), the solutions of equations (61) and (62) satisfying boundary conditions (65) and (88) are given by

$$\frac{v_x}{V_0} = 2\eta - \eta^2 - \frac{K}{Q^2 \cosh Q} [Q(\sinh Q - \sinh Q(1-\eta))$$

$$+ 1-\cosh Q\eta] \ , \tag{89}$$

$$\bar{\nu}_z = -(1-\eta) + [Q \cosh Q(1-\eta) - \sinh Q\eta]/Q \cosh Q \ , \tag{90}$$

where

$$\bar{\nu}_z = \nu_z \ h/V_0 \ , \quad V_0 = \frac{\rho^2 \nu_0 g h^2 \sin \theta}{2\mu+\kappa} \ . \tag{91}$$

The velocity at the free surface then is given by

$$\frac{V_M}{V_0} = 1 - \frac{K}{Q^2 \cosh Q} (Q \sinh Q + 1 - \cosh Q) \ . \tag{92}$$

The shear stress follows a linear profile, i.e.

$$\frac{t_{21}}{\tau_0} = 1-\eta \tag{93}$$

where

$$\tau_0 = t_{21}(0) = (2\mu+\kappa) \ V_0/h \ . \tag{94}$$

For small values of parameter Q the simplified expression for the velocity distribution becomes

$$\bar{v}_x = 2\eta-\eta^2 - \frac{KQ^2}{12(2-K)} (6\eta - 9\eta^2 + 4\eta^3 - \eta^4) \ . \tag{95}$$

Examples of velocity and gyration profiles are shown in figures 4 and 5. The velocity profile corresponding to Bagnold's constitutive equation as is given by equation (86) is also shown in figure 4. Numerical evaluations show that for $Q \leq 1$, expression (95) becomes an accurate approximation for the velocity distribution. Using the approximate velocity profile as given by equation (95) in equation (62) and employing a linearization similar to that of equation (81), we find

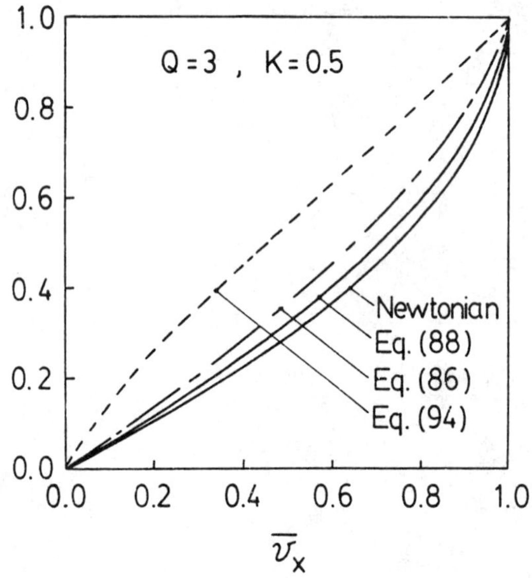

Fig. 4. Velocity profiles for the gravity flow of granular materials.

$$\frac{\partial^3 \nu}{\partial y^3} + \frac{\zeta_1}{4\alpha\nu_0}\frac{\partial v_x}{\partial y}\frac{\partial^2 v_x}{\partial y^2} - \rho^s g \cos\theta = 0 \,. \tag{96}$$

The corresponding linearized form of the normal stress becomes

$$t_{22} = 2\alpha\nu_0 \frac{\partial^2 \nu}{\partial y^2} \,. \tag{97}$$

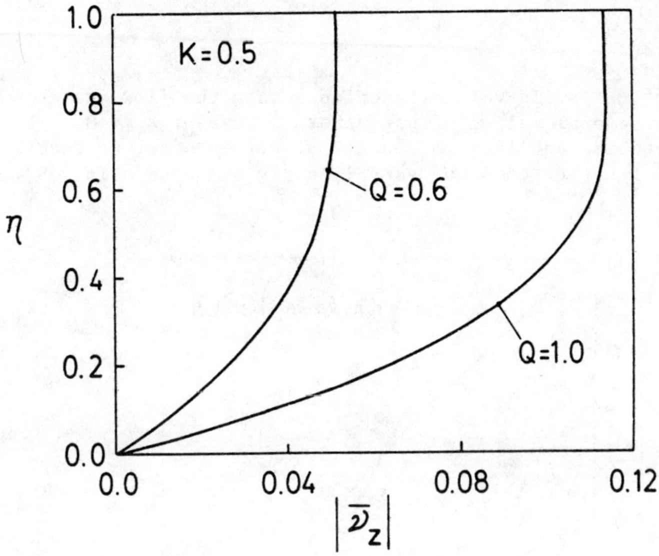

Fig. 5. Variations of granular gyration for the gravity flow.

The expression for the solid volume fraction satisfying equation (96) and boundary conditions (65), (67) and (85) then becomes

$$\bar{\nu}_g = \frac{\nu-\nu_0}{G} = \frac{3}{2} - 6\eta^2 + 2\eta^3 + \bar{\zeta}_1[\frac{11}{20} - \eta - \frac{1}{4}(1-\eta)^4$$

$$+ \frac{KQ^2}{2-K}(3\eta^2 - \frac{1}{2}\eta^3 - \frac{1}{5}\eta^5 + \frac{1}{30}\eta^6 - \frac{711}{840})] , \quad (98)$$

where

$$G = \frac{\rho^s g\, h^3 \cos\theta}{24\alpha} \quad (99)$$

$$\bar{\zeta}_1 = \frac{\zeta_1 V_0^{\,2}(2-\kappa)}{24\alpha\, \nu_0} \quad . \tag{100}$$

Variations of the solid volume fraction across the flow are shown in figure 6 for several values of parameters. Case of $\bar{\zeta}_1 = 0$, corresponds to the equilibrium condition. It is observed that the solid volume fraction does not vary linearly as assumed in [59].

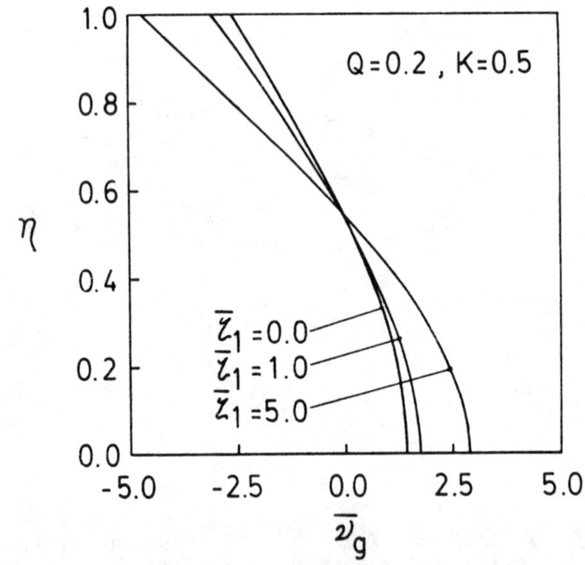

Fig. 6. Solid volume fraction profiles for the gravity flow.

With ν given by equation (98), the equations for the normal stress could be evaluated from equations (97). That is

$$\bar{t}_{22} = \{ \eta - 1 + \frac{\bar{\zeta}_1}{12}[-3(1-\eta)^2 + \frac{KQ^2}{2-K}(6-3\eta-4\eta^3+\eta^4)]\} \tag{101}$$

where

$$\bar{t}_{22} = \frac{t_{22}}{\rho^s gh\, \nu_o \cos\theta} . \tag{102}$$

The variations of the normal stress across the flow field are shown in figure 7. It is observed that the normal stress varies linearly at the equilibrium. However, its variation becomes nonlinear during the motion.

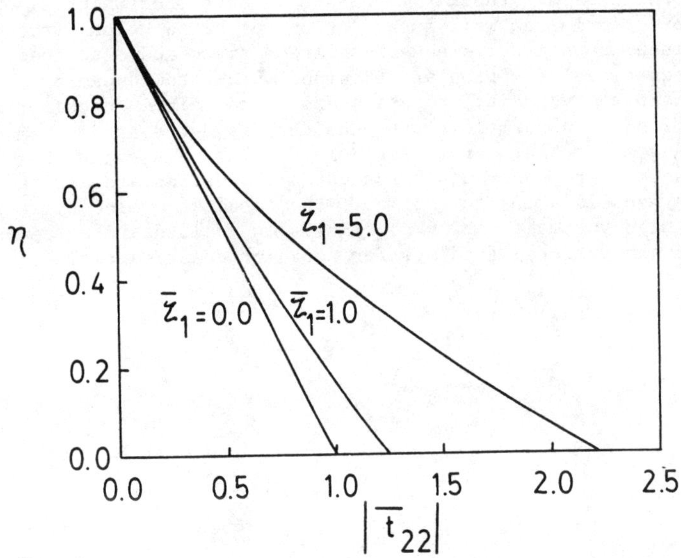

Fig. 7. Variations of normal stress in gravity flow.

The linearized analysis of shear flows presented in this section shows that the consideration of rotational effects of granules leads to the correct qualitative trends of the known granular flow properties with rough granules.

To make the agreement quantitative, the full nonlinear problems must be treated with appropriate ν-dependence of the material parameters being considered. However, that involves elaborated numerical computations and is left for future studies. Note that for smooth particle when the boundary condition $\nu_z = 0$ on a rigid boundary is modified to $\nu_z = \frac{1}{2}\frac{\partial v_x}{\partial y}$, the shape of the velocity profile becomes quite different. Solutions for the linearized equations under that boundary conditions are discussed in [64,65].

6. CONCLUSIONS

A theory for flow of granular materials is developed which allows the rotation of granules as well as variations of solid volume fraction. A set of nonlinear constitutive equations are derived and discussed. The explicit equations of motion are obtained which show dynamic couplings between the velocity, rotation and solid volume fraction fields. The examples of plane shear flows are considered and the special cases of Couette and gravity flows are treated in details. Assuming that the material parameters are constants, closed form solutions are derived from the linearized equations of motion. The resulting velocity, gyration and solid volume fraction profile, show qualitative agreement with the known experimental findings for rough granular particles.

REFERENCES

1. Coulomb, M.A., "Essai sur une application des regles des maximis of minimis a' quelques problemes de statique relative a' 1' architecture," Mem. Acad. Roy. Divers Sav. 5, 383 (1967).

2. Kotter, F., "Die Bestimmung des Druckes an gekrummten Gleitflachen, eine Aufgabe aus der Lehre Vom Eiddruck," Akad. Berichte Berlin (1903).

3. Sokolovskii, V.V., "Statics of Granular Media," Pergamon Press, London (1965).

4. Drucker, D.C. and Prager, W., "Solid Mechanics and Plastic Analysis of Limit Design," Quart. Appl. Math. 10, 157 (1952).

5. Shield, R.T., "On Coulomb's Law of Failure in Solis," J. Mech. Phys. Solids 4, 10 (1955).

6. Biot, M.A., "Theory of Propagation of Elastic Waves in Fluid Saturated Porous Solid-I, Low Frequency Range," J. Acoust. Soc. Am. 28, 168 (1956).

7. Biot, M.A., "The Elastic Coefficients of the Theory of Consolidation," J. Appl. Mech. 24, 594 (1957).

8. Deresiewicz, H., "Mechanics of Granular Matter," Adv. Appl. Mech. 5, 233 (1958).

9. Truesdell, C. and Toupin, R.A., "The Classical Field Theories, Handbuck der Physik. Vol. III/I (Edited by S. Flugge), Springer-Verlag (1960).

10. Truesdell, C., "Mechanical Basis of Diffusion," J. Chem. Phys. 37, 2336 (1962).

11. Eringen, A.C. and Ingram, J.D., "A Continuum Theory of Chemically Reacting Medie-I," Int. J. Engng Sci. 3, 197 (1963).

12. Green, A.E. and Naghdi, P.M., "A Dynamical Theory of Interacting Continua," Int. J. Engineering Sci. 3, 231 (1965).

13. Bown, R.M., "Theory of Mixture, Continuum Physics Vol. III," (Edited by A.C. Eringen). Academic Press (1976).

14. Goodman, M.A. and Cowin, S.C., "Two Problems in the Gravity Flow of Granular Materials," J. Fluid Mech. 45, 321 (1971).

15. Goodman, M.A. and Cowin, S.C., "A Continuum Theory for Granular Materials, Arch. Rat. Mech. Anal. 44, 249 (1972).

16. Jenkins, J.T., "Static Equilibrium of Granular Materials, J. Appl. Mech. 42, 603 (1975).

17. Passman, S.L., "Mixtures of Granular Materials," Int. J. Engng Sci. 15, 117 (1977).

18. Ahmadi, G. and Sohrabpour, S., "Ealsto-Viscoplastic-Viscous Theories of Granular and Porous Media," Int. J. Nonlinear Mech. 14, 133 (1979).

19. Shahinpoor, M. and Ahmadi, G., "Free Energy of Granular Materials in Static Equilibrium," J. Appl. Mech. Trans. ASME 46, 944 (1979).

20. Ahmadi, G., "On Mechanics of Saturated Granular Materials," Int. J. Nonlinear Mech. 15, 251 (1980).

21. Ahmadi, G., "A Generalized Continuum Theory for Porous Elastic Solids," Scientia Sinica 24, 179 (1981).

22. Ericksen, J.L., "Anisotropic Fluids," Arch. Rat. Mech. Anal. 4, 231 (1960).

23. Ericksen, J.L., "Transversely Isotropic Fluids," Kolloidzeitschrift 173, 117 (1960).

24. Leslie, F.M., "Some Constitutive Equations for Liquid Crystals, Arch. Rat. Mech. Anal. 28, 265 (1968).

25. Ericksen, J.L., "Conservation Laws for Liquid Crystals," Trans. Soc. Rheol. 5, 23 (1961).

26. Cosserat, E. and F., "Theories des Crops Deformable," Paris (1909).

27. Toupin, R.A., "Elastic Materials with Couple Stress," Arch. Rat. Mech. Anal. 11, 385 (1962).

28. Midlin, R.D. and Tiersten, H.F., "Effect of Couple Stresses in Linear Elasticity," Arch. Rat. Mech. Anal. 11, 415 (1962).

29. Eringen, A.C. and Suhubi, E.S., "Non-Linear Theory of Simple Micro-Elastic Solids-I," Int. J. Engng Sci. 2, 189 (1964).

30. Midlin, R.D., "Microstructure in Linear Elasticity," Arch. Rat. Mech. Anal. 16, 51 (1964).

31. Eringen, A.C., "Simple Microfluids," Int. J. Engng Sci. 2, 205 (1964).

32. Eringen, A.C., "Linear Theory of Micropolar Elasticity," J. Math. Mech. 15, 909 (1966).

33. Eringen, A.C., "Theory of Micropolar Fluids," J. Math. Mech. 16, 1 (1966).

34. Eringen, A.C., "Foundation of Micropolar Thermoelasticity," UDINE, Springer-Verlag (1970).

35. Ahmadi, G., Koh, S.L. and Goldschmidt, V.W., "A Theory of Non-simple Microfluids, Recent Advances in Engineering Science," Vol. 5, Golden and Breach, 9 (1970).

36. Ahmadi, G., Koh, S.L. and Goldschmidt, V.W., "Mechanics of a Second Order Microfluid," Iranian J. Sci Tech. 1, 223 (1971).

37. Shahinpoor, M. and Ahmadi, G., "Stability of Cosserat Fluid Motions, Arch. Rat. Mech. Anal. 47, 283 (1972).

38. Ahmadi, G. and Firoozbakhsh, K., "First Strain Gradient Theory of Thermoelasticity," Int. J. Solids Struct. 11, 339 (1975).

39. Ahmadi, G., "Mechanics of a Second-Order Micro-Elastic Solids," Rheol. Acta, 14, 710 (1975).

40. Bagnold, R.A., "Experiments on a Gravity-Free Dispersion of Large Spheres in Newtonian Fluid Under Shear," Proc. Roy. Soc. Lond. A 225, 49 (1954).

41. Bagnold, R.A., "The Shearing and Dilatation of Dry Sand and the Singing Mechanism," Proc. Roy. Soc. Lond. A 295, 219 (1966).

42. Savage, S.B., "Gravity Flow of Cohesionless Granular Materials in Chutes and Channels," J. Fluid Mech. 92, 53 (1979).

43. Jenkins, J.T. and Cowin, S.C., "Theories for Flowing Granular Materials," The Joint ASME-CSME Appl. Mech. Fluid Engng. and Bioengng. Conf., AMD - Vol. 31, pp. 79-89 (1979).

44. McTigue, D.F., "A Model for Stresses in Shear Flow of Granular Materials, Proc. US-Japan Seminar on Continuum Mech. and Statist. Approaches in Mech. of Granular Materials, Ed. S.C. Cowin and M. Satake, Sendai, Japan, pp. 266-271 (1978).

45. Blinowski, A., "On the Dynamic Flow of Granular Media," Archives of Mech., Arch. Mech. Stoso, 30, 27 (1976).

46. Ogawa, S., "Multitemperature Theory of Granular Materials," Proc. US-Japan Seminar on Continuum Mech. and Statist. Appraoches in Mech. of Granular Materials, Ed. S.C. Cowin and M. Satake, Sendai, Japan, pp. 208-217 (1978).

47. Kanatani, K.I., "A Micropolar Continuum Theory for the Flow of Granular Materials," Int. J. Engng. Sci. 17, 417 (1979).

48. Kanatani, K.J., "A Continuum Theory for the Flow of Granular Materials, Part I," Theo. Appl. Mech. 27, 571 (1980), part II, Theo. Appl. Mech. 28, 485 (1980), Univ. of Tokyo Press.

49. Shahinpoor, M. and Lin, S.P., "Rapid Couette Flow of Granular Materials," Acta Mechanica 42, 183 (1982).

50. Shahinpoor, M. and Siah, J.S.S., "New Constitutive Equations for the Rapid Flow of Granular Materials," J. Non-Newtonian Fluid Mechanics 9, 147 (1981).

51. Ogawa, S., Umemura, A. and Oshima, N., "On the Equations of Fully Fluidized Granular Materials," ZAMP 31, 483 (1980).

52. Savage, S.B. and Jeffrey, D.J., "The Stress Tensor in a Granular Flow at High Shear Rate," J. Fluid Mech. 110, 225 (1981).

53. Jenkins, J.T. and Savage, S.B., "The Mean Stress Resulting from Interparticle Collisions in a Rapid Granular Shear Flow," Proc. 4th ICCMDS, Stockholm, June (1981).

54. Ahmadi, G. and Shahinpoor, M., "Towards a Turbulent Modeling of Rapid Flow of Granular Materials (to appear).

55. Ahmadi, G. and Shahinpoor, M., "A Kinetic Theory for Rapid Flows and Evolution of Fluctuations," Research Report No. MIE-077 Clarkson College, Potsdam, NY (1982).

56. Brown, R.L. and Richards, J.C., "Principles of Powder Mechanics," Pergamon Press, Oxford (1970).

57. Shahinpoor, M., "On Rapid Flow of Bulk Solids," Bulk Solids Handl. Int. J. 1, 487 (1981).

58. Ahmadi, G., "A Generalized Continuum Theory for Granular Materials," Int. J. Nonlinear Mech. 17, 21 (1982).

59. McTigue, D.F., "A Nonlinear Constitutive Model for Granular Materials: Application to Gravity Flow," J. Appl. Mech. ASME 49, 291 (1982).

60. Ahmadi, G., "A Possible Continuum Theory for Desification and Liquefaction of Saturated Sand," Math. Geology 13, 37 (1981).

61. Spencer, A.J.M. and Rivlin, R.S., "Theory of Matrix Polynomials and Its Application to the Mechanics of Isotropic Continua," Arch. Rat. Mech. Anal. 2, 309 (1959).

62. Spencer, A.J.M. and Rivlin, R.S., "Further Results in the Theory of Matrix Polynominals," Arch. Rat. Mech. Anal. 4, 214 (1960).

63. Spencer, A.J.M., "Theory of Invariants," Ch. 3, Treatise on Continuum Physics, I, Ed. A.C. Eringen, Academic Press, N.Y. (1971).

64. Kirwan, A.D., Jr. and Newman, N., "Plane Flow of Afluid Containing Rigid Structures," Int. J. Engng Sci. 7, 883 (1969).

65. Ahmadi, G., "Self Similar Solution of Incompressible Micropolar Boundary Layer Flow Over a Semi-inifinte Plate," Int. J. Eng. Sci. 14, 639 (1976).

FLUID MECHANIC CHARACTERISTICS OF BUBBLING FLUIDIZED BEDS

Avram Bar-Cohen

Department of Mechanical Engineering, Ben Gurion University
Beer Sheva, Israel

ABSTRACT

In gas fluidized beds, where bubbly fluidization is generally encountered, the characteristics of the gas voids, which rise and grow within the emulsion, exert a profound influence on the fluid dynamics and chemical kinetics of the bed. This study focuses on the application of two-phase theory and gas flow continuity to the prediction of bed expansion and bubble diameter in beds both with and without immersed heat exchange pipes. The derived theoretical formulations are found to offer useful bounds on the bubble fraction, in both the slow and fast bubble regimes, and to provide approximate relations for the prediction of bubble diameter, accounting for bubble coalescence in open beds as well as coalescence and splitting associated with tube/bubble interactions.

1. INTRODUCTION

The primary operating characteristics of gas fluidized beds, including: bed height, heat transfer rate and chemical kinetics, are largely determined by the fluid dynamic mechanisms active in the fluidized medium. At sufficiently low velocities, an imposed gas flow percolates through the tortuous channels, formed by the interparticle spaces, without disturbing the packing of the particles and fixed-bed behavior is encountered. As the flow rate increases, dynamic and viscous forces acting on the particles increase. When these forces become equal to the gravitational force, the particles are suspended in the gas flow and "fluidized" behavior is achieved. For still larger flow rates, the interparticle separation distance grows and distinct bubble-like voids appear in the gas/solids emulsion. These "bubbles" grow with increased gas flow rate but are progressively constrained by the bed side walls and for larger flow rates take on the appearance of gas slugs. For gas flow rates larger than those associated with slugging, turbulent chaotic flow develops in the emulsion and progresses to pneumatic transport as the particle terminal velocity is approached.

The present discussion is limited to the bubbling regime of fluidized bed operation and the superficial gas velocities of interest are bounded from below by U_{mb}, the bubbling incipience value, and from above by U_{slug}, corresponding to the initiation of slug flow in the bed.

While the well-known Stewarts' criterion [1] and its later modification by Bayens & Geldart [2] can serve to define the slugging transition, the precise conditions under which bubbling commences in a gas fluidized bed have yet to be determined. In very small particle systems, where U_{mf} is on the order ot 0.1 to 1.0 cm/sec, the ratio of U_{mb} to U_{mf} varies from approximately 3 to 10. Alternately, in large particle systems with U_{mf} values approaching 0.5 to 1.0 m/sec, the initiation of bubbling is often indistinguishable from the minimum fluidization condition. In the absence of an accepted criterion, Cranfield & Geldhart's [3] measured bubbling incipience excess velocity, $U_{mb}-U_{mf}$, for 1760 μm diameter alumina particles, equal to 5 cm/sec, may serve as a point of reference for large particle beds.

In the analysis of fluid dynamic phenomena in fluidized beds, it is often convenient to distinguish between "slow" bubbles, whose rise velocity is less than the interstitial gas velocity, and "fast" bubbles, which rise at a velocity greater than the interstitial gas velocity [4]. Due to their relatively high minimum fluidization velocities, large-particle beds can be expected to operate both in the slow bubble and fast bubble regime, while the more common, fine-powder fluidized beds generally operate only in the fast bubble regime. In the analyses presented herein, distinct expressions will be derived for each of the two bubbling regimes, though it must be noted that "slow" and "fast" are relative terms and that a bubble of a given diameter could be slow in a bed of 1000-μm quartz particles and fast in a bed of 50 μm.

2. THEORY OF BUBBLING BEDS

In gas fluidized beds, in which bubbly fluidization is generally encountered, the characteristics of the gas voids, which rise and grow within the emulsion, exert a profound influence on the fluid dynamics and chemical kinetics of the bed. Extensive investigations by Davidson et al. [4], have shown that the dynamics of a single, isolated gas void in a fluidized medium are analogous to those of a gas bubble in a homogeneous liquid and that potential flow theory, as first developed by Davies & Taylor [5], can be used to establish the velocity fields in the surrounding emulsion and in the rising bubble. However, the more complex characteristics of bubble swarms have not been thoroughly explored and the diameter, rise velocity and bubble concentration have generally been determined from empirical relations for similar beds, leaving major discrepancies unresolved.

Two-phase theory--particulary with Davidson's bubble model--forms the basis for much of the analysis of fluidized bed behavior [4,6]. In this representation, the emulsion, consisting of both gas and solid particles, is assumed to be a single homogenous phase at the minimally fluidized condition and the bubble gas is viewed as the dispersed second phase. This theory is commonly employed in the analysis of small-particle aggregatively fluidized beds [4,7] and when properly structured has been shown by Bar-Cohen et al. [8] to apply as well to large particle fluidization. However, in very fine particle systems, involving solids belonging to Geldart's type A [9], the emulsion phase is frequently expanded well beyond the value at minimum fluidization and this category will, thus, be excluded from this discussion.

The expansion of the solids/gas emulsion in a fluidized medium, as well as its variation with gas flow rate and other relevant parameters, establishes the height of the bed container and is of critical importance to its design. When the total mass of the solids is fixed, expansion of the fluidized medium results from increased gas fraction and in bubbly or aggregative fluidization this expansion can be related to the emulsion voidage and bubble fraction prevailing in the bed.

Following Kunii & Levenspiel [7] and neglecting wall effects, the average bed voidage, ε, can be expressed as a summation of emulsion voidage and bubble voidage, the former assumed to equal ε_{mf} and the latter to equal unity. Thus

$$\varepsilon = \varepsilon_{mf}(1 - \delta) + \delta \tag{1}$$

where ε_{mf} is the bed voidage at minimum fluidization and δ is the bubble fraction. The bubble fraction can itself be determined from gas flow continuity considerations which dictate that total bed gas flow equal the sum of emulsion flow and bubble flow.

In the slow-bubble regime, when gas flow in the wake and recirculation zones may be neglected, the continuity relation takes the form [7]:

$$U = U_{mf}(1 - \delta)+(u_b + 3U_{mf})\delta \tag{2}$$

where U is the superficial velocity of the fluidizing gas, U_{mf} the minimum fluidization value and u_b the bubble rise velocity. Some uncertainty exists as to the range of applicability of this relation and especially the effects of neighboring bubbles in multiple-bubble flow. An analysis by Bar-Cohen et al. [8] of large particle data obtained by Cranfield & Geldart [3] suggests, however, that (2) is valid for multiple-bubbles throughout at least most of the slow bubble regime, i.e. for $0 < u_{br} < u_f$ (where u_{br} is the rise velocity of an isolate bubble and u_f the interstitial gas velocity taken to equal U_{mf}/ε_{mf}) and is thus applicable to the present analysis.

For very fast bubbles, i.e. when u_{br} is more than five times the interstitial velocity and the clouds surrounding the bubbles are negligible in size, or alternately when $u_{br} >> 3U_{mf}$, the superficial gas velocity can be approximated by [7]:

$$U = U_{mf}(1 -\delta)+u_b\delta. \tag{3}$$

The gas continuity relation for either slow or fast bubbles necessarily involves the bubble rise velocity, u_b. Extensive studies of isolated bubbles have established that the rise velocity of a single bubble in an otherwise undisturbed gas fluidized medium is given by [4]:

$$u_{br} = 0.711\ (gd_v)^{\frac{1}{2}} \tag{4}$$

where d_v is the diameter of the equivalent spherical bubble.

In a bed occupied by a significant bubble fraction, u_b must reflect the "drift flux" contribution of neighboring bubbles and is then commonly expressed as [4]:

$$u_b = u_{br} + U - U_{mf}. \tag{5}$$

Detailed measurements of u_b in bubble swarms (e.g. [10, 11, 12]) appear to suggest that the velocities of individual bubbles may differ considerably from values given by Eq. (5), but, in the absence of a more precise relation, this simple and commonly employed formulation will be used throughout the present discussion.

When the u_b relation is combined with the slow and fast bubble continuity expressions, respectively, the bubble fraction in the bed is found to be given by

$$\delta_{sb} = (U - U_{mf})/(u_{br} + U + U_{mf}) \tag{6}$$

for $u_{br} < u_f$, and

$$\delta_{fb} = (U - U_{mf})/(u_{br} + U - 2U_{mf}) \tag{7}$$

for $u_{br} > 5u_f$, where δ_{sb} is the bubble fraction in the slow regime and δ_{fb} the bubble fraction in the fast bubble regime.

An alternate formulation for bubble fraction can be obtained by noting that the volumetric bubble flow crossing a horizontal plane in a fluidized bed can be expressed as the product of bubble volume and the frequency f_1, with which bubbles cross the reference plane or, alternatively, the product of bubble rise velocity and the area occupied by the bubbles. Thus,

$$(\pi d_v{}^3/6)f_1 = u_b \delta A \tag{8}$$

where f_1 is the frequency with which bubbles cross a given level in the bed and A is the cross-sectional area of the bed. For a random bubble distribution, a simpler measure of bubble frequency, namely the point frequency, defined as the rate at which bubbles strike a single immersed probe, can replace f_1 in (8). These two measures of bubble frequency are related through the ratio of bubble cross-sectional area to total bed area, i.e.

$$f_1/f_p = A/(\pi d_v{}^2/4) \tag{9}$$

with f_p equal to the point frequency.

3. THEORETICAL BOUNDS FOR BUBBLE FRACTION

Equations (6) and (7) reveal that the bubble fraction and, hence, the bed voidage and expansion ratio, in both the slow and fast bubble regimes, are dependent on the bubble rise velocity. Due to bubble growth [3, 12], the rise velocity can be expected to increase with height (i.e. the distance from the distributor plate) but as there is, as yet, no acceptable theory of bubble growth, it is difficult to determine the variation of bubble fraction in the bed.

The relative bubble rise velocity is, however, constrained to vary within prescribed bounds--between zero and the interstitial velocity for slow bubbles and between the interstitial velocity and the slug velocity for fast bubbles, It is, thus, possible to identify several characteristic founds for the bubble fraction, each associated with a limiting value of bubble rise velocity [13].

3.1 Slow Bubble Regime

The maximum bubble rise velocity in the slow bubble regime is, by definition, equal to the interstitial gas velocity, u_f, itself taken equal to U_{mf}/ε_{mf} by the common two-phase assumption. A lower-bound on bed voidage in this regime can thus be obtained by setting u_{br} equal to U_{mf}/ε_{mf} in (6) to yield with ($U' \equiv U/U_{mf}$)

$$\delta_{tr} = (U' - 1)/(U' + 1 + \varepsilon_{mf}^{-1}) \tag{10}$$

where δ_{tr} is the bubble fraction associated with u_{br} equal to u_f. Alternately, the bubble fraction upper bound can be derived from (6) by allowing the relative bubble rise velocity to approach zero, as might be appropriate in the region adjacent to the gas distributor. Thus

$$\delta_{stat} = (U' - 1)/(U' + 1) \tag{11}$$

where δ_{stat} is associated with stationary bubbles. Using similar reasoning the bounding expressions for voidage and expansion ratio can be derived.

In the large-particle fluidized bed study by Cranfield & Geldart [3], local bubble frequency, $f_{C\&G}$, (the frequency with which bubbles strike a point probe) and bubble diameter were measured by independent means over a significant parametric range. These experimental values can be used to determine the local bubble fraction in the bed via Eqs. (8) and (9).

Typical δ values for two different tests in the slow bubble regime, with u_{br}/u_f ranging from 0.4 to 0.67 and 0.58 to 0.94, respectively, are shown in figure 1. As transition is approached, the bubble fraction closely approaches the transition value of δ. The data are seen to display the anticipated variation with bed height and fall within the bounds established by the transition and stationary-bubble δ expressions.

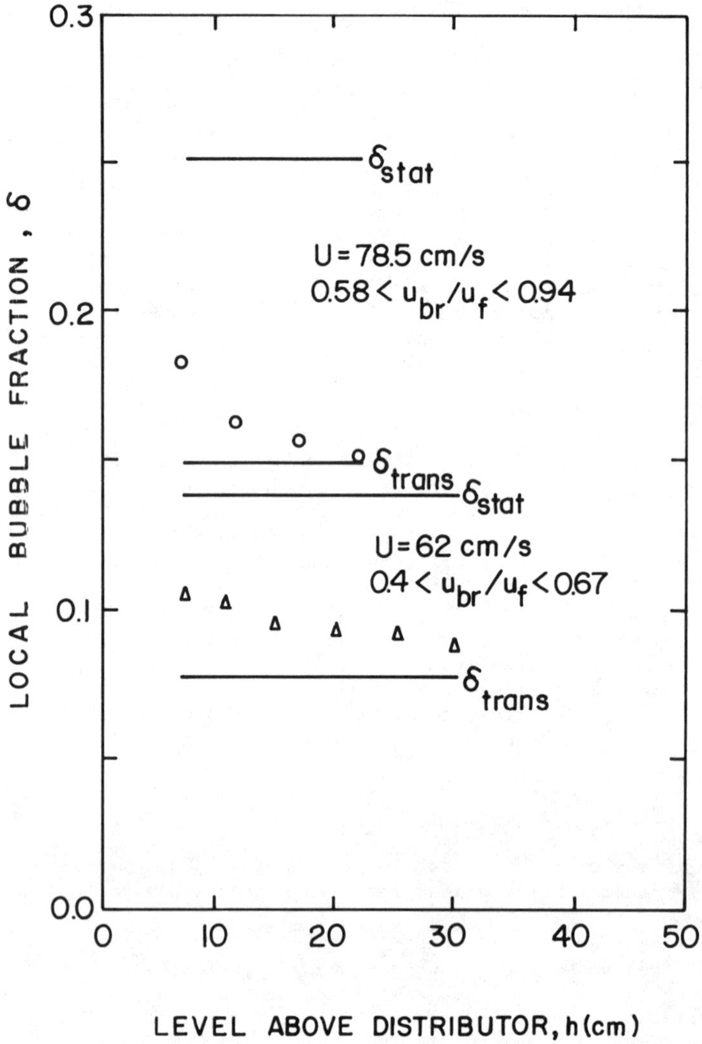

Fig. 1 Variation of local bubble fraction with level above distributor. For [3] [d_p = 1760 um , U_{mf} = 47 cm/s , $\epsilon_{mf} \simeq 0.55$, Bed: 61cm x 61cm]

Similarly, the average bubble fraction recorded in the nineteen Cranfield & Geldart [3] data runs are found to follow the trends suggested by (6) and to be properly bounded by stationary and transition bubble fraction relations.

The paucity of consistent slow-bubble results in the literature makes it most difficult to validate theoretical relations for slow-bubble fluidized beds and often dictates analytical comparisons with incomplete data. In particular, for the bubbling study by McGrath & Streatfield [14], where u_{br}/u_f varied from 0.24 to 0.68 the minimum fluidization voidage, ε_{mf}, is not reported. Nevertheless, when, as shown in figure 2, expansion ratio values are compared with the previously derived analytical expressions with an assumed value of $\varepsilon_{mf} = 0.41$, nearly all the data are found to lie between the stationary and transition bounds.

The large particle data obtained by Canada et al. [15], for relatively high gas flow rates in 30.5 and 61 cm wide beds filled with 2600μ glass particles, offer a further opportunity for examining the applicability of the bounding relations. Canada et al. [15] correlated their results by an expression of the form

$$U/\varepsilon = U + (1 - \varepsilon_{mf})/\varepsilon_{mf} U_{mf}. \tag{12}$$

Equation (12) and the Canada et al. large particle data are shown in figure 3 along with the stationary and transition bounds. Equation (12) is seen to be nearly identical with the stationary-bubble bound for the 31 cm bed and indistinguishable from this upper-bound on slow-bubble voidage in the 61 cm bed.

3.2 Fast Bubble Regime

The onset of slugging in fluidized beds represents the end of simple bubbling behavior, and in the absence of other flow regime transitions, sets an upper-limit on the relative rise velocity of the bubble. Based on the maximum bed expansion ratio in the slugging regime found by Hovmand & Davidson [16], it is possible to establish the desired lower bound for bubble fraction in the fast bubble regime, as

$$\delta_{slug} = 1 - [1 + (U-U_{mf})/0.35\sqrt{gD}]^{-1} \tag{13}$$

The minimum bubble velocity in the fast bubble regime is established by the transition criteria, i.e. $u_{br} = U_{mf}/\varepsilon_{mf}$, and consequently, δ_{tr} sets the upper bound on fast bubble voidage.

In evaluating fast bubble data, it is important to note that, during transition from bubbly to slug flow, the bubble rise velocity can exceed the values associated with the slug by approx. 30 per cent and, furthermore, that an isolated bubble of approx. one-quarter the column diameter can be expected to display a rise velocity equal to the slug velocity [11, 16].

A typical fast-bubble comparison is shown in figure 4 where data,

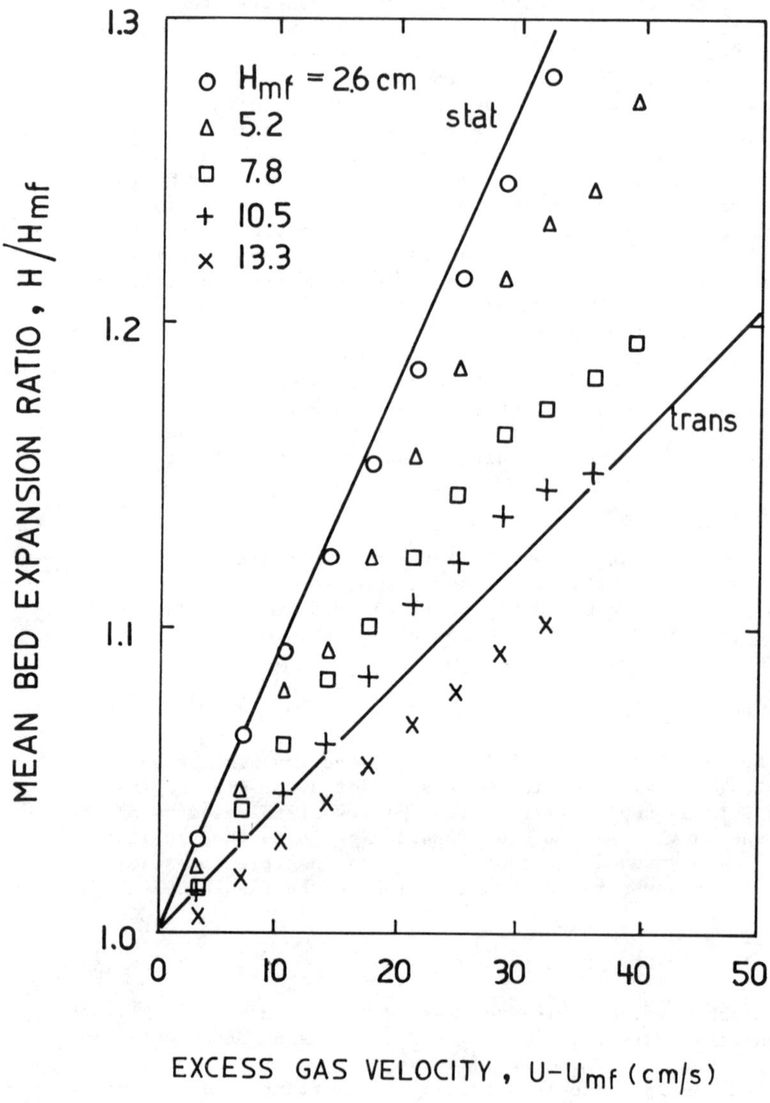

Fig.2 Variation of height ratio with velocity ratio For [14]
[d_p=1540μm, U_{mf}=55cm/s, ϵ_{mf}=0.41 (estimated), Bed: 15.2cm × 30.5cm]

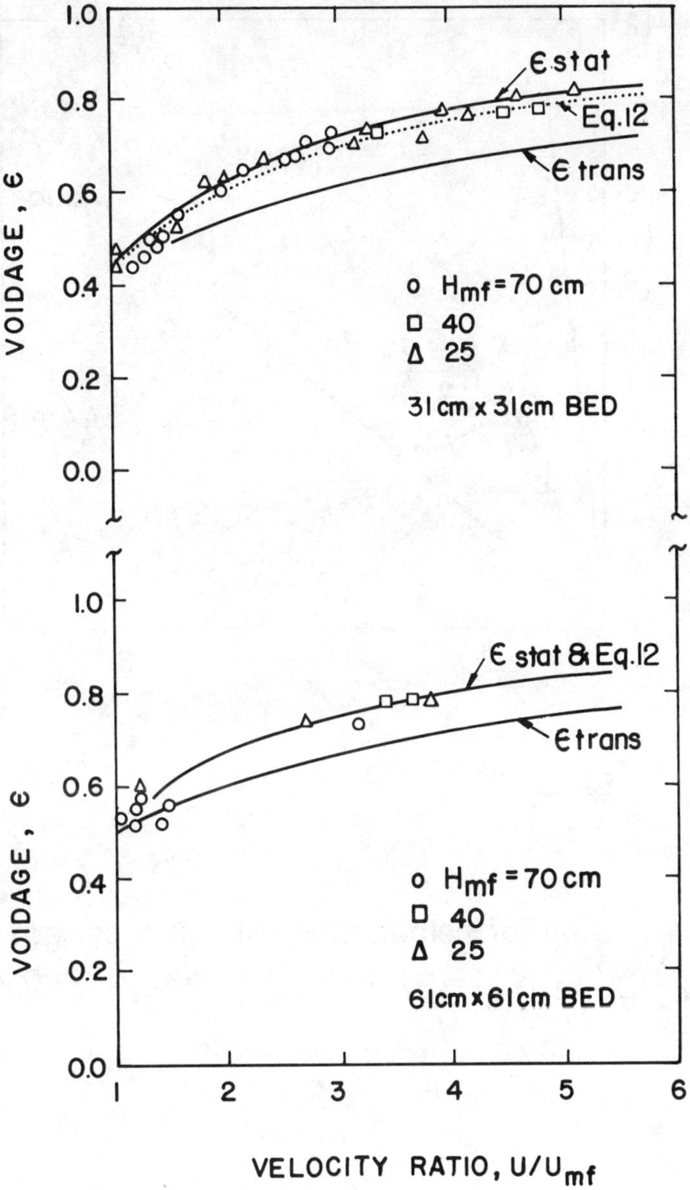

Fig.3(a) Variation of voidage with velocity ratio. For [15]. [d_p = 2600 μm, U_{mf} = 1.3 m/s, $\epsilon_{mf} \simeq$ 0.43, Bed: 30.5 cm x 30.5 cm]

Fig.3(b) Variation of voidage with velocity ratio. For [15] [d_p = 2600 μm, U_{mf} = 1.3 m/s, $\epsilon_{mf} \simeq$ 0.50, Bed: 61 cm x 61 cm]

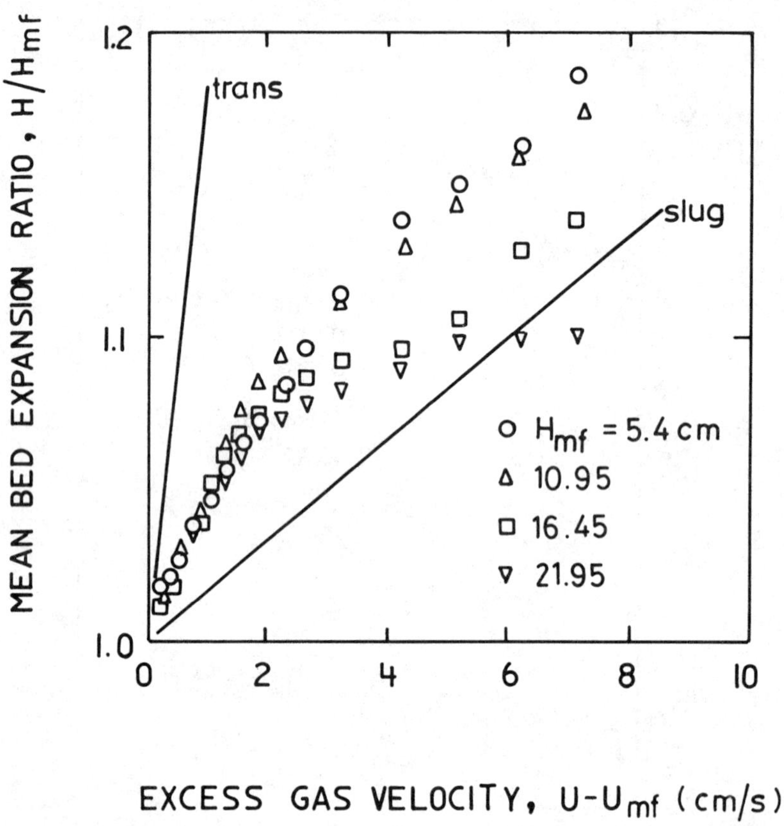

Fig.4 Variation of height ratio with excess gas velocity. For [17] [d_p=101 μm, U_{mf} =1.37 cm/s, ϵ_{mf} =0.49, Bed: 30.8 cm dia.]

obtained by Geldart [17] for 101 μm sand particles, are seen to generally lie between the loci of the transition expression and the slug relation. Despite the absence of slugs in the bed, the reported voidage ratio approaches (and for two data points crosses) the slug relation. This behavior can be related to the possibility of a bubble velocity somewhat larger than the slug velocity in the bubbly to slug flow transition, as mentioned above.

4. PREDICTION OF BUBBLE DIAMETER

4.1 Analysis

Examination of Eqs. (4)-(7) reveals that, at specified values of U and U_{mf}, gas flow continuity can provide a relationship between the bubble fraction, δ, and bubble diameter, d_v. A second relation between bubble fraction and diameter is thus needed to theoretically determine bubble diameter at fixed values of superficial and minimum fluidization velocities and it can be obtained by considering volumetric bubble flow at a given height within a fluidized bed as in Eq. (8).

In the slow bubble regime, the fraction of the bed occupied by bubbles is given by (6). Combining (6) and (8) yields an algebraicly complex relation for slow bubble diameter, as

$$d_{vs} = [6A(U - U_{mf})/\pi f_1(1 + 2U_{mf}/(0.711\sqrt{(gd_v)} + U - U_{mf}))]^{1/3} \tag{14}$$

where d_{vs} is the diameter of a slow bubble. Similarly, the desired expression for bubble diameter in the fast bubble regime can be obtained by solving simultaneously (8) and (9), which yields

$$d_{vf} = [6A(U - U_{mf})/\pi f_1(1 - U_{mf}/(0.711\sqrt{(gd_v)} + U - U_{mf}))]^{1/3} \tag{15}$$

where d_{vf} is the fast bubble diameter. When some measure of bubble frequency is available (14) and (15) can be solved numerically by successive substitution of trial values of d_v, or, with the aid of judicious approximation, can be cast in a form amenable to analytic evaluation.

It is significant to note that, while several bubble diameter expressions are available in the literature, and one due to Darton et al. [18] achieves significant accuracy for small particle beds, none of these formulations attempts to satisfy gas flow continuity. In contrast, both (14) and (15) fully satisfy gas flow continuity as structured by the two-phase equations.

4.2 Comparison with Data

4.2.1 Slow bubble regime. In the large-particle ($d_p \simeq 1760$ μm) fluidized bed investigation by Cranfield & Geldart [3] both bubble frequency and bubble diameter were measured by independent means. The relative predictive capabilities of the present bubble growth model and Darton et al.'s [18] fast-bubble formulation can thus be examined by comparison with Cranfield & Geldart's experimental d_v values. Such a comparison is shown in figure 5, where the values obtained by numerical solution of (14) are seen to be in excellent agreement with the data. Alternately, use of the method proposed by Darton et al. [18] appears to result in a widening disparity as the measured bubble diameter increases, suggesting that it may be inadvisable to use fast-bubble formulations to predict bubble size in large particle beds.

In an earlier large particle ($d_p \simeq 1540$ μm) fluidization study [14], the diameter and frequency of the bubbles erupting from a vigorously bubbling bed were measured, although with less accuracy than in [3]. Examination of figure 6 reveals that the solution of (14) using the measured frequency, yields moderately good agreement between predicted and measured bubble diameters, the latter taken to equal 2/3 of the observed eruption diameters. The modest disparity noted could perhaps be accounted for by experimental inaccuracies. The other two predictions shown in the figure, obtained by solving (14) with the Cranfield & Geldart [3] frequency correlation and the Darton et al. [18] model, respectively, are seen to offer significantly poorer agreement with the data, especially for small bubble diameters. These results can be traced to differences in bubble frequency between the two large particle studies and suggest again that fast-bubble correlation may not be applicable to beds operating in the slow bubble regime.

4.2.2 Fast bubble regime. In a study of small particle ($d_p \simeq 103$ μm) fluidization, Werther [19] measured both volumetric mean bubble diameters and bubble frequency as functions of fluidizing velocity and bed height. The excellent agreement between predicted bubble diameter, obtained by solution of (15) with the empirical frequency, and the measured values is shown in figure 7. Significantly, the Darton et al. [18] correlation, developed in part on the basis of the [19] data, is in this regime seen to offer equally good agreement. Values predicted on the basis of the Cranfield & Geldart [3] frequency, are, on the other hand, seen to diverge widely from the measured fast bubble diameters.

4.3 Modified Darton Formulation

4.3.1 Introduction. Examination of available empirical and semi-empirical correlations for bubble diameter in fluidized beds [3, 9, 18-23] reveals striking differences in both the form and specific dependence of d_v on key variables. Yet, much of the data used to obtain three of the more prominent correlations -- Cranfield & Geldart's [3] for large particles and both Werther's [19] and Darton et al. [18] for small particles -- can be shown to agree with two phase theory [26]. It would thus appear possible to use the two-phase/frequency approach to derive approximate expressions for bubble diameter that are compatible with the present understanding of fluidization phenomena and offer a rational basis for the correlation of future experimental results.

In the previously referred to Darton et al. study, the coalescence of neighboring streams was assumed to be the primary mechanism for bubble growth in fluidized media and coalescence was postulated to occur after each bubble had risen a distance λD_c in the medium, where λ is an

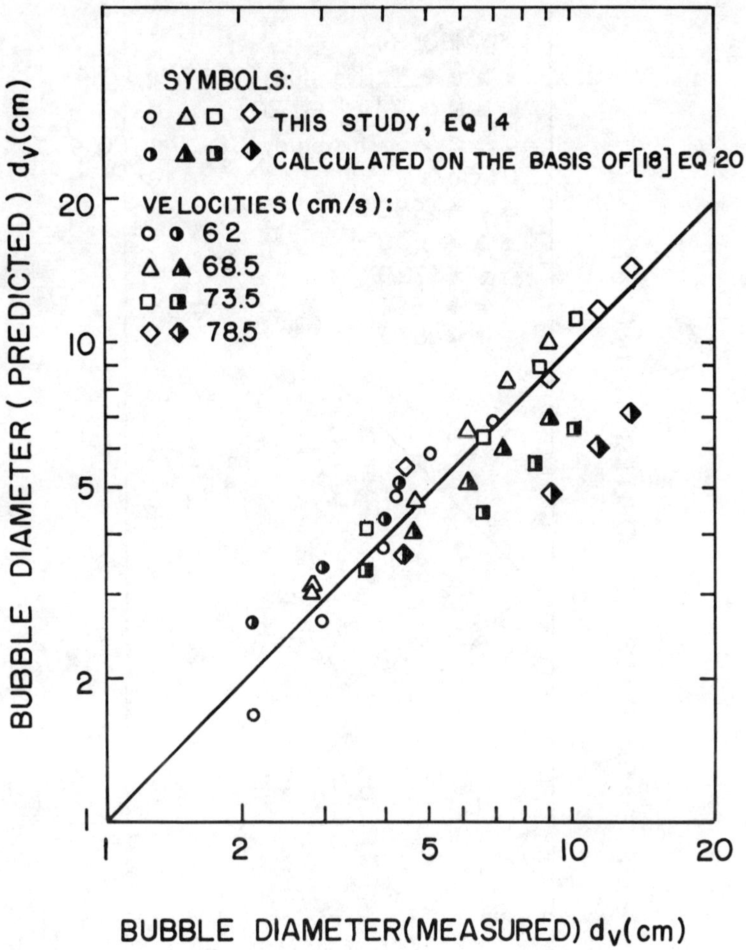

Fig.5 Bubble diameter predictions, data of Cranfield & Geldart [3]

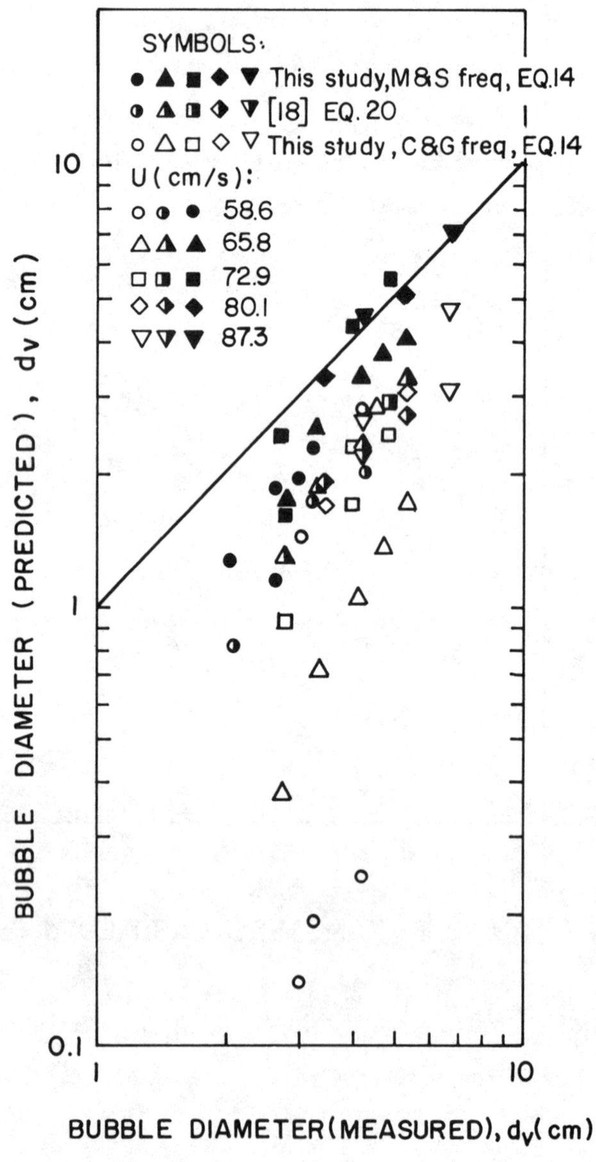

Fig.6 Bubble diameter predictions, data of Mc Grath & Streatfield [14]

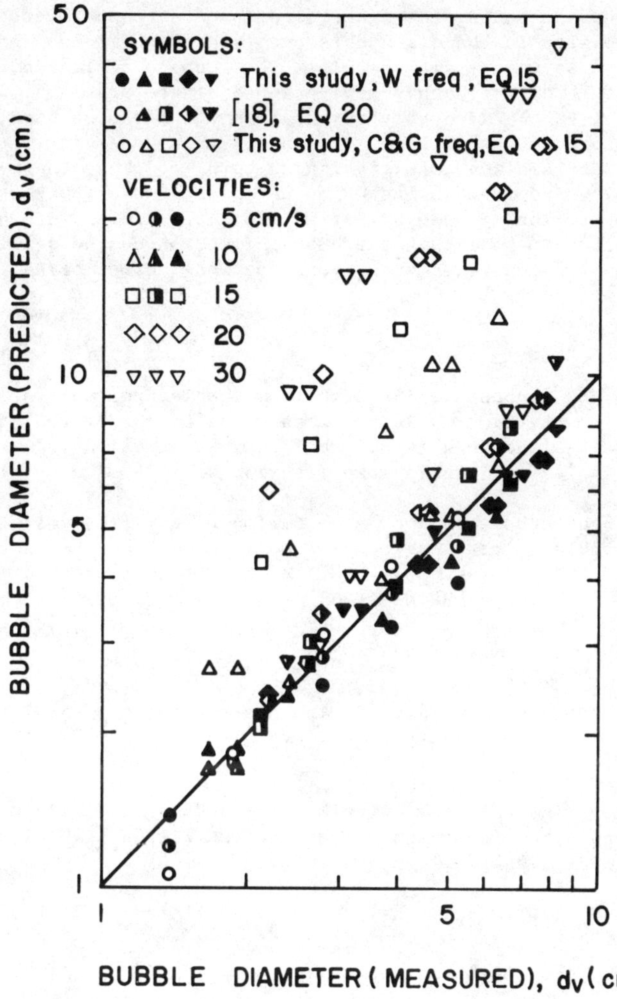

Fig. 7 Bubble diameter predictions, data of Werther [19]

empirical constant and D_c the diameter of the "catchment" area of each bubble and related to the distance separating adjacent bubbles.

The basic Darton et al. postulate that bubble coalescence occurs at a fixed multiple of the bubble separation distance provides, in effect, an alternative bubble frequency formulation which can be used, in the manner described in the earlier sections of this paper, to derive a bubble diameter relation consistent with both two phase theory and gas flow continuity and free of the unproven assumption of bubble volume conservation originally used by Darton [18]. To aid in this process, it is convenient to note that the bubble fraction can be related to the number of bubbles contained in a cross-sectional slice of height kd_v in the bed, according to

$$\delta = N'' \left(\frac{\pi}{6} d_v^{\,3}\right)/kd_v . \tag{16}$$

In (16), N" represents the number of bubbles per unit area, at a given height in the bed, while kd_v equals the height of the bubble and wake or cloud. Although k is not well defined, a value of 1.1 would be consistent with a fast bubble wake fraction of 1/3, as suggested in [24].

4.3.2 Fast bubble regime. The Darton et al. [18] postulate on successive bubble coalescence at increments of D_c can be used to relate the number of bubbles (per unit area) at height h in the bed to the number at height h_0. Following Hughes [25]

$$N''(h) = N''_0[1 + (\sqrt{2} - 1)(h - h_0)/\lambda D_c\,]^{-2} \tag{17}$$

where D_{c0} is the catchment area at the distributor and equal to $\sqrt{(4A_0/\pi)}$ or $\sqrt{4/\pi N''_0}$). Comparison of this relation with the data of [19, 26] shows λ to vary from approx. 0.5 to 2.45 and to average 1.7 for the two studies cited [25].

Combining (16) and (17), with k = 1.1 and λ = 1.7, yields an expression for fast bubble fraction which can be equated with δ_{fb} obtained from (7). This latter operation eliminates δ and yields

$$d_v = 1.284[U-U_{mf}]^{0.5}[0.414(h-h_0)/1.7+D_{c0}]/[0.711\sqrt{gd_v} + U-2U_{mf}]^{0.5} \tag{18}$$

In the fast bubble regime, and especially for the range of large bubble sizes encountered in [19, 27], the denominator of (18) can be shown to approach $(0.711\sqrt{gd_v})^{0.5}$. With this approximation and some algebraic manipulation, (18) yields an explicit expression for bubble diameter

$$d_v = 0.45(U-U_{mf})^{0.4}[h + 4.63\sqrt{A_0}]^{0.8}/g^{0.2}. \tag{19}$$

This expression is very similar to the Darton et al. [18] relation of Eq. 20, and as shown in figure 8, the lower coefficient (0.45 vs 0.54)

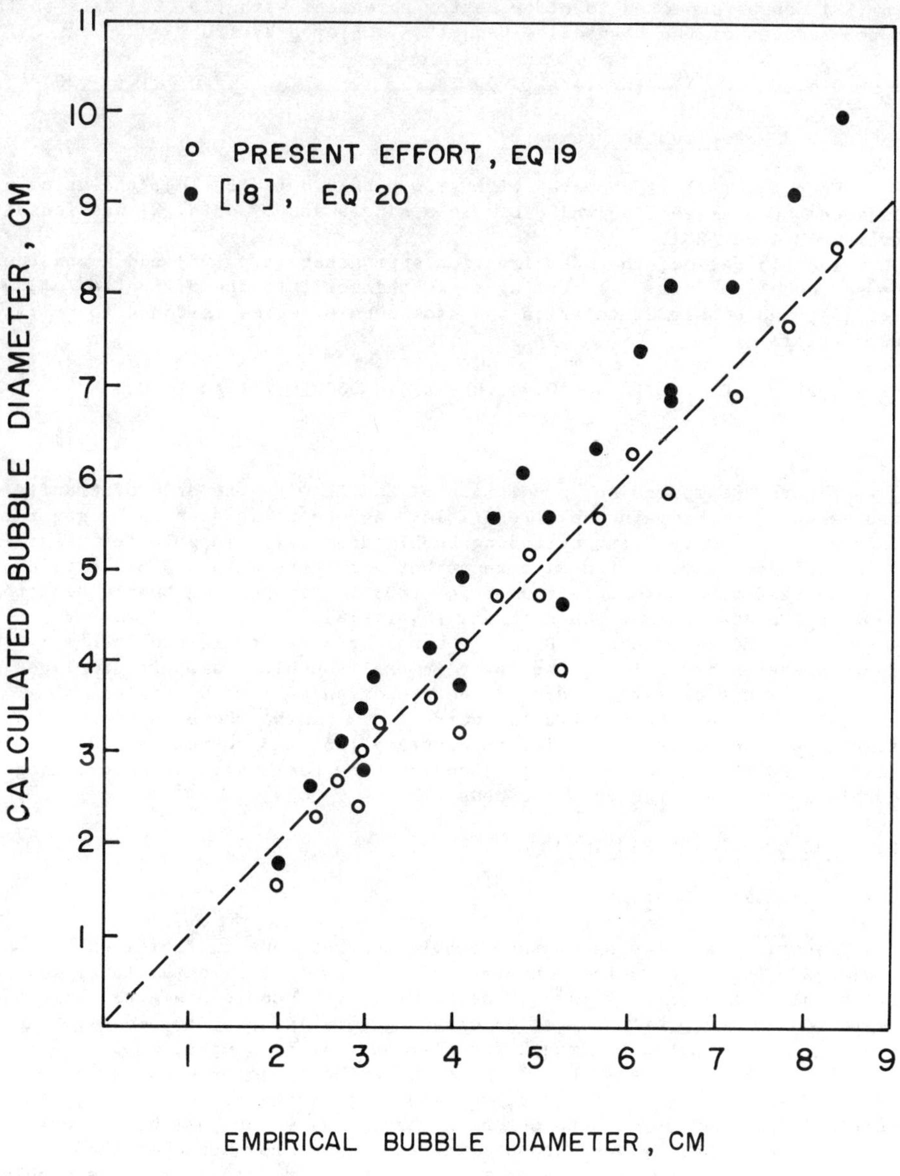

Fig.8 Calculated and empirical bubble diameters in small particle fluidized beds (data of Werther, 1976).

in (19) can be expected to offer better agreement with [19, 27] data in the upper reaches of the respective beds than achieved by (20)

$$d_v = 0.54(U - U_{mf})^{0.4}(h + 4\sqrt{A_0})^{0.8}/g^{0.2} \tag{20}$$

4.3.3 Slow bubble regime

When Eq. (17) is compared with slow bubble data [3, 14], the value of λ is found to average 0.55 while D_{c_0} in a single set of data [3] was found to average 2.4 cm [25].

In this regime, the wake fraction approaches zero [28] and k can thus be set equal to unity. Inserting these changes into the derivation of Eq. (9), the bubble diameter in the slow bubble regime is found to be expressible as

$$d_v = 1.22(U - U_{mf})^{0.5}[2.4 + 0.414(h - h_0)/0.55]/[0.711\sqrt{gd_v} + U + U_{mf}]^{0.5} \tag{21}$$

Figure 9 compares the prediction of Eq. 21 with the data of Cranfield and Geldart and McGrath and Streatfield. Agreement is seen to be generally fair and to improve with increasing bubble diameter. This latter trend is in notable contrast to the divergence between these data and prediction via the fast bubble formulation of Eq. (20) for increasing bubble diameter shown by Hughes and Bar-Cohen et al. [8] (1978).

Since the influence of D_{c_0} on d_v is initially strong and then decreases with bed height, the error in small-bubble diameters displayed in figure 9 may be largely due to the use of an average D_{c_0} value in Eq. (21). Use of the empirically correct D_{c_0} can be shown to yield a significant increase in predictive accuracy [25] and to thus validate reliance on the Darton et al. [18] coalescence postulate for determining bubble growth in large particle beds.

5. TUBE/BUBBLE INTERACTIONS

5.1 Qualitative Description

Experimental studies of tube/bubble interactions in fluidized media [29, 30] suggest that bubbles colliding with horizontal tubes may be split into two or more voids. Consequently, the bubble diameter expressions presented in the previous section cannot generally apply to beds with heat-exchange tubes and an alternate formulation must be sought.

Below the tube bank, bubbling can be expected to proceed as in an open bed and Eqs. (19) and (21), for small and large particle beds, respectively, can be used to determine bubble diameter just below the first row of tubes. Upon encountering this first row, some of the bubbles split -- the exact percentage depending on such factors as bubble size, tube spacing and tube diameter. In the vertical spacing between the tube rows, bubble coalescence can be expected to resume and to follow

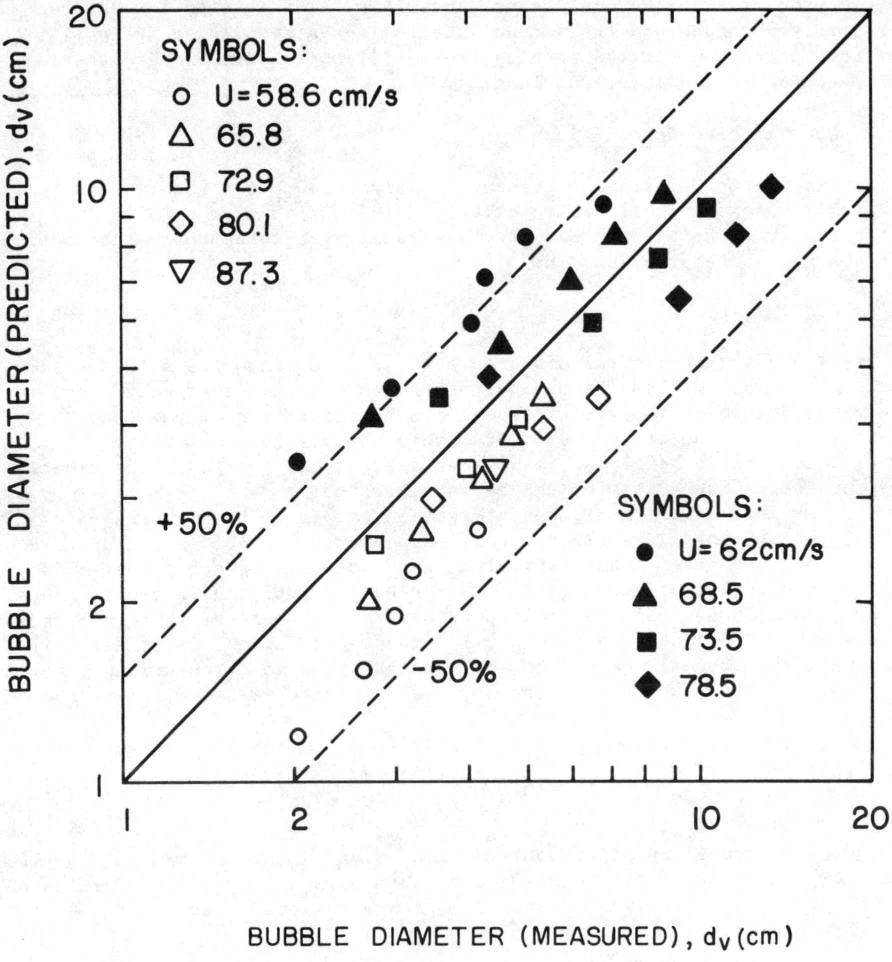

Fig.9 Predicted and empirical bubbe diameters in large particle fluidized beds: solid symbols, data of Cranfield and Geldart [3]; open symbols, data of Mc Grath and Streatfield [14]; predictions via EQ 21

the Darton et al. [18] coalescence postulate. The interaction between the rising bubbles and the second tube row again results in the splitting of some percentage of the bubbles, to be followed by bubble growth via coalescence in the inter-row spacing.

5.2 Quantitative Description

Based upon Glass and Harrison's observation that a bubble tends not to split unless it collides relatively head-on with a tube [30], the probability of splitting, associated with a given tube/bubble interaction, can perhaps be approximated by

$$p_s = 0.5d_v/l_h \tag{22}$$

where l_h is the horizontal distance between tube centerlines in the same tube row. This expression can be interpreted to mean that only those bubbles whose centers pass within half a bubble radius of the tube centerline will split. As empirical data becomes available, the coefficient in Eq. (22) can be altered and, if necessary, an additional dependence on tube diameter can be incorporated.

Numerical solution of the interaction between bubble chains and multiple tube rows in a fluidized bed appears to suggest that a dynamic equilibrium, between bubble splitting and coalescence, can result in an asymptotic bubble diameter [31]. By the use of judicious approximations Bar-Cohen & Hughes [31] found it possible to derive an analytic expression for the asymptotic bubble diameter in tube banks and thus provide a first estimate for thermal combustion and chemical kinetic calculations. For the large particle beds of interest, the requisite relation is given by

$$d_a = 1.49(l_v\, l_h)^{\frac{1}{2}} \left[\frac{U - U_{mf}}{0.711\sqrt{gl_h} + U + U_{mf}}\right]^{1/4} \tag{23}$$

where d_a is the asymptotic bubble diameter, and l_v and l_h are the vertical and horizontal separation distances between pipe centers. The results of a recent study by Whitehead [32] offer tentative confirmation of the validity of this approach.

6. CLOSURE

The desire to provide designers of bubbling fluidized beds with approximate analytical expressions which embody the physics of bubble growth and gas redistribution in the emulsion, motivated much of the present development. The mathematical form of the resulting relations for bubble fraction and diameter, as well as the implicit parametric dependencies, can also serve to establish a structure for the correlation of empirical results.

Application of two-phase theory and gas flow continuity to bubbly fluidized beds suggests that bubble fraction varies inversely with bubble rise velocity. In the absence of appropriate data, an estimate of bubble size can be used to determine the relevant flow regime - slow or fast bubble - and to guide the selection of bounding expressions for bubble fraction and, hence, bed voidage or expansion. When more precise bubble fraction predictions are needed and when transport process or combustion calculations require a detailed knowledge of bubble diameter, the modified Darton relations for open or tubed beds can be used to provide an estimate of average bubble diameter as well as its parametric variation in the fluidized medium.

7. ACKNOWLEDGEMENT

Much of the effort reported herein was funded under DOE Contract E-18-2295 at the MIT Energy Laboratory (1977/78). The author acknowledges with thanks the contributions of L.R. Glicksman and R.W. Hughes to the conceptual framework and the interpretation of data presented in this study.

8. REFERENCES

1. Stewart, P.S.B. and Davidson, J.F., "Slug Flow in Fluidized Beds," Powder Technology, Vol. 1, pp. 61-80 (1967).

2. Bayens, J. and Geldart, P., "An Investigation into Slugging Fluidized Beds," Chem. Eng. Sci., Vol. 29, pp. 255-265 (1974).

3. Cranfield, R.R. and Geldart, D., "Large Particle Fluidization," Chem. Engng. Sci., Vol. 29, pp. 935-947 (1974).

4. Davidson, J.F. and Harrison, D., "Fluidized Particles," Cambridge University Press, Cambridge (1963).

5. Davies, R.M. and Taylor, G.I., "The Mechanics of Large Bubbles Rising through Extended Liquids and through Liquids in Tubes," Proc. R. Soc., Vol. A200, p. 375 (1950).

6. Toomey, R.D. and Johnstone, H.F., "Gaseous Fluidization of Solid Particles," Chem. Engng. Prog., Vol. 48, pp. 220-226 (1952).

7. Kunii, D. and Levenspiel, O., "Fluidization Engineering," Wiley, New York (1969).

8. Bar-Cohen, A., Glicksman, L. and Hughes, R.W., "Fluid Dynamic Modelling of Fluidized Bed Combustors," Proc. 5th Int. Conf. Fluidized Bed Combustion, Washington, DC, Vol. III, pp. 458-474 (1978).

9. Geldart, D., "The Effects of Particle Size Distribution on the Behavior of Gas-Fluidized Beds," Powder Technol., Vol. 6, pp. 201-205 (1972).

10. Godard, D. and Richardson, J.F., "Bubble Velocities and Bed Expansion in Freely Bubbling Fluidized Beds," Chem. Engng. Sci., Vol. 24, pp. 663-670 (1969).

11. Werther, J., "Influence of the Bed Diameter on the Hydrodynamics of Gas Fluidized Beds," Fluidization and Fluid Particle Systems, AIChE, Symp. Series Vol. 70, pp. 63-61 (1975).

12. Werther, J., "Bubble Chains in Large Diameter Gas Fluidized Beds," Int. J. Multiphase Flow, Vol. 3, pp. 367-381 (1977).

13. Bar-Cohen, A., Glicksman, L.R. and Hughes, R.W., "Theoretical Bounding Relations for Void Fraction in Bubbling Fluidized Beds," Int. J. Multiphase Flow, Vol. 6, pp. 319-327 (1980).

14. McGrath, L. and Streatfield, R.E., "Bubbling in Shallow Gas-Fluidized Beds of Large Particles," Trans. Inst. Chem. Engng., Vol. 49, pp. 70-79 (1971).

15. Canada, G.S., McLaughlin, M.H. and Staub, F.W., "Flow Regimes and Void Fraction Distribution in Gas Fluidization of Large Particles in Beds without Tube Banks," Paper 103d, 69th AIChE Meeting, Chicago (1976).

16. Hovmand, S. and Davidson, J.F., "Slug Flow Reactors." In Fluidization, J.F. Davidson and D. Harrison (editors), Academic Press, London, pp. 193-257 (1971).

17. Geldart, D., "1967/68 The Expansion of Bubbling Fluidized Beds," Powder Technol., Vol. 1, pp. 355-368.

18. Darton, R.C., LaNauye, R.D., Davidson, J.F. and Harrison, D., "Bubble Growth due to Coalescence in Fluidized Beds," Trans. Instit. Chem. Engng., Vol. 55, pp. 274-280 (1977).

19. Werther, J., Bubble Growth in Large Diameter Fluidized Beds." In Fluidization Technology, Kearnis, D.L. (editor), Hemisphere, Washington, DC, pp. 215-235 (1976).

20. Rowe, P.N., "1976 Prediction of Bubble Size in a Gas Fluidized Bed," Chem. Engng. Sci., Vol. 31, pp. 285-288 (1976).

21. Yashi, G. and Johanson, L.N., "Characteristics of Gas Pockets in Fluidized Beds," AIChE J., Vol. 4, pp. 445-452 (1958).

22. Kato, K. and Wen, C.Y., "Bubble Assemblage Model for Fluidized Bed Catalytic Reactors," Chem. Engng. Sci., Vol. 24, pp. 1351-1369 (1969).

23. Park, W.H., Kang, W.K., Capes, C.E. and Osbert, G.L., "The Properties of Bubbles in Fluidized Beds of Conducting Particles as Measured by an Electroresistivity Probe," Chem. Engng. Sci., Vol. 24, pp. 851-865 (1969).

24. Rowe, P.N. and Partridge, B.A., "An X-Ray Study of Bubbles in Fluidized Beds," Trans. Inst. Chem. Engng., Vol. 43, pp. 157-175 (1965).

25. Hughes, R.W., "Modelling Bubble Growth in Gas/Solid Fluidized Beds," M.S. Thesis, Dept. of Mech. Engng., MIT, Cambridge, MA (1978).

26. Bar-Cohen, A., Glicksman, L.R. and Hughes, R.W., "Semi-Empirical Prediction of Bubble Diameter in Gas Fluidized Beds," Int. J. Multiphase Flow, Vol. 7, pp. 101-113 (1981).

27. Rowe, P.N. and Everett, D.J., "Fluidized Bed Bubbles Viewed by X-Rays--III. Bubble Size and Number When Unrestrained Three-Dimensional Growth Occurs," Trans. Inst. Chem. Engrs., Vol. 50 (1972).

28. Geldart, D. and Cranfield, R.R., "The Gas Fluidization of Large Particles," Chem. Engng. J., Vol. 3, pp. 211-231 (1972).

29. Cloete, F.L.D., Discussion in Proceedings of the International Symposium on Fluidization, Netherlands University Press, Amsterdam, pp. 305-307 (1967).

30. Glass, D.H. and Harrison, D., "Flow Patterns Near a Solid Obstacle in a Fluidized Bed," Chem. Eng. Sci., Vol. 19, pp. 1001-1002 (1964).

31. Bar-Cohen, A. and Hughes, R.W., "Predicting Asymptotic Bubble Size in Tube/Bundle Interactions Within Fluidized Beds," Israel Journal of Technology, Vol. 18, pp. 137-145 (1980).

32. Whitehead, A.B., "Asymptotic Bubble Size in Tubed Fluidized Beds," submitted to Chem. Eng. Sci. (1981).

ADVANCES IN THE FLUIDIZATION OF COARSE PARTICLES

D. Geldart

Schools of Chemical Engineering and Powder Technology
University of Bradford
Bradford, West Yorkshire, U.K.

ABSTRACT

Fluidization is a technique which has been applied extensively to the processing and use of fine powders but much less to coarse granular materials. There is increasing interest in the fluidization of coarse solids largely because of their growing use in fluid bed dryers and coal combustors.

Recent experimental work on entrainment and segregation at high velocities, and on solids discharge, has led to greater understanding of coarse particle fluidization.

1. INTRODUCTION

The technique of gas fluidization has been applied extensively to the processing of powders or to their use as catalysts, but much less to coarse granular solids, and this is reflected in the relative number of publications - less than 100 dealing with the fluidization of coarse solids compared with many thousands concentrating on small particles. Some workers have reserved the term *teeter bed* for the fluidization of coarse solids to emphasize that their behaviour is somewhat different from that of, for example, cracking catalyst, often considered to be an ideal powder for fluidization. The term *teeter bed* has fallen into disuse as workers have recognized, rightly, that although there are many ways in which coarse solids behave differently from fine powders, they do have sufficient features in common to justify them being called fluidized beds. The development of fluid bed combustion in particular has led to increasing interest in the fluidization behaviour of coarse solids.

In this paper I shall first of all define what is meant by *coarse* particles in the context of fluidization, and then describe some recent advances in understanding their behaviour.

2. WHAT ARE COARSE PARTICLES?

In 1973, Geldart |2| proposed a system which classified solids into four broad groups, designated in order of increasing particle size, C, A, B and D, the last being reserved for fluid dynamically coarse particles. Two criteria were used to distinguish the coarse group D solids from the others; (i) group D solids are those in which a spouted bed at least 30 cm deep can be formed; (ii) group D solids are those in which gas bubbles more upwards more slowly than the interstitial gas at incipient fluidization. Both definitions involve not only the properties of the gas and solid, but also the bed geometry - for example, ease of spouting increases as the jet diameter decreases and bed diameter increases; bubbles move more slowly near the distributor than under otherwise identical conditions near the top of deep beds. In spite of these complications, the two definitions give somewhat similar values, and when simplified by relating them to fluidization by ambient air, the groups can be depicted on Fig.1.

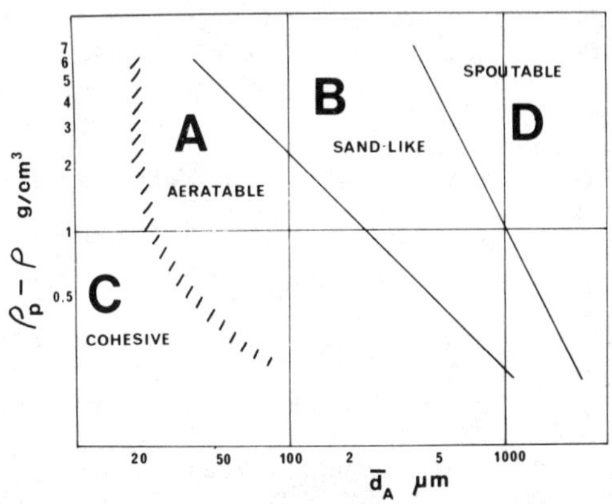

Figure 1. Powder classification diagram for air at ambient conditions |1|.

It can be seen that it is also important to state the particle density when discussing what are coarse particles, but for silica sand ($\rho_p \simeq$ 2600 kg/m^3) beds of particles having a mean size larger than about 600-700 μm would be considered as coarse, or group D, solids.

3. GENERAL FLUIDIZATION CHARACTERISTICS OF COARSE PARTICLES

The general fluidization characteristics of group D solids can be summarised as follows:

All but the largest bubbles rise more slowly than the interstitial fluidizing gas, so that gas flows into the base of the bubbles and out of the top, providing a mode of gas exchange and by-passing different from that observed with smaller particles |2|. The gas velocity in the dense phase is high, solids mixing relatively poor; consequently back-mixing of the dense phase gas is small. The flow regime around the particles is likely to be turbulent, causing some particle attrition |3| with rapid elutriation of the fines produced. Relatively sticky materials can be fluidized since the high particle momentum and fewer particle-particle contacts minimize agglomeration |4|.

Horizontal voids appear close to the distributor; in columns in which the bed depth/diameter ratio; H_s/D_B, is large (>2) these may extend across the tube giving bridging and piston-like solids slugs |5| which slide up the tube and collapse into wall slugs near the surface. In a bed with a low H_s/D_B (>1) at low velocities these voids split into bubbles which appear to grow without coalescence, as though draining gas from the surrounding dense phrase. As the gas velocity is raised the bubbles coalesce into large diameter voids which produce large bed oscillations; this has been called the apparent slug regime |6| in beds of low H_s/D_B since the bed is not deep enough for slugging to become fully developed. At even higher velocities a turbulent or churning bed regime develops in which the slugs break down, there is less vibration and smaller pressure oscillations. The onset of these regimes is a function of bed depth, particle properties, gas velocity and total operating pressure, but no general correlation is available. A typical fluidization flow regime map for 650 μm glass particles is shown in Fig. 2., where the gas velocity, U, is expressed as a fraction of the particle terminal velocity V_t. Further exploration of these regimes is needed but requires very large compressors, particularly for investigation of the turbulent regime.

4. RECENT STUDIES WITH COARSE PARTICLES

An extensive study carried out at the University of Bradford and financed largely by Electric Power Research Institute, Palo Alto, U.S.A., in support of fluid bed combustion development, has given further insights into the fluidization behaviour of coarse particles. Most of the work was done in a column 0.29m diameter x 5m tall, details of which are given elsewhere |7|.

4.1 Segregation by Size

It is generally believed that vigorously bubbling beds of small

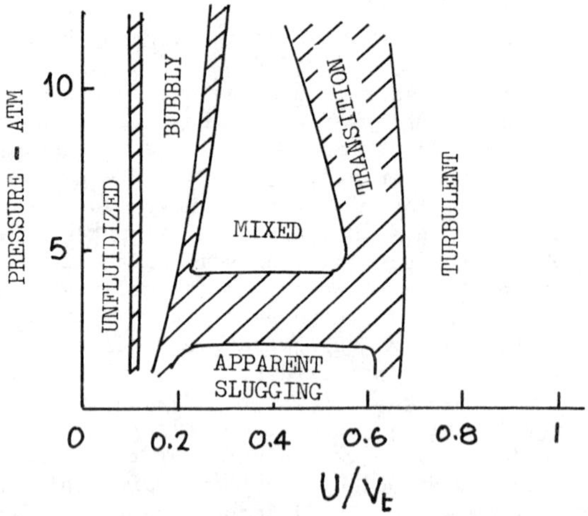

Figure 2. Flow regime map for 650μm particles; bed heights 0.25 - 0.7 m |6|.

particles are well-mixed particularly with respect to size distribution and indeed fluid beds are sometimes used as mixers. However, in our studies using coarse solids, we found evidence of segregation by size.

The work fell into two related, but distinct sets of experiments. In one we concentrated on segregation in beds of (nominally) 1mm and 2.5 mm coarse sands into which "fines" (nominally < 550 μm) were injected and elutriated continuously, separated in a cyclone and re-injected. In the other, the solids (limestone and sand) were so large (0.6 - 6 mm) that virtually no elutriation occurred and so reinjection was not required. The materials had essentially bimodal size distribution as can be seen in Fig. 3. (Sand and limestone have almost identical particle densities). Full experimental details are given in Ref. |8|.

The results are presented as $x/\bar{x}$ against distance above the gas distributor where x is the mass concentration of the size fraction under consideration at a given position and $\bar{x}$ is the average concentration of that fraction in the bed as a whole. Gas velocity has a considerable effect, with segregation increasing as velocity is decreased and this is not unexpected. However there was still significant segregation of sand

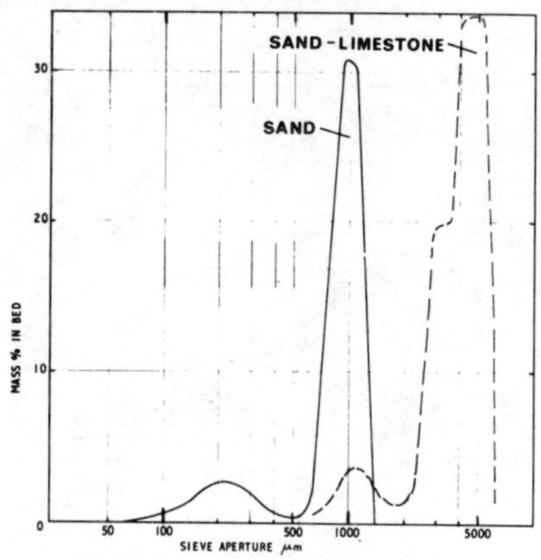

Figure 3. Size distribution of solids used in segregation in tests |8|.

at the top of the bed in the sand/limestone mixture when the gas velocity was 4 m/s ($U - U_{MF}$ = 2.5 m/s) - see Fig. 4.

The size of the fines relative to the coarse is also important, as seen in Fig. 5., where segregation can be seen to increase as the particle size of the fraction decreases.

This is also demonstrated quite strikingly in Fig. 6B and C. Clearly segregation of the fines (in this case all the material less than 550 μm) increases as the size of the coarse solids making up the bulk of the bed is increased from (nominally) 1mm to 2.5mm.

Much more work remains to be done before the degree of segregation can be predicted with any confidence but a qualitative equation of the form below is tentatively suggested to describe segregation in beds containing a wide size distribution of particles having similar densities.

$$\left\{\frac{x_f}{\bar{x}_f}\right\}_{\text{top of bed}} \propto \left\{\frac{\bar{d}_p}{d_f}\right\}^m \times \left\{\frac{U}{U - U_{MF}}\right\}^n$$

where x_f and $\bar{x}_f$ are the concentration and mean concentration of a fines of size d_f; $\bar{d}_p$ is the mean sieve size of particles making up the bulk of the bed, U and U_{MF} are the superficial velocities of the gas and incip-

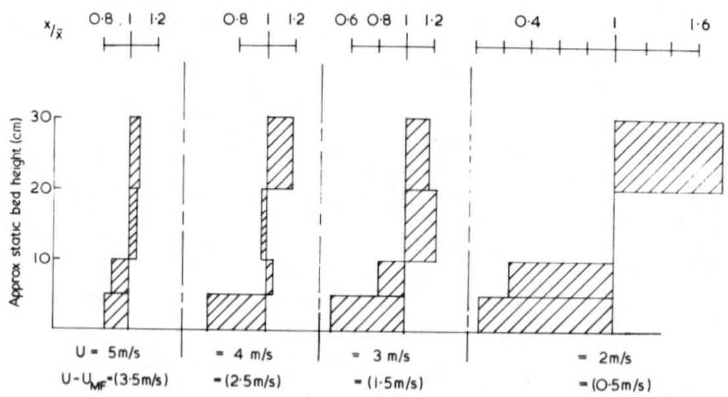

Figure 4. Decreasing the gas velocity increases the segregation of −1.4 + 0.6mm sand (ρ_p = 2650kg/m^3) in −6 + 1.4mm limestone (ρ_p = 2720kg/m^3) |8|.

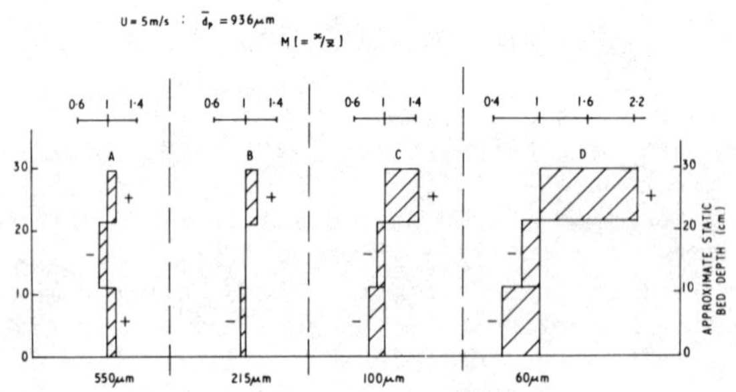

Figure 5. Segregation of various fines fractions in a bed of 1 mm coarse sand at 5m/s |8|.

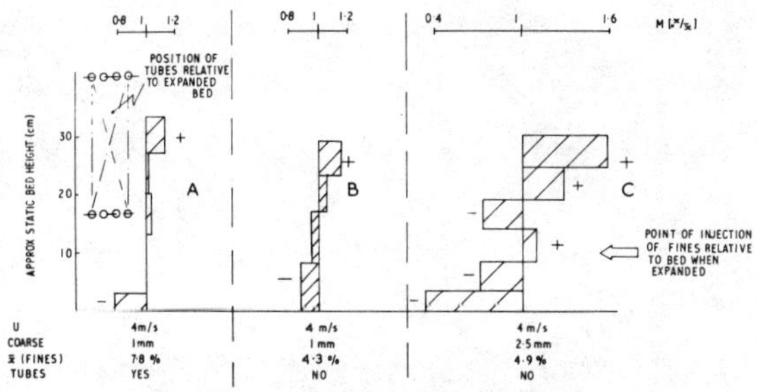

Figure 6. Segregation of >550μm fines bed of 1 mm and 2.5 mm coarse sands |8|.

ient fluidization velocities respectively. The type of size distribution and the bed depth/diameter ratio may also be important.

4.2 Entrainment and Elutriation

One consequence of fines segregation is that for the same average concentration in the bed, the rate of carryover of any particular size fraction of fines is larger when the bed is predominantly 2.5 mm solids than when it consists of 1 mm particles (Fig.7). This is because it is the top region of the bed which acts as a feeder for the freeboard through the bursting of the bubbles. Modelling the carryover of particles for well-mixed fluidized beds has not yet been done successfully and segregation is an extra complication which, though interesting, we could well do without.

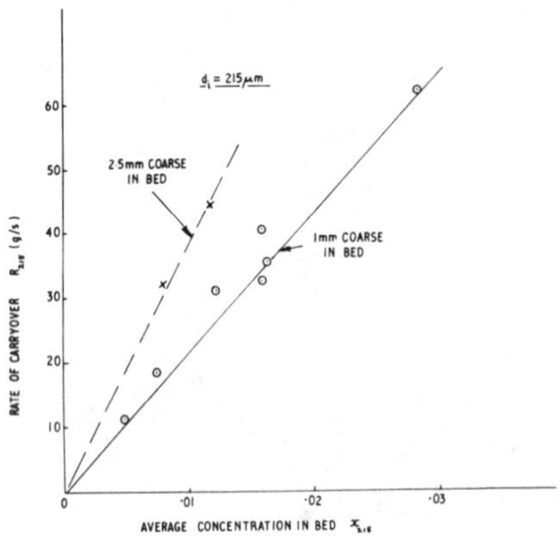

Figure 7. The carryover rate of 215μm sand increases as the mean size of the coarse solids comprising the bed increases: U=4m/s, tubes in bed |8|.

There is, however, yet another complexity which has been discovered recently |7, 9|, and whilst it is not exclusively a coarse particle phenomenon, it was observed during the programme of research into the fluidization behaviour of coarse solids. Providing the height of the column is large (at least 2 m higher than the bed surface) if the particles being fluidized have terminal velocities larger than the superficial gas velocity, they will disengage from the gas and fall back into the bed. Only particles with a terminal velocity, V_t, smaller than the superficial gas velocity, U, might be expected to appear in the offgases. However, in a small column operated batchwise, we observed |10| that the addition of fines to the bed increased the rate of carryover of coarser particles, and resulted in the carryover of particles (having $V_t/U>1$) which previously remained in the column. These experiments were extended to much larger particles in the 0.29 m dia. column x 5 m tall operated under continuous conditions |7,9| and the same effect was observed (Fig.8). The reasons are not

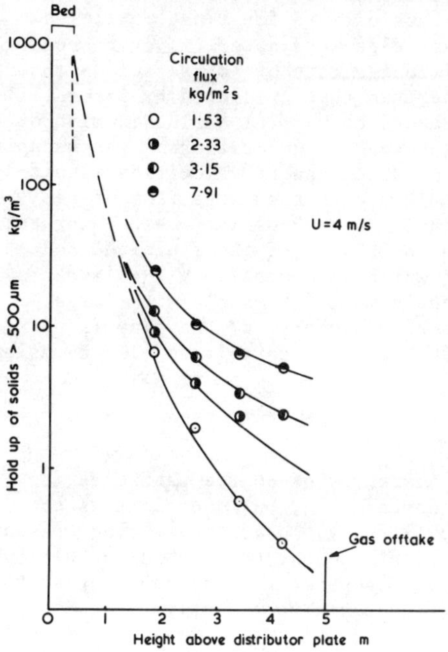

Figure 8. Increasing the circulation rate of fine particles increases the hold up of coarse above the bed |9|.

fully understood, but it appears possible that the fines (<500 μm in this context) collide with the coarse solids, exchange momentum, and cause them to travel further up the column before they disengage and fall back. As the fines entrainment flux is increased, more and more coarse particles are carried out of the column. There is no evidence that the elutriation rate of the fines is affected, though their hold-up in the column is increased slightly.

4.3 Fluidized Discharge of Coarse Solids

The rate at which solids flow out of fluidized beds is of interest in many industrial processes. A fluid bed may be used to feed powder to another part of the process or it may be the reactor from which processed solids or spent catalyst has to be removed, continuously or at regular intervals. In almost all published work the powders used had mean particle sizes less than 500 μm and most were smaller than 300 μm. During a recent development programme on discharge through short vertical pipes |11| it became apparent that the mass discharge rates of coarse solids were significantly lower than values predicted using equations derived for fine particles. Two effects peculiar to coarse particles were noted. (i) even when vigorously fluidized, the mass discharge rate of coarse solids showed a smaller dependency on bed height than that predicted by Bernoulli's equation and this appears to be caused by partial defluidization of the upper portion of the bed due to substantial quantities of gas escaping with the solids. (ii) coarse solids discharge at velocities significantly higher than $(2gH)^{\frac{1}{2}}$, probably due to the large flow of gas. This results in the stream of solids having a bulk density much lower than that of either the dense phase in the bed or of the fluidized bed as a whole. There is one other factor which is common to all solids flowing out of orifices or pipes but which is more noticeable with large particles and small pipes. The entire diameter of the pipe is not available for flow and the results had to be correlated |11| by using an effective pipe diameter $D_o-4\bar{d}_p$.

5. CONCLUSIONS

Our understanding of the behaviour of coarse particles is considerably better than it was 10 years ago, but much remains to be learnt. Progress may well be slow largely because any further research which is likely to be of use to industry ought to be done in relatively large diameter beds using pressures, temperatures and gases other than ambient air; such research is difficult and very costly.

REFERENCES

1. Geldart, D., "Types of Gas Fluidization", Powder Technol. Vol.7, pp.285-292, (1973).

2. Catipović, N.M., Jovanović, G.N., and Fitzgerald, T.J., "Regimes of Fluidization for Large Particles", A.I.Ch.E.J. Vol. 24, pp.543- , (1978).

3. Vaux, W.G., and Keairns, D.L., "Particle Attention in Fluid-Bed Processes", in Fluidization, J.R.Grace and J.M.Matsen (editors), Plenum Press, New York and London, pp.437-444, (1980).

4. Gluckman, M.J., Yerushalmi, J., and Squires, A.M., "Defluidization Characteristics of Sticky or Agglomerating Beds", in Fluidization Technology, Vol.II., D.L.Keairns (editor), Hemisphere Pub.Corp., Washington, pp.395-422, (1976).

5. Geldart, D., Hurt, J.M. and Wadia, P.M., "Slugging in Beds of Large Particles", A.I.Ch.E. Symp. Ser. Vol.74, No.176, pp.60-66, (1978).

6. Canada, G.S. and McLaughlin, M.H., "Large Particle Fluidization and Heat Transfer at High Pressures", A.I.Ch.E., Symp.Ser. Vol. 74, No.176, pp.27-37, (1978).

7. Geldart, D., "Behaviour of Fines in a Fluidized Bed of Coarse Solids", Final Report Oct.1981, Research Project 839-1, Electric Power Research Institute, Palo Alto. CS-2094

8. Geldart, D., Baeyens, J., Pope, D.J. and Van de Wijer, P., "Segregation in Beds of Large Particles at High Velocities", Powder Technol. Vol.30, pp.125-205, (1981).

9. Geldart, D. and Pope, D.J., "Interaction of Fine and Coarse Particles in the Freeboard of a Fluidized Bed", Powder Technol. (in Press).

10. Geldart, D., Cullinan, J., Georghides, S., Gilvray, D. and Pope, D.J., "Effect of Fines on Entrainment From Gas Fluidized Beds", Trans.Inst. Chem.Engrs. (London), Vol.57, pp.269-275, (1979).

11. Geldart, D. and Haesebrouck, M., "Studies on the Intermittant Discharge of Coarse Solids from Fluidized Beds", Paper presented at A.I.Ch.E. Annual Meeting, Los Angeles, Nov. (1982).

THE FLOW OF SOLID PARTICLES IN AN AERATED INCLINED CHANNEL

M. Ishida and H. Hatano

Research Laboratory of Resources Utilization
Tokyo Institute of Technology, Yokohama 227, Japan

ABSTRACT

The dynamic behavior of the flow of solid particles in an inclined open channel with air flow introduced through a porous base plate was examined. The solid particle velocity distributions were measured in detail using an optical fiber probe.

Five types of flow patterns were found depending on the air velocity and the slope of the channel. They were: sliding flow, immature sliding flow, splashing flow, bubbling flow, and gliding flow.

In the sliding flow, the velocity distribution normal to the bottom plate appeared linear. These observed velocity distributions were analyzed based on the variational principle, by which the velocity distribution was obtained as the solution which minimized a certain integral consisting of several energy terms.

1. INTRODUCTION

In the previous papers [1,2], the authors measured the velocity of solid particles within an aerated inclined channel in detail by a fine optical fiber probe. Based on those experiments, they classified the flow in an inclined open channel into five types depending on the gas velocity and the slope of the channel. They were: sliding flow, immature sliding flow, splashing flow, bubbling flow, and gliding flow. The first two could be observed when the air velocity was less than the minimum fluidization velocity, U_{mf}, for the particles. The last two were observed when it became higher than U_{mf}.

In the sliding flow, linear velocity distribution was observed and the velocity gradient in this region was almost independent of the thickness of the particle layer and increased as the slope of the channel became steep.

To analyze these velocity distributions, the authors developed a variational analysis and derived the result that the solid particles moved in such a manner that a certain integral term became a minimum.

This paper summarizes the main feature of those works by adding some new experimental results.

2. EXPERIMENTAL

Figure 1 shows the outline of the apparatus. The inclined channel (4) was 39 mm wide and 954 mm long. To make the flow in it similar to that on an inclined flat plate, the side walls (4-a) were smooth glass plates. The bottom air distributor (4-b) was made of a porous plastic plate of 20 mm in thickness. Two perforated plates below it were installed to distribute the air flow uniformly.

The upstream end of the main channel was connected to a guide channel (3) which was slightly steeper than the main one. It was installed to accelerate the particles initially in order to minimize the velocity

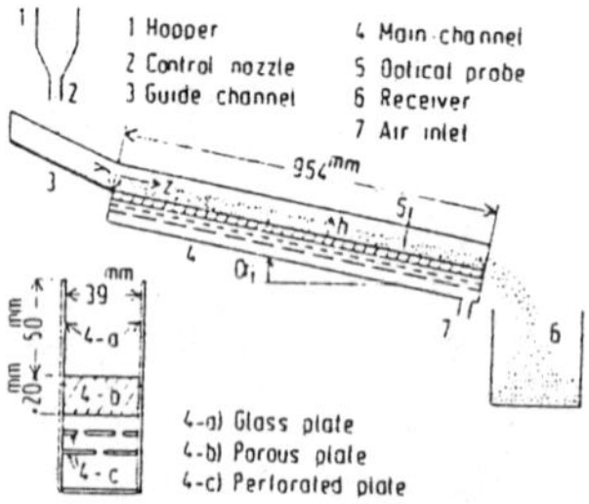

Fig. 1. Outline of the experimental apparatus.

TABLE 1
Properties of solid particles

Particles	d_p [mm]	ρ_B [g/cm^3]	ϕ_r [deg]	α_{ic} [deg]	U_{mf} [cm/s]
Glass beads (L)	0.39	1.57	23.0	23.0	14.0
Glass beads (S)	0.16	1.48	24.5	24.2	2.8
Porous alumina	0.23	0.95	31.0	30.7	2.8
Sand particles	0.19	1.31	35.0	36.2	5.9
F.C.C.	0.05	0.80	32.0	32.4	0.2

change along the longitudinal direction in the main channel. A hopper (1) which could store 0.03 m^3 of particles was located above the guide channel. A nozzele (2) was provided at the bottom of the hopper to control the feed rate of particles.

The velocity of the particles at each local point within the layer was then measured using a fine optical fiber probe. Also the bulk density of particles of the flow was obtained for five kinds of non-adhesive particles shown in Tabele 1. The details of the experimental procedure were shown elsewhere [1,2].

2. EXPERIMENTAL RESULTS

2.1 Effect of gas velocity and slope of the channel

Figure 2 shows typical examples of the velocity distribution for the large glass beads when the inclination angle of the channel, α_i, was as small as 14.4°, which was less than their angle of repose. Particles flowed when the air flow rate through the porous plate exceeded 8 cm/s. As the air velocity was further increased, the particle velocity increased rapidly. The horizontal bar in Fig. 3 represents the position of the free surface measured by a cathetometer.

For such a low bed height, the distributions of the particle velocity within the layer appeared linear when the gas velocity was less than or equal to the minimum fluidization velocity U_{mf}. When U_{mf} was exceeded, the distribution became gradually curved.

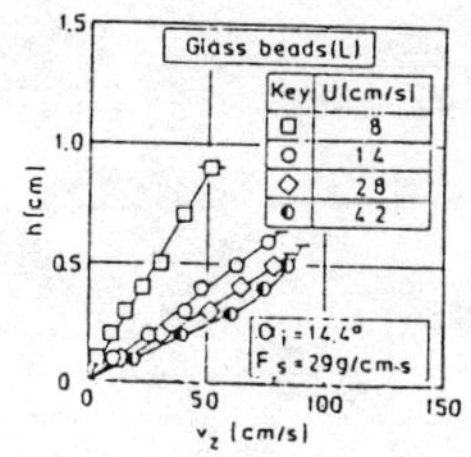

Fig. 2. Distribution of particle velocity for a low inclination angle.

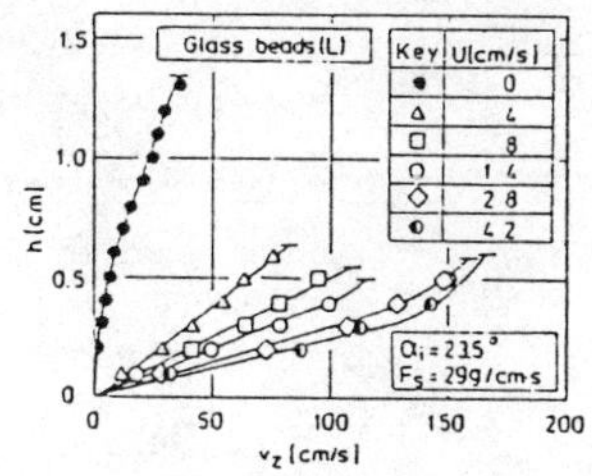

Fig. 3. Distribution of particle velocity for a high inclination angle.

Figure 3 shows the velocity distributions for a greatly inclined channel. The flow of particles then took place even without the air flow. However, the velocity profile without the air flow was curved near the bottom plate. As the air velocity was increased, the distribution of the particle velocity was higher than U_{mf}, however, the upper part of the profile became curved. By comparing Figs. 2 and 3, it was found that the velocity distributions for $\alpha_i = 23.5°$ in Fig. 3 were much steeper than those for $\alpha_i = 14.4°$ in Fig. 2.

2.2 Effect of layer height

The effect of the feed rate of the particles on those velocity profiles is shown in Figs. 4 and 5.

Figure 4 shows the cases for a low inclination angle with the air velocity kept at 14 cm/s ($U/U_{mf} = 1.0$). As the feed rate of particles F_s became high, the layer height was increased. But the velocity gradient was found to be almost independent of the layer height. When the feed rate became as great as 107 g/(cm·s), however, the profile became extremely curved near the free surface.

Figure 5 shows the case for a high inclination angle with the air velocity kept also at 14 cm/s. The velocity profile became curved even for a lower feed rate was 57 g/(cm·s), for example, the distribution of the particle velocity appeared nearly linear in Fig. 5, whereas it became significantly curved in Fig. 5. When the upper part of the distribution became curved, the velocity gradient in the lower linear portion was found to be affected by the feed rate of particles.

2.3 Classification of flow types

The velocity gradients, dv_z/dh, for the linear part in the observed velocity profiles were calculated. Figure 6 and 7 show data for the large glass beads at $F_s = 29$ g/(cm·s) and $U = 14$ cm/s, respectively.

The region (I) in Fig. 6 was the sliding region and the particles showed laminar movement, as shown schematically in Fig. 8 (I). Linear velocity distributions were observed in this region and the velocity gradients were independent of the layer height, as shown in the left part of Fig. 7.

When the inclination angle became extremely low, however, the velocity of the particles near the bottom plate was retarded, as

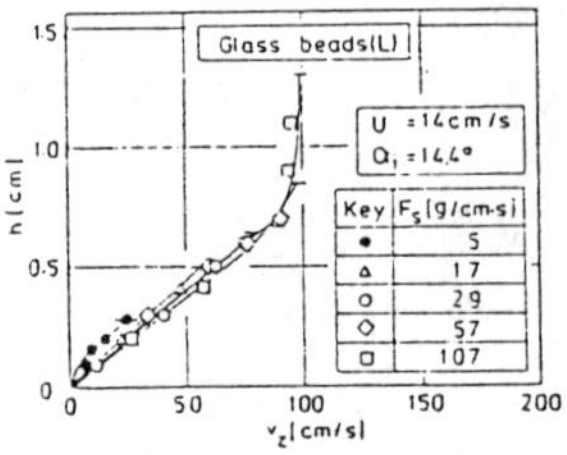

Fig. 4. Effect of feed rate of particles on the velocity distribution for a low inclination angle.

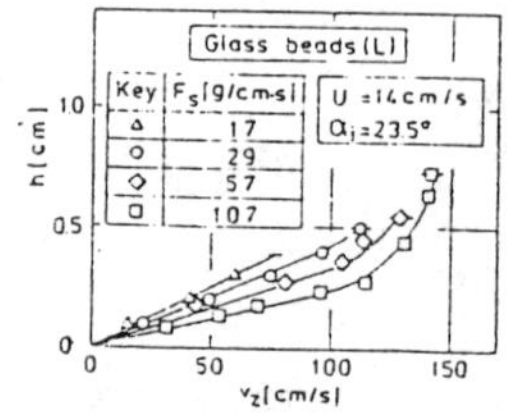

Fig. 5. Effect of feed rate of particles on the velocity distribution for a high inclination angle.

schematically shown in Fig. 8 (II), and some particles near the bottom plate in the upstream side remained stationary. Such profiles were observed in the immature sliding region (II) in Figs. 6 and 8.

The critical angle α_{ic} at which each line in Fig. 6 intersected the abscissa decreased with the gas velocity and became zero at the minimum fluidization velocity for such non-adhesive particles.

The region (III) in Fig. 6 was the splashing region. In this region the particles, especially near the free surface, showed bounding movement as schematically shown in Fig. 8 (III). The velocity gradients calculated from the layer height increased, as shown in Fig. 7. The shift from the sliding region (I) to the splashing region (III) took place when the particle velocity at the free surface reached about 1 m/s for the large glass beads.

When the air velocity increased beyond U_{mf} and with very low angle of inclination, bubbling was observed. The velocity profile then became significantly curved, as schematically shown in Fig. 8 (IV). Corresponding velocity gradients are shown by dotted lines in Fig. 6, but only a short portion adjacent to the bottom plate gave linear profiles.

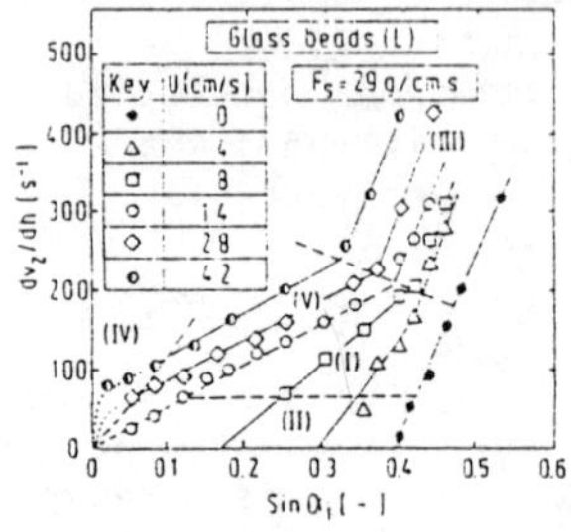

Fig. 6. Velocity gradients in five flow regions.

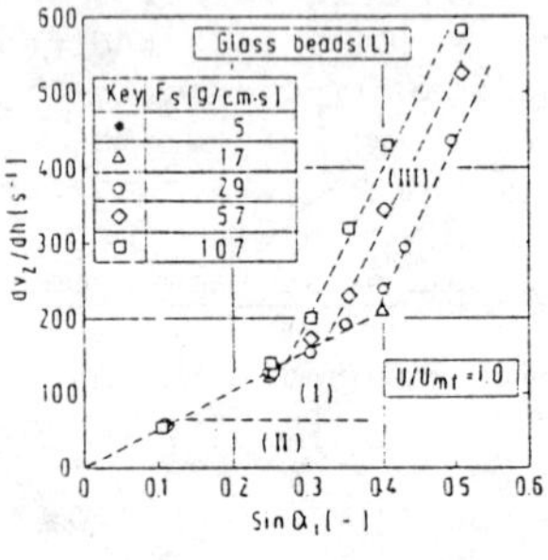

Fig. 7. The effect of the feed rate of particles on the velocity gradient, indicating the difference between the sliding region (I) and the splashing region (III).

As the inclination angle was increased, the rapid stream of particles extinguished the bubbles, as remarked by Mcguigan and Pugh [3] and

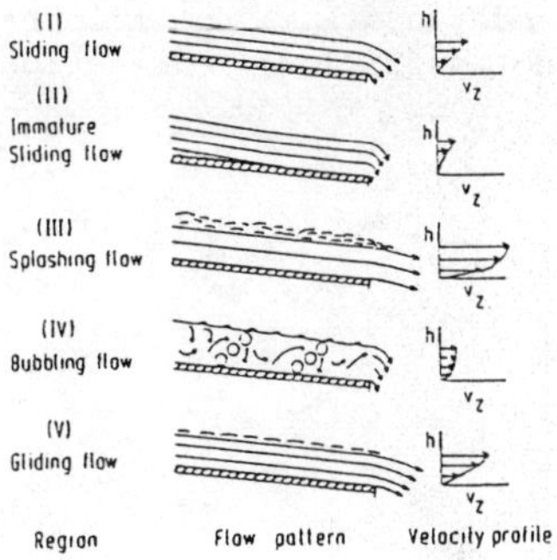

Fig. 8. Five types of flow pattern.

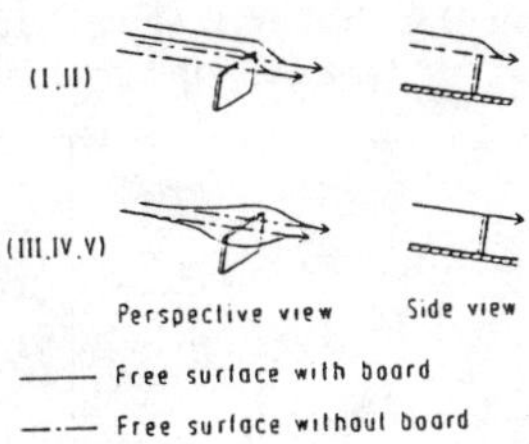

Fig. 9. Overpassing and roundabout movement of particles when a board is set in the flow.

Botterill and Abdul-Halim [4]. Although the bed expanded, the particles were distributed much more homogeneously than those in the bubbling region (IV) and glided as schematically shown in Fig 8 (V). In this gliding region, (v) in Fig. 6, the velocity gradient increased slightly as the layer height was increased.

The distinction between the sliding region (I) and the gliding region (V) could be made by installing an obstacle in the flow. Thus, if a board of similar height to the layer height and with width a third of the channel width was set in the flow, as shown in Fig. 9, in the sliding region (I), the layer height in the upstream side increased and then the particles passed over the board. On the other hand, for flow in the gliding region (V), the particles passed round the board. These overpassing and roundabout movements represent the important characteristics of each flow.

2.4 Bulk density of particles in the flow

The bulk density of the particles of each flow are shown in Fig. 10. In the sliding flow, the density was as close as the bulk density at the stationary state. However in the gliding flow, it decreased with the air flow rate increased, but are almost independent of the slope angle. In the splashing flow, however, it decreased significantly with the slope of the angle increased.

2.5 VARIATIONAL ANALYSIS

When these observed velocity distributions, especially in the sliding region, are compared with those of fluids, one may find out several distinctive characteristics of the particle flow. Especially, it is noticed that the velocity distribution appeared linear and that the velocity gradient at the free surface was not zero.

Although the size of particles is much greater than that of fluid molecules, a throng of particles may be treated macroscopically as a continnum, and the equation of continuity and the stress equation of motion can be applied. However, they are not sufficient to describe the motion of matter under given boundary conditions. In addition, one must

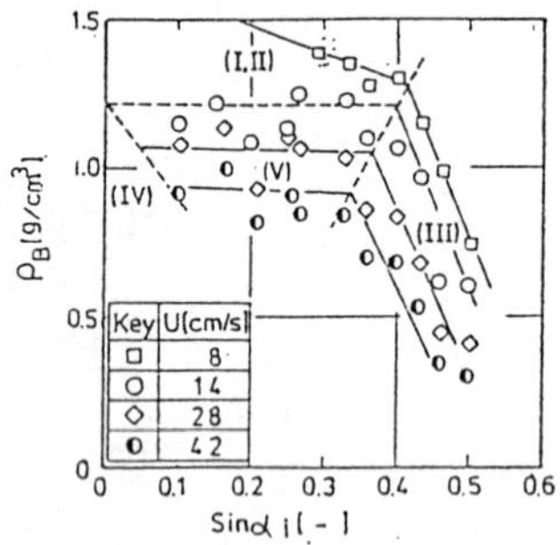

state the relation between stress and deformation, in order to describe the behavior of the particular matter to be considered. For a Newtonian fluid, this relation gives the definition of viscosity. Also for a non-Newtonian fluid, several relations such as Bingham model, Ostwald-de Waale model, and Reiner-Philipoff model have been proposed. For the motion of particles, however, such relations have not been suggested.

Consequently, an attempt was made to find such a relation for particle flow based on the experimental results by applying the variational principle.

The parabolic velocity distribution for a Newtonian fluid on an inclined flat plate can be obtained by finding the velocity distribution to minimize the following integral.

$$J = \int_0^H \left[\frac{\mu}{2}\left(\frac{dv_z}{dh}\right)^2 - \rho g v_z \sin\alpha_i\right] dh \tag{1}$$

The integrated consists of two energy terms: The first one is viscous dissipation, $(1/2)(-\tau:\nabla v)$, and the second one is rate of work done by gravity on the volume element, $\rho(v\cdot g)$ [5].

The velocity distribution to minimize the value of J can be found by Pontryagin's maximum principle [6,7], as summarized in Appendix, and the following parabolic distribution is obtained.

$$v_z = \frac{\rho g H^2 \sin\alpha_i}{2\mu}\left[1-\left(\frac{Y}{H}\right)^2\right] \tag{2}$$

Similarly, the velocity distribution for a Bingham fluid on an inclined flat plate can be obtained by finding the velocity distribution to minimize the following integral.

$$J = \int_0^H \left[\tau_B \frac{dv_z}{dh} + \frac{\mu_B}{2} \left(\frac{dv_z}{dh}\right)^2 - \rho g v_z \sin \alpha_i \right] dh \tag{3}$$

The numerical coefficient of the first term in the integrand is twice that of the second term, because these terms are derived from [6]

$$\int_0^{(dv_z/dh)} \left[\tau_B + \mu_B (dv_z/dh) \right] d(dv_z/dh) \tag{4}$$

And the following well-known relation can be obtained as outlined in Appendix.

$$v_z = \frac{\rho g (H-H_C)^2 \sin \alpha_i}{2\mu_B} \qquad \text{for } y < H_C \tag{5}$$

$$v_z = \frac{\rho g (H-H_C)^2 \sin \alpha_i}{2 \quad B} \left[1 - \left(\frac{y-H_C}{H-H_C}\right)^2 \right] \qquad \text{for } y \geqq H_C \tag{6}$$

where the critical thickness for a Bingham fluid is given by

$$H_C = \tau_B / (\rho g \sin \alpha_i) \tag{7}$$

Meanwhile, the shear stress τ_{yz} at depth y is given by the following equation:

$$\tau_{yz} = \rho g y \sin \alpha_i \tag{8}$$

By combining Eq. (8) and Eqs. (2) and (6), one may derive the following famous relation between stress and rate of deformation:

$$\tau_{yz} = -\mu (dv_z/dy) \qquad \text{for a Newtonian fluid} \tag{9}$$

$$\tau_{yz} = \tau_B - \mu_B (dv_z/dy) \qquad \text{for a Bingham fluid} \tag{10}$$

Therefore it may be induced that the velocity distribution for particle flow on an inclined flat plate can be obtained also bv minimizing a certain integral. As a first-step, the linear distribution for the sliding region cosidered and the following integral which explains it is proposed.

$$J = \int_0^H \left[k_\tau y \left(\frac{dv_z}{dh}\right) + \frac{k_\mu y}{2} \left(\frac{dv_z}{dh}\right)^2 - \rho g v_z \sin \alpha_i\right] dh \tag{11}$$

where k_τ and k_μ are constants.

Then the so-called Ishida-Shirai equation can be obtained, as outlined in Appendix.

$$v_z = 0 \qquad \text{for } \alpha_i < \alpha_{ic} \tag{12}$$

$$v_z = \frac{\rho g \sin \alpha_i - k_\tau}{k_\mu} \cdot h \qquad \text{for } \alpha_i \geqq \alpha_{ic} \tag{13}$$

where the critical angle α_{ic} is given by

$$\sin \alpha_{ic} = k_\tau / (\rho g) \tag{14}$$

Equations (11) and (14) give the physical meaning of the angle of repose. Equation (13) supports that the velocity gradient is independent of layer thickness H in the sliding flow.

By combining Eq. (8) and Eq. (13), one obtains the following relation between stress and rate of deformation for the flow of particles on an inclined plate.

$$\tau_{yz} = k_\tau y - k_\mu y \, (dv_z/dy) \tag{15}$$

Consequently, Equation (15) may be considered as a simple model which can explain the important features of particle flow in an inclined channel.

Although Eqs. (11) and (15), respectively, are very similar to Eqs. (3) and (10) for a Bingham fluid, both yield stress $k_\tau y$ and apparent viscosity $k_\mu y$ for particles are proportional to y. This may be related closely to the fact that buoyancy does not act in the layer of solid particles. This depth y, which is proportional to the normal stress σ, plays a very important role and gives a velocity distribution which is entirely different from that for a Bingham fluid. Much more effort should be made to establish a more general relation which is applicable to various kinds of flow of solid particles.

CONCLUSIONS

The velocity distributions of non-adhesive particles flowing in an inclined open channel with air flow was investigated and the following conclusions were reached.

(1) The flow patterns could be classified into five types depending on

the fluid velocity and the inclination angle. These were sliding, immature sliding, splashing, bubbling, and gliding flow.
(2) In the sliding flow, the velocity distribution appeared linear and its gradient was independent of the layer height. This fact was explained by the variational analysis.
(3) In the splashing and gliding flow, the velocity gradient increased as the layer height was increased.
(4) In the sliding and immature sliding flow particles could only flow over an obstruction, whereas they flowed round in the splashing, bubbling, and gliding flow.

APPENDIX: Velocity Distributions to Make the Integral J Minimum

A-1. Newtonian fluid

The state equation for this flow system is

$$dv_z/dh = p \qquad (p \geqq 0) \tag{A-1}$$

Then, try to find p which will maximize the following Hamiltonian function H.

$$H = -[(\mu/2)p^2 - \rho g v_z \sin \alpha_i] + \psi_1 [p]$$

$$= -\frac{\mu}{2}\left(p - \frac{\psi_1}{\mu}\right)^2 + \frac{\psi_1{}^2}{2\mu} + \rho g v_z \sin \alpha_i$$

where ψ_1 is a multiplier and satisfies the following equation.

$$\frac{d\psi_1}{dh} = -\frac{\partial H}{\partial v_z} = -\rho g v_z \sin \alpha_i \tag{A-2}$$

Hence,

$$\Psi_1 = -\rho g h \sin \alpha_i + c_1 \tag{A-3}$$

H becomes maximum when

$$p \equiv \frac{dv_z}{dh} = \frac{\psi_1}{\mu} = \frac{-\rho g h \sin \alpha_i + c_1}{\mu}$$

Hence,

$$v_z = (-\rho g h^2 \sin \alpha_i + 2c_1 h)/2\mu + c_2$$

The integral constants c_1 and c_2 are determined by

$$v_z = 0 \qquad \text{at } h = 0 \tag{A-4}$$

$$\psi_1 = 0 \qquad \text{at } H = 0 \tag{A-5}$$

Equation (A-5) holds because the velocity v_z at $h=H$ is not given. Then, Eq. (3) is obtained. Note that the condition, $dv_z/dh=0$ at $h=H$, has not been used to derive the velocity distribution.

A-2. Bingham fluid

Hamiltonian function H is given as

$$H=-[\tau_B p+(\mu_B/2)p^2-gv_z\sin\alpha_i]+\psi_1[p]$$

$$=-\frac{\mu_B}{2}\left(p-\frac{\psi_1-\tau_B}{\mu_B}\right)^2+\frac{(\psi_1-\tau_B)^2}{2\mu_B}+\rho g v_z\sin\alpha_i$$

Hence, Eqs. (A-2), (A-3), and (A-5) hold also in this case. Since p must be zero or positive, the following two cases are considered.

(Case 1) $\psi_1-\tau_B<0 \qquad y<H_c$

$$p\equiv dv_z/dh=0 \qquad\qquad p\equiv dv_z/dh=0 \tag{A-6}$$

(Case 2) $\psi_1-\tau_B\geqq 0 \qquad y\geqq H_c$

$$p\equiv\frac{dv_z}{dh}=\frac{\psi_1-\tau_B}{\mu_B}=\frac{\rho g\sin\alpha_i(H-h)-\tau_B}{\mu_B} \tag{A-7}$$

Equation (6) is obtained by integrating Eq. (A-7) with the boundary condition, Eq. (A-4). Equation (6) is derived from Eq. (6) and Eq. (A-6).

A.3 Particles

Hamiltonian function H is given as

$$H=-[k_T(H-h)p+(k_\mu/2)(H-h)p^2-\rho g v_z\sin\alpha_i]+\psi_1[p]+\psi_2$$

$$=-\frac{k_T(H-h)}{2}\left(p-\frac{\psi_1-k_T(H-h)}{k_\mu(H-h)}\right)^2+\frac{[\psi_1-k_T(H-h)]^2}{2k_\mu(H-h)}+\rho g v_z\sin\alpha_i+\psi_2$$

Hence, Eqs. (A-2), (A-3), and (A-5) hold also in this case. Then, following two cases are considered.

(Case 1) $\psi_1-k_T(H-h)<0 \qquad \sin\alpha_i<k_T/(\rho g)$

$$p\equiv dv_z/dh=0 \qquad\qquad v_z=0$$

(Case 2) $\psi_1-k_T(H-h)\geqq 0 \qquad \sin\alpha_i\geqq k_T/(\rho g)$

$$p\equiv\frac{dv_z}{dh}=\frac{(\rho g\sin\alpha_i-k_T)(H-h)}{k_\mu(H-h)} \tag{A-8}$$

Equation (13) is obtained by integrating Eq. (A-8) with the boundary condition, Eq. (A-4).

NOMENCLATURE

d_p	= diameter of particles	[mm]
F_s	= feed rate of particles per unit width	[g/cm·s]
H	= height of the layer	[cm]
H_c	= critical thickness for a Bingham fluid	[cm]
h	= height measured from the bottom plate	[cm]
k_μ	= coefficient for apparent viscosity of particles	[$g/cm^2.s$]
k_τ	= coefficient for yield stress of particles	[$g/cm^2.s^2$]
U	= air velocity	[cm/s]
U_{mf}	= minimum fluidization velocity	[cm/s]
v_z	= longitudinal velocity	[cm/s]
y	= depth measured from the free surface	[cm]
z	= longitudinal distance	[cm]
α_i	= inclination angle of the channel	[deg]
α_{ic}	= critical inclination angle of the channel	[deg]
μ	= viscosity	[g/cm.s]
μ_B	= apparent viscosity for a Bingham fluid	[g/cm.s]
ρ	= density	[g/cm^3]
ρ_B	= bulk density	[g/cm^3]
τ_B	= shear stress for a Bingham fluid	[$g/cm.s^2$]
τ_{yz}	= shear stress	[$g/cm.s^2$]
ϕ_r	= angle of repose	[deg]

REFERENCES

1. Ishida, M. and Shirai, T., "VELOCITY DISTRIBUTIONS IN THE FLOW OF SOLID PARTICLES IN AN INCLINED OPEN CHANNEL," J. Chem. Eng. Jpn., vol. 12, pp. 46-50 (1979).

2. Ishida, M., Hatano, H. and Shirai, T., "The Flow of Solid Particles in an Aerated Inclined Channel," Powder Technol., vol. 27, pp. 7-12 (1980).

3. McGuigan, S.J. and Pugh, R.R., Proc. 3rd Int. Conf. on Pneumatic Transport of Solids in Pipes, BHRA, U.K., paper E2,(1976).

4. Botterill, J.S.M. and Abdul-Halim,B.H., "The Open-Channel Flow of Fluidized Solids," Powder Technol., vol. 23, pp. 67-78 (1979).

5. Bird, R.B., Stewart, W.E. and Lightfoot, E.N., "Transport Phenomena," John Wiley & Sons, Inc. (1960).

6. Pontryyagin, L.S., Et al., "The Mathematical Theory of Optimal Processes," Interscience Publishers, Inc. (1962).

7. Uno, T. and Kikuchi, T., "Saidai Genri Nyumon," Kyoritsu Shuppan (1967).

SOME MATHEMATICAL MODELS OF SNOW AVALANCHES

E. Margarita Eglit

Department of Hydromechanics
Moscow State University
Moscow 117234, USSR

1. INTRODUCTION

Snow avalanches, in spite of great variety of their types, are flows of continuum media which often may be considered as flows of granular materials. Elaboration of mathematical models for calculation of dynamical parameters of snow avalanches (snow velocity, flow depth, run-out distance - path till stopping and so on) is an actual problem because of the necessity to organize reliable protection in mountain regions. Description of an avalanche flow in detail is an extremely complicated problem especially because of the fact that avalanche flows are almost always turbulent.

At present there are some models of avalanche flows which are called hydraulics. In some of these models the structures of a flow in direction perpendicular to the slope is not considered at all and only average meanings of density and velocity in a cross section of the flow are taken into account. In other models the flow is conventionally being devided into layers and average characteristics for these layers are considered. The single-layer models are used for avalanches from dense dry or wet snow when mixing up with surrounding air is inessential and also for dust snow avalanches if a cloud of snow dust is long enough. Two-layer models are used for "mixed" avalanches consisting of dense part - nucleus and a layer of snow dust moving upon it. In two-layer models postulation of laws defining friction and mass exchange between the layers is the most important part of the theory.

In the first part of this paper a single-layer model of an avalanche flow is described. In the second part of the paper we will consider some two-layer models of an avalanche flow.

2. GOVERNING EQUATIONS

The system of equations for the simplest case of one-dimensional motion along a cylindrical surface when the ratio of characteristic flow thickness h to characteristic curvature radius of the surface R is small but is taken into account has the following form:

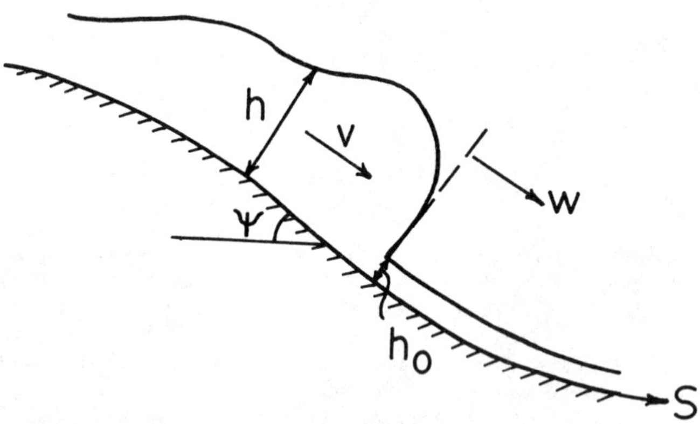

Fig. 1. Diagram of a moving snow avalanche.

The continuity equation

$$(1 - \frac{h}{R}) \frac{\partial h}{\partial t} + \frac{\partial hv}{\partial S} = 0 \tag{1}$$

The equation of motion

$$\frac{\partial v}{\partial t} + (1 + \frac{h}{2R}) v \frac{\partial v}{\partial S} = g \sin \psi - \frac{1+\frac{h}{2R}}{2h} \frac{\partial}{\partial S} [h^2(g \cos \psi + \frac{v^2}{R})] - (f_1 + f_2) \tag{2}$$

In this equation f_1 and f_2 are "dry" and "hydraulic" frictions. Assume the following notations to hold:

$$f_{11} = \mu(\frac{v^2}{R} + g \cos \psi)(1 + \frac{h}{R})$$

$$f_{12} = (1 + \frac{h}{R})\tau^*/\rho h$$

$$f_{13} = g \sin \psi - \frac{1+\frac{h}{2R}}{2h} \frac{\partial}{\partial S} [h^2(g \cos \psi + \frac{v^2}{R})] \tag{3}$$

Then the expression for f_1 is

for $v \neq 0$

$$f_1 = \begin{cases} 0 & \text{at} \quad f_{11} < 0 \;, \\ f_{11} \cdot \text{sign } v & \text{at} \quad 0 \leqq f_{11} \leq f_{12} \;, \\ f_{12} \cdot \text{sign } v & \text{at} \quad f_{11} > f_{12} \;. \end{cases} \tag{4}$$

For $v = 0$

$$f_1 = \begin{cases} f_{13} & \text{at} \quad |f_{13}| < \min(|f_{11}|, |f_{12}|) \;, \\ f_{11} \cdot \text{sign } f_{13} & \text{at} \quad |f_{13}| \geq f_{11}, \quad f_{11} < f_{12} \;, \\ f_{12} \cdot \text{sign } f_{13} & \text{at} \quad |f_{13}| \geq f_{12}, \quad f_{12} < f_{11} \;. \end{cases} \tag{5}$$

The formulas defining hydraulic friction f_2 are

$$f_2 = \begin{cases} \dfrac{\lambda v}{h^2}\left(1 + \dfrac{h}{R}\right) & \text{at} \quad \dfrac{vh}{\nu} < Re_{crit}, \quad \lambda = 3\nu \\ \dfrac{kv^2}{h}\left(1 + \dfrac{h}{R}\right) & \text{at} \quad \dfrac{vh}{\nu} \geq Re_{crit} \end{cases} \tag{6}$$

In the equations (1-6) t and S are time and coordinate along the mountain slope, v and h - velocity and flow thickness (flow depth), ψ - a slope angle, R - curvature radius of the avalanche path, $R > 0$ for convex slope, μ - "dry" friction coefficient, k and λ - "hydraulic" turbulent and laminar friction coefficients, ν - viscosity coefficient, Re_{crit} - the critical value of Reynolds' number defining the transition from laminar to turbulent regime, τ^* - minimum of shear strengths of avalanche snow and underlying material (ground for ground avalanches and underlying layer of snow for surface avalanches).

The equations (1-6) are analogous to those used in hydraulics or in shallow water theory. Here terms of the order of h/R are taken into account which is not usual in hydraulics. Viscous - plastic nature of flow is shown in special laws of friction which are different for moving and resting material. For moving snow friction force consists of hydraulic friction connected with an exchange of momentum between neighboring parts of a flow and friction of solids being equal to Coulomb's friction unless Coulomb friction is not larger than some limiting value τ^* and to τ^* in the other case. The inflow of mass is not included into the equations (1-6). It is because of the assumption that: involving additional portions of snow from the slope occurs only in narrow zone adjoining the boundaries of an avalanche flow, in particular the avalanche leading front. This narrow zone is represented in this theory by a "shock wave" (hydraulic jump) where the snow cover lying on the mountain slope in front of an avalanche is being destroyed and then evolve into an avalanche. For one-dimensional motion along wide slopes the equations of conservation of mass and momentum at the avalanche leading front are

if

$$0.5\ \rho\bar{h}^2\ (g \cos\psi + \frac{\bar{v}^2}{R}) \geq \sigma_* h_o \quad , \qquad (7')$$

then

$$\rho\bar{h}\ (w - \bar{v}) = \rho_o h_o w \quad ,$$

$$\rho_o h_o w\bar{v} = 0.5\ \rho\bar{h}^2\ (g \cos\psi + \frac{v^2}{R}) - \sigma_* h_o \quad , \qquad (7)$$

if

$$0.5\ \rho\bar{h}^2\ (g \cos\psi + \frac{v^2}{R}) < \sigma_* h_o \quad . \qquad (8')$$

Then instead of (7) it is possible to use the following conditions (8a), or (8b), or (8c)

$$w = 0\ ,\ \bar{v} = 0 \quad , \qquad (8a)$$

$$w = 0\ ,\ \bar{v} = 0,\ h_o = 0.5\ \rho\ g \cos\psi\ \bar{h}^2/\sigma_* \quad , \qquad (8b)$$

$$w = \bar{v}\ ,\ \bar{h} = 0,\ h_o = 0 \quad . \qquad (8c)$$

In the equations (7), (8) $\bar{h}$ and $\bar{v}$ are a height of front and a snow velocity at the front, w is the front speed, σ_* is a contraction strength of snow layer which begins to move with the avalanche, h_o is a thickness of this layer, ρ and ρ_o - snow density in an avalanche and in a snow cover. In some mathematical models of avalanches it is assumed that an avalanche involve additional portions of snow not only in a front zone but all along the flow from it's bottom. In such models the differential equations of continuity and motion contain some additional terms connected with the inflow of mass and the front conditions are of the form (8c). An example of such a model is given in the second part of this paper.

The equations (1-8) and corresponding equations for two-dimensional flows upon wide nonhomogeneous slopes and also for flows in channels of various forms of cross-sections were studied analytically and numerically [1-5]. The aim of analytical investigations was to get approximate analytical formulas for speed and height of an avalanche front and also for velocity and flow depth distribution along an avalanche. Such formulas were obtained for motion along wide long homogeneous slopes and for motion in long homogeneous channels of trapezoidal (in particular rectangular and three - cornered) cross-sections [3]. The aim of numerical computations was 1) to estimate the influence of various terms in the equations (1-8) and also the influence of variation of magnitude of model coefficients on computed dynamical parameters of an avalanche flow, 2) to choose values of model coefficients by means of comparison of computed and observed data. Qualitative description of an influence of various terms in the equations (1-8) and results of numerical computations are given below.

3. NUMERICAL RESULTS

1. The influence of slope curvature. The equations (1-8) are the equations of motion for shallow layer of continuum media on curvilinear surface small terms of order of $\frac{h}{R}$ being taken into account. Among the terms connected with the slope curvature those containing centrifugal force are the most important. The centrifugal force influences onto the motion of two ways. For the same slope it is both accelerating and braking on. If the dry friction increase when the pressure on the bottom increase (i.e. it does not reach its top magnitude yet) then the dry friction will be greater on concave slope because of centrifugal force, so breaking will be greater. But at the same time the force connected with the pressure of deeper parts of a flow into less deep parts increases and there occurs acceleration of an avalanche front and snow on the front. The effect of acceleration of the front due to increasing of pressure gradient on concave part of a slope (and in winding part of a channel - if an avalanche moves in a channel) may be stronger then the effect of braking due to increasing friction. The final result depends on the velocity at the beginning of curvilinear region, curvature radius and value of dry friction coefficient. In testing computations not very significant increasing of an avalanche run-out distance (i.e. the whole path till stop) was observed when the initial velocity was 10-12 m/sec., curvature radius was 50 m and 100 m and $\mu = 0.1$, $\mu = 0.25$. For $\mu = 0.5$ insignificant decreasing of the run - out distance was obtained comparatively with the results of computations without taking centrifugal force into account.

2. Changing of friction law at stopping. The flow of granular material in contrast with the flow of fluid may stop even on an inclined plane. It is important that mechanical properties of stopped material are different from those of moving material. In particular laws defining friction in motion differ from laws for friction at rest. This fact is taken into account in formulas (3-5). When computing the process of stopping of an avalanche one must compare magnitudes of active force and friction force calculated from the law for friction in motion, for those parts of a flow where $v = 0$. If this friction force is larger than the active force then it must be substituted by a force whose magnitude is equal and sign is opposite to an active force. As long as an active force is less than dry friction in motion, the motion in stopped part of an avalanche can not begin. Yet in general secondary motions may appear because the other parts of the avalanche are in motion so the form of the avalanche changes and active force connected with interaction between different parts of the avalanche changes. Testing numerical computations show that secondary motions of previously stopped parts of an avalanche don't lead to any significant increasing of avalanche path till final step.

3. Replacement of turbulent friction by laminar friction. Formulas (8) give different expressions for hydraulic friction in turbulent and laminar zones of a flow. Numerical and analytical solutions of (2-8) show that even when an avalanche moves down a very steep slope a zone with small velocity and depth exists in an avalanche flow, usually a tail of an avalanche represents such zone. Before final stopping all the flow becomes laminar. Numerical computations were done to compare results obtained when it was supposed that hydraulic friction in all the flow was turbulent and results obtained when replacement of turbulent friction by laminar one for those parts where $Re < Re_{crit}$ was taken into account. Replacement of turbulent friction which is proprotional to the square of the velocity, by laminar friction, which is proportional to the first power of the velocity, in zones where velocity is very small leads to growth of friction and intensification of braking in these zones. But on a transition zone where the Reynolds' numbers have an order of Re_{crit} this replacement of friction laws diminishes the friction force so that this zone begins to move with larger velocity. It is important that the length of this zone present the greater part of the avalanche length the greater co-efficient of laminar friction is. That is why the greater the laminar friction coefficient the greater the accelerating influence of the transition zone. The final result depends on the proportion between laminar $\lambda(\lambda = 3\nu)$ and turbulent k friction coefficients and also on the value of Re_{crit}. Testing computations were done for $Re_{crit} = 500$, $\nu = 1 \div 0.001\ m^2/sec$, $k = 0.001 \div 0.1$. It appears that when ν is not too large so that turbulent zone does exist in the flow if one takes into account the difference in friction laws for

different parts of the avalanche the smaller velocity and depth occur in the tail of the avalanche, maximum depth increases and the final path until stop does not practically change. Therefore for practical calculations it is possible to use the simplified model in which hydraulic friction everywhere is assumed to be turbulent.

4. Effect of involving additional parts of snow from the slope and stopping criterion: The quantity and the law of involving new portions of snow into an avalanche is essential especially for motion down long slopes. Terminal front height and speed for avalanches moving down long homogeneous slopes are proportional to the square root of the thickness of snow layer stripping off by an avalanche. For starting and transit phases of motion the thickness of the involving snow layer depends chiefly on the properties of snow cover which is situated on the mountain slope along the path of an avalanche and so is known in advance. But in the time of braking and stopping, as it was noticed by observers, the involving of additional portions of snow in front zone ceases and an avalanche snow overflows upon the snow cover. In the model (1-8) it is supposed that all additional snow comes into an avalanche only at its front and as long as a force acting onto the layer of snow cover in front of an avalanche is large enough to destroy it then all the layer is stripped off by an avalanche. At the moment when this force becomes insufficient, i.e., the condition (8') becomes true it may be assumed either that an avalanche stops in accordance with the condition (8a), or that only smaller part of the snow layer is stripped off. The part which can be destroyed by present force, i.e. the condition (8b) holds, at the moment the involving of additional portions of snow ceases at all and an avalanche snow overflows over the snow cover in accordance with (8c). It appears that all the conditions (8a), (8b), (8c) give approximately equal values of avalanche run-out distance, but the maximum run out distance is usually obtained when one uses the condition (8c).

Some models of avalanche flows have been developed in which seizing of additional snow is supposing to take place in general along all the length of an avalanche from snow layer situated beneath it. In the paper [4] a formula is proposed defining the magnitude of mass being seized at unit square of contact of an avalanche and underlying snow. This magnitude is proportional to the flow depth h. In models which will be described below it is postulated that this magnitude is proportional to the flow velocity.

4. EQUATIONS FOR THE TWO-LAYER MODELS

Let us consider now some two-layers mode of an avalanche. Here only the equations for one-dimensional motion down wide slope will be written, terms connected with the curvature of the path being neglected.

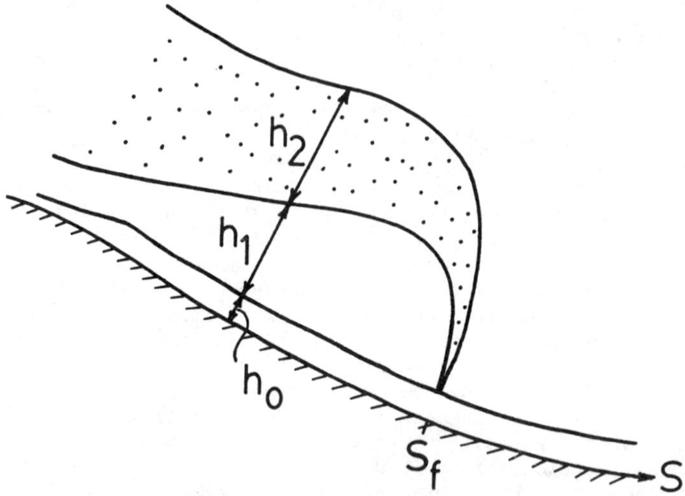

Fig. 2. An Avalanche as a Two-Layer Flow.

Let $\rho_1(S_1t)$, $v_1(S_1t)$, $h_1(S_1t)$ be the density, the velocity and the thickness of the lower dense layer which may be called a nucleus of an avalanche of mixed type. Upon the dense nucleus and perhaps in front of it there moves a powder layer - a flow of snow dust with density $\rho_2(S_1t)$, the velocity $v_2(S_1t)$ and the thickness $h_2(S_1t)$. The dense compact nucleus moves upon some layer of snow cover whith thickness $h_0(S_1,t)$ changing because of transfer of some amount of snow to an avalanche. Above the powder layer there is an air with the density ρ_a and a snow dust flow interact with it (Fig. 2). The flow is considered to be incompressible but ρ_1 and ρ_2 are not constant because of mass exchange between the neighboring layers.

The equations for snow dust layer may be written in the following form.

The equation of variation of volume

$$\frac{\partial h_2}{\partial t} + \frac{\partial h_2 v_2}{\partial S} = V_{2a} + V_{21} - V_S \tag{9}$$

Here V_{2a}, V_{21}, V_S are the rates of variation of volume for unit width and unit length of an avalanche occurring respectively due to the entrance of air, dense snow from a nucleus and due to s sedimentation of snow dust.

The equation of continuity

$$\frac{\partial \rho_2 h_2}{\partial t} + \frac{\partial \rho_2 h_2 v_2}{\partial S} = \rho_a V_{2a} + \rho_1 V_{21} - (\rho_2 - \rho_a) V_S \tag{10}$$

The equation of motion

$$\frac{\partial \rho_2 h_2 v_2}{\partial t} + \frac{\partial \rho_2 h_2 v_2^2}{\partial S} = g(\rho_2 - \rho_a) h_2 \sin \psi$$

$$- 0.5 \frac{\partial}{\partial S} [(\rho_2 - \rho_a) h_2^2 \; g \cos \psi] + \tau_{2a} + \tau_{21} + \rho_1 V_{21} v_2 \tag{11}$$

Here τ_{2a}, τ_{21} are friction forces between a powder layer and air and dense layer respectively related to a unit of mass.

The equations for dense nucleus are analogous to those for powder layer

$$\frac{\partial h_1}{\partial t} + \frac{\partial h_1 v_1}{\partial S} = -V_{21} + V_{10} + V_S \tag{12}$$

$$\frac{\partial \rho_1 h_1}{\partial t} + \frac{\partial \rho_1 h_1 v_1}{\partial S} = -\rho_1 V_{21} + \rho_0 V_{10} + (\rho_2 - \rho_a) V_S$$

$$\frac{\partial \rho_1 h_1 v_1}{\partial t} + \frac{\partial \rho_1 h_1 v_1^2}{\partial S} = g \; \rho_1 h_1 \sin \psi - \frac{\partial \rho_2 g h_2 \cos \psi}{\partial S} h_1$$

$$- \frac{\partial}{\partial S} \left[\frac{\rho_1 h_1^2 g \cos \psi}{2}\right] - \rho_1 \; g \cos \psi \; h_1 \frac{\partial h_0}{\partial S}$$

$$- \tau_{21} + \tau_{10} - \rho_1 V_{21} v_2 \tag{14}$$

Here V_{10}, τ_{10} are the rate of volume variation and friction force connected with an interaction with snow layer upon which an avalanche moves, ρ_0 and h_0 - the density and the thickness of this layer.

The equation describing variation of h_0 has the form

$$\frac{\partial h_0}{\partial t} = -V_{10} \tag{15}$$

Starting from the analysis of the theory of intermixing of turbulent jets and the theory of stability of the boundaries between two flows with different densities and velocities A.G. Kulikovskiy had suggested the following formulas for mass exchange between layers in a mixed avalanche

$$V_{2a} = m_{2a} \cdot v_2 \cdot \frac{\sqrt{\rho_2 \, \rho_a}}{\rho_2 + \rho_a}$$

$$V_{21} = \begin{cases} m_{21} \cdot |v_2 - v_1| \cdot \sqrt{\rho_2 \rho_1}/(\rho_2 + \rho_1) \text{ at } h_1 > 0,\ S < S_f \ , \\ m_{10} \cdot v_2 \cdot \sqrt{\rho_2 \rho_0}/(\rho_2 + \rho_0) \text{ at } h_1 = 0,\ S > S_f \ , \end{cases}$$

$$V_{10} = \begin{cases} m_{10} \cdot v_1 \cdot \sqrt{\rho_1 \rho_0}/(\rho_1 + \rho_a) \text{ at } h_0 > 0 \ , \\ 0 \text{ at } h_0 = 0 \ . \end{cases} \qquad (16)$$

Here m_{2a}, m_{21}, m_{10} are some empirical coefficients, S_f is a coordinate of the leading front of a dense layer of an avalanche.

The formulas for friction forces in this model are analogous to (3-6). It follows from the preceeding analysis that if the motion in the main part of the flow is turbulent then the hydraulic friction may be assumed to be proportional to the square of the relative velocity of the flow. In addition to this it seems plausible that for a powder layer it is possible to neglect dry friction and friction with air.

$$\tau_{2a} = 0 \ ,$$

$$\tau_{21} = -K_{21}(v_2 - v_1) \cdot |v_2 - v_1| \ , \qquad (17)$$

$$\tau_{10} = -K_{10} \cdot v_1 \cdot |v_1| - f_1 \ ,$$

$$f_1 = \begin{cases} f_{11} \cdot \text{sign } v_1 \equiv \mu \, g \cos \psi (\rho_2 h_2 + \rho_1 h_1) \text{sign } v_1 \text{ at } |f_{11}| < |f_{12}|,\ v_1 \neq 0 \ , \\ f_{12} \cdot \text{sign } v_1 \equiv \tau^* / (\rho_2 h_2 + \rho_1 h_1) \text{sign } v_1 \text{ at } |f_{11}| > |f_{12}|,\ v_1 \neq 0 \ . \end{cases}$$

$$f_1 = \begin{cases} f_{13} = \rho_1 h_1 g \sin\psi - h_1 \frac{\partial}{\partial S}(\rho_2 h_2 g \cos\psi) - 0.5 \frac{\partial}{\partial S}(\rho_1 h_1{}^2 g\cos\psi) \\ \quad - \rho_1 h_1 g \cos\psi \frac{\partial h_0}{\partial S} \text{ at } |f_{13}| < \min(|f_{11}|, |f_{12}|) \text{ and } v_1=0 \\ f_{11}\cdot \text{sign } f_{13} \text{ at } |f_{13}| \geq f_{11}, f_{11} < f_{12} \text{ and } v_1=0 , \\ f_{12}\cdot \text{sign } f_{13} \text{ at } |f_{13}| \geq f_{12}, f_{12} < f_{11} \text{ and } v_1=0 . \end{cases}$$

The mathematical model of snow avalanche described by equations (9-17) is a generalization of hydraulic models used for avalanches of dense compact snow [1-5] and of the model of moving snow - dust cloud considered in [6,7]. All the latter models have been used for computation of the dynamical parameters of snow avalanches and gave results consistent with data of observations in nature for physically realistic magnitudes of friction and mass - exchange coefficients.

5. REFERENCES

1. Bachvalov, S.N., Eglit, E.M., Issledovanie odnomernogo dvizhenija snezhnoj lavini po ploskomu sklonu (The investigation of one-dimensional motion of snow avalanche upon a plane slope) Isvestija AN SSSR, Mechanika zhidkosti i gasa, no. 5, pp. 7-14 (1973).

2. Kulikovskij, G.A., Eglit, E.M., Dvumernaja zadacha o dvizhenii snezhnoj lavini po sklonu s plavno menjajuschimisja svojstvami. (Two-dimensional problem of snow avalanche motion upon a slope with smoothly varying properties) Prikladnaja Matematika i mechanika, t 37, No. 5, pp. 837-848 (1973).

3. Danilova, M.E., Eglit, E.M., Dvizhenije lotkovish lavin (Motion of channel avalanches.) Data of Glaciological Studies Chronicle, Moscow, Discussion, 31, pp. 65-73 (1977).

4. Grigorian, S.S., Ostroumov, V.A., Mathematicheskoje modelirovanije dvizhenija snexhnich lavin (Mathematical simulation of motion of snow avalanches) International Congress on applied Glaciology Cambridge, 1976.

5. Danilova, M.E., Eglit, E.M., Dvizhenije snezhnikh lavin v uslovijakh predelonogo trenija. (Motion of snow avalanches in limiting friction condition) Izvestija AN SSSR, Mekhanika zhidkosti i gasa, No. 5, pp. 30-37 (1977).

6. Kulikovskij, G.A., Sveschnikova, I.E., Model dlja rascheta dvizhenija pilevoj snezhnoj lavini (A model for calculation of motion of powder snow avalanche) Data of Glaciological Studies Chronicle Moscow, Discussion, 31, pp. 74-80 (1977).

7. Eglit, E.M., Sveschnikova, I.E., Mathematicheskoje modelirovanije snezhnikh lavin (Mathematical modeling of snow avalanches) Data of Glaciological Studies Chronicle, Moscow, Discussion, 38, pp. 79-84 (1980).

SIMULATION OF SOLID PARTICLE MOTION IN TURBULENT GAS-SOLID PIPE FLOW AND RELATED TOPICS

H. Masuda[1], H. Yoshida[1] and K. Iinoya[2]

[1]Department of Chem. Eng., University of Hiroshima,
Higashi-Hiroshima 724, Japan
[2]Department of Ind. Eng., Aichi Institute of Technology,
Toyota 470-03, Japan

ABSTRACT

Solid-particle motion in turbulent flow controls the performance of powder handling processes and also causes various phenomena such as particle electrification. In relation to the particle motion, a number of works and findings have been reported in Japan. Here we will give a brief survey of the works with particular attention to the simulation method of solid-particle motion in turbulent pipe flow. First, we will discuss the simulation of coarse-particle motion, then simulation of fine-particle motion and we will give further comments on the calculation of particle trajectories.

1. SIMULATION OF COARSE-PARTICLE MOTION

Coarse-particle trajectories are affected by collisions between particles and a pipe wall. The number of collisions were measured by Masuda et al.[4] in relation to the particle electrification. Microscopic analysis was carried out on the scars made on a plastic film covering the inside wall of a pipe. Some scars showed that particles might collide with small incident angle and rebounded with larger angle.

In the study by Matsumoto et al., an irregular bouncing model has been proposed based on the assumption that the particle is not spherical but ellipsoidal[8] or the assumption that the particle is spherical but the wall surface is rough[9]. The wall roughness is given by a sine curve. Calculation of particle bounce based on the rough wall model is easier than the ellipsoidal particle model. The calculation by these models well explaines the suspension of particles in a horizontal pneumatic transport line without particle deposition. Figure 1 shows the distribution of particle concentration for polystyrene beads transported in a rectangular horizontal duct. The assumed amplitude of wall roughness, Ar = (1/70)Dp, explaines the experimental distribution of particle concentration well.

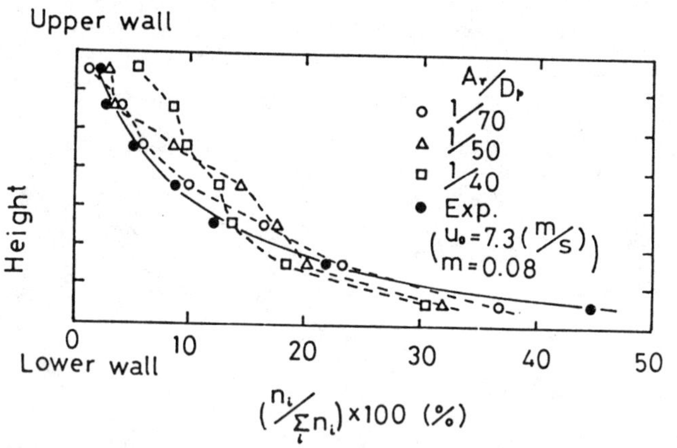

Fig. 1 Effect of wall roughness on the concentration distribution of particles. (at X = 500, polystyrene beads of 0.94 mm diameter)

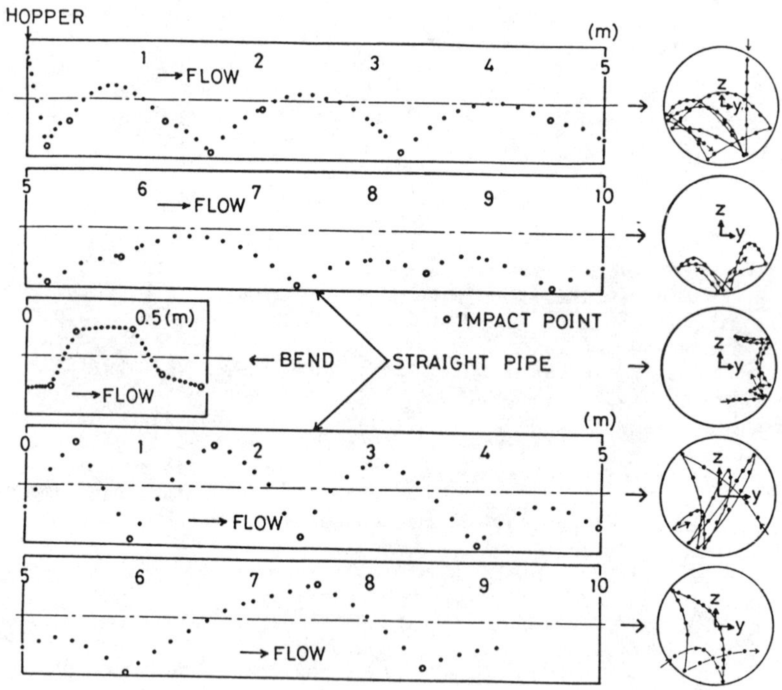

Fig. 2 Calculated particle-trajectories in a 90°-bend. (particle rebounds on the wall)

Morikawa and Tsuji[10] calculated solid particle motion in a circular bend. They solved three dimensional equation of particle-motion using polar coordinates and localized orthogonal coordinates to represent sliding particle-motion and rebounding particle-motion, respectively. Figure 2 shows the claculated particle trajectories in a 90°-bend. The calculated pressure drop and particle velocity agree well with the experimental results. The simulation does not, however, account the irregular bounce of particles, and therefore the particles ultimately deposit on the wall.

These two simulations of coarse-particle motion neglected the effects of turbulence. The calculated results, however, well coinside with the corresponding experimental results, because the particle-inertia is large enough for the particles to pass through the eddies in the fluid.

2. SIMULATION OF FINE-PARTICLE MOTION

When a particle is small, particle trajectory in turbulent flow depends on the turbulence intensity. Yuu[16] calculated the distributions of particle concentration in a horizontal circular pipe assuming the constant turbulence intensity during the Lagrangian diffusion time scale (integral time scale). To calculate fluid turbulent diffusivity, Reichart's[10] and Deissler's[1] empirical equations are used for the turbulent core-region and the wall-region, respectively. Then the particle turbulent diffusivity is calculated by solving the equation of motion of a particle. The distributions of particle concentration are obtained by substituting the particle diffusivity into the three-

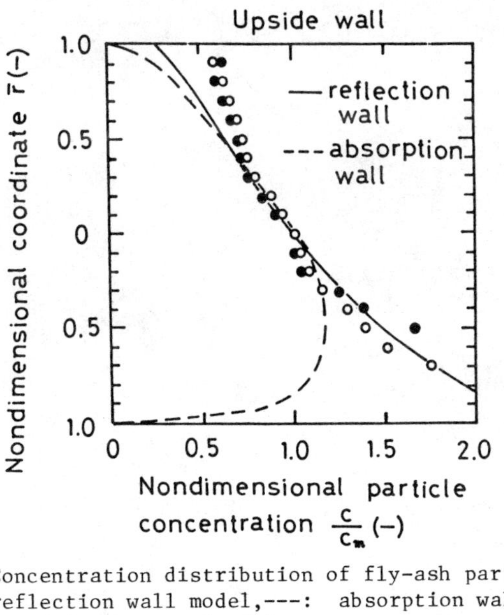

Fig. 3 Concentration distribution of fly-ash particles. (—: reflection wall model,---: absorption wall model, L = 8.5 m, u_{av} = 5 m/s)

dimensional diffusion equation of particles and solving it numerically. He made the calculation for two cases; one assuming a perfect absoving wall and the other, a perfect reflection wall. Figure 3 shows the comparison between the calculated and experimental distribution of particle concentration. The reflection wall model well explains the experimentally obtained concentration distribution of fly-ash particles.

Yoshida, Masuda, and Iinoya[12] simulated particle motion in turbulent vertical upward flow. Here we will introduce the procedure briefly.

Equations of motion of a particle, in non-dimensional form, are given by

a) axial direction;

$$P \frac{d^2\bar{z}}{d\bar{t}^2} = - \frac{C_D Re_p}{24} \left(\frac{d\bar{z}}{d\bar{t}} - \bar{u}_z\right) + \frac{\rho_p D_p^2 g}{18\mu u_o} \tag{1}$$

b) radial direction;

$$P \frac{d^2\bar{r}}{d\bar{t}^2} = - \frac{C_D Re_p}{24} \left(\frac{d\bar{r}}{d\bar{t}} - \bar{u}_r\right) \tag{2}$$

where the inertia parameter (Stokes number) P is defined by

$$P = \frac{\rho_p D_p^2 u_o}{18\mu R} \quad (-) \tag{3}$$

The drag coefficient C_D is give by

$$C_D = \frac{24}{Re_p} (1.0 + 0.15\, Re_p^{0.687}) \tag{4}$$

$\bar{r}$, $\bar{z}$, $\bar{t}$, $\bar{u}_z$, etc. represent the dimensionless quantities (see also Nomenclature). Reynolds number of particle Re_p is expressed by

$$Re_p = Re_{po} \sqrt{\left(\frac{d\bar{r}}{d\bar{t}} - \bar{u}_r\right)^2 + \left(\frac{d\bar{z}}{d\bar{t}} - \bar{u}_z\right)^2} \tag{5}$$

$$\text{where} \quad Re_{po} = \frac{D_p u_o \rho}{\mu} \tag{6}$$

The time-mean axial flow-velocity is assumed by 1/7th power law;

$$(\bar{u}_z)_{av} = \bar{u}_{max} (1 - \bar{r})^{1/7} \tag{7}$$

On the other hand, root-mean square values of fluctuating fluid velocities are given by Laufer[3], and are represented by

a) radial direction ($\bar{r} < 0.8$);

$$\frac{\sqrt{(u'^2_r)_{av}}}{u^*} = -18.89(1-\bar{r})^4 + 43.51(1-\bar{r})^3 - 33.43(1-\bar{r})^2 + 9.17(1-\bar{r}) + 0.35 \tag{8a}$$

b) axial direction ($\bar{r} < 0.8$);

$$\frac{\sqrt{(u'_z)_{av}}}{u^*} = -3.05(1-\bar{r})^4 + 5.18(1-\bar{r})^3 - 1.54(1-\bar{r})^2 + 1.98(1-\bar{r}) + 2.11 \tag{8b}$$

In the above equations, u^* represents friction velocity.
Now, the instantaneous fluid velocities are expressed as follows:

a) radial direction; as the mean air veloctiy is zero, we have

$$\bar{u}_r = \bar{u}'_r \tag{9a}$$

b) axial direction;

$$\bar{u}_z = (\bar{u}_z)_{av} + \bar{u}'_z \tag{9b}$$

Further, we need a life time of fluid eddy which is represented by

$$\Delta\bar{t}^* = \frac{\bar{\eta}}{\bar{u}'} = \frac{\bar{\eta}}{\sqrt{(\bar{u}'^2_r)_{av} + (\bar{u}'^2_z)_{av}}} \tag{10}$$

In Eq. (10), $\eta(= R\bar{\eta})$ is the Kolmogorov's scale of fluid eddy, and is given by[2]

$$\eta = \left(\frac{\mu^3}{\varepsilon\rho^3}\right)^{1/4} \tag{11}$$

The energy dissipation, ε, of unit mass of fluid par unit time is expressed by:

$$\varepsilon = \frac{2\pi RL\cdot\frac{1}{2}\rho u_o^2 f\cdot u_o}{\pi R^2 L\cdot\rho} = \frac{u_o^3 f}{R} \tag{12}$$

Substituting Eq. (12) into Eq. (11), the Kolmogorov's eddy scale is represented by the following equation:

$$\eta = 2R\left(\frac{1}{2fR_e^{\,3}}\right)^{1/4} \tag{13}$$

Velocity fluctuations $\bar{u}'_r$ and $\bar{u}'_z$ are normally distributed randam numbers, and $(\bar{u}'_r)_{av} = (\bar{u}'_z)_{av} = 0$. Standard deviations $\sqrt{(\bar{u}'^2_r)_{av}}$ and $\sqrt{(\bar{u}'^2_z)_{av}}$ are given by Eqs. (8a) and (8b), respectively. These randam numbers were produced by a computer. It is assumed that the velocity fluctuations $\bar{u}'_r$ and $\bar{u}'_z$ have constant values during the life time Δt^*. After the time Δt^* elapsed, they have new randam values. Thus we can calculate the velocities $\bar{u}_r$ and $\bar{u}_z$ from Eqs. (9a) and (9b). Substituting Eqs. (9a) and (9b) into Eqs. (1) and (2), particle trajectories are numerically calculated.

Figures 4(a) and (b) show the particle trajectories plotted by a computer. The radial motions are exaggerated in the figure, because different scales are adopted in a radial and axial directions. Figure 4(a) shows the particle movements of relatively large-inertia particles (24 μm diameter). Fluid fluctuations are not strong enough for the particles to change the direction of motion in a short distance of flow. However, overall trajectories are affected by the fluctuations. Figure 4(b) shows the trajectories of small-inertia particles (3.5 μm diameter), where the inertia is about one order smaller than in Fig. 4(a). Comparing Fig. 4(a) and (b), it is clear that the trajectories of particles with small inertia are more susceptible to fluid fluctuations. These figures show successive dense and sparse particle-concentration distribution resulting from random motion of particles. The tendency becomes more pronounced as the air velocity and/or the particle diameter decreases.

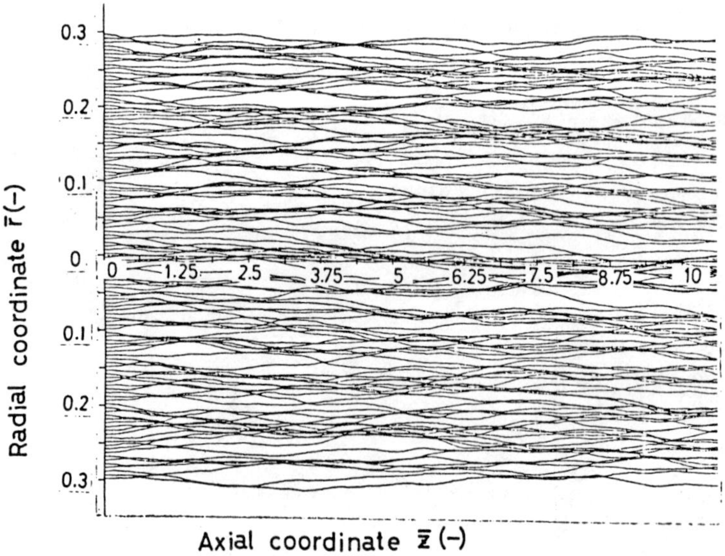

Fig. 4 Particle trajectories simulated by a computer for (a) large-inertia particles

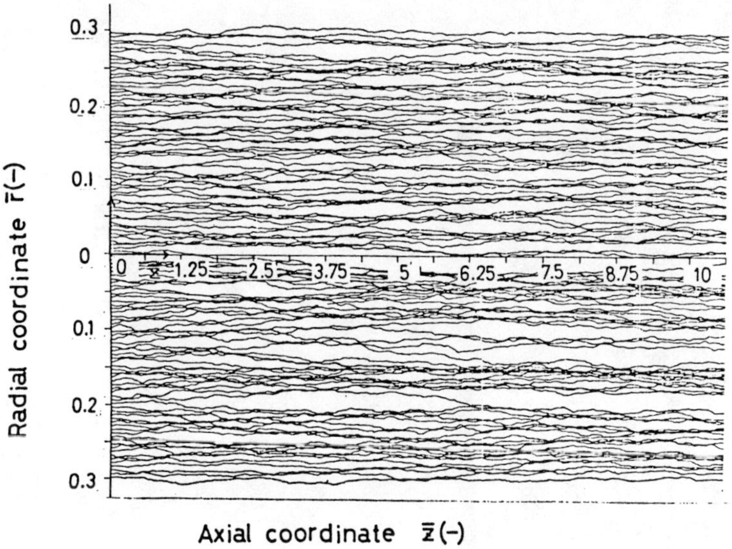

Fig. 4 Particle trajectories simulated by a computer for (b) small-inertia particles

Such uneven distribution of particles causes non-uniform distribution of electrically charged particles, resulting in pulsed electric current in gas-solid pipe flow[6]. Also, in the measurement of fine particle concentration, similar fluctuations may be observed in the instantaneous output of the detectors based on a light scattering or light extinction method, because the particle concentration in their sampling volume fluctuates with time.

3. FURTHER COMMENTS ON THE CALCULATIONS OF PARTICLE TRAJECTORIES

When a small particle is suspended in turbulent air flow, the instantaneous trajectory is affected by the turbulence. Nevertheless, the performance of particle classifier or the efficiency of particle sampling is calculated by a deterministic procedure, and the calculated results agree well with the experiment. Figure 5 shows the particle-trajectories calculated near the sampling probe without consideration for the turbulence-effect[15]. From these trajectories, we can determine the critical trajectory, where the particles are divided into two groups; one flows into the sampling probe and the other does not.

Following the simplified deterministic procedure, Yoshida et al.[17] calculated the sampling efficiency of fine particles in stationary air, and Masuda et al.[5] calculated electrostatic effects on isokinetic dust sampling. The same procedure was applied to calculate the performance of fine particle classifier called virtual impactor[15]. Based on the calculated results, Masuda et al.[7] developed a precise classifier.

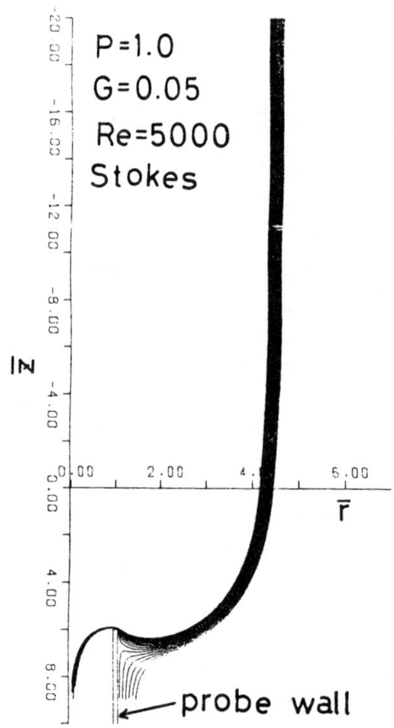

Fig. 5 Calculated particle-trajectories near the sampling probe (stationary air, viscous flow field)

These works suggest the effectiveness of the critical-trajectory sumulation.

NOMENCLATURE:

A_r; amplitude of roughness (m)

D_p; particle diameter (m)

D'; nozzle diameter (m)

D(=2R); pipe diameter (cm)

f; friction factor (-)

L; pipe length (cm)

$P(=\rho_p D_p^2 u_o/18\mu R)$; inertia parameter (-)

$\bar{r}(=r/R)$, $\bar{z}(=z/R)$; dimensionless radial and axial coordinates (-)

R'; radius of curvature for bend (m)

$R_e(=Du_o\rho/\mu)$; pipe Reynolds number (-)

$Re_p(=D_p u_p \rho/\mu)$; particle Reynolds number (-)

$Re_{po}(=D_p u_o \rho/\mu)$; particle Reynolds number for free stream (-)

$\bar{t}(=tu_o/R)$; dimensionless time (-)

Δt^*; dimensionless lifetime of fluid eddy (-)

u_o; main flow velocity (m/s)

u_p; particle velocity (m/s)

$(\bar{u}_z)_{av}$; dimensionless average fluid velocity (-)

$\bar{u}_{max}$; dimensionless fluid velocity at $\bar{r} = 0$ (-)

$\bar{u}_r(=u_r/u_o)$, $\bar{u}_z(=u_z/u_o)$; dimensionless radial and axial fluid velocity (-)

$\sqrt{(u_r'^2)_{av}}$, $\sqrt{(u_z'^2)_{av}}$; root-mean-square values of velocity fluctuations in $\bar{r}$ and $\bar{z}$ directions (m/s)

$\bar{u}_r'(=u_r'/u_o)$, $\bar{u}_z'(=u_z'/u_o)$; dimensionless radial and axial velocity fluctuations

$\bar{u}'\ (=\sqrt{(\bar{u}_r'^2)_{av} + (\bar{u}_z'^2)_{av}})$; dimensionless velocity fluctuation (-)

$u^*(=(u_z)_{av}\sqrt{f/2})$; friction velocity (m/s)

X; dimensionless distance from feed point (-)

η; kolmogorov's scale of fluid eddy (cm)

μ; fluid viscotiy (g/cm·s)

ρ, ρ_p; fluid and particle density (g/cm)

ε; energy dissipation of fluid per unit mass and unit time (erg/g·sec)

$(\ )_{av}$; time-averaged value

REFERENCES:

1. Deissler, R.G. and Taylor, M.F., NASA Tech. Note, 3451 (1955)

2. Hinze, J.O., "Turbulence", New York, McGraw-Hill (1959)

3. Laufer, J., "The Structure of Turbulence in Fully Developed Pipe Flow", NASA Tech. Note, 1174, pp. 417-434 (1954)

4. Masuda, H., Komatsu, T. and Iinoya, K., "The Static Electrification of Particles in Gas-Solids Pipe Flow", AIChE J., Vol. 22, pp.558-564 (1976)

5. Masuda, H., Yoshida, H. and Iinoya, K., "Theoretical Study of Electrostatic Effects on Isokinetic Dust Sampling", J. Chem. Eng. Japan, Vol. 13, pp. 467-472 (1980)

6. Masuda, H., Mitsui, N. and Iinoya, K., "Experimental Investigation of the Pulsed Electric Current Generated by a Two-Phase Flow Consisting of a Gas-Solid Suspension of Low Concentration", Int. Chem. Eng., Vol. 19, pp.287-292 (1979)

7. Masuda, H., Hochrainer, D. and Stöber, W., "An Improved Virtual Impactor for Particle Classification and Generation of Test Aerosols with Narrow Size Distributions", J. Aerosol Sci., Vol. 10, pp. 275-287 (1979)

8. Matsumoto, S. and Saito, S., "On the Mechanism of Suspension of Particles in Horizontal Pneumatic Conveying: Monte Carlo Simulation Based on the Irregular Bouncing Model", J. Chem. Eng. Japan, Vol. 3, pp. 83-92 (1970)

9. Matsumoto, S. and Saito, S., "Monte Carlo Simulation of Horizontal Pneumatic Conveying Based on the Rough Wall Model", J. Chem. Eng. Japan, Vol. 3, pp. 223-230 (1970)

10. Morikawa, Y. and Tsuji, Y., "Computer Simulation for the Pneumatic Transport in Pipes with Bends" , Transactions of the Japan Soc. of Mechanical Engineers, Vol. 43, pp. 4158-4167 (1977) Proc. 4th International Conference on the Pneumatic Transport of Solids in Pipes, BHRA Fluid Engineering, Cranfield, Bedford, England, Vol. 1, pp. B1-1∿20 (June 1978)

11. Reichardt, M., ZAMM, Vol. 31, pp. 208 (1951)

12. Yoshida, H., Masuda, H. and Iinoya, K., "Model Simulation of Particle Motion in Turbulent Gas-Solid Pipe Flow", Powder Technology, Vol. 26, pp. 217-220 (1980)

13. Yoshida, H., Ohsugi, T., Masuda, H., Yuu, S. and Iinoya, K., "Anisokinetic Sampling Errors for Aerosols", Kagaku Kogaku Ronbunshu,

Vol. 2, pp. 336-340 (1976)

14. Yoshida, H., Urgami, M., Masuda, H. and Iinoya, K., "Sampling Efficiency of Particles in Stationary Air", Kagaku Kogaku Ronbunshu, Vol. 4, pp. 123-128 (1978)

15. Yoshida, H., Fujii, K., Yonemoto, Y., Masuda, H. and Iinoya, K., "Classification Performance of Virtual Impactor", Kagaku Kogaku Ronbunshu, Vol. 4, pp. 419-424 (1978)

16. Yuu, S., "Particle Concentration Distribution in a Turbulent Stream in a Horizontal Circular Pipe", Kagaku Kogaku Ronbunshu, Vol. 6, pp. 70-78 (1980)

ON THE STATISTICAL APPROACHES TO THE DYNAMICS OF FULLY FLUIDIZED GRANULAR MATERIALS

S. Ogawa

Computer Center, National Aerospace Laboratory, Chofu, Tokyo, Japan

ABSTRACT

Statistical structure of homogeneous granular particles colliding with each other is investigated. The interaction relation of inelastic spherical particle collision is obtained by using two empirical constants, and the Boltzmann equation for inelastic particles is derived. Using the Monte Carlo method, the time-dependent energy distributions are calculated, and it becomes clear that the distribution of kinetic energy approaches to χ^2-distribution more rapidly than that of rotational energy, and that the distribution of rotational energy gradually seperates from χ^2-distribution.

1. INTRODUCTION

The analysis of the dynamical behaviour of granular materials has been one of the most interesting subjects in the fields of civil engineering and chemical engineering. However, granular materials are very complicated in structure and composition, and a consistent treatment of the dynamical problem is extremely difficult. From the view point of continuum physics, the rotational motion of particles[1,2], the micromotion[3,4] and the motion of void[5] have been introduced as the internal variables to characterize the motion of granular materials. Comparing to the extensive studies of continuum physics approach, the statistical or kinetic approaches are few. At present it is no exaggeration to say that nothing is known about the statistical structures, i.e., the distribution function, for the granular particles colliding with each other. Whence a certain doubtful hypothesis for the statistical structures is needed in deriving the constitutive equations of fluidized granular particles. In [6], we have deduced the constitutive equations of fully fluidized granular materials by using a simple kinematical model for the collision of particles, where the random velocities of granular particles are assumed that the directions of velocities are statistically isotropic and the absolute values of velocities takes a same constant. In recent paper by Savage et al.[7], the distribution function is assumed to be normal distribution. However, it will be natural to consider that a certain distribution function will be defined for the ensemble of granular particles and it should be determined by some roles.

If we intend to obtain the constitutive relations by statistical procedures, we must know the statistical structures of the materials.

In conservative systems as gas molecules, it is well known that the distribution function satisfies the Boltzmann equation[8-10] under some conditions, and it approaches to the normal distribution(Maxwell-Boltzmann distribution) in thermal equilibrium condition. Whence without solving the equation of distribution function, we can easily calculate the physical properties[11], the results of which are in fairly good agreement with experiments. The validity to use the normal distribution in calculating the physical properties of fluidized granular materials will not be justified, since the collision processes of granular particles are inelastic and the energies of motion are not conserved, i.e., change to internal energy.

The purpose of this paper is to construct the foundations of kinetic theory for fluidized granular materials, and we analyze the energy dissipation processes in homogeneous system of granular particles colliding with each other. In 2, we derive the relations of states of particles before and after a binary inelastic collision by using the similar procedures to [6]. The rotational motion of particles is also considered in this paper. It is known that the relations are expressed by two empirical constants, the coefficient of restitution and that of solid friction. In 3, the equation similar to the Boltzmann equation is derived, which takes a very simple form in the case that the collisional relations are assumed to be linear. Since the analytical solution is hardly obtainable, the Monte Carlo method[12,13] is used to calculate the distribution function in 4, the results of which are shown in 5. In order to examine the shape of the distribution function, the Hellinger distance[14] is used, and several interesting features of the statistical structures of granular particles become evident.

2. INELASTIC BINARY COLLISION OF SPHERICAL GRANULAR PARTICLES

Let us consider the interaction relation of inelastic binary collision of spherical particles. In kinetic theory[8,9], the collision of such particles as granules has scarecely been studied. Here we deduce the relation of inelastic collision by using two empilical physical constants; the coefficient of restitution and that of solid friction. We are interested only in inelastic binary collision of particles, and the procedures in deriving the relation are similar to [6](See also [15]).

Consider a collision between two particles of masses m_1 and m_2, radii a_1 and a_2 and inertia moments I_1 and I_2, respectively(Fig.1). The linear and angular velocities of each particle are denoted by $\mathbf{v}_\alpha$ and $\mathbf{w}_\alpha$ (α=1,2). In general, two particles contact at face in a collision, however in our case, the impact is so small that we can neglect the distortion of particles. Whence we assume the following:

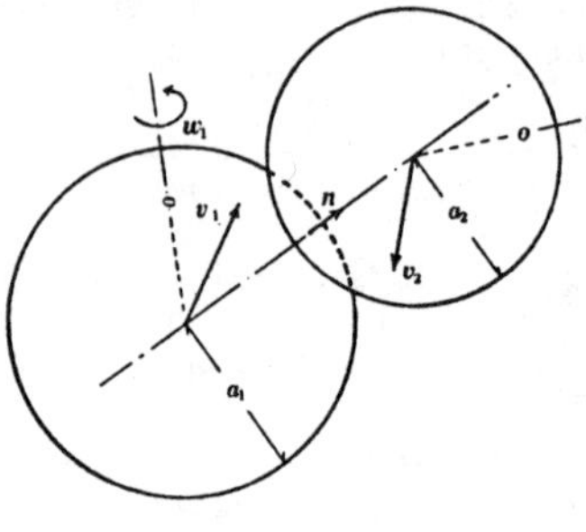

Fig.1 Inelastic binary collision

Assumption 1 Particles collide at one point, and during the collision the point does not move. In other words, the outward normal unit vector $\mathbf{n}$ is kept constant during the collision.

Particles contact during an infinitesimal time, and the collision process satisfies the laws of conservation of linear momentum and of angular momentum. The energies of motions are not conserved, and dissipate to the internal energy in inelastic collision. Let $\mathbf{F}$ be the force acting on particle (1) at the contact point, then the equations of linear and angular velocities become as follows:

$$m_\alpha d\mathbf{v}_\alpha/dt = (-1)^{\alpha+1}\mathbf{F}, \qquad (\alpha=1,2) \tag{1}$$

$$I_\alpha d\mathbf{w}_\alpha/dt = a_\alpha \mathbf{n}\dagger\mathbf{F}, \tag{2}$$

where † denotes the vector product. Since the collisional relations in $\mathbf{n}$ direction and $\mathbf{n}$-normal direction are different with each other, we decompose Eqn.(1) in two directions:
In $\mathbf{n}$ direction, we have

$$[m_\alpha d(\mathbf{v}_\alpha\cdot\mathbf{n})/dt]\mathbf{n} = (-1)^{\alpha+1}(\mathbf{F}\cdot\mathbf{n})\mathbf{n}, \tag{3}$$

and in $\mathbf{n}$-normal direction

$$m_\alpha d[\mathbf{v}_\alpha\cdot(\boldsymbol{\delta}-\mathbf{nn})]/dt = (-1)^{\alpha+1}\mathbf{F}\cdot(\boldsymbol{\delta}-\mathbf{nn}), \tag{4}$$

where $\boldsymbol{\delta}$ denotes Khronecker's delta, and $\cdot$ the inner product. Since $\mathbf{F}$ can be written as

$$\mathbf{F} = (\mathbf{F}\cdot\mathbf{n})\mathbf{n} + \mathbf{F}\cdot(\boldsymbol{\delta}-\mathbf{nn}) \tag{5}$$

the equation of angular velocity (2) is rewritten as

$$I_\alpha d\mathbf{w}_\alpha/dt = a_\alpha \mathbf{n}\cdot[\mathbf{F}\cdot(\boldsymbol{\delta}-\mathbf{nn})]. \tag{6}$$

In deriving the relations between the quantities before and after collision, we must know the interaction force $\mathbf{F}(t)$, and it is not determined without more microscopic considerations. However, we can deduce the relations by making use of two empilical relations as follows. Let us obtain the relative velocity at the point of collision. Since the velocity of each particle at the point of collision is given by $\mathbf{v}_\alpha + (-1)^{\alpha+1}a_\alpha \mathbf{w}_\alpha\dagger\mathbf{n}$ (α=1,2), the relative velocity at the point is

$$\mathbf{V} = V\mathbf{e} = (\mathbf{V}\cdot\mathbf{n})\mathbf{n} + \tilde{V}\tilde{\mathbf{e}} = (\mathbf{v}_1 - \mathbf{v}_2) + \sum(a_\alpha \mathbf{w}_\alpha\dagger\mathbf{n}), \tag{7}$$

where $\mathbf{e}$ is the unit vector in the direction of $\mathbf{V}$, and $\tilde{\mathbf{e}}$ the unit vector in the direction of $\mathbf{n}$-normal component of $\mathbf{V}$, and is given by

$$\tilde{\mathbf{e}} = \mathbf{e}\cdot(\boldsymbol{\delta}-\mathbf{nn})/\sqrt{1-(\mathbf{e}\cdot\mathbf{n})}. \tag{8}$$

Two empilical assumptions are used:

<u>Assumption 2</u> In n direction, the ratio of the relative velocities before and after collision is constant

$$(\mathbf{v}_1' - \mathbf{v}_2')\cdot\mathbf{n}/(\mathbf{v}_1 - \mathbf{v}_2)\cdot\mathbf{n} = -\varepsilon \tag{9}$$

where ε is the coefficient of restitution, and (') denotes the corresponding quantity after collision.

<u>Assumption 3</u> The force acting in $\mathbf{n}$-normal direction, $\mathbf{F}\cdot(\boldsymbol{\delta}-\mathbf{nn})$, is in propotion to $|\mathbf{F}\cdot\mathbf{n}|$, and the direction of the force is the opposite direction to $\tilde{\mathbf{V}}$:

$$\mathbf{F}\cdot(\boldsymbol{\delta}-\mathbf{nn}) = -\eta[\mathbf{F}\cdot(-\mathbf{n})]\tilde{\mathbf{e}}, \tag{10}$$

where η is the coefficient of solid friction.

Let us assume that the particles contact during an infinitesimal time τ, and integrate Eqn.(3) with respect to time 0 to τ. Subtracting the equation of particle (2) from that of particle (1), we obtain the following relation,

$$[(\mathbf{v}_1' - \mathbf{v}_2')\cdot\mathbf{n}]\mathbf{n} - [(\mathbf{v}_1 - \mathbf{v}_2)\cdot\mathbf{n}]\mathbf{n} = \int_0^\tau (1/\mu)(\mathbf{F}\cdot\mathbf{n})\mathbf{n}dt, \tag{11}$$

where μ is the reduced mass, and is defined by $1/\mu = 1/m_1 + 1/m_2$. Subtracting Eqn.(9) from the above relation, we obtain

$$\int_0^\tau (\mathbf{F}\cdot\mathbf{n})dt = -\mu(1+\varepsilon)(\mathbf{v}_1 - \mathbf{v}_2)\cdot\mathbf{n}. \tag{12}$$

Substituting Eqn.(12) into Eqn.(11), the following relation in $\mathbf{n}$ direction is obtained

$$(\mathbf{v}_\alpha'\cdot\mathbf{n})\mathbf{n} = (\mathbf{v}_\alpha\cdot\mathbf{n})\mathbf{n} + (-1)^\alpha(\mu/m_\alpha)(1+\varepsilon)(\mathbf{U}\cdot\mathbf{n})\mathbf{n}, \tag{13}$$

where $\mathbf{U} \equiv \mathbf{v}_1 - \mathbf{v}_2$

Now let us derive the relation in $\mathbf{n}$-normal direction. The slip velocity $\tilde{\mathbf{V}}$ is defined by the projection of the relative velocity $\mathbf{V}$ on the tangent space at the point of collision, and the governing equation of $\tilde{\mathbf{V}}$ is, from Eqns.(4) and (6), and using the identity

$$\mathbf{n}\dagger[\mathbf{F}\cdot(\boldsymbol{\delta}-\mathbf{nn})]\dagger\mathbf{n} = \mathbf{F}\cdot(\boldsymbol{\delta}-\mathbf{nn}), \tag{14}$$

given by

$$d\tilde{\mathbf{V}}/dt = K\mathbf{F}\cdot(\boldsymbol{\delta}-\mathbf{nn}), \tag{15}$$

where $K \equiv 1/\mu + \sum a_\alpha^2/I_\alpha$, and substituting the relation (10) we obtain

$$d\tilde{\mathbf{V}}/dt = -\eta K\mathbf{F}\cdot(-\mathbf{n})\tilde{\mathbf{e}}. \tag{16}$$

The slip velocity after collision can be obtained by integrating Eqn.(16) from 0 to τ, however, we notice that $\tilde{V}$ never takes a negative value in case of friction force acting on the interface. Though we have assumed η as a constant in relation (10), η must have a property of step function in the integration of Eqn.(16):

$$\begin{aligned} \eta(\tilde{V}) &= \eta = \text{constant} && \text{if } \tilde{V} > 0 \\ &= 0 && \text{if } \tilde{V} \leqq 0 \end{aligned} \tag{17}$$

It would be irrational that the slip velocity is reversed by the friction force, and such a nature of solid friction makes the problem extremely difficult. Whence let us consider the following way. Integrating Eqn.(16) formally with respect to time, and using Eqn.(12), we obtain

$$\tilde{\mathbf{V}}' - \tilde{\mathbf{V}} = -\eta\mu(1+\varepsilon)K\tilde{\mathbf{e}}(\mathbf{U}\cdot\mathbf{n}). \tag{18}$$

The right hand side of the above equation is considered to give the maximum value of the deceleration of slip velocity. Since $\mathbf{U}\cdot\mathbf{n} = \mathbf{V}\cdot\mathbf{n}$ is satisfied, rewritten Eqn.(18) by using (8), we obtain

$$\begin{aligned} \tilde{\mathbf{V}}' &= R\tilde{\mathbf{V}} \\ R &\equiv 1 - \eta\mu(1+\varepsilon)K(\mathbf{e}\cdot\mathbf{n})/\sqrt{1-(\mathbf{e}\cdot\mathbf{n})^2}. \end{aligned} \tag{19}$$

As mentioned above, R must satisfy the inequality $0 \leqq R \leqq 1$. Let θ be the angle between $\mathbf{e}$ and $\mathbf{n}$, then R satisfies

$$\begin{aligned} R(\theta) &= [1 - \Lambda\cot(\theta)] && \text{if } \theta^* \leqq \theta \leqq \pi/2 \\ &= 0 && \text{if } 0 \leqq \theta < \theta^* \end{aligned} \tag{20}$$

where Λ is defined by $\Lambda = \eta\mu(1+\varepsilon)K$, and θ is the angle defined by

$\tan(\theta^*) = \Lambda$, and the range of θ is confined $(0, \pi/2)$ from the condition of collision; $\cos(\theta) = \mathbf{e}\cdot\mathbf{n} > 0$.

In use of R, the impulse of $\mathbf{F}\cdot(\boldsymbol{\delta}-\mathbf{nn})$ is given by

$$\int_0^\tau \mathbf{F}\cdot(\boldsymbol{\delta}-\mathbf{nn})dt = -(1 - R)\tilde{\mathbf{V}}/K, \tag{21}$$

and substituting the above relation into Eqns.(4) and (6), we obtain

$$\mathbf{v}'_\alpha\cdot(\boldsymbol{\delta}-\mathbf{nn}) = \mathbf{v}_\alpha\cdot(\boldsymbol{\delta}-\mathbf{nn}) + (-1)^\alpha(1 - R)\tilde{\mathbf{V}}/K/m_\alpha, \tag{22}$$

$$\mathbf{w}'_\alpha = \mathbf{w}_\alpha - a_\alpha(1 - R)(\mathbf{n}\dagger\mathbf{V})/K/I_\alpha. \quad (\alpha=1,2) \tag{23}$$

Summarizing Eqns.(13) and (22), the velocity of the particle after collision is given in the following form:

$$\mathbf{v}'_\alpha = \mathbf{v}_\alpha + (-1)^\alpha[(\mu/m_\alpha)(1+\varepsilon)(\mathbf{U}\cdot\mathbf{n})\mathbf{n} + (1 - R)\tilde{\mathbf{V}}/K/m_\alpha]. \tag{24}$$

Note that, in Eqns.(23) and (24), $\varepsilon = R = 1$ corresponds to the collisional relations of elastic collision(hard sphere model), and $\varepsilon = -R = 1$ to those of rough sphere model[12], the essential feature of which is that the relative velocity at the point of collision is reversed.

3. BOLTZMANN EQUATION FOR GRANULAR PARTICLES

Let us consider the circumstances that the granular particles move in space colliding with each other, and in order to simplify the problem, we assume that the system is homogeneous and the radii of granular particles are identical, the radius of which is denoted by a. When the energy dissipation does not occure, the distribution function $f = f(\mathbf{v}, \mathbf{w}, t)$ satisfies the Boltzmann equation for hard sphere molecules under sufficiently dilute condition[8]. Since the energy does not conserved in our problem, the collision integral of Boltzmann equation must slightly be rewritten.

Following to [9], the Boltzmann equation for homogeneous system with no external force can be written as

$$\partial f/\partial t = \Gamma^{(+)} - \Gamma^{(-)}, \tag{25}$$

where $\Gamma^{(+)}d\mathbf{v}d\mathbf{w}dt$ is the number of particles which in a time dt join the group of points which started from $(\mathbf{v}, \mathbf{w})$ because of collision, and $\Gamma^{(-)}d\mathbf{v}d\mathbf{w}dt$ is the number of particles which are removed from the volume element $d\mathbf{v}d\mathbf{w}$ by collision during the time interval dt. For hard sphere particles, $\Gamma^{(+)}$ and $\Gamma^{(-)}$ are given as follows:

$$\Gamma^{(+)}d\mathbf{v}d\mathbf{w}dt = d\mathbf{v}'d\mathbf{w}'dt\int\cdots\int f(\mathbf{v}', \mathbf{w}', t)f(\mathbf{v}'_1, \mathbf{w}'_1, t)\mathbf{g}'\cdot\mathbf{n}(2a)^2 d\Omega d\mathbf{v}'_1 d\mathbf{w}'_1,$$

$$\Gamma^{(-)}d\mathbf{v}d\mathbf{w}dt = d\mathbf{v}d\mathbf{w}dt\int\cdots\int f(\mathbf{v}, \mathbf{w}, t)f(\mathbf{v}_1, \mathbf{w}_1, t)\mathbf{g}\cdot\mathbf{n}(2a)^2 d\Omega d\mathbf{v}_1 d\mathbf{w}_1, \tag{26}$$

where $d\Omega = \sin(\theta)d\theta d\phi$ is the differential solid angle, and $\mathbf{g}$ the relative velocity defined by $\mathbf{g} = \mathbf{v} - \mathbf{v}_1$.

Now, let us calculate the relation connecting the pre-collisional relative velocity to post-collisional value, and the Jacobian J of the transformation: $(\mathbf{v}, \mathbf{w}) \to (\mathbf{v}', \mathbf{w}')$. When the potential force acts between the particles, $|\mathbf{g}| = |\mathbf{g}'|$ and $J = 1$ are satisfied, and the usual expression for collision integral is obtained(For example, see [9]).

In Eqn.(25), the post-collisional values must be rewritten by the pre-collisional values, and we must calculate the Jacobian:

$$dv'dw'dv_1'dw_1' = J\ dvdwdv_1dw_1. \tag{27}$$

However, the transformation $(\mathbf{v}, \mathbf{w}) \to (\mathbf{v}', \mathbf{w}')$ is nonlinear and is not differentiable as R has the property of step function(See Eqn.(20)). If an effective mean value of R be determined by taking a certain statistical average, the transformation becomes linear. In this case, J is the determinant of 12 x 12 matrix, and after rather complicated calculations

$$J = \varepsilon R^2 \tag{28}$$

is obtained. The relation of relative velocity is easily obtained from Eqn.(9) as

$$|\mathbf{g}'| = \varepsilon|\mathbf{g}|. \tag{29}$$

Rewritten Eqn.(25) by using the above relations, we obtain the Boltzmann equation for inelastic particles as

$$\partial f/\partial t = (2a)^2\int\cdots\int[(\varepsilon J)f'f_1' - ff_1]\mathbf{g}\cdot\mathbf{n}d\Omega d\mathbf{v}_1 d\mathbf{w}_1. \tag{30}$$

As mentioned in previous section, Eqn.(30) becomes the Boltzmann equation of hard sphere particles in the case that $\varepsilon = R = 1$, and of rough sphere particles in the case that $\varepsilon = -R = 1$, respectively.

It will be better to non-dimensionalize Eqn.(30). Following to [8], non-dimensional variables are defined by

$$\begin{aligned} \bar{\mathbf{v}} &= \mathbf{v}/v_o, \quad \bar{\mathbf{w}} = \mathbf{w}/w_o, \\ \bar{f} &= f/[N/(v_o^3 w_o^3)], \\ \bar{t} &= t[\pi N(2a)^2 v_o], \end{aligned} \tag{31}$$

where v_o and w_o are the characteristic velocity and angular velocity, respectively, and N the number density of particles. Using these non-dimensional variables, the non-dimensionalized Boltzmann equation is written as follows:

$$\partial\bar{f}/\partial\bar{t} = \int\cdots\int[(\varepsilon J)\bar{f}'\bar{f}_1' - \bar{f}\bar{f}_1]|\bar{\mathbf{g}}|d\bar{\Omega}d\bar{\mathbf{v}}_1 d\bar{\mathbf{w}}_1, \tag{32}$$

where $d\bar{\Omega} \equiv \cos(\theta)d\Omega/\pi$.

4. MONTE CARLO METHOD

The Boltzmann equation for granular particles derived in the previous section is essentially non-linear, and we cannot solve that equation analytically. We are interesting in the change of distribution function with time, and the Monte Carlo method[12,13] is suitable for our purpose. In this paper we use the procedures proposed by Koura[13], the essential feature of which is that the collision frequency is calculated in Monte Carlo procedures. Using Koura's scheme, the distribution function $\bar{f}(\bar{t}+\Delta\bar{t})$ at time $\bar{t}+\Delta\bar{t}$ is obtained from $\bar{f}(\bar{t})$ at time $\bar{t}$ as follows:

Simulated particles, the number of which is N_s, are considered, and the distribution function $\bar{f}$ is replaced by the set of states of N_s particles.

The time interval $\Delta\bar{t}_c$ is assigned so that $\Delta\bar{t}_c << \Delta\bar{t}$, and we calculate whether each of simulated particles may collide or not in the time interval $\Delta\bar{t}_c$.

1) A particle is assigned, then the mean collision frequency in unit time interval is given by

$$\gamma = \int\cdots\int \bar{f}(\bar{v}_1, \bar{w}_1, \bar{t})|\bar{v}_a - \bar{v}_1| d\bar{\Omega} d\bar{v}_1 d\bar{w}_1, \tag{33}$$

where $\bar{\mathbf{v}}_a$ is the velocity of assigned particle. If the density of particles is sufficiently low and the collision frequency takes small value, the probability that the first collision occures in the time interval $(\bar{t}, \bar{t}+\Delta\bar{t})$ is known to satisfy the Poisson distribution[16]

$$p(\bar{t}) = \gamma\exp(-\gamma\bar{t}), \tag{34}$$

and integrating with respect time from 0 to $\Delta\bar{t}_c$, we obtain the probability that the particle collides in the time interval $\Delta\bar{t}_c$ as

$$P(\Delta\bar{t}_c) = 1 - \exp(-\gamma\Delta\bar{t}_c). \tag{35}$$

2) On the condition that the collision occures in $\Delta\bar{t}_c$, the probability of collision with a particle, the velocity of which belonges in the interval $(\bar{\mathbf{v}}_1, \bar{\mathbf{v}}_1+\Delta\bar{\mathbf{v}}_1)$, is given by

$$\omega(\Delta\bar{\mathbf{v}}_1) = [\int\bar{f}_1 d\bar{w}_1]|\bar{v}_a - \bar{v}_1|\Delta\bar{v}_1/\gamma. \tag{36}$$

3) The states of particles after collision are calculated by assigning a unit normal vector at the point of collision, $\mathbf{n} = \mathbf{n}(\Theta, \Phi)$, by the probability

$$P_n(\Delta\Theta, \Delta\Phi) = \sin(\Theta)\cos(\Theta)\Delta\Theta\Delta\Phi/\pi, \tag{37}$$

where (Θ, Φ) is the polar coordinates, the polar axis of which is fixed with $(\bar{\mathbf{v}}_a - \bar{\mathbf{v}}_1)$.

4) The procedures 1)-3) are carried out for each of simulated particles until the accumulated time $\sum\Delta\bar{t}_c$ exceeds $\Delta\bar{t}$, and the distribution function $\bar{f}(\bar{t}+\Delta\bar{t})$ at time $\bar{t}+\Delta\bar{t}$ is obtained from the velocities and angular velocities of simulated N_s particles at time $\bar{t}+\Delta\bar{t}$.

In computer simulation, we use 10^4 particles, and calculate the change of distribution function. Two initial distributions are considered; the normal distribution(Case 1), and the distribution of equi-energies (Case 2). Since the random numbers of the standard form of normal distribution$(1/\sqrt{2\pi}\exp(-x^2/2))$ are usedfor the initial velocities and angular velocities in Case 1, it is convenient to define the characteristic velocity and angular velocity by

$$\begin{aligned} v_o^2(t) &\equiv (1/3)\int(\mathbf{v}^2 f)d\mathbf{v}d\mathbf{w}, \\ w_o^2(t) &\equiv (1/3)\int(\mathbf{w}^2 f)d\mathbf{v}d\mathbf{w}. \end{aligned} \tag{38}$$

To calculate the mean collision frequency γ, N_c random number are drawn $(N_c < N_s)$, and (33) is replaced by

$$\gamma = \sum_{\beta=1}^{N_c}|\bar{\mathbf{v}}_a - \bar{\mathbf{v}}_\beta|/N_c. \tag{39}$$

Then one random number is drawn which determine the particle may collides or not comparing with the probability Eqn.(35). In order to determine the particle, with which a collision has occured, we draw a random number with the probability density Eqn.(36). Then we draw further two random numbers which determine the collision parameters. If Θ and Φ are determined with the probability Eqn.(37), then the normal unit vector at the point of collision $\underline{n}$ is calculated(See [15]). In calculation, we have put $\Delta\bar{t}_c = 0.025$ and $N_c = 10$, considering the computational time and the errors caused by the difference of these values.

5. RESULTS AND DISCUSSIONS

In our problem, three parameters appear, namely the coefficient of restitution, solid friction and the ratio of the rotational energy to the kinetic energy in initial condition C_e, which is defined by $C_e = I\langle w_o^2\rangle/m\langle v_o^2\rangle$. Since one of the main aims of this paper is to examine the forms of distribution function, a certain criterion which defines the distance between two distributions will be needed. Though several criterions are proposed[14], the Hellinger distance will be suitable considering that the distribution obtained by Monte Carlo simulation is discrete. The Hellinger distance is defined by

$$D_H = \int[1 - \sqrt{P(\mathbf{x})Q(\mathbf{x})}]d\mathbf{x}, \qquad (40)$$

where $P(\mathbf{x})$ and $Q(\mathbf{x})$ are the probability distributions. In conservative systems as gas molecules, it is well known that the distribution function becomes the normal distribution(Maxwell-Boltzmann distribution) in thermal equilibrium condition, hence it will have a significant meaning to examine the distance between the simulated distribution of granular particles and the normal distribution with same mean and variance. To calculate the distance, we construct a histgram of simulated distribution, in which thirty class intervals of same length each are laid off on the horizontal, and considering the fact that if $x_i(i=1,2,3)$ are drawn from $N(0,1)$, the normal distribution with mean 0 and variance 1, then $\{x_i x_i\}$ satisfies the χ^2-distribution with 3 degrees of freedom, the distribution of energies of simulatedparticles are compared with the χ^2-distribution with 3 degrees of freedom.

The time-dependent energy distributions are calculated for two initial distributions, the normal distribution(Case 1), and the equi-energies distribution(Case 2). In Case 1, the simulated distribution of kinetic energy always stays near χ^2-distribution, while that of rotational energy gradually separates from χ^2-distribution. Fig.2 shows the distribution of rotational energy for $\varepsilon = 0.5$, $\eta = 0.3$ and $C_e = 1.0$ at time 5.0, 10.0 and 15.0, and it indicates that the populations over χ^2-distribution in both low energy level and high energy tail increase gradually with time lapth. Fig.3 shows the Hellinger distance between the simulated distribution of rotational energy and χ^2-distribution, where the influence of for $= 0.5$ and $C_e = 1.0$ is illustrated. From the calculations, it becomes clear that the distance takes a large value in $\varepsilon = 0.4 \sim 0.7$ and $\eta = 0.3 \sim 0.6$. In Case 2, the distribution of kinetic energy approaches to χ^2-distribution more rapidly than that of rotational energy. In $\bar{t} >$

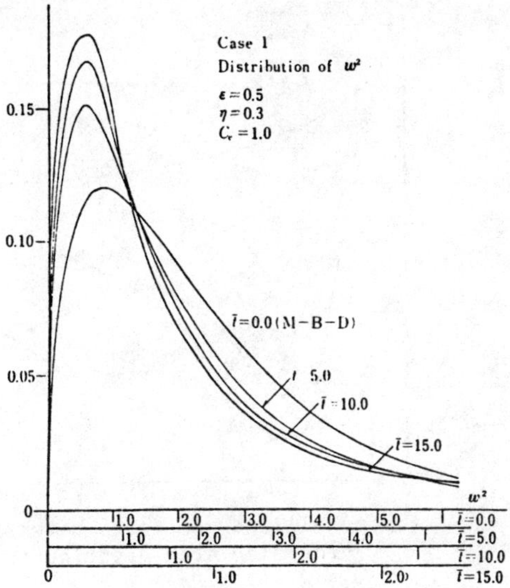

Fig.2 Distribution of rorational energy in Case 1

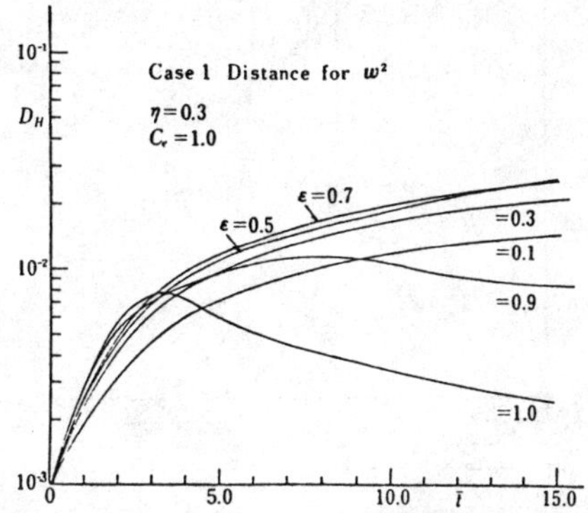

Fig.3 Hellinger distance between the simulated distribution and χ^2-distribution

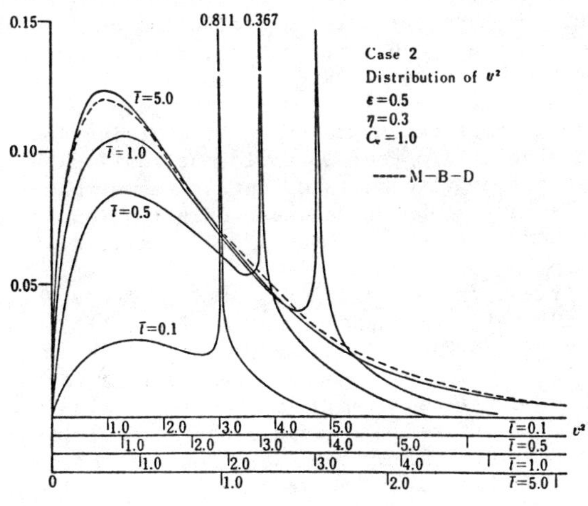

Fig.4 Distribution of kinetic energy in Case 2

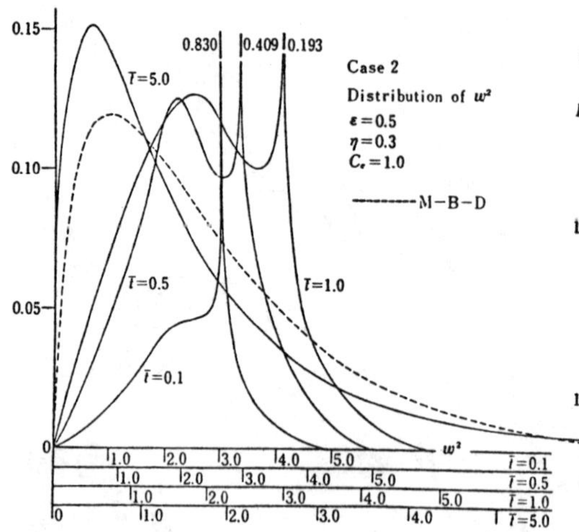

Fig.5 Distribution of rotational energy in Case 2

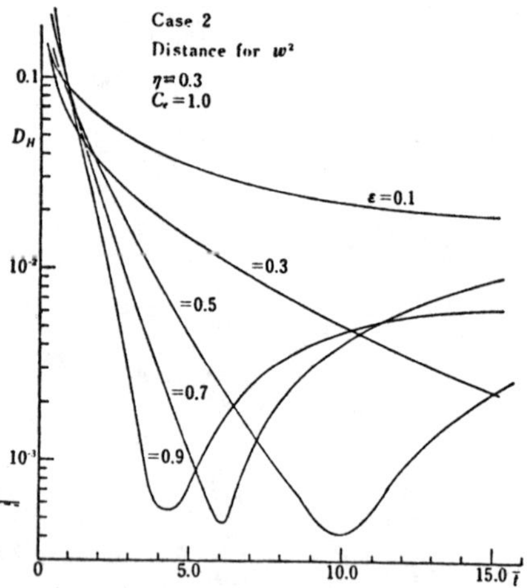

Fig.6 Hellinger distance between the simulated distribution and χ^2 in Case 2

5.0, the distribution of kinetic energy almost coincides with χ^2, however the distribution of rotational energy gradually separates from χ^2 similar to Case 1, see Fig.4 and Fig.5. After several collisions, the distance between the simulated kinetic energy distribution and χ^2 becomes $\mathcal{O}(10^{-3})$, which is involved in the range of error caused by the random numbers. The distance for the rotational energy is shown in Fig.6, where the parameters are similar to Case 1(Fig.3). In any cases the distribution of rotational energy does not approach to χ^2-distribution. Since the computational time being limited, we could not confirm the asymptotical behaviour of distance as time elapses, but it seems that the distribution of rotational enegy approaches to another distribution corresponding to the parameters ε and η, which is about $\mathcal{O}(10^{-2})$ distant from χ^2-distribution.

The mean collison frequency is presented in Fig.7. In initial conditions, the mean collision frequencies are easily calculated as $\gamma = 4/\sqrt{\pi}$ for Case 1, and $\gamma = 4/\sqrt{3}$ for Case 2, respectively. For $C_e = 1.0$, the change of collision frequency is shown by solid line for Case 1 and broken line for Case 2. The influence of η is shown by chain line for $\varepsilon = 0.5$. It is known that the difference between both cases diminishes with time lapth. Recently, we have calculated the effect of volume fraction of granular particles on the collision frequency[17]. Using Monte Carlo method, numerical experiments are done for 10^2 particles, moving in cubic box, where the elastic binary collision and the specular reflection from the wall are assumed. Fig.8 shows the effect of volume fraction on collision frequency. The horizontal coordinate denotes the volume fraction of particles, and the perpendicular coordinate is the calculated collision frequency normalized by the collision frequency theoretically derived in dilute case[18]. The values presented in the figure are calculated by using the average of 50000 particle-particle collisions in the box. Differing from the usual gases, the volume fraction of particles is not negligibly small, and we believe that Fig.8 will become a very useful data in the numerical simulations of flow problems of granular particles.

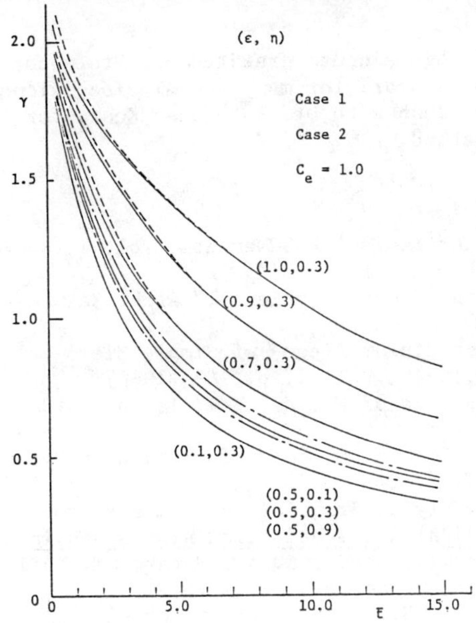

Fig.7 Mean collision frequency

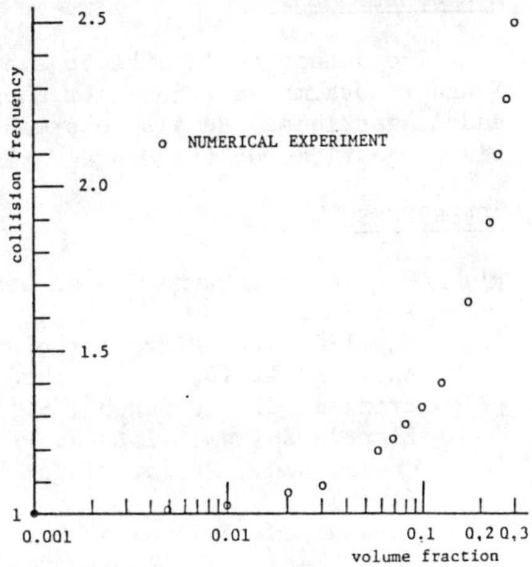

Fig.8 Effect of Volume fraction of particles on mean collision frequency

6. CONCLUSIONS

In order to construct the foundations of statistical theory for fluidized granular materials, we have investigated the statistical structures of granular particles colliding with each other. The relations of inelastic binary collision were obtained by using two empilical constants, the coefficient of restitution and of solid friction, and the Boltzmann equation for inelastic particles was obtained. To investigate the change of distribution function of granular particles, the Boltzmann equation was solved by Monte Carlo mehtod in homogeneous case. Two initial distributions were assumed, and it was known that the distribution of kinetic energy approaches to χ^2-distribution with 3 degrees of freedom more rapidly than that of rotational energy, and that the distribution of rotational energy gradually seperates from χ^2-distribution. The effect of the volume fraction of particles is also calculated by Monte Carlo method, which will become the most important foundation of statistical treatments of granular materials.

Considering that the granular materials are discrete medis, we believe, the Monte Carlo procedures will become the most useful and important methods in investigating the dynamics of granular materials. It may be said that the analysis of dynamical behaviours of granular materials has just started and many interesting subjects will be left in the future.

ACKNOWLEDGEMENT

The author would like to express his sincere gratitude to Professor Nobunori Oshima and Professor Shun-ichi Amari for many useful discussions and suggestions. He also express his thanks to Dr. Kathuhisa Koura for his suggestions of the Monte Carlo method.

REFERENCES

[1] Oshima, N., "Dynamics of Granular Medis," RAGGMemoirs, Vol. 1, p. 111 (1955)
[2] Mindlin, R., "Micro-Structure in Linear Elasticity," Arch. Rat. Mech. Anal., Vol. 16, p. 51 (1964)
[3] Eringen, C. and Suhubi, S., "Non-linear Theory of Simple Micro-Elastic Solids," Int. J. Engng. Sci., Vol. 2, p. 189 (1964)
[4] Cowin, S.C., "Polar Fluids," Physics of Fluids, Vol. 11, p. 1919 (1968)
[5] Goodman, M. and Cowin, S.C., "A Continuum Theory for Granular Materials," Arch. Rat. Mech. Anal., Vol. 44, P. 249 (1972)
[6] Ogawa, S., "Multitemperature Theory of Granular Materials," Proc. US-Japan Semi. Continuum-Mechanical and Statistical Approaches in the Mechanics of Granular Materials, Cowin, S. and Satake, M. (editors), Gakujutsu Bunken Fukyukai, Tokyo, Japan, p. 208 (1978)
Ogawa, S., Umemura, A. And Oshima, N., "On the Equations of Fully Fluidized Granular Materials," ZAMP, Vol. 31, p. 483 (1980)
[7] Savage, S.B. and Jeffrey, D.J., "The Stress Tensor in a Granular Flow at High Shear Rates," J. Fluid Mech., Vol. 110, p. 255 (1981)
[8] Kogan, K., "Rarefied Gas Dynamics", Plenum Press, New York (1969)
[9] Hirschfelder, J., Curtiss, C. and Bird, R., "Molecular Theory of Gases and Liquids," John Wiley and Sons, New York (1954)
[10] Harris, S., "The Boltzmann Equation," Holt, Rinehart and Winston, Inc., New York (1971)
[11] Jeans, J., "The Dynamical Theory of Gases," Dover Pub., Inc., New York (1954)
[12] Bird, R., "Molecular Gas Dynamics," Oxford Univ. Press, Oxford (1976)
[13] Koura, K., "Nonequilibrium Velocity Distributions and Reaction Rates in Fast Highly Exothermic Reactions," J. Chemical Physics, Vol. 59, p. 691 (1973)
[14] Amari, S., "Differential Geometry of Curved Exponential Families," Annals of Statistics, Vol. 10, No. 2, p. 357 (1982)
[15] Ogawa, S., "On the Statistical Approaches for Inelastic Granular Particles," ISAS Report No. 3 (1982), or NAL TR-740T (1982)
[16] Feller, W., "An Introduction to Probability Theory and Its Applications," Vol. 2, John Wiley & Sons, Inc., New York (1966)
[17] Ogawa, S., "Monte Carlo Simulations for Fluidized Granular Materials," (to appear)
[18] Kubo, R., "Statistical Mechanics," North-Holland Pub. Co., Amsterdam (1965)

ON SHEAR FLOW OF COHESIONLESS GRANULAR MATERIALS DOWN AN INCLINED CHUTE

Thomas Schweiwiller and Kolumban Hutter

Laboratory of Hydraulics, Hydrology
and Glaciology, ETH Zurich, Switzerland

ABSTRACT

The continuum model, describing rapid flow of granular materials, which was recently deduced by Jenkins & Savage [11] from considerations of statistical mechanics, is taken up here to formulate and solve numerically gravity driven flow down a chute. In analysing this problem a sliding condition at the basal boundary is conjectured. We present the differential equations and boundary conditions and suggest a procedure of non-dimensionalization which yields the dimensionless parameters governing the problem. The solutions of the nonlinear boundary value problem depend on the coefficient of restitution, the ratio of the particle diameters to the thickness of the shear layer, a parameter characterizing the shear, two sliding coefficients and the inclination angle of the chute. Velocity profiles show the qualitative features observed by Savage [17] but density profiles differ from those reported by others. It is shown that, on occasion, solutions may critically depend on boundary conditions. This points at the need for a deduction of boundary conditions from principles which are as rigorous as those leading to the field equations. This paper extends work of a complementary article by Hutter & Scheiwiller [8].

1. INTRODUCTION

Two idealized flow regimes exist in cohesionless granular materials, one appropriate for slow flow and governed by interparticle friction, the other valid under rapid flow conditions and dominated by particle collisions. Bagnold [1] was the first to clearly identify the two regimes and to perform viscosimetric experiments in both, the viscosity dominated and the inertia dominated regimes. Since Bagnold rapid deformations of dry, relatively dense granular materials have extensively been studied, both theoretically and experimentally, but until recently no systematic theoretical model was available in which the structure of the model and the phenomenological equations would have been deduced with equal rigor from a common fundament. Savage [17] has extensively discussed the literature on various gravity flows of cohesionless materials and has mentioned the difficulties that are encountered in the Goodman-Cowin [4][5] continuum theory. Nunziato et al. [13] present a rectification of that model, but certain features in shear flows between two rigid walls and in plane flows down a chute seem not to be confirmed by experiments. Typical for these models is that the balance laws of mass, momentum (and energy if temperature variation is thought significant) are complemented by a balance law for the volume distribution function of the particles. Methods of continuum thermodynamics are used to deduce the final system of equations. Coefficient functions of the constitutive equations remain largely undetermined, however.

An alternative to deduce a continuum model for the rapid flow of granular materials proceeds along lines of statistical mechanics. Blinowski [2] takes advantage of the fact that the particle motion may be divided into a smooth, mean component and a fluctuation component and deduces balance laws for the mean fields similar to those used in turbulence theory. Statistical ideas were also used by Ogawa et al. [14], Jenkins & Cowin [9] and Jenkins & Savage [10]. In all these formulations a balance law for the fluctuation energy complements the mechanical balance equations; each term in this balance law has a sound physical explanation. Presentation of constitutive relations in these works is, however, still ad-hoc.

A first attempt to deduce a continuum model for the rapid flow of granular materials from statistical mechanics is due to Savage & Jeffrey [18], but it is incomplete as only the stress tensor was analysed. A complete statistical model for hard, tucky identical spherical balls was developed later by Jenkins & Savage [11]. It includes the derivation of balance laws for mass, momentum and collisional fluctuation energy including deduction of constitutive relations for the stress tensor, the flux of fluctuation energy and its annihilation. Physically, Jenkins and Savage's model is simple: only three parameters enter the field equations, namely the particle diameter, the coefficient of restitution in binary collisions and a coefficient characterizing the anisotropy of the particle distribution function at collision whose value is known.

Explicit demonstration whether this model is capable of producing the correct velocity and density distributions in typical boundary value problems has so far not been given to our knowledge. Available experimental evidence, see Savage [17], indicates that for granular flow down a chute velocity profiles deviate substantially from those of Newtonian fluids as they may have a point of inflexion. Furthermore, the density of the bulk material seems to be largest in a layer midway between the bottom and the free surface.

Formulating the field equations is one step of a sucessful description of a physical problem. However, boundary conditions are crucial in the solution of a typical boundary value problem. For granular materials they are not yet clearly understood. At a rigid wall, depending on its roughness, sliding may be assumed. In fluid mechanics of polymers (see e.g. Pearson & Petrie [16], Uhland [21] [22]) and other nonclassical fluid applications (Hutter [6][7]) the sliding law consists of a (nonlinear) relation between shear traction and sliding velocity (usually a power law). This law will also be adopted here, but must be complemented by another relationship involving the fluctuation energy. We propose a relation between the fluctuation energy and the sliding velocity and study its significance in a typical boundary value problem, namely the rapid flow of a granular material down a chute. In particular it will be demonstrated that nearly identical flow configurations may exist for different choices of the phenomenological coefficients in the field equations and boundary conditions. Boundary conditions at a free surface are better understood; they require the vanishing of the fluxes of momentum and fluctuation energy perpendicular to the surface.

Analytical solutions of the complicated nonlinear two-point boundary value problem being inaccessible, we perform a numerical integration. To this end, the governing equations are non-dimensionalized. Thereby the physically important scales become explicitly apparent and the significance of the various different scales can be studied by performing the integration for a wide range of the non-dimensional numbers which govern the problem. We investigate in detail the variations of the coefficient of restitution, of the inclination angle and of the sliding conditions. Results indicate that velocity profiles with an inflexion point emerge when the density is non-uniformely distributed over depth. For a uniform density profile the velocity profile is blunt and flow conditions are near Newtonian. Which of these flow features is attained depends on the parameters which govern the problem. Both can be obtained by either varying the parameters arising in the field equations and keeping those of the boundary conditions fixed, or vice versa. In other words, the experimental separation of the coefficients in the field equations and boundary conditions is difficult if not impossible. This would not be so if the boundary conditions were derived from a statistical model say, similar to the one leading to the field equations. We regard this as the next major step in the proper understanding of the rapid flow of granular materials.

Complementary results to this article are also given by Hutter & Scheiwiller [8].

2. THE BOUNDARY VALUE PROBLEM

Consider steady plane shear flow down a flat plate with angle of inclination α^*, see Figure 1. Let (x,y) be Cartesian coordinates in the direction of flow and perpendicular to it and let $u(y)$ and $\Theta(y)$ be streamwise velocity and fluctuation energy per unit mass, respectively, which in this steady gravity flow are merely functions of the coordinate y. For this situation, using the field equations of Jenkins & Savage [11], it is easy to see that balance of momentum and fluctuation energy reduce to the equations (see Scheiwiller & Hutter, [8]).

$$\frac{2}{5}(2+\alpha\sqrt{\pi})(\kappa k)' + \rho g \sin\alpha^* = 0,$$

$$\frac{\sqrt{\pi}}{\sigma}(\kappa\sqrt{\Theta})' + \rho g \cos\alpha^* = 0, \qquad (1)$$

$$(\kappa\Theta')' - \frac{6(1-e)}{\sigma^2}\kappa\Theta + \frac{4}{5}(2+\alpha\sqrt{\pi})\kappa k^2 = 0,$$

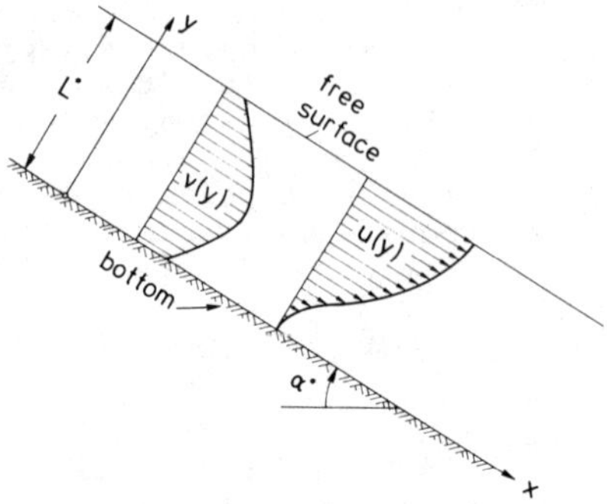

Fig. 1

Definition of steady flow geometry: Gravity driven flow of a layer with thickness L* and inclination angle α*. Typical velocity profile u(y) and density distribution as surmised by the literature (see Scheiwiller & Hutter [19].

in which $(\ )' = d(\ )/dy$ and

$$k \equiv \frac{1}{2}u', \quad \kappa = \frac{2\hat{\rho}\sigma(1+e)}{\sqrt{\pi}}\ \phi_0(\nu)\sqrt{\Theta}, \quad \rho = \nu\hat{\rho}. \tag{2}$$

Here k is the shear rate, κ a conductivity, ρ the bulk density of the material under consideration, g the gravity acceleration, ν the volume distribution function and $\hat{\rho}$ the density of granuls. Furthermore, $\phi_0(\nu)$ is related to the isotropic statistical correlation function $g_0(\nu)$ at contact, which according to Carnahan & Starling [3] is given by

$$\begin{aligned} g_0(\nu) &= \frac{1}{1-\nu} + \frac{3\nu}{2(1-\nu)^2} + \frac{\nu^2}{2(1-\nu)^3}, \\ \phi_0(\nu) &= \nu^2\, g_0(\nu), \end{aligned} \tag{3}$$

and is a good approximation for $\nu \lesssim 0.5$. The remaining parameters in equations (1), σ, e and α are the diameter of the spherical particles, the coefficient of restitution $(0 \ll e \leq 1)$ and a parameter describing the anisotropy of the collisional distribution function. Generally, $\alpha = \alpha(\nu)$ but we shall assume $\alpha = 1/\sqrt{\pi}$, as suggested by Jenkins & Savage [11].

Boundary conditions, to which equations (1) are subject are a viscous sliding law at the bottom.

$$u = f(\tau^2)\tau, \quad \Theta = g(u^2)u^2, \quad \text{at} \quad y = 0, \tag{4}$$

with $\tau = \frac{2}{5}(\kappa + \alpha\sqrt{\pi})\kappa k$ and f and g depending on surface roughness, and no flux of momentum and fluctuation energy perpendicular to the free surface,

$$k = 0, \quad \Theta = 0, \quad \Theta' = 0, \quad \text{at} \quad y = L^*. \tag{5}$$

As far as sliding is concerned, we shall restrict ourselves to the power laws

$$f(x^2) \equiv \mathcal{C}[x^2]^{(n-1)/2}, \quad g(x^2) \equiv \mathcal{c}[x^2]^{(m-1)/2}, \tag{6}$$

with sliding coefficients $\mathcal{C}$ and $\mathcal{e}$ and power law coefficients n and m. The limit $\mathcal{C} \to 0$ corresponds to the no slip condition. The first two of conditions (5) are the expressions of no shear traction and no surface pressure, the third expresses the fact that there is no flux of fluctuation energy.

Equations (1), (4) and (5) with the specifications (6) form a nonlinear two-point-boundary-value problem for the variables ν, u and Θ. The construction of its solutions is the aim of this article. To this end, the equations will first be non-dimensionalized.

Let Θ^*, U^*, ν^* be reference values for Θ, u, ν and L^* a typical length (the thickness of the layer) and let

$$\Theta = \Theta^* \, \bar{\Theta}, \qquad u = U^* \, \bar{u}, \qquad \nu = \nu^* \, \bar{\nu}, \qquad y = L^* \, \bar{y}, \tag{7}$$

so that overbarred quantities are dimensionless. With these the field equations assume the form

$$\begin{aligned} &(\phi_0 \sqrt{\Theta}\, \bar{u}')' + A \sin\alpha^* \, \bar{\nu} = 0, \\ &(\phi_0 \, \Theta)' + B \cos\alpha^* \, \bar{\nu} = 0, \\ &(\phi_0 \sqrt{\bar{\Theta}}\, \bar{\Theta}')' - 6(1-e)\left(\frac{L^*}{\sigma}\right)^2 \bar{\Theta}^{3/2} + \frac{(2+\alpha\sqrt{\pi})U^{*2}}{5\,\Theta^*} \, \phi \, \sqrt{\bar{\Theta}}\, \bar{u}'^2 = 0, \end{aligned} \tag{8}$$

with coefficients A and B as listed in equations (13). Primes now denote differentiation with respect to $\bar{y}$. Orders of magnitude for the coefficient $6(1-e)(L^*/\sigma)^2$ are typically from 10^{-1} to 10^4. This indicates that equations (8) may, on occasion exhibit boundary layer structure. On the other hand the second and third term in the third of equations (8) represent the annihilation of fluctuation energy by collisions and its dissipation due to shearing of the mean motion. Both terms are regarded to be of equal importance, and thus we choose the scales U^*, Θ^* and L^* such that

$$\frac{U^{*2}}{\Theta^*} = (1-e)\left(\frac{L^{*2}}{\sigma}\right). \tag{9}$$

With this choice and with the new variables

$$X = \phi_0 \, \bar{u}'/\sqrt{\bar{\Theta}}, \qquad Y = \phi_0 \, \bar{\Theta}, \qquad Z = \phi_0 \sqrt{\bar{\Theta}}\, \Theta', \tag{10}$$

straightforward manipulations show that the boundary value problem (1)-(5) can be replaced by the following new dimensionless first order nonlinear two-point-boundary-value problem for X, Y, Z, $\bar{u}$, $\bar{\nu}$

$$\begin{aligned} &X' = -A \sin\alpha^* \, \bar{\nu}, \qquad Y' = -B \cos\alpha^* \, \bar{\nu}, \\ &Z' = \frac{1}{\varepsilon^2}\left[(Y/\phi_0)^{3/2} - c\,X\right], \qquad \bar{u}' = X/\sqrt{\phi_0 \, Y}, \\ &\bar{\nu}' = -(\psi_0/\phi_0)\, B \cos\alpha^* \, \bar{\nu} - \psi_0 \, Z/\sqrt{\phi_0 \, Y}, \end{aligned} \tag{11}$$

for $\bar{y} = (0,1)$, subject to the boundary conditions

$$\bar{u} - \mathbb{C}\, X^n = 0, \qquad Y - \mathbb{D}\, \phi_0\, \bar{u}^{m+1} = 0, \qquad \text{at} \quad \bar{y} = 0,$$
$$X = Y = Z = 0, \qquad \text{at} \quad \bar{y} = 1. \tag{12}$$

In equations (11), A, B, C and ε are dimensionless coefficients involving typical physical parameters and the scales Θ^*, U^*, ν^*, L^*, and are defined by

$$A = \frac{5\sqrt{\pi}\, g\, \sqrt{1-e}}{2(2+\alpha\sqrt{\pi})(1+e)} \; \frac{\nu^* L^{*3}}{U^{*2}\,\sigma^2}, \qquad C = \frac{2+\alpha\sqrt{\pi}}{30} = 0.1$$
$$B = \frac{g(1-e)}{2(1+e)} \; \frac{\nu^* L^{*3}}{U^{*2}\,\sigma^2}, \qquad \varepsilon^2 = \frac{1}{6\,(1-e)} \left(\frac{\sigma}{L^*}\right)^2. \tag{13}$$

and the function ψ_0 is given by

$$\psi_0(\bar{\nu}) \equiv \frac{1}{\nu^*} \; \frac{\phi_0^2\,(x)}{d\,\phi_0(x)/dx}\Bigg|_{x=\nu^*\bar{\nu}} . \tag{14}$$

Furthermore, $\mathbb{C}$ and $\mathbb{D}$ are dimensionless sliding coefficients, defined by

$$\mathbb{C} = \frac{\mathcal{C}\, D^{*n}}{U^*}, \qquad \mathbb{D} = \frac{\mathcal{e}}{2\,\varepsilon^2}\, U^{*m-1}, \tag{15}$$

in which

$$D^{*2} = \frac{4(1+e)^2\, \hat{\rho}^2\, (2+\alpha\sqrt{\pi})^2}{25\,\pi\,(1-e)} \left(\frac{U^*\,\sigma}{L^*}\right)^4 . \tag{16}$$

In order that sliding is significant these must differ from zero.

Orders of magnitude for the various coefficients (13) and (15) will be given below. Here it may suffice to mention that ε, on occasion, may be small pointing at possible boundary layer effects of equations (11).

3. NUMERICAL SOLUTION PROCEDURE

Standard numerical integration of the two-point-boundary-value problem (11)-(12) might involve quasilinearization combined with forward shooting techniques, solving a series of initial value problems (see Szidarovski & Yakowitz [20]). In a certain domain of the physical parameters this approach is not necessary, however, because by using a simple iteration scheme, the two-point-boundary-value problem transforms to a series of initial value problems as follows: Assume that an iterate of the volume distribution function $\bar{\nu}(\bar{y})$ is known. The first, second and third of equations (11) may then be integrated, subject to the boundary conditions $(12)_{3-5}$. This yields

$$(X,Y) = (A \sin \alpha^*, B \cos \alpha^*) \int_{\bar{y}}^{1} \bar{v}(\xi)\, d\xi ,$$

$$\varepsilon^2 Z = - \int_{\bar{y}}^{1} \left[\frac{B \cos \alpha^*}{\phi_0(\xi)} \int_{\xi}^{1} \bar{v}(\phi) d\phi \right]^{3/2} d\eta - CA \sin \alpha^* \int_{\bar{y}}^{1} (\bar{y}-\eta)\, \bar{v}(\eta)\, d\eta . \qquad (17)$$

Once X, Y and Z are known as functions of $\bar{y}$ the integration of equation $(11)_4$ can be performed subject to the boundary condition $(12)_4$. Having found $\bar{u}(\bar{y})$, equation $(11)_5$ will yield a new, hopefully improved, distribution for $\bar{v}(\bar{y})$. This requires another forward integration in which $\bar{v}$ on the right hand side of $(11)_5$ is treated as known and the initial value of $\bar{v}$ at $\bar{y} = 0$ is deduced from the boundary condition $(12)_2$, or

$$\phi_0(\bar{v}) = Y/(\mathbb{D}\, \bar{u}^{-m+1}), \quad \text{at} \quad \bar{y} = 0. \qquad (18)$$

Since the function $\phi_0(\nu)$ is monotone for $0 \le \nu < 1$ this equation has a unique solution and thus yields a unique initial condition for the ensuing integration. With new values for the distribution function $\bar{v}$ the integration scheme re-commences and continues until a suitably chosen residue falls below a certain prescribed value.

The above integration procedure goes without difficulties, but needs special consideration near $\bar{y} = 1$ when X, Y and Z vanish simultaneously. Here the rule of Bernoulli de L'Hôpital yields

$$X/\sqrt{\phi_0 Y} = 0, \quad Z/\sqrt{\phi_0 Y} = 0, \quad \text{at} \quad \bar{y} = 1. \qquad (19)$$

Experience also shows that convergence of the iteration scheme is generally without difficulty provided that initially $\bar{v}(\bar{y}) = 0.4$ is chosen and $\alpha^* < 90^o$ and $\varepsilon^2 \ge 0.3$. Further scrutiny of the iteration scheme is needed to achieve convergence outside this range.

4. NUMERICAL VALUES

For a linear sliding law (n = 1, m = 0) the boundary value problem (11)-(12) is governed by the coefficients A, B, C, ε^2, α^*, $\mathbb{C}$ and $\mathbb{D}$. Since $C = 0.1$ there remain six quantities which can be varied. Setting $\mathbb{R} = g\, \nu^*\, L^{*3}/(U^* \sigma)^2$, $\mathbb{A} = \sigma/L^*$, $\alpha = 1/\sqrt{\pi}$ and asuming e to be near unity we obtain from equations (13)

$$A \sim 7.5 x \sqrt{1-e}\, \mathbb{R}, \quad B \sim 2.5 x\, (1-e)\, \mathbb{R},$$

$$\varepsilon^2 = \frac{1}{6(1-e)} \mathbb{A}^2. \qquad (20)$$

The physically relevant independent dimensionless parameters are e, α^*, $\mathbb{R}$, $\mathbb{A}$, $\mathbb{C}$ and $\mathbb{D}$. Realistic maximal and minimal values for the scalings are, perhaps, $\nu^* = (0.1, 0.75)$, $L^* = (0.5, 5)$ m, $U^* = (1, 100)$ m/s,

$\sigma = (10^{-3}, 10^{-1})$ m, so that $\mathbb{R}$ varies, roughly, between 10^{-3} and 10^{6}. Thus, with $(1-e) = (10^{-2}, 10^{-6})$ we obtain

$$10^{-5} < A < 10^{6}, \quad 10^{-8} < B < 10^{4}, \quad 10^{-5} < \varepsilon^2 < 10. \tag{21}$$

These indicate a wide range of variability of the coefficients in the governing equations.

We focus attention here on a variation of $\mathbb{A}$, $\mathbb{R}$, e, α^*, $\mathbb{C}$ and $\mathbb{D}$ in the ranges

$$10^{-3} < \mathbb{A} < 2\text{x}10^{-1}, \quad 5\text{x}10^{2} < \mathbb{D} < 10^{4}, \quad 10^{-6} < 1 - e < 10^{-4}$$

$$10^{\circ} < \alpha^* < 90^{\circ}, \quad 10^{-3} < \mathbb{C} < 10^{-1}, \quad 10^{0} < \mathbb{D} < 10^{4}$$

and shall analyse the sliding law (6) for $-2 < (n-1) < 2$ and $-2 < (m-1) < 2$. Outside the indicated range of the coefficient of restitution convergence of the iteration scheme was uncertain. Values of $\mathbb{A}$ larger than 10^{-1}, roughly, are physically meaningless. Further, with small $\mathbb{C}$ and O(1) values of $\mathbb{D}$ it is ascertained that sliding velocities remain small yet the non-classical boundary condition is likely to play a significant role.

5. DISCUSSION OF RESULTS

Our primary interest is in profiles of the volume distribution function, of the velocity and the fluctuation energy and in orders of magnitudes of the surface velocity and maximal values of the fluctuation energy as function of the above mentioned parameters. In particular, we shall emphasize the significance of boundary conditions. The analysis is performed mostly for a chute with $\alpha^* = 45^{\circ}$, and a layer of the granular material of 0.5 m thickness. Deviations from it will be indicated. Also, until further noticed, the sliding law exponents will be chosen to be $n = 1$ and $m = 0$, corresponding to linear relationships.

1) Variation of the coefficient of restitution

For $\mathbb{A} = 2\text{x}10^{-2}$, $\mathbb{R} = 925$, $\mathbb{C} = 10^{-3}$, $\mathbb{D} = 10$, $\alpha^* = 45^{\circ}$, the profiles of the volume distribution, velocity and fluctuation energy are shown in Figure 2, from which the following facts are evident:

(i) The density distribution function is a monotonic function of depth; density is largest at the bottom and smallest at the free surface. This is contrary to the qualitative results displayed in Figure 1 and reported by Savage [17] to be obtained obervationally and also disagrees with the findings of Nunziato et al. [13], which are based on a different theoretical model.

(ii) With decreasing values of e, that is increasing plasticity of the collisions the non-uniformity of the volume distribution

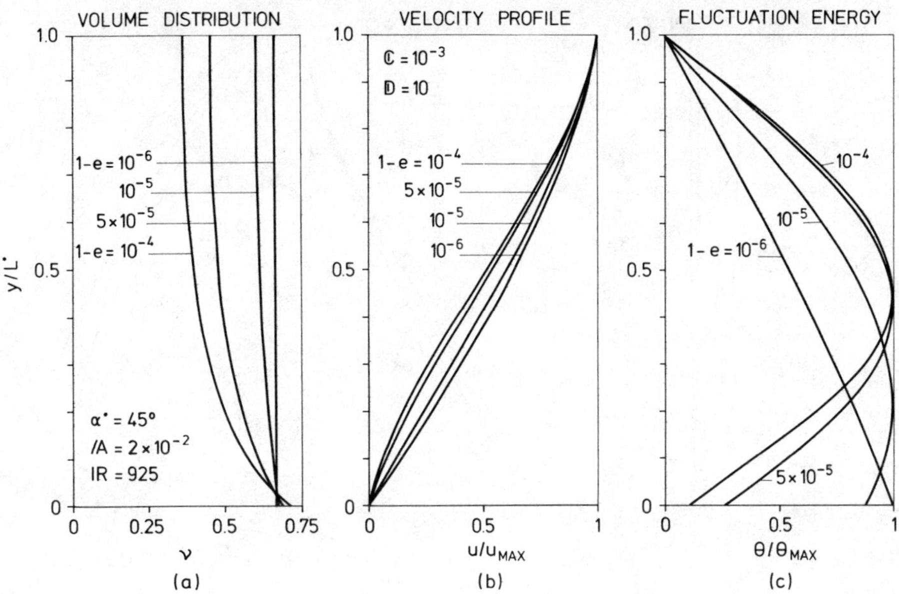

Fig. 2 Profiles of volume distribution (a), velocity (b) and fluctuation energy (c) as functions of the coefficient of restitution, e. Calculations were performed for a chute with $L^* = 0.5$ m, $\alpha^* = 45^\circ$ and $\mathbb{A} = 2\mathrm{x}10^{-2}$, $\mathbb{R} = 925$, $\mathbb{C} = 10^{-3}$, $\mathbb{D} = 10$.

function increases. For $(1\text{-}e) \leq 10^{-6}$ uniformity is attained.

(iii) The velocity profile is blunt (close to a parabola) when $(1\text{-}e) \leq 10^{-6}$, but has an inflexion point for values $(1\text{-}e) > 10^{-5}$. This latter fact is in agreement with the observations reported by Savage [17].

(iv) When $(1\text{-}e) \leq 10^{-6}$ the fluctuation energy is monotonic; it increases from top to bottom. With increasing plasticity of the collisions this monotonicity breaks down; the maximum of the fluctuation energy moves into the layer, and values at the boundaries are now smallest.

Figures 3 and 4 show the same profiles but for values of $\mathbb{D} = 10^2$, 10^3 while the remaining constants are kept the same. Densities tend to become smaller with increasing $\mathbb{D}$ according to (18) and their distribution is more uniform. Correspondingly, the formation of the inflexion point in the velocity profile is less pronounced, and so is the distribution of fluctuation energy. Thus the figures suggest that the flow is Newtonian-like when the density profile is nearly constant or the fluctuation energy linearly distributed over depth,

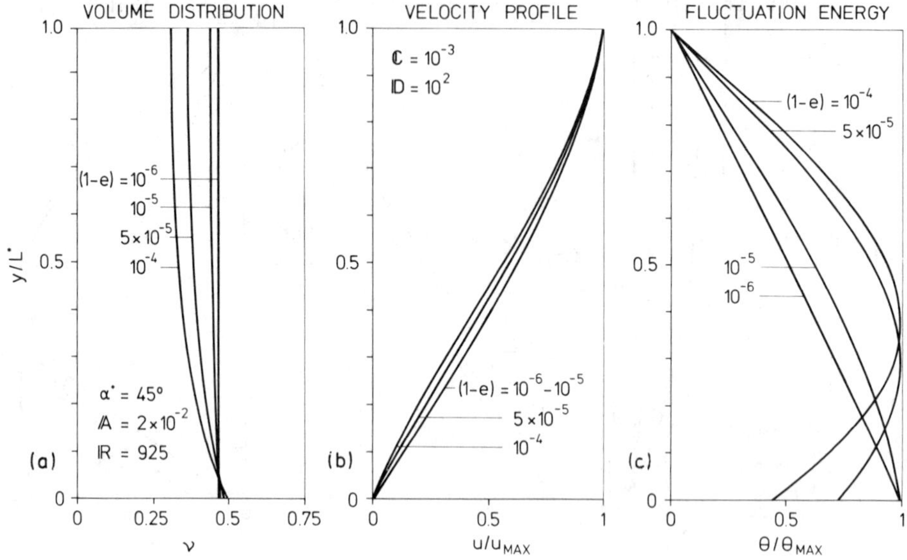

Fig. 3 Same as Figure 2 but for D = 100.

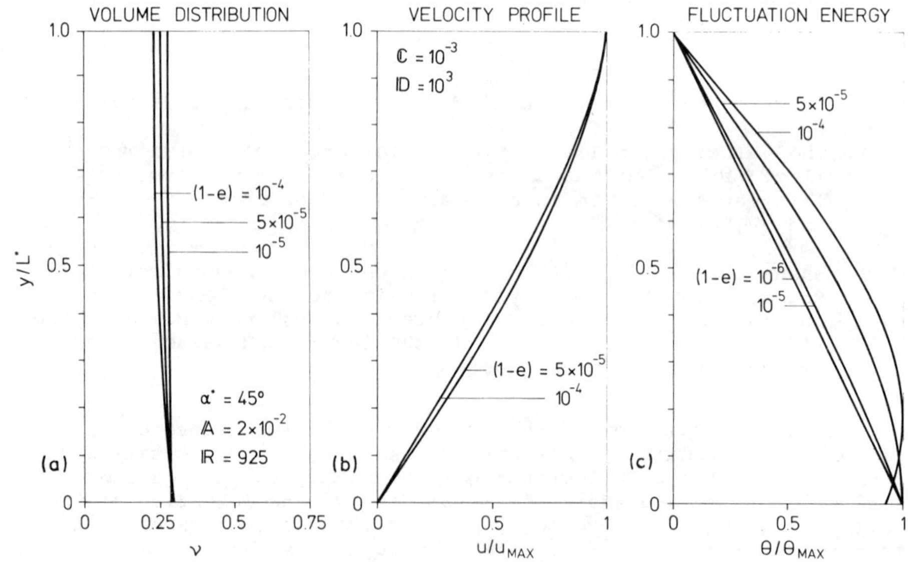

Fig. 4 Same as Figure 2 but for D = 1000.

which arises when $(1-e) \lesssim 10^{-5}$. Later figures will broadly corroborate this behavior.

Surface velocities and maximal values of the fluctuation energy for a chute with $\alpha^* = 45^o$, $L^* = 0.5$ m are shown in Figure 5. Notice that changes in u_{Max} and Θ_{Max} are appreciable for $(1-e) > 10^{-5}$, but negligible when $(1-e) \leq 10^{-6}$. This brings about very clearly the significance of collisional inelasticity. For e very near unity collisions are elastic and flow behavior apparently Newtonian-like. To obtain a velocity profile with inflexion point a certain inelasticity of the collisions is necessary. However, it need not be large and $(1-e) = O(10^{-4})$ suffices to obtain velocity profiles which qualitatively agree with those observed by Savage [17]. It should be recognized that, depending on the values of $\mathbb{D}$, u_{Max} may be very large, while $\Theta^{1/2}_{Max}$ is one to two orders of magnitude smaller compared to it. The strong dependence of u_{Max} and Θ_{Max} on the frictional coefficient $\mathbb{D}$ also points at the importance of a clear understanding of boundary conditions. This will further be emphasized below. Finally, it should be stressed that for values $(1-e) > 10^{-4}$ no convergence of the numerical integration scheme was achieved. This requires further scrutiny because Figure 5 suggests this to be an interesting range of the coefficient of restitution.

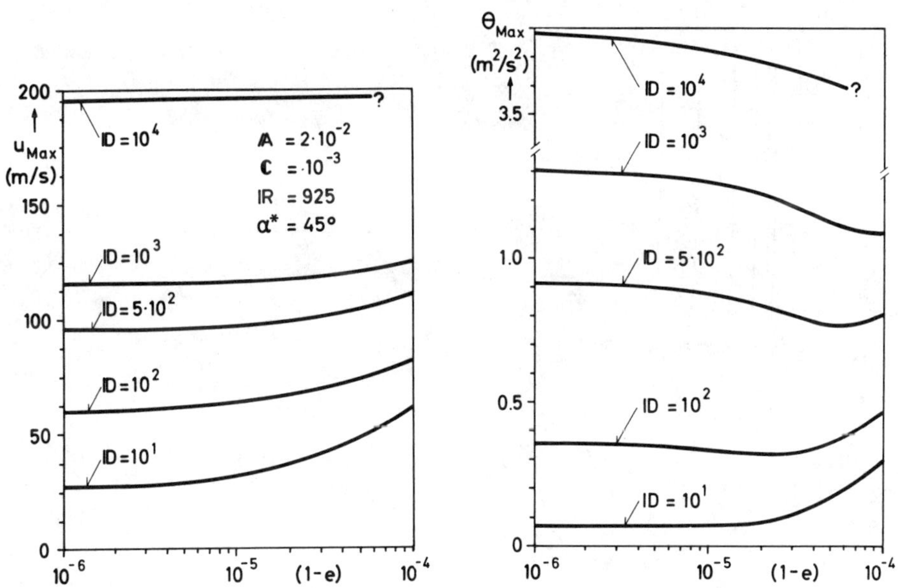

Fig. 5 Surface velocity (a) and maximal values of the fluctuation energy (b) as functions of coefficient of restitution, e, and for various values of the friction parameter $\mathbb{D}$, valid for a chute with $L^* = 0.5$m, $\alpha^* = 45^o$, $\mathbb{A} = 2\times10^{-2}$, $\mathbb{R} = 925$ and $\mathbb{C} = 10^{-3}$. Question marks indicate domains where numerical integration procedure did not converge.

2) Variation of the inclination angle

In Figure 2, profiles with a coefficient of restitution satisfying $(1-e) = 10^{-4}$ agree, qualitatively, with those of Savage [17]. Ensuing discussions will therefore mostly be limited to this value.

Figures 6 and 7 show the same profiles as do Figures 2 and 3, but now for α^* varying in the range $10^\circ \lesssim \alpha^* \lesssim 85^\circ$. Values of other parameters are shown in the figure captions. Profile forms are largely independent of the inclinating angle α^* but magnitudes of ν, u and Θ may vary appreciably. Density profiles are again monotonic with largest particle concentrations at the base and smallest values at the free surface; and for small inclination angle they tend to be more uniform than for large ones. Other theoretical models (Nunziato et al., [13]) tend to agree with this, but the monotonicity is not borne out by these other models. Density decreases with increasing α^* in Figures 6 and 7 as one would expect intuitively.

Not surprising either, the values ν, u_{Max} and Θ_{Max} depend critically on the inclination angle. Figure 8 collects some results for u_{Max} and Θ_{Max} as functions of inclination angle; u_{Max} increases monotonically with increasing α^*, but Θ_{Max} does not. In fact the

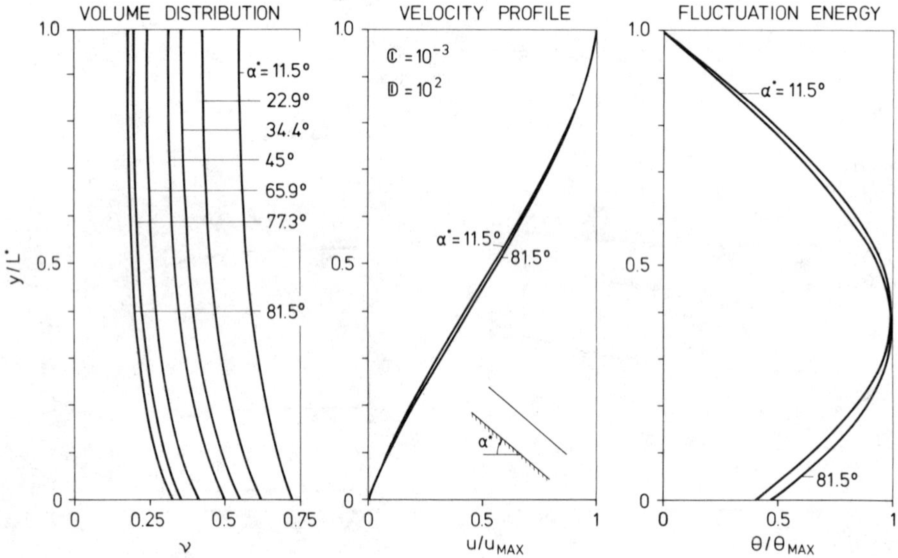

Fig. 6 Profiles of the volume distribution function (a), velocity (b) and fluctuation energy (c) as functions of the inclination angle α^*. Calculations were performed for a chute with $L^* = 0.5$ m, $\mathbb{A} = 2\times10^{-2}$, $\mathbb{R} = 925$, $\mathbb{C} = 10^{-3}$, $\mathbb{D} = 10^2$.

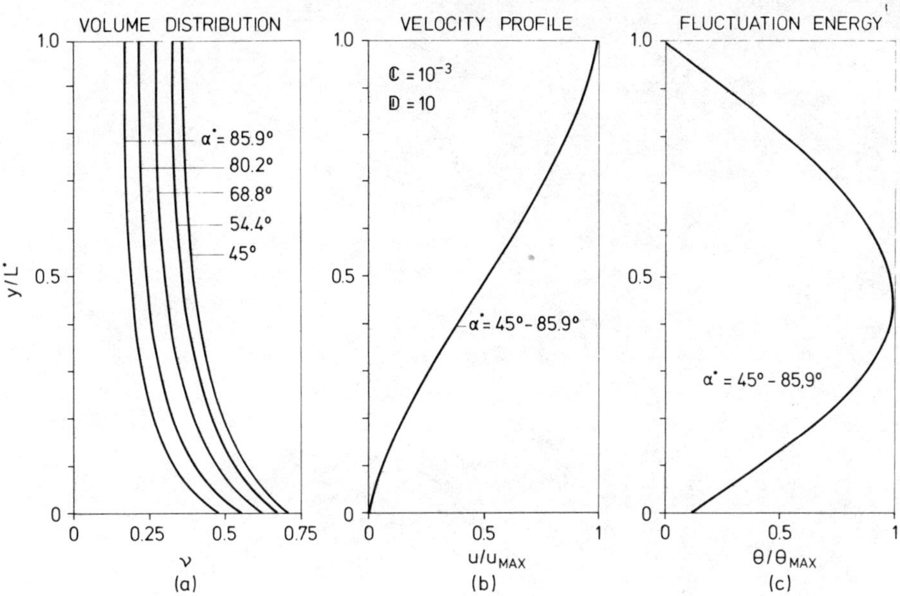

Fig. 7 Same as Figure 6 but calculated with a value of the sliding coefficient $\mathbb{D}$ = 10.

function $\Theta_{Max}(\alpha^*)$ is not bijective. Notice also the conspicuous dependence of both variables, u_{Max} and Θ_{Max} on $\mathbb{D}$, and the unusually large surface velocity at large inclination angles. For these the model equations are likely to be invalid because of the neglect of air resistance.

3) Variation of the sliding conditions

Because deduction of boundary conditions at a rigid, rough wall is largely conjectural, a sensitivity analysis of the solutions of the boundary value problem to the postulated sliding conditions is particularly instructive. Figure 9 shows the three transverse profiles of density, velocity and fluctuation energy using a linear basal sliding law according to equation (12), varying $\mathbb{C}$ and keeping $\mathbb{D}$ fixed. Figure 10 summarizes the same results when $\mathbb{D}$ is varied but $\mathbb{C}$ is held fixed. Both sliding coefficients are significant and apparently may have similar effects. In Figure 9 $\mathbb{C}$ varies between 10^{-3} and $2x10^{-2}$ and $\mathbb{D}$ = 10. In view of equation (12) one may thus be tempted to conclude that basal sliding is small. This is indeed so, as the sliding component in Figures 9 and 10 is not visible in comparison to the gliding part. However, varying $\mathbb{C}$ and $\mathbb{D}$ in the indicated ranges changes velocity profiles from near Newtonian to the form with inflexion point. Near-Newtonian behavior of the velo-

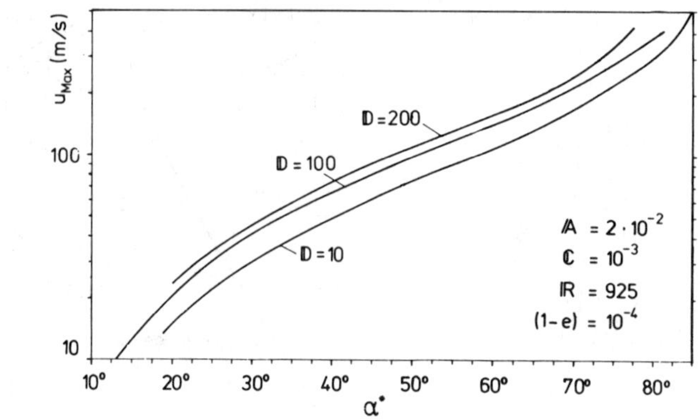

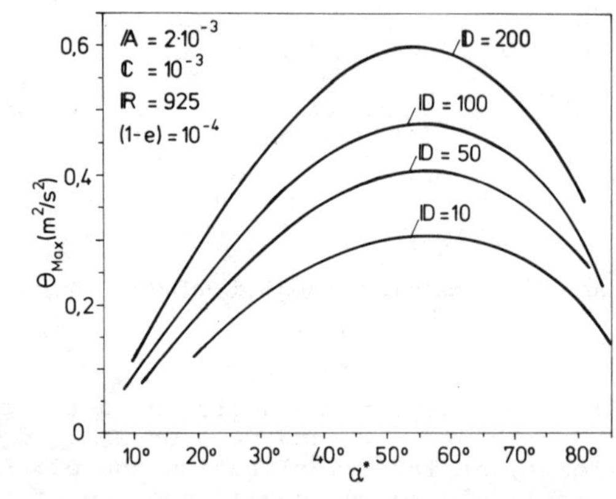

Fig. 8 Surface velocities (a) and maximum fluctuation energy (b) as functions of inclination angle α^* and several values of the friction parameter $\mathbb{D}$, valid for a chute with $L^* = 0.5\,m$, $\mathbb{A} = 2x10^{-2}$, $\mathbb{R} = 925$ and $\mathbb{C} = 10^{-3}$.

city profile is accompanied by both, a uniform distribution and a linear distribution of the fluctuation energy, as before. The same results are also observed in Figure 10. Hence, nearly identical density, velocity and fluctuation energy profiles can be obtained with two different choices of the sliding coefficients. Figure 11 shows the surface velocity and the maximal value of the fluctuation energy. Evidently, u_{Max} and Θ_{Max} are increasing functions of both $\mathbb{C}$ and $\mathbb{D}$, and both may assume quite large values. Question marks indicate domains where the integration scheme did not converge. Further, to prescribed values of u_{Max} and Θ_{Max}, there are an infinity of values for the pair $(\mathbb{C}, \mathbb{D})$, and there are certainly two pairs with nearly the same density, velocity and fluctuation energy profiles. On the other hand, comparing the profiles of the volume distribution of Figures 9 or 10 with those of Figure 2 it is recog-

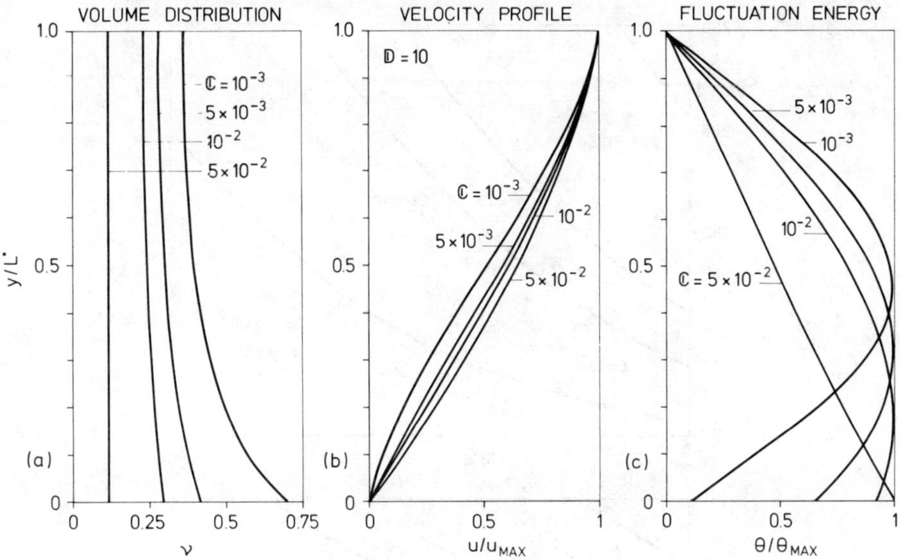

Fig. 9 Profiles of the volume distribution function (a), velocity (b) and fluctuation energy (c) as function of the sliding parameter ℂ, keeping $(1-e) = 10^{-4}$, 𝔻 = 10, A = $2x10^{-2}$, ℝ = 925, $\alpha^* = 45^o$ fixed. A linear sliding law with n = 1, m = 0 was used.

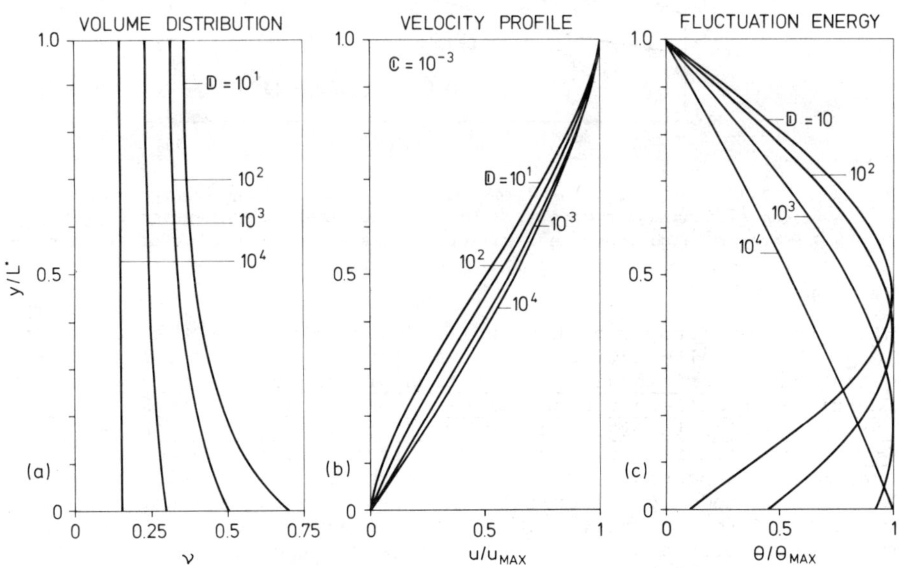

Fig. 10 Same as Figure 9 but now 𝔻 is varied and ℂ is held fixed.

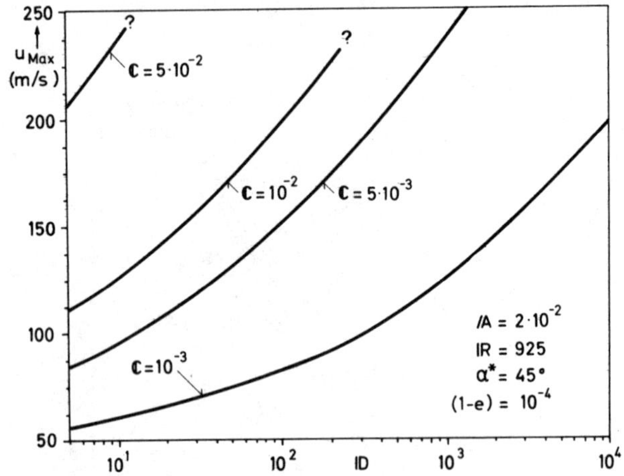

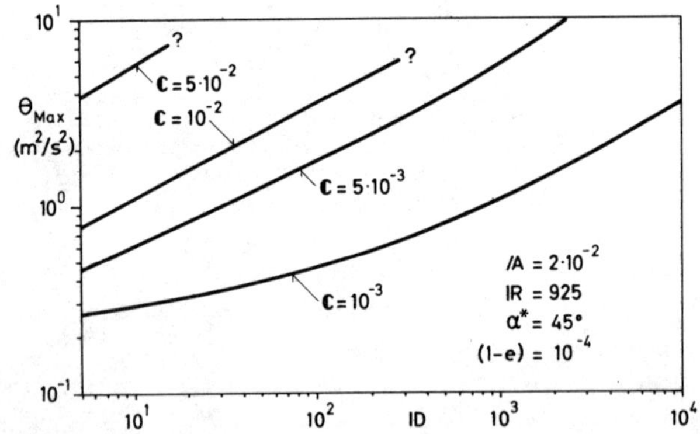

Fig. 11 Surface velocity (a) and maximum fluctuation energy (b) as functions of the sliding coefficients. Values of the other non-dimensional parameters are also indicated. Question-marks indicate domains of the parameters, where convergence of our integration procedure was not achieved.

nized that uniformity of the profile is obtained for the large $\mathbb{C}$ or large $\mathbb{D}$ and, alternatively, for small values of (1-e). This suggests that experimental determination of the volume distribution function for different basal sliding conditions is necessary to separate the effects of inelasticity of collisions and of basal sliding. On the other hand, experimental identification of $\mathbb{C}$ and $\mathbb{D}$ does not seen possible. It requires a theoretical model, perhaps a statistical model of spherical particles bouncing against a rough wall. Proof for the difficulty of the separation of parameters e, $\mathbb{C}$ and $\mathbb{D}$ is provided by Figure 12. It shows the profiles for two different chutes characterized by $\alpha^* = 45^o$, $\mathbb{A} = 2x10^{-2}$, $\mathbb{R} = 925$, and the remaining parameters as indicated in Figure 12. As evident, pro-

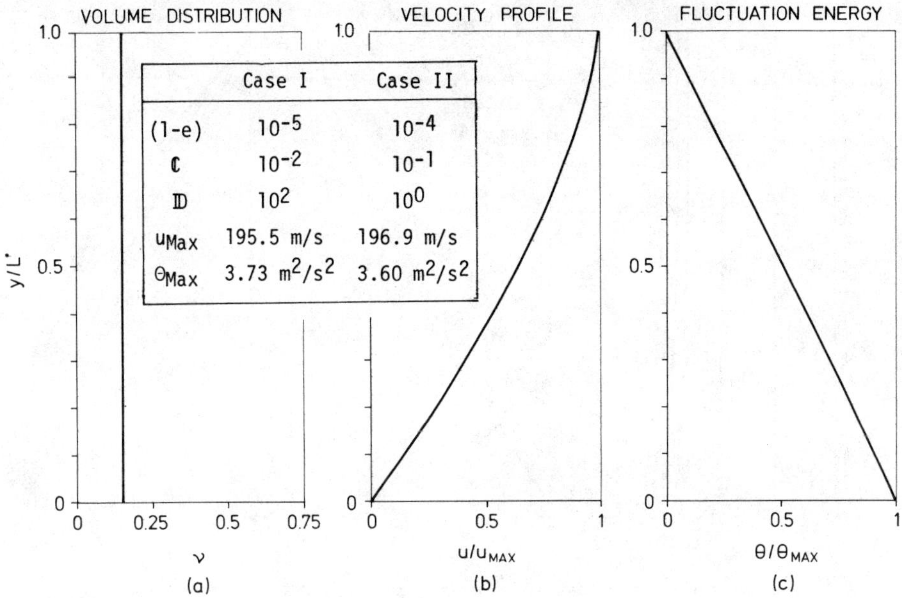

Fig. 12 Profiles of the volume distribution function (a), velocity (b) and fluctuation energy (c) for a chute with $\alpha^* = 45^o$, $L^* = 0.5\,m$, $\mathbb{A} = 0.02$, $\mathbb{R} = 925$ and e, $\mathbb{C}$, $\mathbb{D}$, u_{Max}, Θ_{Max} as given above.

files hardly differ at all and neither is there a difference in u_{Max} and Θ_{Max}. There is no experiment separating the two cases.

The above considerations ignore nonlinear sliding laws which may or may not follow the postulated power laws. Nevertheless, it is interesting to see how results vary with the variation of the exponents in the power laws of the sliding conditions. Calculations were performed for $\mathbb{A} = 2x10^{-2}$, $\mathbb{C} = 10^{-3}$, $\mathbb{D} = 10$, $\mathbb{R} = 925$, $(1-e) = 10^{-5}$ and

$$a = \frac{n-1}{2} = 0 \qquad -0.8 < b = \frac{m-1}{2} < 1$$
$$-0.4 < a < 0.4 \qquad b = 0. \tag{21}$$

In the first case the classical sliding condition $(12)_2$ is under-linear or over-linear; in other words, the exponent in equation $(12)_2$ is either larger than unity or smaller. In the second case both sliding laws are non-linear. However, $b = 0$ corresponds to the relation $\Theta = \mathfrak{e}u^2$ which physically is the most natural relation.

In figure 13 we have summarized the profiles for the conditions as listed in relations $(21)_1$ and $b < 0$ (under-linear); Figure 14 contains the same results when $b > 0$ (over-linear). In the first

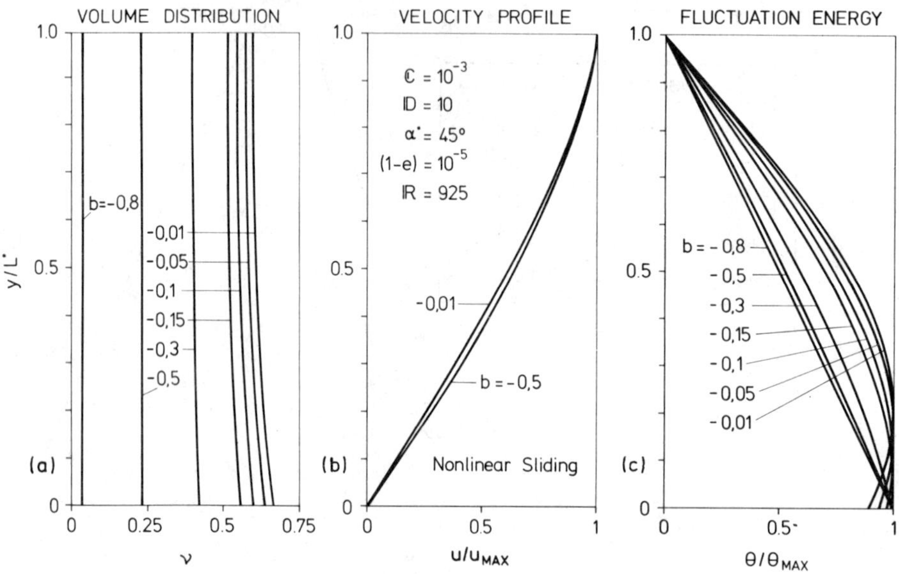

Fig. 13 Profiles of volume distribution (a), velocity (b) and fluctuation energy (c) as functions of the sliding law exponent $b = (m-1)/2$, keeping $(1-e) = 10^{-5}$, $\mathbb{A} = 2\times10^{-2}$, $\mathbb{C} = 10^{-3}$, $\mathbb{D} = 10$, $\mathbb{R} = 925$, $\alpha^* = 45°$ and $n = 1$ $(a = 0)$ fixed, $b \leq 0$.

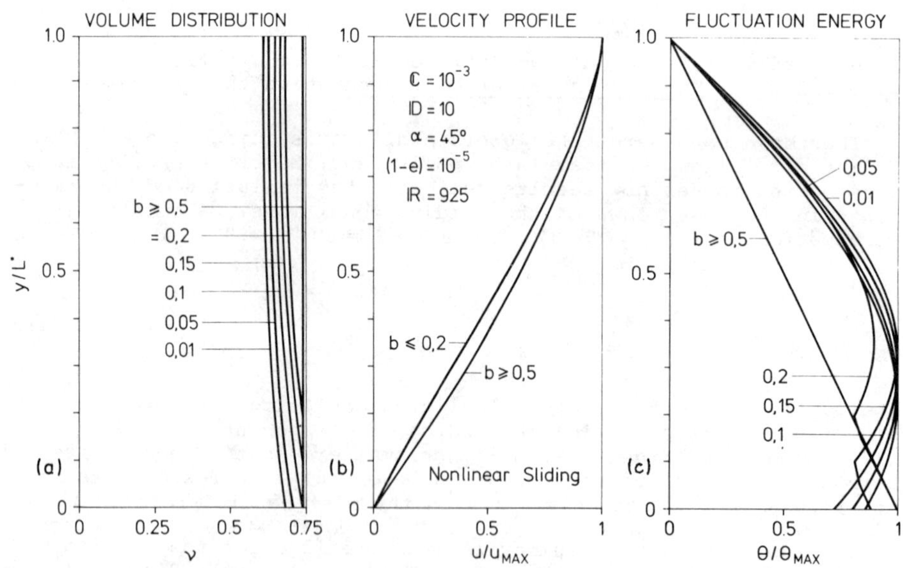

Fig. 14 Same as Figure 13, but now $b \geq 0$.

case, density profiles are nearly uniform but the absolute value for the volume distribution of particles depends crucially on the exponent b. With decreasing values of it the density decreases. Velocity profiles are nearly insensitive to changes in b but the distribution of fluctuation energy again depends on it. A similar observation also holds when $b > 0$ (Figure 14), but calculations for $b > 0.1$ can hardly be trusted here (even through the numerical scheme converged) because for $b \geq 0.2$ a kink arises in the ν- and Θ-distributions. This kink occurs because the volume distribution function reaches the value corresponding to the densiest packing. For this value $\nu \cong 0.74$ rigid body motion would have to result which is not the case in Figure 14 when $b \geq 0.2$.

Figures 15a,b indicate, how the surface velocity and the maximum fluctuation energy vary when the power law exponent $(m-1)/2 = b$ varies between -0.8 and 0.5. The transition from a continuous differentiable solution to a flow where parts of the layer should move as a rigid body can clearly be identified.

When the power law exponent of the other, first equation of the sliding law is varied (case $(21)_2$), results are as shown in Figures 16 - 17. The first displays the profiles for $a > 0$, indicating a definite but not pronounced dependence of the density and fluctuation energy profiles on the exponent a. For $a > 0.3$ convergence of our

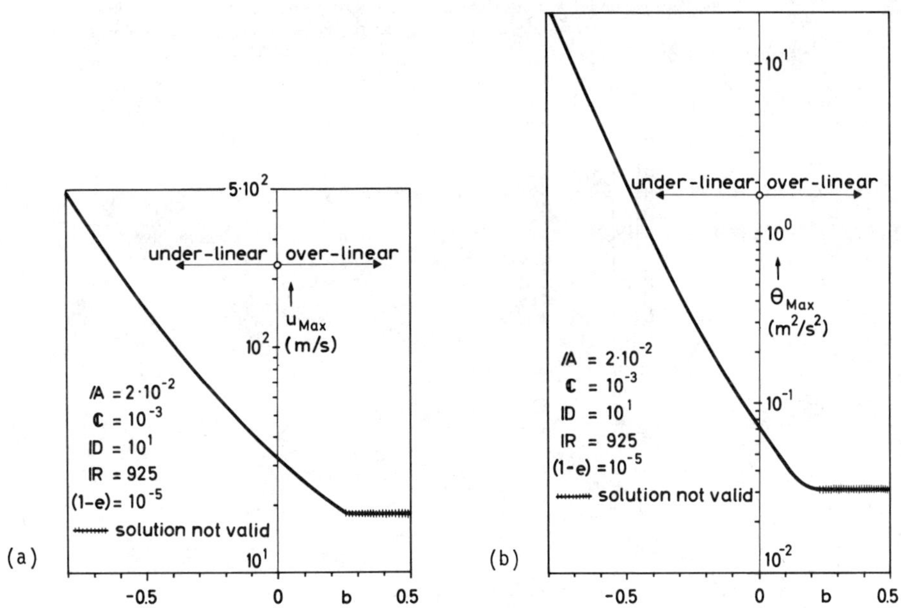

Fig. 15 Surface velocity (a) and maximum fluctuation energy (b) as functions of the sliding law exponent $b = (m-1)/2$.

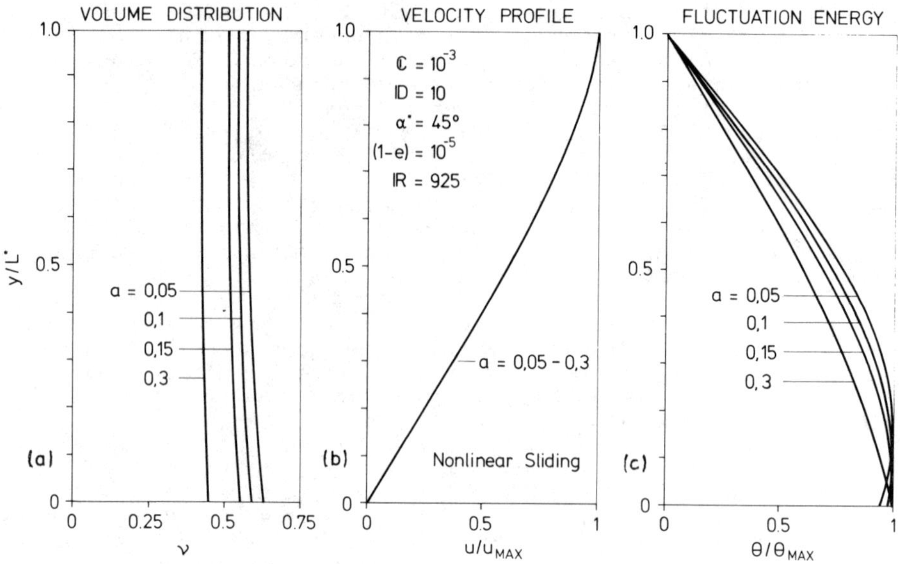

Fig. 16 Profiles of volume distribution (a), velocity (b) and fluctuation energy (c) as functions of the sliding law exponent $a = (n-1)/2$, keeping $(1-e) = 10^{-5}$, $\mathbb{A} = 2\times10^{-2}$, $\mathbb{C} = 10^{-3}$, $\mathbb{D} = 10$, $\mathbb{R} = 925$, $\alpha^* = 45^{\circ}$ and $m=1$ fixed. $a \geq 0$.

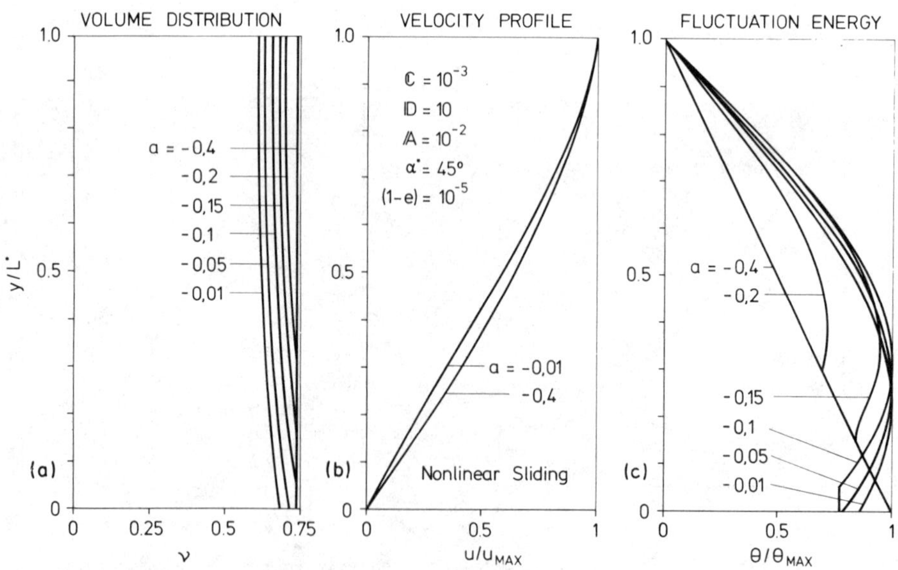

Fig. 17 Same as Figure 16, but now $a \leq 0$.

integration scheme was not achieved. The second (Figure 17) contains again profiles with kinks. Their cause can again be traced to the volume distribution function reaching its maximum value $\nu \cong 0.74$. Profiles are again false solutions because they are not differentiable throughout the layer depth. In both graphs there is virtually a zero sliding velocity, however, surface velocities u_{Max} and maximal fluctuation energies Θ_{Max} nevertheless depend substantially on the value of the exponent a (Figure 18). This is proof for the fact that knowledge of the form of the sliding law is important.

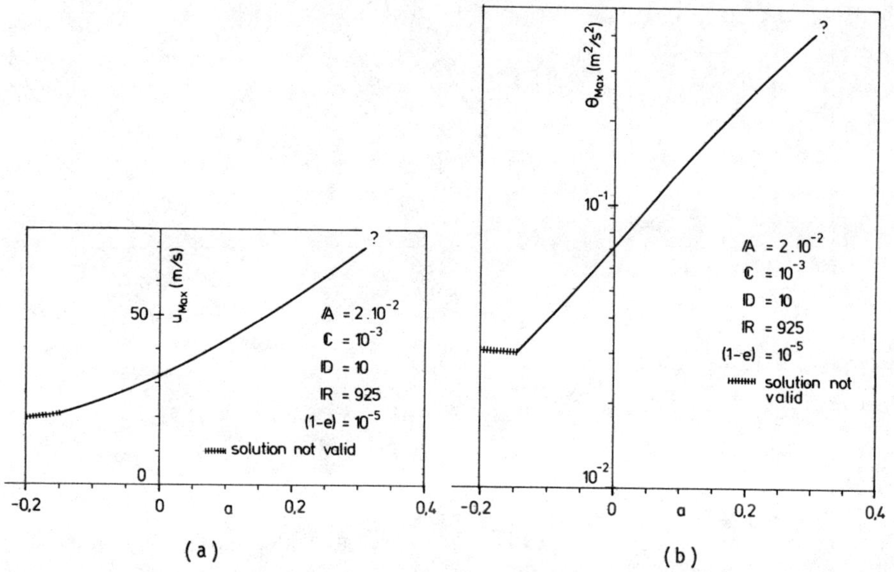

Fig. 18 Surface velocity (a) and maximum fluctuation energy (b) as functions of the sliding exponent $a = (n-1)/2$.

4) Variation of the particle size and the parameter $\mathbb{R}$

The parameter governing the particle size is $\mathbb{A} = \sigma/L^*$, representative values of which are smaller than 0.2, approximately. Keeping all other parameters fixed the results for this case are shown in Figure 19. Calculations were also performed for $\mathbb{A} = 0.5$ even though the model can not be valid in this range. Evidently, the aspect ratio has a sizeable effect on the profiles of the volume distribution, velocity and fluctuation energy. The smaller $\mathbb{A}$, the less "Newtonian-like" is the flow and the more does the velocity profile resemble the shape with inflexion as observed by Savage [17].

There remains the variation of the parameter $\mathbb{R}$. This parameter contains the gravity constant, L^*, U^* and σ, and can be written as

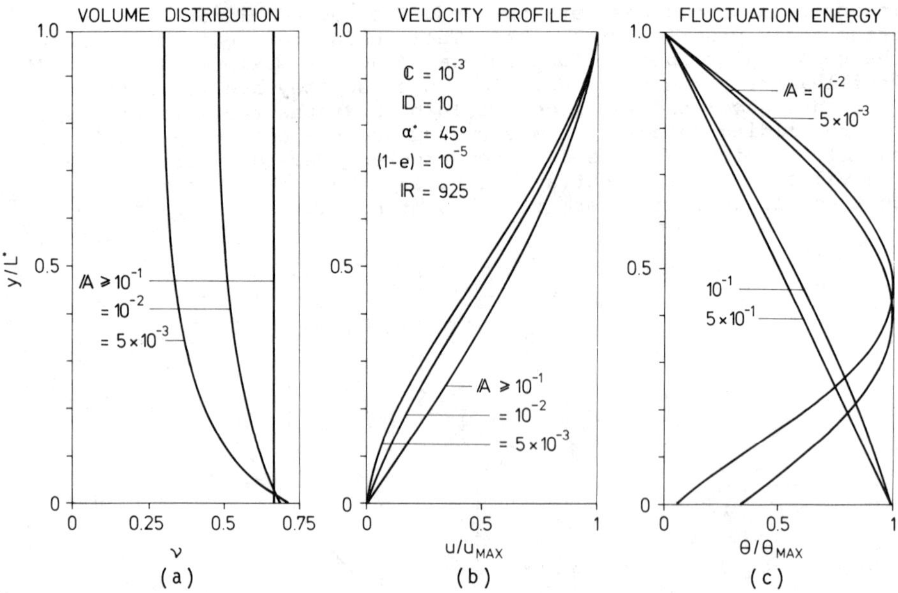

Fig. 19 Profiles of volume distribution (a), velocity (b) and fluctuation energy (c) as functions of the aspect ratio $\mathbb{A}$, keeping $(1-e) = 10^{-5}$, $\mathbb{C} = 10^{-3}$, $\mathbb{D} = 10$, $\mathbb{R} = 925$, $\alpha^* = 45^o$ and $n = 1$, $m = 0$.

$(g\,\nu^* L^*/U^{*2})\cdot\mathbb{A}^{-2}$. The first is a Richrdson number (inverse of a Froude number). Thus, $\mathbb{R} = Ri\,\mathbb{A}^{-2}$. Varying $\mathbb{R}$ and keeping the aspect ratio constant thus means that the Richardson number is varied. Calculations were performed for $5x10^2 < \mathbb{R} \leq 10^4$ and profiles are obtained as shown in Figure 20. For $\mathbb{R} = 5x10^2$, ν reaches the largest possible value $\nu = 0.74$ in the lower portion of the layer. Numerical results cannot be true solutions of the differential equations in this case. For the other values of $\mathbb{R}$ considered by us this situation does not arise so that results are trustworthy. Large $\mathbb{R}$ means large Richradson numbers or small Froude numbers, but since U^* was chosen independently (and does not agree with the surface velocity) this does not necessarily correspond to large surface velocity. Figure 20 also shows that by varying $\mathbb{R}$ profiles for the volume distribution function, velocity and fluctuation energy can be obtained which are as rich in their variability as they are for many of the parameters investigated before. A similar statement also applies for the surface velocity and the maximum values of the fluctuation energy. Figure 21 shows how these vary with the aspect ratio $\mathbb{A}$ but fixed $\mathbb{R}$; in Figure 22 $\mathbb{R}$ is varied while $\mathbb{A}$ is held constant. The dependence of both on $\mathbb{A}$ or $\mathbb{R}$ is conspicuous; u_{Max} and Θ_{Max} both grow with increasing $\mathbb{R}$ and decrease with decreasing $\mathbb{A}$.

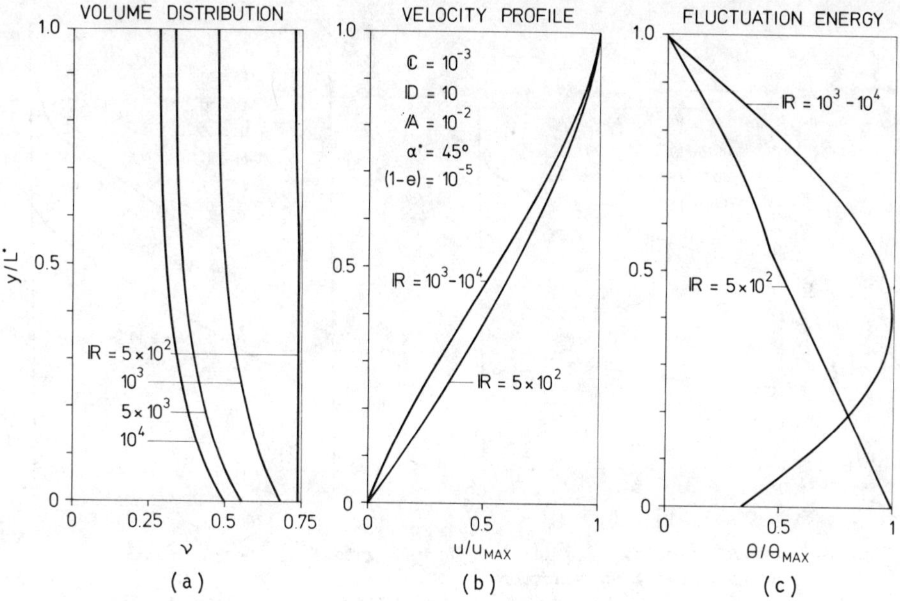

Fig. 20 Profiles of volume distribution (a), velocity (b) and fluctuation energy (c) as functions of the parameter IR, keeping $(1-e) = 10^{-5}$, $\mathbb{A} = 10^{-2}$, $\mathbb{C} = 10^{-3}$, $\mathbb{D} = 10$, $\alpha^* = 45^{\circ}$ and $n = 1$, $m = 0$. Keeping $\mathbb{A}$ constant means that IR is proportional to a Richardson number.

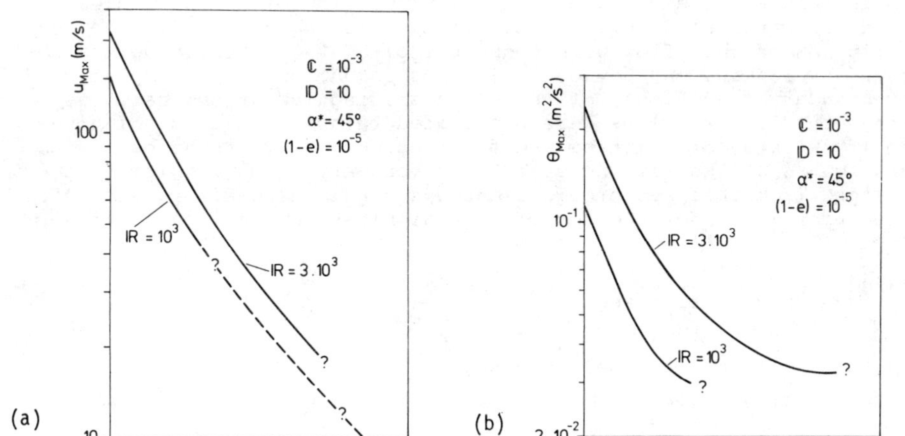

Fig. 21 Surface velocity u_{Max} (a) and maximum values of the fluctuation energy Θ_{Max} (b) as functions of the aspect ratio.

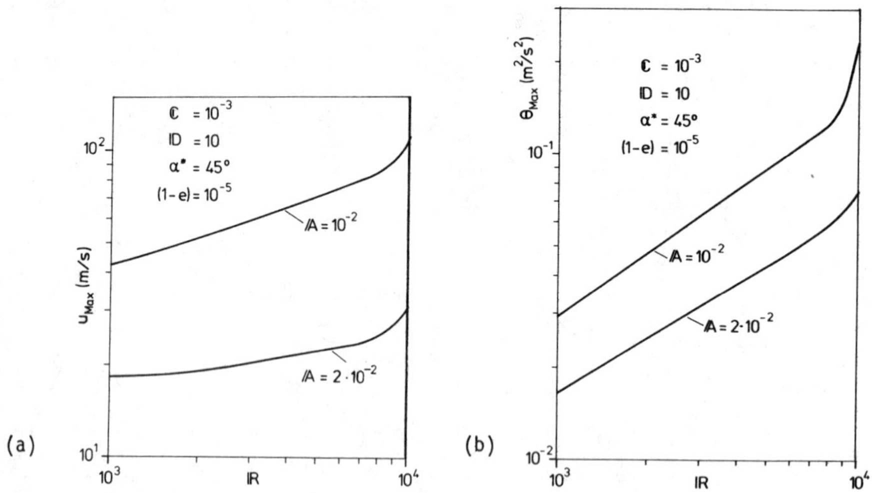

Fig. 22 Surface velocity u_{Max} (a) and maximum values of the fluctuation energy Θ_{Max} (b) as functions of $\mathbb{R}$.

6. CONCLUSIONS

In this article we have analysed plane flow of a cohesionless granular material down a chute. Differential equations for the description of this flow were deduced from a statistical model as derived by Jenkins and Savage [11]. These equations were complemented by boundary conditions; of these the sliding law at the base was conjectural. Power laws were postulated relating basal sliding velocity with shear traction and fluctuation energy at the base with the square of the sliding speed. The boundary value problem that emerged from this set of equations was nondimensionalized and solved numerically for the set of dimensionless physical parameters,

e	coefficient of restitution,
α^*	inclination angle,
$\mathbb{A}$	aspect ratio,
$\mathbb{C},\mathbb{D},m,n$	sliding law coefficients,
$\mathbb{R}$	flow number ($\mathbb{R} = Ri\,\mathbb{A}^{-2}$ where Ri is the Richardson number),

arising in the system. The numerical results were displayed graphically and the plots show how the particle density, velocity and fluctuation energy vary across the moving layer of granuls and also indicate orders of magnitude for the surface velocity and the maximum value of the fluctuation energy. Results may, perhaps, be summarized as follows:

1. By varying any one of the parameter e, $\mathbb{A}$, $\mathbb{C}$, $\mathbb{D}$ or $\mathbb{R}$ (and keeping the remaining ones fixed) velocity profiles could be obtained which are either blunt (near parabolic) or having a form with an inflexion point.

2. The formation of a profile with an inflexion point is always accompanied by a rather non-uniform density distribution and a distribution of the fluctuation energy having a maximum value within the gliding layer. Uniform particle density distributions are always accompanied by blunt velocity profiles. In this regard the effective value of the volume distribution function does not seem to play a decisive role.

3. On occasion, almost identical profiles for ν, u and Θ can be obtained with two choices of the sliding law parameters and, correspondingly, two different values of the coefficient of restitution, e. This demonstrates that there is no experiment using chutes alone by which the sliding parameters and the coefficient of restitution could uniquely be determined.

4. Profiles for ν, u and Θ, and values for u_{Max} and Θ_{Max} also depend conspicuously on the power law exponents of non-linear sliding laws. This gives even more freedom in the selection of the parameters e, $\mathbb{C}$, $\mathbb{D}$, m and n and makes determination of these by experiments even more difficult.

To remedy this situation a relationship for the sliding law should be found from theoretical considerations. This model will hopefully yield relationships for the sliding law. in which the values of the velocity, fluctuation energy, shear traction etc. are related by equations involving fewer parameters than in the conjectural power law above. Only in this case is there a hope to quantify the phenomenological coefficients with one experimental setup (the chute) alone.

This theoretical model for the boundary condition is also needed for the simple reason that sliding laws others than ours are possible. One could for instance relate Θ with its derivative normal to the wall.

Before finishing these conclusions it should be mentioned that in an actual problem $\mathbb{R}$ cannot be freely varied, once the other parameters are selected. This is because $\mathbb{R}$ can be calculated as soon as U^* is selected. The best choice is $U^* = u_{Max}$ but u_{Max} is only known once the solution has been determined. In an actual problem, therefore, given α^*, e, $\mathbb{A}$, $\mathbb{C}$, $\mathbb{D}$, m and n $\mathbb{R}$ is calculated from a first estimate of U^*. With it, a new value for $\mathbb{R}$ can be calculated yielding a new value for u_{Max}, and so on. This procedure was not necessary here, even though it is physically more appropriate. But with our procedure integration is easier and $\mathbb{R}$ can be picked with a certain freedom. However, u_{Max} and Θ_{Max} must then explicitly be calculated.

7. ACKNOWLEDGMENTS

While performing this work Th. Scheiwiller was supported by the Swiss National Science Foundation.

8. REFERENCES

[1] Bagnold, R.A., 1954. Experiments on a gravity-free dispersion of large spheres in Newtonian fluid under shear. Proc. Roy. Soc. London, Vol. A 225, pp. 49-63.

[2] Blinowski, A., 1978. On the dynamic flow of granular media. Arch. of Mech., Arch. Mech. Stoso., Vol. 30, No. 1, pp. 27-34.

[3] Carnahan, N.F. and Starling, K.E., 1969. Equations of state for non-attracting rigid spheres. J. Chem. Phys., Vol. 51, pp. 625-636.

[4] Goodman, M.A. and Cowin, S.C., 1971. Two problems in the gravity flow of granular materials. J. Fluid Mech., Vol. 45, No. 2, pp. 321-339.

[5] Goodman, M.A. and Cowin, S.C., 1972. A continuum theory for granular materials. Arch. Rational Mech. Anal., Vol. 44, No. 4, pp. 249-266.

[6] Hutter, K., 1982a. Glacier Flow. American Scientist, Vol. 70, No. 1, pp. 26-34.

[7] Hutter, K., 1982b. Dynamics of Glaciers and large ice masses. Ann. Rev. Fluid Mech., Vol. 14, pp. 87-130.

[8] Hutter, K. and Scheiwiller, Th., 1983. Rapid plane flow of granular materials down a chute. US-Japan Seminar on New Models and Constitutive Relations in the Mechanics of Granular Materials. (J.T. Jenkins Ed.). Elsevier Scientific Publ. Comp. Amsterdam.

[9] Jenkins, J.T. and Cowin, S.C., 1979. Theories for flowing granular materials. The Joint ASME-CSME App. Mech. Fluid Engng. and Bioengng. Conf., AMD - Vol. 31, pp. 79-89.

[10] Jenkins, J.T. and Savage, S.B., 1981. The mean stress resulting from interparticle collisons in a rapid granular shear flow. In: Proceedings of the Fourth International Conference on Continuum Models of Discrete Systems. (Eds. O. Bruhin and R.K.T. Hsien), North Holland.

[11] Jenkins, J.T. and Savage, S.B., 1983. A theory for the rapid flow of identical, smooth, nearly elastic spherical particles. J. Fluid Mech. (in press).

[12] Kanatani, K.I., 1979. A micropolar continuum theory for the flow of granular materials. Int. J. Engng. Sci., Vol. 17, pp. 419-432.

[13] Nunziato, J.W., Passmann, S.L. and Thomas, J.P.Jr., 1980. Gravitational flow of granular materials with incompressible grains. J. Rheology, Vol. 24, No. 4, pp. 395-420.

[14] Ogawa, S., Umemura, A. and Oshima, N., 1980. On the equations of fully fluidized granular materials. ZAMP 31, pp. 483-493.

[15] Passmann, S.L., Nunziato, J.W., Bailey, P.B. and Thomas J.P. Jr., 1980. Shearing flows of granular materials. ASCE J. Engng. Mech. Div., Vol. 106, No. EM 4, pp. 773-783.

[16] Pearson, J.R.A. and Petrie, C.J.S., 1968. On melt flow instability of extruded polymers. In: Polymer Systems, Deformation and Flow. Proceedings of the 1966 Annual Conference of the British Society of Rheology. Ed. R.E. Whetton and R.W. Whorolow, MacMillan, London.

[17] Savage, S.B., 1979. Gravity flow of cohesionless granular materials in chutes and channels. J. Fluid Mech., Vol. 92, pp. 53-96.

[18] Savage S.B. and Jeffrey, D.J., 1981. The stress tensor in a granular flow at high shear rates. J. Fluid Mech., Vol. 110, pp. 255-272.

[19] Scheiwiller, Th. and Hutter, K., 1982. Lawinendynamik. Uebersicht über Experimente und theoretische Modelle von Fliess- und Staublawinen. Mitteilung Nr.56 der Versuchsanstalt für Wasserbau, Hydrologie und Glaziologie an der ETH, Zürich.

[20] Szidarovszki, F. and Yakowitz, S., 1978. Principles and procedures of numerical analysis. Plenum Press, New-York & London.

[21] Uhland, E., 1979. Stratified two-phase flow of molton polymers in circular dies. Polym. Eng. & Sci., Vol. 17, p. 671.

[22] Uhland, E., 1979. Das anomale Fliessverhalten von Polyäthylenen hoher Dichte. Rheol. Acta, Vol. 18, 1.

A KINETIC THEORY FOR THE RAPID FLOW OF ROUGH INELASTIC SPHERICAL PARTICLES AND THE EVOLUTION OF FLUCTUATIONS

M. Shahinpoor and G. Ahmadi

Department of Mechanical and Industrial Engineering
Clarkson College of Technology
Potsdam, NY 13676, USA

ABSTRACT

A kinetic statistical theory for the rapid flow of rough and linearly inelastic identical spherical particles is developed. This, in particular, embodies evolutionary equations for the conservation and diffusion of random fluctuation energy in such rapid flows. By a rapid flow is meant a flow that is primarily maintained by exchange of momenta between the colliding particles. Further, this theory clearly demonstrates the effect of intergranular friction and grain inelasticity on the nature of the flow regimes. The Krook equation is solved by the Chapman-Enskog iterative expansion for the collision operator. A particle distribution density is then derived that is anisotropic and general enough to include the effect of inelasticity of grains and intergranular friction. Based on this particle velocity distribution the pertinent constitutive equations are derived for the stress tensor and the flux vector of random fluctuation energy. Detail of a binary collision of rough inelastic particles is then analyzed. Furthermore, a complete set of five governing equations are derived which contains an equation for the evolution of random fluctuation energy in such rapid flows. These equations are then applied to two problems of practical interest. One deals with a rapid gravity flow of granular materials down an inclined plane and clearly shows how interparticle friction and inelasticity influences the flow profile. The other one deals with a rapid plane Couette flow of granular materials and again shows the effect of intergranular friction and inelasticity on the flow regime. In both cases no-slip condition is applied at flow solid boundaries. Furthermore, explicit expressions are derived for the dynamic pressure, root mean square of random velocity fluctuations and fluctuation energy in terms of the bulk granular density, the mean shear rate and other flow characteristics such as the applied shear stress or the boundary velocities.

1. INTRODUCTION

Rapid transport of granular materials occurs in many industrial applications and geophysical processes. Its understanding has become extremely important in recent years. Thus, there has been created a great need for a plausible kinetic statistical theory for the rapid flow of such materials. Realistically, these materials are rough, slightly inelastic and more orless rotund in shape. When in rapid motion they create a field of random fluctuations on all pertinent variables such as velocity, bulk density, and interparticle's dispersive and frictional forces. The dispersive forces are generally created by pair collision of grains which continuously transfer moment among colliding particles. The frictional forces are created both as a result of pair collision as well as particle sliding at points where shear rate is small. The random fluctuation fields are somewhat analogous to the molecular velocity fluctuations which ultimately define an absolute temperature for a gas. Similarities between the statistical mechanical description of granular materials and the statistical geometric theory of dense fluids have recently been discussed by Shahinpoor [1], [2]. In the molecular description of gases, the mean flow velocity is usually small compared with the velocity associated with thermal molecular excitations. The difference between the mean velocity and the thermal fluctuation velocity becomes small for sonic and transonic flows and the mean velocity may be considerably bigger than the random velocity associated with thermal molecular excitation for supersonic and hypersonic flows(*). In a rapid granular flow of practical interest the velcoity scales associated with the mean shear and the random fluctuation field are almost equal and thus such flows somewhat resemble transonic flows of gases. However, because of dissipative mechanisms associated with rapid granular flows a continuous source of energy must feed the field of fluctuation energy by whatever means or otherwise such fields would immediately die out. Therefore, an essential feature of such flows must be an evolution equation for the random fluctuation energy, their mode of diffusion and their interaction with other modes of energy. Thus, it appears that a correct formulation of governing equations for the behavior of granular materials in rapid flows must have an equation for the evolution of fluctuation energy coupled with other field equations. This point was first appreciated by Jenkins and Cowin [4]. Apparently, the first attempts in this direction were initiated by Blinowski [5] and Ogawa [6], who proposed equations pertaining to the balance of fluctuation energy. Guided by the similarities between the rapid flows of granular materials and the turbulent flows of Newtonian fluids Jenkins and Savage [7] have recently discussed a balance law for the average specific kinetic energy associated with the particle velocity fluctuations. These fluctuations arise due to a continuous process of binary collisions among the particles. Their formulation

---

* - A similar situation exists in the dynamics of planetary rings where velocity fluctuations are much smaller than the mean velocity (see Goldreich [3]).

is guided by the direct modeling of the evolution of random turbulent energy as described by Launder and Spalding [8].

Guided by the results obtained in [5], [6] and [7] Ahmadi and Shahinpoor [9] have recently presented a turbulent model for the rapid flow of granular materials by a fundamental recomputation of statistical averages. Due to irregular and stochastic motions created by particle collisons a similar situation such as in a turbulent flow of fluids is created in a rapid granular flow. There, however, exists a major difference in the sense that granular media in motion are macroscopically discontinuous (particulate, discrete) while the fluid medium is considered a continuum. Never the less, the technique of dealing with hydrodynamic turbulence can in principle be applied to the rapid flow of granular materials. In fact, it was Blinowski [5] who first attempted this line of approach with relative success. He used a special averaging technique that is different from the standard turbulent flow type averaging. As discussed by Ahmadi and Shahinpoor [9] his averaging technique may create complexities in the calculations of statistical quantities. Further, he derives no equation for the fluctuating stress tensor in an explicit form. However, Blinowski's work [5] should be considered as a pioneering work in this direction. Ahmadi and Shahinpoor [9] on the other hand use the integral balance laws and a different averaging (including a mass weighted average) technique and derive an explicit expression for the fluctuation stress tensor that is in remarkable agreement with the experimental results.

Attempts in developing a kinetic theory involving particle collision and transfer of momenta were first initiated by Bagnold [10] and later expanded by McTigue [11]. Neither of them considered the fluctuations in their formulation. Ogawa, Omemura and Oshima [12] have considered both random fluctuations as well as interparticulate friction to derive a set of constitutive equations for the stresses. Their treatment, however, does not include the fluctuation diffusion terms. Lun, Savage and Jeffrey [13] have recently shown that stresses given in [12] are much smaller than those measured by Savage and Sayed [14], [15]. Savage and Jeffrey [16], [17], on the other hand, employed the kinetic theory of dense gases to derive certain constitutive equations for the rapid granular flows which depend on the value of a parameter which signifies the ratio of the mean shear rate to the root mean square of random velocity fluctuations. Their theory does not give an equation for the evolution of random fluctuation energy.

Guided by a program outlined by Jenkins and Savage [7], Jenkins and Savage [18] and Ahmadi and Shahinpoor [19] have developed a kinetic theory for rapid granular flows. They derive explicit expressions for the balance laws for mass, linear momentum, and fluctuation energy as well as the flux and production of linear momentum and fluctuation energy. The two approaches are, however, fundamentally different.

Jenkins and Savage [18] employ a two particle distribution function (see Chapman and Cowling [20]) for identical, smooth, nearly elastic spherical particles. They model the collisional pair distribution function to be linear in the mean rate of deformation. This anisotropy is compatible with the numerical predictions of Ashurst and Hoover [21] and Evans and Watts [22]. Furthermore, for the associated radial distribution function they employ an empirical model suggested by Carnahan and Starling [23]. The inelasticity of grains is assumed to be given in terms of a linear coefficient of restitution. On the other hand, Ahmadi and Shahinpoor [19] have considered rough particles and the associated dissipative mechanisms. In their formulation they [19] employed the Chapman-Enskog [24] iterative expansion of the Krook's equation (see Bhatnagar, Gross and Krook [25], Vincenti and Kruger [26]). They derived explicit expressions for the production and diffusion of random fluctuation energy in such rapid flows. They [19] however, disregarded the frictional dissipation of energy in applying their theory to some flow situations of interest. This created some ambiguities in the correct interpretation of the relevant boundary conditions. In the present paper we intend to circumvent the aforementioned difficulties by including the effect of intergranular friction and roughness of spherical particles in our formulation and solution of two flow problems of practical interest. In this endeavor we will use an expression for the frictional and inelastic energy dissipation in particle collisions as derived by Shahinpoor and Siah [27] and Shahinpoor [28], [29]. Thus, the Krook equation [25] will be solved by the Chapman-Enskog iterative expansion for the collision operator. A particle distribution density will be derived which is general enough to include anisotropy as well as intergranular friction. Consequently, the relevant constitutive equations for the stress tensor and flux vector of random fluctuation energy will be derived. The results will also include an equation for the production of random fluctuation energy. Finally two specific problems are solved. The first one deals with a rapid gravity flow of rough grains down an inclined plane. The second one deals with a rapid plane Couette flow of rough grains.

For a survey of the literature of rapid granular flows the reader is referred to Shahinpoor [29], Shahinpoor and Siah [27], [30], and Shahinpoor and Lin [31].

2. GOVERNING KINETIC EQUATIONS

Let the distribution function $f(v_i,x_i,t)$ of a granular material be defined on a six-dimensional phase space (v_i,x_i). The probable number of granules which are situated in the volume element(*) dx^3 around x_i and have velocities within the range v_i, v_i+dv_i, is equal to $f(v_i,x_i,t)dv_1dv_2dv_3dx_1dx_2dx_3$. The number density of the granules then is given by

* - With respect to a Rectangular Cartesian Coordinate System $x_i \equiv (x_1,x_2,x_3)$.

$$n = \int_V f(v_i, x_i, t) d^3v \, , \quad d^3v = dv_1 dv_2 dv_3 \quad . \tag{1}$$

Consider a rapid flow of equal rough elastic spherical grains in which each granule of mass m is subjected to an external force mF_i. At time t + dt, the distribution function becomes $f(v_i + F_i dt, x_i + v_i dt, t + dt)$. Here the change in the number of particles in a volume element of phase space is due to the collisions between the granules. Therefore

$$[f(v_i + F_i dt, x_i + v_i dt, t+dt) - f(v_i, x_i, t)] d^3v d^3x = \Delta f_c d^3v d^3x dt \, , \tag{2}$$

where Δf_c is the change in the distribution function due to collision, and $d^3v \stackrel{c}{=} dv_1 dv_2 dx_3$, $d^3x = dx_1 dx_2 dx_3$. As dt approaches zero equation (2) becomes*

$$\frac{\partial f}{\partial t} + v_i \frac{\partial f}{\partial x_i} + F_i \frac{\partial f}{\partial v_i} = \Delta f_c \, , \tag{3}$$

which is the celebrated Boltzmann equation. For a comprehensive discussion of the Boltzmann equation the reader is referred to the monographs by Chapman and Cowling [21] and Vincenti and Kruger [26](**). Multiplying equation (3) by a function $\psi(v_i, x_i)$ and integrating over the velocity space we find

$$\frac{\partial \bar{\psi}}{\partial t} - \overline{\frac{\partial \psi}{\partial t}} + \frac{\partial}{\partial x_i} (\overline{v_i \psi}) - \overline{v_i \frac{\partial \psi}{\partial x_i}} - F_i \overline{\frac{\partial \psi}{\partial x_i}} = \Delta \psi_c \, , \tag{4}$$

where

$$\bar{\psi} = \int_V \psi \, f \, dv_1 dv_2 dv_3 \tag{5}$$

and $\Delta\psi_c$ is the change in ψ due to granular encounters. Equation (4) is known as the Enskog's equation of change.

If there exist particle-particle statistical correlation such as for example in pairs of particles and the evolution of pair collisions then statistical averages must be accordingly adjusted. Thus, one may define n-particle averages such as $\langle A \rangle_n$ such that

(*) - Einstein's summation convention is applied throughout on repeated indices.

(**) - see also Truesdell and Muncaster [32]

$$\langle A\rangle_n = \int_n f^{(n)}(v_i, x_i, t) A \, d^n v_i \quad , \tag{6}$$

where $f^{(n)}(v_i, x_i, t)$ is generally an n-particle velocity distribution function. This function must then satisfy the generalized Boltzmann equation or the BBGKY (Bogoliubov, Born, Green, Kirkwood, Yvon) equation, i.e.,

$$\frac{\partial f^{(n)}}{\partial t} + \sum_{i=1}^{n} \{ \underset{\sim}{v}_i \cdot \frac{\partial f^{(n)}}{\partial \underset{\sim}{r}_i} + [\underset{\sim}{F}_i + \sum_{j=1}^{n} \underset{\sim}{X}_{ij}] \cdot \frac{\partial f^{(n)}}{\partial \underset{\sim}{v}_i} \}$$

$$= - \iint (\sum_{i=1}^{n} \underset{\sim}{X}_{i,n+1} \cdot \frac{\partial f^{(n+1)}}{\partial \underset{\sim}{v}_i}) \, d\underset{\sim}{r}_{n+1} d\underset{\sim}{v}_{n+1} \quad , \tag{7}$$

where $\underset{\sim}{F}_i$, $\underset{\sim}{X}_{ij}$, are the acceleration of the particles, due respectively to external forces and to the forces exerted by particles.

Let

$$v_i = u_i + v_i' \quad , \tag{8}$$

where u_i is the average velocity and v_i' is the random fluctuating velocity. If ψ is assumed to be only a function of velocity, then equation (4) becomes

$$\frac{d}{dt}(\bar{\psi}) + \bar{\psi} \frac{\partial u_i}{\partial x_i} + \frac{\partial}{\partial x_i}(\overline{\psi v_i'}) - F_i \overline{\frac{\partial \psi}{\partial v_i}} = \Delta\psi_c \quad , \tag{9}$$

where the material time derivative is defined by:

$$\frac{d}{dt} = \frac{\partial}{\partial t} + u_i \frac{\partial}{\partial x_i} \quad . \tag{10}$$

For $\psi = m$ equation (9) becomes simply

$$\frac{\partial \rho}{\partial t} + \frac{\partial}{\partial x_i}(\rho u_i) = 0 \quad , \tag{11}$$

where $\rho = nm$ is the mass density. In the derivation of equation (9) from equation (7), $\Delta\psi_c$ was set equal to zero, since the total mass of the granules participating in a collision is conserved. The properties which are conserved in granular encounters are called "summation invariants" or "collisional invariants." The total mass and momentum of granules are collisional invariants. For elastic frictionless granules, the total kinetic energy also becomes one of collisional invariants. However, for rough elastic particles $\Delta\psi_c$ is different from zero.

For $\psi = mv_i$, equation (9) becomes

$$\rho \frac{du_j}{dt} = \rho F_j + \frac{\partial \tau_{ij}}{\partial x_i} , \tag{12}$$

where

$$\tau_{ij} = - \overline{mv_i'v_j'} , \tag{13}$$

is the stress tensor.

Taking $\psi = \frac{1}{2} mv^2$, $v^2 = v_k v_k$ in equation (9) and making use of equation (11) and (12), the result becomes

$$\rho \frac{de}{dt} = - \frac{\partial Q_j}{\partial x_j} + \tau_{ji} \frac{\partial u_i}{\partial x_j} - \rho \Delta e_c \tag{14}$$

where $v'^2 = v_k'v_k'$, $e = \frac{1}{2} \overline{v'^2}$ and

$$Q_j = \frac{1}{2} \overline{mv'^2 v_j'} , \tag{15}$$

is the flux vector of random fluctuation energy. Further, Δe_c is the statistically averaged collisional loss of energy per unit mass due to both friction and inelasticity of grains. In order to find an explicit expression for Δe_c one must consider the details of a binary collision of two particles in the presence of friction and inelasticity. This is discussed in the next section.

3. DETAIL OF A BINARY COLLISION OF ROUGH INELASTIC PARTICLES

For a collection of dense elastic spheres with binary collision it was shown in [20] that

$$\left(\frac{\partial f}{\partial t}\right)_c = \iint_{\underset{\sim}{v}_2 \underset{\sim}{n}} \{\chi(\underset{\sim}{x} + \frac{1}{2} \sigma \underset{\sim}{n}) f(\underset{\sim}{v}_1', \underset{\sim}{x}) f(\underset{\sim}{v}_2', \underset{\sim}{x} + \sigma \underset{\sim}{n})$$ - (refer to next page)

$$- \chi(\underset{\sim}{x} - \frac{1}{2}\sigma\underset{\sim}{n}) f(\underset{\sim}{v}_1,\underset{\sim}{x}) f(\underset{\sim}{v}_2,\underset{\sim}{x} - \sigma\underset{\sim}{n})\}\sigma^2 \underset{\sim}{g}\cdot\underset{\sim}{n}d\underset{\sim}{n}d\underset{\sim}{v}_2 \quad , \tag{16}$$

where $\underset{\sim}{n}$ is unit vector in the direction of normal collision (Fig. 1) and $\underset{\sim}{g} = \underset{\sim}{v}_1 - \underset{\sim}{v}_2$, $\underset{\sim}{g}' = \underset{\sim}{v}_1' - \underset{\sim}{v}_2'$.

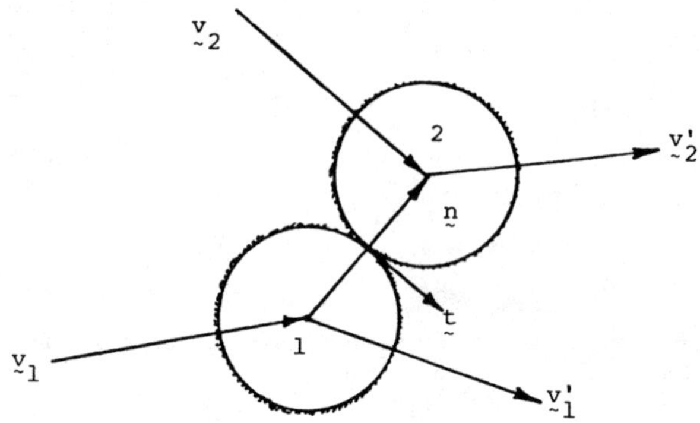

Fig. 1 - Binary Collision of Two Rough Particles

$\underset{\sim}{v}_1'$ and $\underset{\sim}{v}_2'$ are the velocities of a pair of colliding particles just after a binary collision, σ is the diameter of particles, and χ is a function of position only and takes into account the reduction in total volume due to the presence of finite size particles.

Consideration of simple elastic collisions could hardly be applicable to rapid flows of granular materials. Inelasticity and interparticle friction are the significant and distinguishing features of granular encounters.

For collision of two rough and inelastic identical spherical particles, the conservation of momentum yield,

$$\underset{\sim}{v}_2' + \underset{\sim}{v}_1' = \underset{\sim}{v}_2 + \underset{\sim}{v}_1 \ , \tag{17}$$

$$m\underset{\sim}{v}_1 + \int_{\Delta t} \underset{\sim}{F}\, dt = m\, \underset{\sim}{v}_1' \quad , \tag{18}$$

$$m\mathbf{v}_2 - \int_{\Delta t} \mathbf{F}\, dt = m\, \mathbf{v}_2' \quad , \tag{19}$$

where $\mathbf{F}$ is the force felt by each particle due to a collision and Δt is the duration of impact, such that the impulse $\mathbf{I}$ per unit mass is given by

$$\mathbf{I} = \frac{1}{m} \int_{\Delta t} \mathbf{F}\, dt \ . \tag{20}$$

Thus, equations (18) and (19) can be written as

$$\mathbf{v}_1 + \mathbf{I} = \mathbf{v}_1' \quad , \tag{21}$$

$$\mathbf{v}_2 - \mathbf{I} = \mathbf{v}_2' \quad . \tag{22}$$

The relative velocities $\mathbf{g} = \mathbf{v}_1 - \mathbf{v}_2$, $\mathbf{g}' = \mathbf{v}_1' - \mathbf{v}_2'$ are then obtained from equations (21) and (22) to be related such that

$$\mathbf{g} + 2\mathbf{I} = \mathbf{g}' \quad . \tag{23}$$

Due to the inelasticity of particles it is also true that

$$(\mathbf{n} \cdot \mathbf{g}') = -e(\mathbf{n} \cdot \mathbf{g}) \quad . \tag{24}$$

Thus, from (23) and (24) the normal impulse is found to be:

$$I_n = -\frac{1}{2}(1 + e)(\mathbf{n} \cdot \mathbf{g}) \ , \tag{25}$$

where I_n is the component of $\mathbf{I}$ in the $\mathbf{n}$ direction, i.e. $I_n = \mathbf{I} \cdot \mathbf{n}$. The components of the impulse equations in the normal direction $\mathbf{n}$ and the tangential direction $\mathbf{t}$ ($\mathbf{t}$ is a unit vector in the plane of $\mathbf{g}$, $\mathbf{n}$ perpendicular to $\mathbf{n}$) are given by

$$v_{1n} + I_n = v_{1n}' \quad , \qquad v_{1t} + I_t = v_{1t}' \quad , \tag{26), (27}$$

$$v_{2n} - I_n = v'_{2n} \quad , \quad v_{2t} - I_t = v'_{2t} \quad , \qquad (28), (27)$$

where I_n and I_t are respectively the normal and the tangential components of the impulse $\underset{\sim}{I}$ such that

$$I_n = \frac{1}{m} \int_{\Delta t} F_n \, dt \ , \quad I_t = \frac{1}{m} \int_{\Delta t} F_t \, dt \quad , \qquad (30), (31)$$

where F_n and F_t are the normal and the tangential components of the force between the two particles. Here we assume that in the period Δt spent for a binary collision the particles may slide on each other, hence,

$$F_t = \mu F_n \ , \quad I_t = \mu I_n \quad , \qquad (32), (33)$$

where μ is the kinematic coefficient of friction between particles. Therefore

$$v'_{1t} - v_{1t} = I_t = -\mu I_n = -\mu(v'_{1n} - v_{1n}) \ . \qquad (34)$$

$$v'_{2t} - v_{2t} = -I_t = \mu I_n = -\mu(v'_{2n} - v_{2n}) \ , \qquad (35)$$

Thus from equations (25), (26), (27), (28), (29), (34) and (35) we obtain that

$$v'_{1t} - v_{1t} = -\frac{1}{2}\mu(1 + e)(\underset{\sim}{n} \cdot \underset{\sim}{g}) \qquad (36)$$

$$v'_{2t} - v_{2t} = +\frac{1}{2}\mu(1 + e)(\underset{\sim}{n} \cdot \underset{\sim}{g}) \quad . \qquad (37)$$

One notes that

$$\underset{\sim}{I} = I_n \underset{\sim}{n} + I_t \underset{\sim}{t} = I_n(\underset{\sim}{n} + \mu\underset{\sim}{t}) \quad , \qquad (38)$$

where $\underset{\sim}{t}$ is a unit tangent to the colliding particles. The actual direction of $\underset{\sim}{t}$ can be found by noting that it must be perpendicular to $\underset{\sim}{n}$ and it must also lie in the surface formed by $\underset{\sim}{n}$ and $\underset{\sim}{g}$. Thus

$$\underset{\sim}{t} = \underset{\sim}{n} \times (\hat{\underset{\sim}{g}} \times \underset{\sim}{n}) \tag{39}$$

where $\hat{\underset{\sim}{g}} = (\underset{\sim}{g}/|\underset{\sim}{g}|)$.

From equations (25) and (38) one finds that

$$\underset{\sim}{I} = -\frac{1}{2}(1 + e)(\underset{\sim}{n} \cdot g)(\underset{\sim}{n} + \mu\underset{\sim}{t}) \; .$$

Finally from equations (21), (22), (24), (38) and (40) it is found that

$$\underset{\sim}{v}_1' = \underset{\sim}{v}_1 - \frac{1}{2}(1+e)(\underset{\sim}{n}\cdot\underset{\sim}{g})[\underset{\sim}{n} + \mu\underset{\sim}{n} \times (\hat{\underset{\sim}{g}} \times \underset{\sim}{n})] \quad , \tag{41}$$

$$\underset{\sim}{v}_2' = \underset{\sim}{v}_2 + \frac{1}{2}(1 + e)(\underset{\sim}{n} \cdot\underset{\sim}{g})[\underset{\sim}{n} + \mu\underset{\sim}{n} \times (\hat{\underset{\sim}{g}} \times \underset{\sim}{n})] \; . \tag{42}$$

Note that $\underset{\sim}{t}$ in equation (39) can be decomposed in terms of $\underset{\sim}{n}$ and $\hat{\underset{\sim}{g}}$ such that

$$\underset{\sim}{t} = \hat{\underset{\sim}{g}} - (\underset{\sim}{n} \cdot \hat{\underset{\sim}{g}})\underset{\sim}{n} \quad , \tag{43}$$

Equations (41) and (42) are then reduced to

$$\underset{\sim}{v}_1' = \underset{\sim}{v}_1 - \frac{1}{2}\,(1+e)(\underset{\sim}{n}\cdot\underset{\sim}{g})\{\underset{\sim}{n}[1-\mu(\underset{\sim}{n}\cdot\hat{\underset{\sim}{g}})]+\mu\hat{\underset{\sim}{g}}\} \quad , \tag{44}$$

$$\underset{\sim}{v}_2' = \underset{\sim}{v}_2 + \frac{1}{2}(1+e)(\underset{\sim}{n}\cdot\underset{\sim}{g})\{\underset{\sim}{n}[1-\mu(\underset{\sim}{n}\cdot\underset{\sim}{g})]+\mu\hat{\underset{\sim}{g}}\} \quad . \tag{45}$$

From equations (44) and (45) it is clear that intergranular friction creates second order effect in velocity distribution in collisional encounter.

The energy dissipation per mass of particles m can now be found. Let this energy be denoted by $(\frac{\Delta E}{m})$ such that

$$\zeta^{-1}\Delta e_c = (\frac{\Delta E}{m}) = \frac{1}{2}\left[(v_1'^2 + v_2'^2) - (v_1^2 + v_2^2)\right] . \tag{46}$$

By means of equations (44) and (45) it may be shown that equation (46) can be reduced to

$$\zeta^{-1}\Delta e_c = (\frac{\Delta E}{m}) = \frac{1}{4}(\underset{\sim}{n}\cdot\underset{\sim}{g})^2[1-e^2 + \mu^2(1+e)^2] \tag{47}$$

Expressions (44), (45) and (47) will be reduced to the ones recently derived by Jenkins and Savage [18] if the friction μ is set equal to zero. We can now study the expression for the collision integral in the presence of friction and inelasticity of granular particles. It is well known that [20],

$$\begin{aligned}(\frac{\partial f}{\partial t})_c \, d\underset{\sim}{x}\, d\underset{\sim}{v}_1\, dt = &-dt\, d\,\underset{\sim}{v}_1' \iint_{\underset{\sim}{n}\underset{\sim}{v}_2'} f_1' f_2'\, \sigma^2\, \underset{\sim}{g}' \cdot \underset{\sim}{n}\, d\underset{\sim}{n}\, d\underset{\sim}{v}_2' \\ &- dt\, d\,\underset{\sim}{v}_1 \iint_{\underset{\sim}{n}\underset{\sim}{v}_2} f_1 f_2\, \sigma^2\, \underset{\sim}{g} \cdot \underset{\sim}{n}\, d\underset{\sim}{n}\, d\underset{\sim}{v}_2 ,\end{aligned} \tag{48}$$

where the conventional notation $f_2' = f(\underset{\sim}{v}_2', \underset{\sim}{x})$, $f_2 = f(v_2, \underset{\sim}{x})$ have been used. Changing variables in the first integral, using equation (24) and noting that

$$d\underset{\sim}{v}_1'\, d\underset{\sim}{v}_2' = |J|\, d\underset{\sim}{v}_1\, d\underset{\sim}{v}_2 , \tag{49}$$

where J is the Jacobian of transformation, equation (48) becomes

$$(\frac{\partial f}{\partial t})_c = \iint_{\underset{\sim}{n}\underset{\sim}{v}_2} [e|J| f_1' f_2' - f_1 f_2]\sigma^2\, \underset{\sim}{g} \cdot \underset{\sim}{n}\, d\underset{\sim}{n}\, d\underset{\sim}{v}_2 \tag{50}$$

For elastic frictionless particles, $e=1$, $\mu=o$, $|J|=1$, and equation (37) reduces to the classical Boltzmann Collision integral. For granular encounter with inelasticity and friction, the Jacobian of velocity transformation as calculated from equations (44) and (45) is given by

$$|J| = e\{1 - \frac{[\mu(1+e)(\underset{\sim}{n}\cdot\hat{\underset{\sim}{g}})]}{[1-(\underset{\sim}{n}\cdot\underset{\sim}{g})^2]^{1/2}}\}^2, \quad o \le \cos^{-1}(\underset{\sim}{n}\cdot\underset{\sim}{g}) \le \theta^* \ , \tag{51}$$

$$|J| = e, \qquad \theta^* < \cos^{-1}(\underset{\sim}{n}\cdot\hat{\underset{\sim}{g}}) \le 1 \ , \tag{52}$$

where $\theta^* = \frac{\pi}{2} - \tan^{-1}\mu$, so that as $\mu \to o$, $|J| \to e$, and as $\mu \to o$, $e \to 1$, $|J| \to 1$. For a relatively dense flow of granular material, the collision integral must be somewhat modified. Following the method discussed in [20], we find

$$(\frac{\partial f}{\partial t})_c = \iint_{\underset{\sim}{n}\underset{\sim}{v}_2} [\chi(\underset{\sim}{x}+\frac{1}{2}\sigma\underset{\sim}{n})e|J|f(\underset{\sim}{v}_1',\underset{\sim}{x})f(\underset{\sim}{v}_2',\underset{\sim}{x}+\sigma\underset{\sim}{n})$$

$$- \chi(\underset{\sim}{x}-\frac{1}{2}\sigma\underset{\sim}{n})f(\underset{\sim}{v}_1,\underset{\sim}{x})f(\underset{\sim}{v}_2,\underset{\sim}{x}-\sigma\underset{\sim}{n})]\sigma^2\underset{\sim}{g}\cdot\underset{\sim}{n}\ d\underset{\sim}{n}\ d\underset{\sim}{v}_2 \ . \tag{53}$$

Equation (53) gives the general expression for the collision term for granular encounter with friction and inelasticity. When e and J are set equal to one, equation (53) reduces to the classical collision integral for a dense gas given by equation (16). In the next section an approximate technique for the derivation of a one-particle velocity distribution function and the associated balance equations are presented.

4. AN ITERATIVE SOLUTION FOR THE DISTRIBUTION FUNCTION

From the preceding section we noted that even for the simplest collision models, the Boltzmann equation becomes a nonlinear integrodifferential equation. On the other hand, several approximate expressions for the collision operator have been successfully employed in the kinetic theory of dense gases. One simple solution which also retains many of the qualitative features of the true collision term was originally proposed by Bhatnagar, Gross and Krook [25]. This is now known [26] as the BGK model and the resulting pdf equation is known as the Krook's equation. The model essentially asserts that collisions tend to change the distribution f such that they result in a change proportional to (f_o-f) with f_o being the fluctuation equilibrium distribution. Thus, in this case one has

$$\frac{\partial f}{\partial t} + v_i \frac{\partial f}{\partial x_i} + F_i \frac{\partial f}{\partial v_i} = \zeta(f_o - f) \ , \tag{54}$$

where ζ is a collision frequency and f_o is an equilibrium Maxwellian distribution function which arises when fluctuations are at equilibrium much the same way as thermal equilibrium in dense gases at translation. As shown by Shahinpoor and Ahmadi [33] f_o is given by

$$f_o = n\left(\frac{4\pi e}{3}\right)^{-3/2} \exp.\left\{-\frac{3(v_i - u_i)^2}{4e}\right\} \tag{55}$$

where $e = \frac{1}{2}\overline{v'^2}$ was introduced before and is called the energetic. A Chapman-Enskog iterative solution for the Krook's equation (54) leads to the following solution for the distribution function f:

$$f = f_o - \frac{1}{\zeta}\left(\frac{\partial f_o}{\partial t} + v_i \frac{\partial f_o}{\partial x_i} + F_i \frac{\partial f_o}{\partial v_i}\right) \ . \tag{56}$$

Figure 2 shows the possible variation in the profile of the velocity distribution function near the boundary of a flow where small fluctuations and velocities dominate and away from it where large fluctuations and velocities dominate

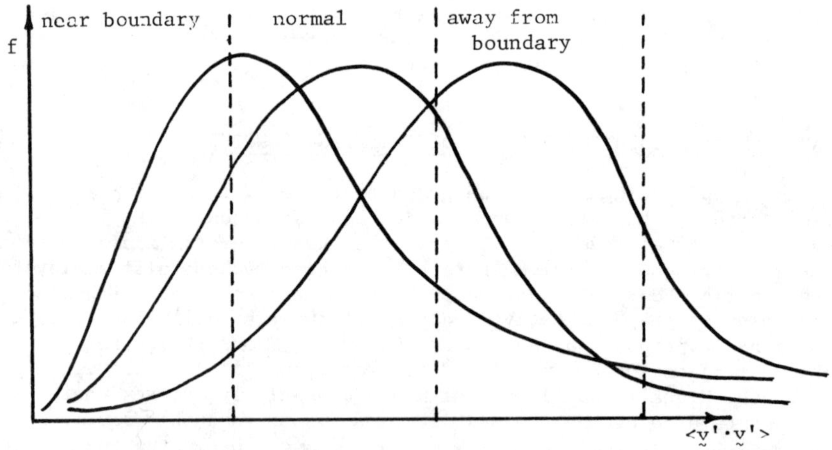

Fig. 2. Possible variation in the velocity distribution function in granular flows and formation of Maxwellian tails [35], [36].

We will employ the above equation in order to derive our governing balance equations and constitutive relations. However, before elaborating further, it must be mentioned that there also exist other exact solutions of the Boltzmann equation as described recently by Krook and Wu [34], [35], Truesdell and Muncaster [32] and Muncaster [36]. These solutions do possess anisotropic features such as Maxwellian tails [34] and [35]. For the purpose of the present paper we will be content with f as given by (56) and proceed to derive the balance equations. Employing equation (55) in (56) and making use of the balance equations (11), (12) and (14) to express $\frac{dn}{dt}$, $\frac{du_i}{dt}$ and $\frac{de}{dt}$ in terms of their space derivatives and other average quantities the expression for f becomes

$$f = f_o\{1 - \frac{1}{\zeta}[\frac{\partial n}{\partial x_i}\frac{v_i'}{n} + \frac{\partial u_i}{\partial x_j}(-\delta_{ij} + \frac{3}{2e}v_i'v_j') + \frac{\partial e}{\partial x_j}(\frac{3v'^2v_j'}{4e^2} - \frac{3v_j'}{2e}) + (\frac{1}{\rho}\frac{\partial \tau_{ji}}{\partial x_j})(\frac{3v_i'}{2e}) + \frac{1}{\rho}(\tau_{ji}\frac{\partial u_i}{\partial x_j} - \frac{\partial Q_j}{\partial x_j} - \rho\Delta e_c)(\frac{3v'^2}{4e^2} - \frac{3}{2e})]\} \quad . \tag{57}$$

Equation (57) gives the iterative solution for the particle random fluctuation velocity distribution in rapid flows of granular materials with rough inelastic grains. The expressions for the stress tensor and the energy flux vector as given by equations (13) and (15) now become

$$\tau_{ij} = (-\frac{2\rho e}{3} + \frac{2}{3\zeta}\tau_{k\ell}\frac{\partial u_\ell}{\partial x_k} - \frac{2}{3\zeta}\frac{\partial Q_k}{\partial x_k} - \frac{2\rho}{3\zeta}\Delta e_c)\delta_{ij} + \frac{2\rho e}{3\zeta}(\frac{\partial u_i}{\partial x_j} + \frac{\partial u_j}{\partial x_i}) \ , \tag{58}$$

$$Q_i = \frac{5}{9\zeta}[2e^2\frac{\partial \rho}{\partial x_i} + 4\rho e\frac{\partial e}{\partial x_i} + 3e\frac{\partial \tau_{ji}}{\partial x_j}] \ . \tag{59}$$

The appearance of $\tau_{k\ell}$ and Q_k on the right hand side of equations (58) and (59) shows the iterative nature of these equations. The first order iterative constitutive equations are obtained by setting

$$\tau_{ij} = -p\delta_{ij} \quad , \quad p = \frac{2}{3}\rho e \quad , \tag{60}$$

$$Q_i = 0 \quad , \tag{61}$$

on the right hand side of equations (58) and (59). Thus we find

$$\tau_{ij} = \left(-\frac{2\rho e}{3} - \frac{4\rho e}{9\zeta}\frac{\partial u_k}{\partial x_k} - \frac{2\rho}{3\zeta}\Delta e_c\right)\delta_{ij} + \frac{2\rho e}{3\zeta}\left(\frac{\partial u_i}{\partial x_j} + \frac{\partial u_j}{\partial x_i}\right) , \tag{62}$$

$$Q_i = -\frac{10\rho e}{9\zeta}\frac{\partial e}{\partial x_i} \quad . \tag{63}$$

Similar substitutions could be carried out in the original expression for the distribution function as given in equations (57). Employing the constitutive equations (62) and (63) into equations (12) and (14), the equations of motion become

$$\rho\frac{du_j}{dt} = \rho F_j - \frac{\partial}{\partial x_j}\left(\frac{2\rho e}{3} + \frac{4\rho e}{9\zeta}\frac{\partial u_k}{\partial x_k} + \frac{2\rho}{3\zeta}\Delta e_c\right)$$

$$+ \frac{\partial}{\partial x_i}\left[\frac{2\rho e}{3\zeta}\left(\frac{\partial u_i}{\partial x_j} + \frac{\partial u_j}{\partial x_i}\right)\right] \tag{64}$$

$$\frac{de}{dt} = \frac{\partial}{\partial x_i}\left(\frac{10\rho e}{9\zeta}\frac{\partial e}{\partial x_i}\right) - \left(\frac{2\rho e}{3} + \frac{4\rho e}{9\zeta}\frac{\partial u_k}{\partial x_k} + \frac{2\rho}{3\zeta}\Delta e_c\right)\frac{\partial u_i}{\partial x_i}$$

$$+ \frac{2\rho e}{3\zeta}\left(\frac{\partial u_i}{\partial x_j} + \frac{\partial u_j}{\partial x_i}\right)\frac{\partial u_j}{\partial x_i} - \rho\Delta e_c \tag{65}$$

The collision frequency must also be evaluated from the expression for the distribution function as given in equation (57). However, at this stage we employ the equilibrium distribution function f_o as given by equation (55) for an approximate evaluation of the collision frequency. Proceeding similar to [33] it follows that

$$\zeta = 2\left(\frac{2\pi}{3}\right)^{\frac{1}{2}} nd^2 e^{\frac{1}{2}} \tag{66}$$

where d is the diameter of the spherical granules. Equation (66) can be written as

$$\zeta = \frac{\rho e^{\frac{1}{2}}}{\alpha} \quad , \tag{67}$$

where

$$\alpha = \tfrac{1}{2}\, m(3/2\pi)^{\frac{1}{2}} d^{-2} \tag{68}$$

is a constant.

With ζ given by equation (67), equations (64), (65) and (11) form five equations for finding the five unknowns u_i, ρ, and e. Therefore, the present mathematical formulation is complete in the sense that the number of equations matches the number of the unknown functions. It is expected that for suitably prescribed initial and boundary conditions these equations could be solved to give the mean velocity distribution, density and energy density profiles for a rpaid flow of granular materials, with rough and inelastic grains. Mathematical analysis of the problems of existence and uniqueness of the solutions of these systems of equations are left for future investigations.

In the next section we shall discuss two simple and yet typical rapid granular flow situations for which some concrete and rather fundamental results can be identified.

5. SHEAR AND GRAVITY FLOWS OF GRANULAR MATERIALS

Here we consider two simple steady flows in which the flows are maintained in the x-direction (Figs. 3 and 4) with field variations with respect to y-direction only; y being perpendicular to the direction of the flow; here (x,y) is a rectangular Cartesian coordinate system that

$$u_i \equiv (u(y),0,0), \quad \rho = \rho(y) \quad , \qquad e = e(y) \quad , \tag{69 1,2,3}$$

Under these circumstances one may show [27], [28], that the energy dissipation per unit mass is given by

$$\Delta e_c = \tfrac{1}{4}(1-\alpha_r^2)\bar{k}^{-2} d^2 \left(\frac{du}{dy}\right)^2 + A(\mu;\xi)\ \bar{k}^{-3} d^2 \left(\frac{du}{dy}\right)^3 \quad , \tag{70}$$

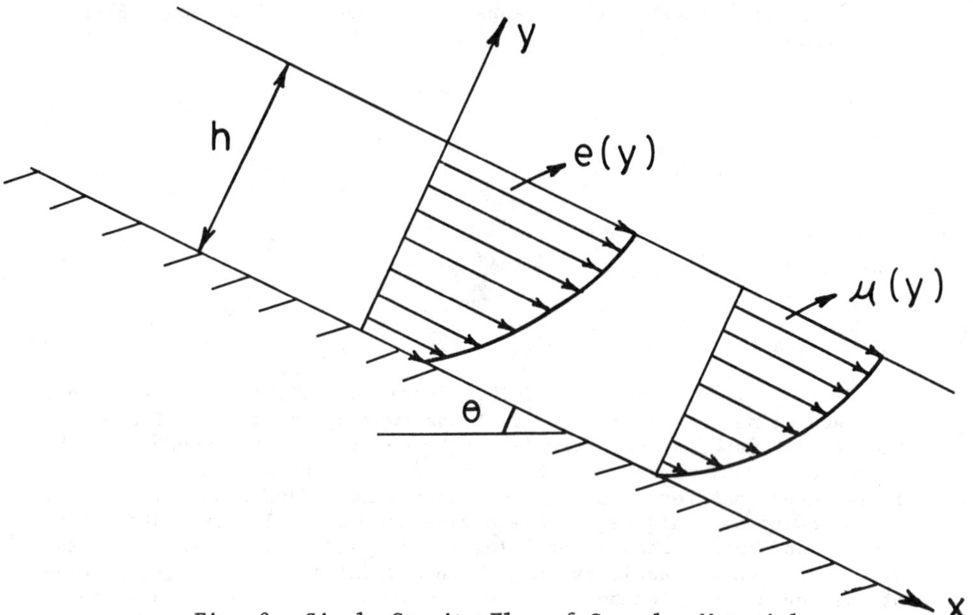

Fig. 3. Simple Gravity Flow of Granular Materials

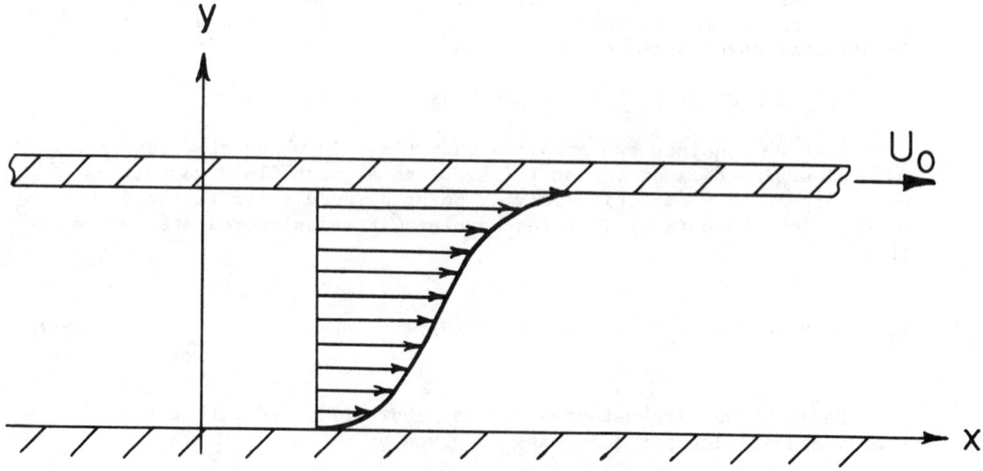

Fig. 4. Plane Couette Flow of Granular Materials.

where $\bar{k}$ is a statistically averaged flow parameter and

$$A(\mu_k,\rho,\beta,\bar{\alpha}) = 2\rho d^2 (\pi/3)^{\frac{1}{2}} \beta \mu_k m^{-1} B , \quad (71)$$

$$B = \left\{\frac{\mu_k \cos^2\bar{\alpha} - \mu_k(1+\beta)\sin^2\bar{\alpha} + [\mu_k^2(1+\beta)-1]\sin\bar{\alpha}\cos\bar{\alpha}}{[1+(1+\beta)\mu_k^2]^2}\right\} , \quad (72)$$

where $\beta(0 \leq \beta \leq 1)$ is the fraction of time that the pair of colliding spheres slip with respect to each other in the entire period of collision (see [27], [28]), $\bar{\alpha}$ is an average angle that the direction of collision makes with the x axis, and μ_k is the kinematic coefficient of friction. In deriving equation (31) we have employed the equilibrium distribution f_o to estimate the collision frequency ζ. Furthermore, for such simple two-dimensional flows we have assumed in conformity with Bagnold [10], and [27], [28] that the relative velocity of grains at collision is related to the mean shear rate such that $k_i \delta U_i = \bar{k}d(\frac{du}{dy})$. In the next sections we will discuss the nature of the solutions to these simple two-dimensional flows.

5.1 Simple Gravity Flow Down an Inclined Plane

Referring to Fig. 3 consider a steady simple gravity flow of a granular materials down an inclined plane in the x direction with field variation in the y direction with respect to a rectangular Cartesian system (x,y) such that

$$u \equiv (u(y),0,0), \quad \rho = \rho(y) , \quad e = e(y) . \quad (73)_{1,2,3}$$

In this case the governing equation (64) and (65) will be simplified to:

$$\rho g \sin\theta + \frac{d}{dy}\left[\frac{2}{3}\frac{\rho e}{\zeta}\left(\frac{du}{dy}\right)\right] = 0 , \quad (74)$$

$$\rho g \cos\theta + \frac{d}{dy}\left(\frac{2}{3}\rho e + \frac{2\rho}{3\zeta}\Delta e_c\right) = 0 , \quad (75)$$

$$\frac{d}{dy}\left[\frac{10\rho e}{9\zeta}\frac{de}{dy}\right] + \frac{2}{3}\frac{\rho e}{\zeta}\left(\frac{du}{dy}\right)^2 - \rho\Delta e_c = 0 \quad (76)$$

and the continuity equation (11) is automatically satisfied.

Equations (74), (75), and (76) comprise a set of coupled nonlinear second order ordinary differential equations for the gravity flow of granular materials down an inclined plane, in terms of the unknown field variables ρ, u, and e. From equations (74) and (75) one can easily show that

$$\frac{\rho e}{\zeta}\frac{du}{dy} - \tan\theta(\rho e + \frac{\rho}{\zeta}\Delta e_c) = C_1 \tag{77}$$

where C_1 is an integration constant. Note that in this case the shear stress is given by

$$\tau_{12} = \frac{2}{3}[C_1 + \tan\theta(\rho e + \frac{\rho}{\zeta}\Delta e_c)] \ . \tag{78}$$

Requiring the shear stress $\tau_{12}(h) = 0$ then dictates C_1 to be

$$C_1 = -\tan\theta[\rho(h)e(h) + \alpha e^{-\frac{1}{2}}(h)\Delta e_c(h)] \ . \tag{79}$$

From equations (67) and (77) it then follows that

$$\frac{du}{dy} = \alpha^{-1}(\tan\theta)\rho e^{\frac{1}{2}} + C_1\alpha^{-1}e^{-\frac{1}{2}} + (\tan\theta)(\Delta e_c/e) \ . \tag{80}$$

From equation (80) certain estimates on e can be found. Note that from (80) one obtains

$$e^{\frac{1}{2}} \approx \frac{1}{2}\alpha\rho^{-1}\text{cotag}\theta[\frac{du}{dy} - \tan\theta(\frac{\Delta e_c}{e})](1+\sqrt{1+R^*}) \ , \tag{81}$$

where

$$R^* = \left[\frac{4\rho\tan^2\theta[\rho(h)e(h) + \alpha e^{-\frac{1}{2}}(h)\Delta e_c(h)}{\alpha^2[(\frac{du}{dy}) - \tan\theta(\frac{\Delta e_c}{e})]^2}\right] \tag{82}$$

Note that when $R^* \ll 1$ then

$$e^{\frac{1}{2}} = \alpha\rho^{-1}(\text{cotag}\theta)[(\frac{du}{dy}) - \tan\theta(\frac{\Delta e_c}{e})] \ , \tag{83}$$

$$\alpha = \tfrac{1}{2}\, m(3/2\pi)^{\frac{1}{2}} d^{-2} \quad , \tag{84}$$

which shows how random fluctuation energy is related to the mean shear rate, the inclination angle θ and the bulk density ρ. On the other hand, if R*>>1 then

$$e^{\frac{1}{2}} \approx \tfrac{1}{2}\, \alpha\rho^{-1} \mathrm{cotag}\theta [(\frac{du}{dy}) - \tan\theta(\frac{\Delta e_c}{e})]$$

$$+\rho^{-\frac{1}{2}}[\rho(h)e(h) + \alpha e^{-\frac{1}{2}}(h)\Delta e_c(h)] \ . \tag{85}$$

One may further integrate equation (76), using equation (67), to arrive at

$$e^{3/2} = \left[C_3 + C_2 y - (\frac{27}{20\alpha})\int_o^y [\int_o^\zeta [\tau_{12}(s)(\frac{du(s)}{ds})ds - \rho(s)\Delta e_c(s)]ds]d\zeta\right] \quad , \tag{86}$$

where C_1 and C_3 are integration constants. Therefore, we have managed to come up with a set of nonlinear integro differential equations for $u(y)$, $\rho(y)$, and $e(y)$ that can in principle be solved numerically. Of course appropriate boundary conditions have to be prescribed before explicit solutions are constructed. In order to construct two approximate solutions for both $e(y)$ and $u(y)$ let us assume that the energy dissipation per unit volume, i.e., $\tau_{12}(\frac{du}{dy})$, can be approximated by

$$\tau_{12}(\frac{du}{dy}) \approx \mu_k \rho_{av.} g\cos\theta u_{av.} \quad , \tag{87}$$

where $\rho_{av.}$ and $u_{av.}$ are, respectively, the average density and velocity. Thus, from equation (86)

$$e^{3/2} = \{C_3 + C_2 y - (\frac{20}{40\alpha})[\mu_k \rho_{av.} g\cos\theta u_{av.} - \rho_{av.}\Delta e_{c(av.)}]y^2\} \quad , \tag{88}$$

where $\Delta e_{c(av.)}$ is the average collisional dissipation energy due to inelasticity and intergranular friction. A profile resembling this $e(y)$ is depicted in Fig. 2. Further, from equation (80) we obtain

$$u(y) = \left[\alpha^{-1}\rho_{av.}\tan\theta\int_o^y e^{\frac{1}{2}}(\zeta)d\zeta + C_1\alpha^{-1}\int_o^y e^{-\frac{1}{2}}(\zeta)d\zeta + \Delta e_{c(av.)}\tan\theta\int_o^y e^{-1}(\zeta)d\zeta\right], \quad (89)$$

where we have used the no-slip boundary condition of $u(0) = 0$, at the lower boundary. Fig. 1 approximately depicts the nature of $u(y)$.

5.2 Simple Shear Flows: Plane Couette Flows

Referring to Fig. 4 consider a steady simple plane Couette two-dimensional flow problem of granular materials in the x-direction with variation in the y-direction with respect to a rectangular Cartesian system (x,y) such that

$$u \equiv (u(y),0,0), \quad \rho = \rho(y) \quad , \quad e = e(y) \ . \quad (90)$$

In this case the continuity equation (11) is automatically satisfied and governing equations (64) and (65) will be simplified to

$$\frac{2}{3}\left(\frac{\rho e}{\zeta}\right)\left(\frac{du}{dy}\right) = \tau_o \quad , \quad (91)$$

$$\rho g + \frac{d}{dy}\left(\frac{2}{3}\rho e + \frac{2}{3}\frac{\rho}{\zeta}\Delta e_c\right) = 0 \ , \quad (92)$$

$$\frac{d}{dy}\left(\frac{10\rho e}{9\zeta}\frac{de}{dy}\right) + \frac{2}{3}\left(\frac{\rho e}{\zeta}\right)\left(\frac{du}{dy}\right)^2 - \rho\Delta e_c = 0 \ , \quad (93)$$

where τ_o is a constant shear stress applied to the upper plate to maintain the flow. From equations (67), (91), (92), and (93) we obtain

$$\frac{2}{3}\alpha e^{\frac{1}{2}}\frac{du}{dy} = \tau_o \ , \quad (94)$$

$$\frac{2}{3}\rho e + \frac{2}{3}\alpha e^{-\frac{1}{2}}\Delta e_c = g\int_o^y \rho(\zeta)d\zeta \quad , \quad (95)$$

$$\frac{d}{dy}\left(\frac{10}{9}\alpha e^{\frac{1}{2}}\frac{de}{dy}\right) + \left(\frac{3}{2\alpha}\right)\tau_o^{\,2} e^{-\frac{1}{2}} - \rho\Delta e_c = 0 \ . \qquad (96)$$

From equation (94) one finds that

$$u(y) = \left(\frac{3\tau_o}{2\alpha}\right)\int_o^y Z^{-1/3}(\zeta)\,d\zeta \ , \qquad (97)$$

where $Z(y) = e^{3/2}$.

From equation (96) one finds that

$$\frac{d^2Z}{dy^2} + \left(\frac{81}{40\alpha^2}\right)\tau_o^{\,2} Z^{-1/3} = \left(\frac{27}{20\alpha}\right)\rho\Delta e_c \ , \qquad (98)$$

Note that in equation (97) we have used the no-slip condition $u(0) = 0$ at the stationary wall at $y = 0$. Furthermore, the fact that $u(h) = U_o$ (Fig. 4), implies that τ_o can be found from the following equation

$$U_o = \left(\frac{3\tau_o}{2\alpha}\right)\int_o^h Z^{-1/3}(\zeta)\,d\zeta \ . \qquad (99)$$

Thus, the only thing to do is to solve equation (98) for $Z(y)$. Since Δe_c in terms of ρ and $\frac{du}{dy}$ is given by equation (70) one in principle may solve the above system of nonlinear integro differential equation numerically. Thus, one appears to have an acceptable solution provided one is able to prescribe correctly and in an acceptable manner the two remaining boundary conditions on $Z(y)$ or the average fluctuation energy e. From equation (94) it is clear that

$$Z(y) = \left(\frac{3}{2\alpha}\right)^3 \tau_o^{\,3}/(du/dy)^3 \ , \qquad (100)$$

so that the values of $Z(0)$ and $Z(h)$ are inversely related to the velocity gradients at the two walls. Although there appears to exist some abmiguities regarding these boundary conditions one may assume that

$$Z(0) = Z(h) = e(0) = e(h) = 0 , \qquad (101)$$

so that, based on equation (94) and the fact that shear stresses are to reamin finite and different from zero, the velocity gradients at the walls acquire infinite values. In this case the velocity and the fluctuation energy profile should look like what is depicted in Fig. 4. Otherwise, it is not yet clear how one would prescribe such fluctuation quantities at the boundaries of this flow. It is possible to envision situations in which the flow regime is quite rapid even very close to the boundaries (Savage [37]). Here, the grains collide with the boundary and give rise to a continuous process of specular reflection. Thus, on the boundaries wall because of changes in magnitude and direction of grain velocities the amount of fluctuation e may differ from zero. This would be certainly true if the grains are allowed to slip momentarily before bouncing off the boundaries.

6. ACKNOWLEDGMENTS

The work of MS is partially supported by a grant from the National Science Foundation.

7. REFERENCES

1. Shahinpoor, M., "Statistical Mechanical Considerations on the Random Packing of Granular Materials," J. Powder Tech., Vol. 25, No. 2, pp. 163-176 (1980).

2. Shahinpoor, M., "Statistical Mechanical Considerations on Storing Bulk Solids," Bulk Solids Handl. Int. J., Vol. 1, No. 1, pp. 31-36 (1981).

3. Goldreich, P., "Dynamics of Planetary Rings," Proc. IXth US Natnl. Cong. Appl. Mech., Cornell Univ., Ithaca, N.Y., June (1982).

4. Jenkins, J.T. and Cowin, S.C., "Theories for Flowing Granular Materials," The Joint ASME-CSME Appl. Mech. Fluid Engng. & Bioengng. Conf., AMD, Vol. 31, pp. 79-89, Niagara Falls (1979).

5. Blinowski, A., "On the Dynamic Flow of Granular Media," Arch. Mech. Stoso., Vol. 30, No. 1, pp. 27-34 (1978).

6. Ogawa, S., "Multitemperature Theory of Granular Materials," Proc. U.S.-Japan Seminar on Continuum Mech. and Statist. Approaches in Mech. of Granular Materials, (Ed: S.C. Cowin and M. Satake), Sendai, Japan. Pub., Gakujutsu Bunken Fukyu-Kai, pp. 208-217, June (1978).

7. Jenkins, J.T. and Savage, S.C., "The Mean Stress Resulting From Interparticle Collisions in a Rapid Granular Shear Flow," Proc. 4th Int. Conf. Continuous Models of Discrete Systems, Stokholm, August (1981).

8. Launder, B.E. and Spalding, D.B., "Lectures in Mathematical Models in Turbulence," Academic Press, London and New York (1972).

9. Ahmadi, G. and Shahinpoor, M., "Towards a Turbulent Modeling of Rapid Flow of Granular Materials," Powder Tech. J. (in press).

10. Bagnold, R.A., "Experiments on a Gravity Free Dispersion of Large Spheres in Newtonian Fluid Under Shear," Proc. Roy. Soc. Lond. A, 225, pp. 49-63 (1954).

11. McTigue, D.F., "A Model for Stresses in Shear Flow of Granular Materials," in Proc. US-Japan Seminar (see Ref. 6 of this paper), pp. 266-271 (1978).

12. Ogawa, S., Omemura, A. and Oshima, N., "On the Equations of Fully Fluidized Granular Materials," ZAMP, Vol. 31, pp. 483-493 (1980).

13. Lun, C., Savage, S.B. and Jeffrey, D.J., "The Stresses Developed During the Simple Shear of a Granular Material Comprised of Smooth, Uniform, Inelastic, Spherical Particles," J. Fluid Mech. (to appear) (1982).

14. Savage, S.B. and Sayed, M., "Experiments on Dry Cohesionless Materials in an Annular Shear Cell at High Strain Rates," Proc. Euromech. 133, Statics and Dynamics of Granular Materials, Oxford Univ., Cambridge (1980).

15. Savage, S.B. and Sayed, M., "Stresses Developed by Dry Cohesionless Granular Materials Sheared in an Annular Shear Cell," J. Fluid Mech. (in press) (1982).

16. Savage, S.B. and Jeffrey, D.J., "The Stress Tensor in the Granular Flow," J. Fluid Mech., 110, pp. 255-272 (1981).

17. Savage, S.B. and Jeffrey, D.J., "Errata: The Stress Tensor in a Granular Flow," J. Fluid Mech., 110, pp. 255-272 (1982).

18. Jenkins, J.T. and Savage, S.B., "A Theory for the Rapid Flow of Identical, Smooth, Nearly Elastic Spherical Particles," (in press) (1982).

19. Ahmadi, G. and Shahinpoor, M., "A Kinetic Theory for Rapid Granular Flows and Evolution of Fluctuations," Int. J. Nonlinear, Mech. (to appear).

20. Chapman, S. and Cowling, T.G., "The Mathematical Theory of Non-Uniform Gases," Cambridge University Press, Cambridge (1970).

21. Ashurst, WT. and Hoover, W.G., "Dense Fluid Shear Viscosity Via Nonequilibrium Molecular Dynamics," Phys. Rev. A., 11, pp. 658-678 (1975).

22. Evans, D.J. and Watts, R.O., "Shear-Dependent Viscosity in Simple Fluids," Chem. Phys., 48, pp. 321-327 (1980).

23. Carnahan, N.F. and Starling, K.E., "Equations of State for Non-Attracting Rigid Spheres," J. Chem. Phys. 51, pp. 635-636 (1969).

24. Cercignani, C., "Theory and Application of the Boltzmann Equation," Elsevier Pub. Comp., New York (1975), also see Hochstim, A.R., "Kinetic Processes in Gases and Plasmas," Academic Press, New York and London (1969).

25. Bhatnagar, P.L., Gross, E.P. and Krook, M., "A Model for Collision Processes in Charged and Neutral One-Component Systems," Phys. Rev., 94, 3, pp. 511-535 (1954).

26. Vincenti, W.G. and Kruger Jr., C.H., "Introduction to Physcial Gas Dynamics," John Wiley and Sons, Inc., New York, London, Sydney (1965).

27. Shahinpoor, M. and Siah, J.S.S., "New Constitutive Equations for the Rapid Flow of Granular Materials," J. Non-Newtonian Fluid Mech., 9, pp. 147-156 (1981).

28. Shahinpoor, M., "On Rapid Flow of Bulk Solids," Bulk Solids Handl. Int. J., Vol. 1, No. 3, pp. 487-500 (1981).

29. Shahinpoor, M., "New Constitutive Equations for Rapid Flow of Granular Materials - II," J. Non-Newtonian Fluid Mech. (to be published) (1982).

30. Shahinpoor, M. and Siah, J.S.S., "On A Two-Dimensional Mean Flow of Granular Materials," Dev. Theo. Appl. Mech., Vol. XI, pp. 499-505 (1982).

31. Shahinpoor, M. and Lin, S.P., "Rapid Couette Flow of Granular Materials," Acta Mechanica, 42, pp. 183-196 (1982).

32. Truesdell, C. and Muncaster, R.G., "Fundamentals of Maxwell's Kinetic Theory of a Simple Monotomic Gas," Academic Press, Pure and Appl. Math. Series, No. 83, New York (1980).

33. Shahinpoor, M. and Ahmadi, G., "Fluctuation Equilibrium in Rapid Flow of Granular Materials," (in press) (1982).

34. Ahmadi, G. and Shahinpoor, M., "On Collision Operators in Rapid Granular Flows," Powder Tech. J. (to appear) (1983).

35. Krook, M. and Wu, T.T., "Formation of Maxwellian Tails," Phys. Rev. Letts., 36, 19, pp. 1107-1109 (1976).

36. Krook, M. and Wu, TT., "Exact Solutions of the Boltzmann Equation," Phys. of Fluids, 20, 10, pp. 1589-1595 (1977).

37. Muncaster, R.G., "On Generating Exact Solutions of the Maxwell-Boltzmann Equation," Arch. Ration. Mech. Anal., 70, 1, pp. 79-90 (1979).

38. Savage, S.B. (private communication) (1982).

GEOMETRICAL FOUNDATION OF THE ANALYSIS OF GRANULAR MATERIALS FLOWS

M. Shimbo

Division of Information Engineering, Hokkaido University
Sapporo, Japan

ABSTRACT

A theoretical approach to the problem of the flows of granular materials is shown in terms of differential geometry. A heuristic treatment of asymmetric fields in flows originated from the intrinsic rotation of particles affords a possible interpretation of the tubular pinch effect in suspensions passing through a tube.

1. INTRODUCTION

Granular materials are treated as an imperfect continuum with microstructure, in which anomalies are distributed, and microscopic local configurations are considered. Each element can deform and rotate freely, so that asymmetric strains and stresses are likely to be entailed. A typical example of such asymmetric features will be found in the dilatancy effect in granular materials [1], where a volumetric strain is caused by shear stresses. It could be due to mutual actions between particles, and the geometrical background has been shown in the previous articles [2] in the framework of the non-Riemannian theory of plasticity [3]. A recognition obtained there was that there could be a certain distribution of asymmetric stress fields, which brought about the gradient of a volumetric strain.

In what follows, an attempt is made to describe the rheological aspects of granular materials along the same lines of thought and with reference to the theory of moving dislocations [4]. For the sake of convenience of analysis, we shall confine ourselves to the case of small deformations.

2. MATERIAL SPACE-TIME

The flow of granular materials can be treated in a four-dimensional material space-time, as in a similar manner as in [4]. Each material point is represented by the coordinates x^a $(a=0,1,2,3)$, where $x^o=t$ is the time coordinate and x^i $(i=1,2,3)$ the space coordinates.

The imperfection is defined by the distortion of a certain reference configuration. Corresponding material elements in two configurations are assumed to be related in a linear form

$$(dx)^{a'} = A^{a'}_{a} dx^{a} , \quad dx^{a} = A^{a}_{a'}(dx)^{a'} \qquad (a,a'=0,1,2,3) \tag{1}$$

where $A^{a'}_{a}$ etc. are transformation tensors, given by a function of x^a, and Einstein's summation convension is adopted with regard to repeated indices.

A metric and a Euclidean connexion are also introduced with regard to this transformation. If the transformation tensor defined by (1) be only slightly deviated from the Euclidean condition, i.e. $A^{a'}_{a}=\delta^{a'}_{a}+e^{a'}_{a}$, the metric and the torsion tensors are given respectively by

$$g_{ba} = \delta_{ba} + 2e_{(ba)} \quad \text{and} \quad S_{cba} = S_{cb}^{\cdot\cdot d} g_{da} = \partial_{[c} e_{b]a} \tag{2}$$

for an affine connexion with distant parallelism [3], where $\delta_{b\,a}$ is Kronecker's delta, $e_{ba}=e^{b'}_{b}\delta_{b'a}$ and $\partial_b=\partial/\partial x^b$. The parentheses ( ) and the brackets [ ] in (2) indicate the mixing and the alternation over indices, e.g., $e_{(ba)}=(e_{ba}+e_{ab})/2$ and $e_{[ba]}=(e_{ba}-e_{ab})/2$. The Riemann-Christoffel curvature tensor vanishes identically in this case.

The spatial component e_{ji} (i,j=1,2,3) in the first equation of (2) represents deformed and rotated states of a material element, which is not generally assured to be symmetric with respect to the indices i and j. Its antisymmetric part does not represent the metric imperfection, but the rotational characteristics of material elements. The mixed component of torsion tensor represents the motion of dislocations. It is given by

$$S_{oi}^{\cdot\cdot k} = -S_{ji}^{\cdot\cdot k} v^{j} \qquad (i,j,k=1,2,3) \tag{3}$$

for a constant dislocation velocity v^j, where $S_{ba}^{\cdot\cdot o}=0$ is assumed since the distortion of time is not considered [4].

3. ASYMMETRIC FIELDS IN FLOW

The density characteristics of flows of granular materials are represented mathematically by the torsion tensor. By contracting the indices i and k in (3) and defining $S_b=[2/(n-1)]S_{ba}^{\cdot\cdot a}$ (n=4), we have

$$S_o = - S_j v^j , \quad S_j = 2S_{ji}^{\cdot\cdot i}/3 .$$

The S_o represents the time rate of dislocation density and S_j the spatial gradient of volumetric strain along the axis of x^j. Since the former depends upon the latter, the instantaneous spatial component S_j is mainly concerned with hereafter.

In the so-called Cosserat continuum [5], an asymmetric strain e_{ji} is given, e.g., by

$$e_{ji} = \partial_j u_i + w_{ji}\ , \quad w_{ji} = - w_{ij}\ ,$$

where u_i is a displacement vector and w_{ji} a rotation tensor independent of the rotation originating from the displacement. The torsion tensor is then given by

$$S_{kji} = \partial_{[k} w_{j]i} \quad \text{and hence} \quad S_i = \partial^j w_{ji}/2\ , \tag{4}$$

where $\partial^j = g^{ji}\partial_i$.

We introduce here a linear coupling relation between the anti-symmetric part of stress and the intrinsic rotation, i.e.

$$\sigma_{[ji]} = C_{[ji]}^{\cdot\cdot[km]} w_{[mk]}\ ,$$

which is reduced to

$$\sigma_{[ji]} = C w_{[ji]} \tag{5}$$

for an isotropic case, where $C_{ji}^{\cdot\cdot km}$ and C are the coefficients of friction [6]. The mixed component σ_{oi} represents the density of momentum 7]. It follows from the second equation of (4) and (5) that

$$S_i = \partial^j \sigma_{[ji]}/2C\ , \tag{6}$$

which means that the gradient of volumetric strain is entailed, if asymmetric stress field is not non-divergent.

It sometimes happens that particles in a flow of initially uniform suspension through a straight tube collect into the central region or a thin annular region in a cross-section [8]. This is known as the tubular pinch effect. If interactions among particles, liquid and the tube wall be the cause, a microscopic rotation field w_{ji} of particles should play an important role for the characteristics of dislocation density of particles.

For the simplicity's sake, a flow of granular materials passing along x^3-direction bounded by a pair of parallel walls at $x^1 = \pm a$ with infinite length in x^2-direction is considered in an orthogonal coordinate system x^i (i=1,2,3). In a laminar flow, an asymmetric stress field is responsible not for the rotation field originating from the displacement of particles, but for the microscopic intrinsic rotation field. The distribution of stresses along the axis of x^2 is assumed to be in equilibrium and is neglected here.

If an unbalance of shear stress operating particles may originate at the wall and transmit through suspensions, it is assumed to be approximately given by

$$\sigma_{[31]} = \exp[-k(1-x^1/a)^2] f(x^3) \tag{7}$$

except near the region at $x^1=0$, where k is a constant and $f(x^3)$ an arbitrary function of x^3. There is no reason to assume the distribution (7), except an observation of the transversal stress of particles in a moving bed has been reported to have such a form near the wall [9].

From (6) and (7) we have the gradient of volumetric stress along the axis of x^3 at a cross-section, given by

$$S_3 = k(1-x^1/a)\exp[-k(1-x^1/a)^2]f(x^3)/C$$

for an isotropic case. The S_3 gives a finite value near the tube axis, zero at the wall, and the extremals at $x^1 = \pm(1-1/\sqrt{2k})$, giving a measure of the rate of packed- or unpackedness of particles in suspensions. The effect will be accumulated according to the deviation from the initial configuration. The particles will collect finally into thin regeions. Thus it is expected that the lateral migration of particles in a cross-section happens in two ways from the wall to the center and from the center to the wall, which explains the centripetal and centrifugal migrations of the tubular pinch effect.

4. CONCLUDING REMARKS

The density characteristics of granular materials flows is analyzed based on the intrinsic rotation field of particles with reference to the theory of moving dislocations. It was shown that the existence of asymmetric stress field, which is not non-divergent, entails the gradient of the density of particles along the direction of suspension flow. It is also suggested that a certain distribution of asymmetric stress causes bilateral migrations of particles in the cross-section of the flow. The distribution of asymmetric stress as well as the estimation of the parameters in the formulation are still in a conjecture. It is necessary to discuss more in detail of the change of the density distribution as a function of time along this lines of thought.

REFERENCES

1. Reynolds, O., "On the Dilatancy of Media Composed of Rigid Particles in Contact. With Experimental Illustrations", Phil. Mag., S. 5, Vol. 20, pp. 469-481 (1885).

2. Shimbo, M., "A Geometrical Formulation of Granular Media", Theor. & Appl. Mech., Vol. 26, pp. 249-256 (1978): "A Geometrical Foundation of the Analysis of Granular Behavior", Proc. U.S.-Japan Seminar on Continuum-Mechanical and Statistical Approaches in the Mechanics of Granular Materials, S. C. Cowin and M. Satake (editors), Gakujutsu Bunken Fukyukai, Tokyo, Japan, pp. 272-281 (1978).

3. Kondo, K. (editor), "Non-Holonomic Geometry of Plasticity and Yielding", Memoirs, Vol. 1, pp. 453-572 (1955); RAAG Memoirs, Vol. 2, pp. 199-232 (1958); ibid., Vol. 3, pp. 91-214 (1962); ibid., Vol. 4, pp. 137-238 (1968).

4. Amari, S., "A Geometrical Theory of Moving Dislocations and Anelasticity", RAAG Memoirs, Vol. 4, pp. 142-152 (1968).

5. Cosserat, E. and F., "Théorie des corps déformables", Paris, A. Hermann et fils (1909).

6. Oshima, N., "Dynamics of Granular Media", Memoirs, Vol. 1, pp. 565-572 (1955).

7. Minagawa, S., "Physical Implications of Identities for the Geometrical Terms Related to the Strain and Stress-Function Space-Times", RAAG Research Notes, Third Series, No. 135, pp. 1-19 (1968).

8. Segré, G. and Silberberg, A., "Radial Particle Displacements in Poiseuille Flow of Suspensions", Nature, Vol. 189, pp. 209-210 (1961).

9. Takahashi, H. and Yanai, H., "Discussion on the Distribution of Radial Stress and Vertical Velocity of Particles in a Moving Bed by Plasticity Theory", Kagaku Kogaku, Vol. 35, pp. 1372-1379 (1971).

DYNAMIC INTERACTIONS OF GRANULAR MATERIALS*

S. L. Soo

Department of Mechanical and Industrial Engineering
1206 West Green Street
University of Illinois at Urbana-Champaign
Urbana, IL 61801, USA

I. ABSTRACT

Interactions of particulates in a moderately dense to a dense suspension are treated according to aspects of momentum, energy, mass, and electrical charge transfers. In terms of applications, they include drag, agglomeration, heat transfer, erosion, attrition, and charge or high voltage generation. Basic interacting parameters are identified and methods illustrated can be extended to various systems.

*Study supported in part by the National Science Foundation

1. INTRODUCTION

Mutual interactions including collision among solid particles in a fluid suspension occur by their difference in responses to a field force or by random excitation. Collisions with the wall of a body or container may be considered as a special case. The field forces include those due to fluid flow field, gravitational field or electro-magnetic field. Random excitations may arise via turbulence, particle induced wakes, or sonic waves. The influence of particle-particle interactions on an overall suspension system depends on whether a suspension is dilute or dense. A suspension is considered to be dilute when the particle-fluid interaction length L_p is smaller than the mean free-path of particle-particle collision Λ_p. The are given by

$$L_p = \langle(\Delta u)^2\rangle^{1/2}/F \tag{1.1}$$

and

$$\Lambda_p = 1/2^{1/2}\, n\, \pi(a_1 + a_2)^2 \tag{1.2}$$

where $\langle(\Delta u)^2\rangle^{1/2}$ is the intensity of relative motion of particle to fluid, F is the inverse relaxation time for fluid-to-particle momentum transfer, n is the number density of particles, $\pi(a_1 + a_2)^2$ is the collision cross-section of particles of radii a_1 and a_2, assuming spherical particles. It has been shown that depending on the above and additional parameters, the upper limit of a dilute suspension is for Λ_p to approach the mean inter-particle spacing or when the volume fraction of particles is 5.5 percent [1]. In pipe flow, a dilute suspension is further characterized by the fact that particle-wall collision is more frequent than particle-particle collision.

Using the idealization of spherical particles, transport processes due to particle-particle collision are treated and updated (from [2] for instance) on a common basis. They include:

Momentum Transfer--impaction and the influence of inertia force, viscous force, and field forces; characterized by the fraction impacted and a drag coefficient.

Heat Transfer--heat transfer by contact, characterized by an accommodation coefficient and a heat transfer coefficient.

Mass Transfer--attrition, erosion, and agglomeration; characterized by a mass removal coefficient.

Charge Transfer--electrostatic charging and charge distribution, characterized by a charge transfer coefficient.

Further, all these transfer processes are characterized by their respective relaxation times or their inverse, the time constants.

Items treated here are intended to be illustrative rather than exhaustive. Besides those treated, extensions to cases such as agglomeration and electric current to ground due to impacts with a pipe wall are readily made.

2. FRACTION IMPACTED

When one considers a stationary sphere 1 of radius a_1 bombarded by a cloud of spheres 2 of radius a_2 and material ensity $\bar{\rho}_{p2}$ and conveyed by a fluid of material density $\bar{\rho}$, viscosity $\bar{\mu}$, and freestream velocity u_o, only a fraction of the particles 2 whose centers lie within the cylinder formed by the velocity u_o and the projected area $\pi(a_1 + a_2)^2$ will hit sphere 1 because the drag exerted by the fluid as balanced by the inertia of the particles. This fraction is called fraction impacted. The analysis of Langmuir and Blodgett [3] gives results which are correlated by

$$\psi_{12} = \frac{\bar{\rho}_{p2}\, u_o (2a_2)^2}{18\bar{\mu}\; 2(a_1 + a_2)} = \frac{u_o}{F_{2f}\; 2(a_1 + a_2)} = N_{m2} \qquad (2.1)$$

$$\phi_{12} = \frac{(N_{Re})_2^2}{2\Psi_{12}} = 9(\bar{\rho}/\bar{\rho}_{p2})\, [\frac{u_o\; 2(a_1 + a_2)\, \bar{\rho}}{\bar{\mu}}] \qquad (2.2)$$

where F_{2f} is the inverse relaxation time for momentum transfer from the fluid to particles 2, $2(a_1 + a_2)$ is the radius of exclusion, N_{m2} is the momentum transfer number from the fluid to particle 2, $(N_{Re})_o = 2a_2\, u_o\, \bar{\rho}/\bar{\mu}$, which gives the local particle Reynolds number as $N_{Re}/(N_{Re})_o = |u - u_{p2}|/u_o$. Further,

$$F_{2f} = (3/4)\, C_D (\bar{\rho}/\bar{\rho}_{p2})\, |u - u_{p2}|/2a_2 \qquad (2.3)$$

and the drag coefficient is given by

$$C_D = 24\, F^*/N_{Re} \qquad (2.4)$$

with

$$F^* \sim 1 + 0.0975\, N_{Re} - 0.00636\, N_{Re}^2 \qquad (2.5)$$

for $N_{Re} < 100$ and

$$F_{2f} = (9/2)\, F^*\, \bar{\mu}/\bar{\rho}_{p2}\, a_2^2. \qquad (2.6)$$

The fraction impacted η as given by Langmuir and Blodgett [3] can be curve-fitted to give

$$-\ln \eta_{12} = \left(\frac{3.243 \times 10^5}{\Phi_{12} + 1.617 \times 10^5}\right) \ln \left[1 + \frac{7.5 \ln(\Phi_{12} + 1.617 \times 10^5) - 89.16}{(\Psi_{12})^{1/2}}\right] \tag{2.7}$$

Modifications are known for the effect of diffusion [4], and for the case of particles in a rarefied gas [5], but neither of these are significant to the range of granular materials.

When concerned with mutual collision of particle clouds of species 1 and 2, both moving, in the center of mass frame of reference, impaction of species 1 by 2 is correlated via

$$\psi_{12} = \left(1 + \frac{m_2}{m_1}\right)^{-1} \frac{\bar{\rho}_{p2}\, \Delta u_{12}}{2(a_1 + a_2)\, 18\bar{\mu}} (2a_2)^2$$

$$= [(m_2/m_1)^{1/2} + (m_1/m_2)^{1/2}]^{-1} [(a_2/a_1)^{1/2} + (\bar{a}_1/\bar{a}_2)^{1/2}]^{-1}$$

$$\cdot\ [2(a_1\ a_2)^{1/2}\ \bar{\rho}\ \Delta u_{12}/18\bar{\mu}]\ [(\bar{\rho}_{p1}\ \bar{\rho}_{p2})^{1/2}/\bar{\rho}]\ [a_1/a_2]^{1/2} \tag{2.8}$$

$$\Psi_{12} = [1 + (m_2/m_1)]\ \{9\bar{\rho}^2\ \Delta u_{12}[2(a_1 + a_2)]/\bar{\rho}_{p2}\ \bar{\mu}\}$$

$$= [(m_2/m_1)^{1/2} + (m_1/m_2)^{1/2}][(a_2/a_1)^{1/2} + (a_1/a_2)^{1/2}]$$

$$\cdot\ \left[\frac{18(a_1\ a_2)^{1/2}\ \bar{\rho}\ \Delta u_{12}}{\bar{\mu}}\right] \frac{\bar{\rho}}{(\bar{\rho}_{p1}\ \bar{\rho}_{p2})^{1/2}} (a_2/a_1)^{3/2} \tag{2.9}$$

with η_{12} given by Eq. (2.7) and $\Delta u_{12} = |u_{p2} - u_{p1}|$.

3. COLLISION OF ELASTIC SPHERES

Another basic interaction is the phenomenon of impacting elastic spheres. The basic formulation was given by Herz (1881) and Rayleigh (1900) [6]. For two spheres 1 and 2 of elastic materials of moduli E_1, E_2 and Poisson ratios ν_1 and ν_2, impacting with relative motion represented in Fig. 1, the interacting force is P when the contacting surface has a radius a_c while the two centers approaching each other by a distance α_1 smaller than the sum of radii a_1 and a_2. The masses of the spheres are m_1 and m_2 of apparent (including hollow spheres) material densities $\bar{\rho}_{p1}$ and $\bar{\rho}_{p2}$ with relative motion denoted by angle θ and relative speed $\Delta u = |u_{p2} - u_{p1}|$. Deviation from perfect elastic collision is represented by incoming relative velocity Δu and reflecting relative velocity $r^*\Delta u$. The duration of contact is given by $\Delta t_{12} = 2.94\ \alpha_1/(\Delta u)\cos\theta$ and

$$\alpha_1 = [(9\pi^2/16)\ P^2(k_1 + k_2)^2\ (a_1 + a_2)/a_1 a_2]^{1/3} \tag{3.1}$$

for static contacts where $k = (1 - \nu^2)/E$ for each material. Taking into consideration of reduced mass in relative motion,

$$\alpha_1 = [\frac{5}{4}(\frac{1 + r^*}{2})(\Delta u)^2 \cos\theta \frac{m_1 m_2}{m_1 + m_2} \frac{3\pi}{4} (\frac{a_1 + a_2}{a_1 a_2})^{1/2} (k_1 + k_2)]^{2/5} \tag{3.2}$$

and

$$P = \frac{2^{4/5}\pi^{1/5}}{3} (1 + \frac{m_2}{m_1})^{-3/5} (k_1 + k_2)^{-2/5} (1 + \frac{a_2}{a_1})^{-1/3} a_2^2$$

$$\cdot [5(\Delta u)^2 (\frac{1 + r^*}{2}) \cos\theta\ \bar{\rho}_{p2}]^{3/5} \tag{3.3}$$

We note that the recoverable elastic work is given by $(1/2)\ P\alpha_1$ approximately which can be expressed as

$$P\ \alpha_1/2 = (5/16)\ (1 + r^*) \cos\theta\ (\Delta u)^2 \frac{m_1 m_2}{m_1 + m_2} \tag{3.4}$$

Hence, the dissipated kinetic energy is

$$(5/16)\ (1 - r^*) \cos\theta\ (\Delta u)^2 \frac{m_1 m_2}{m_1 + m_2} \tag{3.5}$$

when $r^* < 1$, giving additional physical meaning of r^*, the ratio of the reflected velocity to incoming velocity. $r^* = 0$ corresponds to agglomeration.

The radius of maximum area of contact is given by

$$a_c = [\frac{3\pi}{4} \frac{P(k_1 + k_2)\ a_1 a_2}{(a_1 + a_2)}]^{1/3}$$

$$= 2^{-2/5} [(a_1/a_2)^{1/2} + (a_2/a_1)^{1/2}]^{-2/5} [(m/m_2)^{1/2} + (m_2/m)^{1/2}]^{-1/5}$$

$$\times (\cos\theta)^{1/5} (a_1/a_2)^{1/2} N_{Im}^{1/5} \tag{3.6}$$

where N_{Im} is the impact number relating dynamic force and elastic force, or

$$N_{Im} = 5\pi^2(\Delta u)^2 (\bar{\rho}_{p1} \bar{\rho}_{p2})^{1/2} (k_1 + k_2)(1 + r^*)/2 \tag{3.7}$$

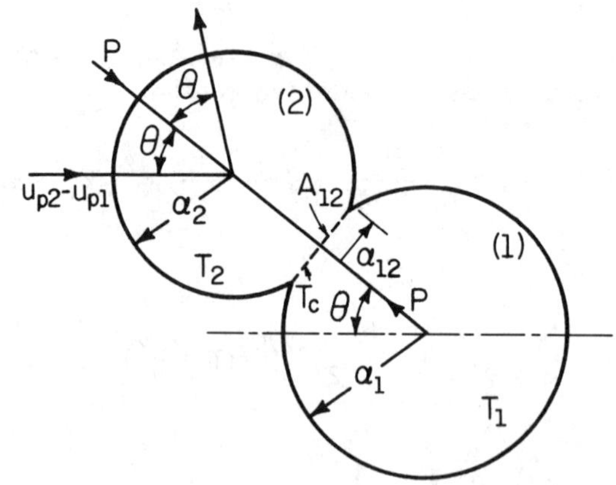

Figure 1 Collision and contact of two elastic spheres

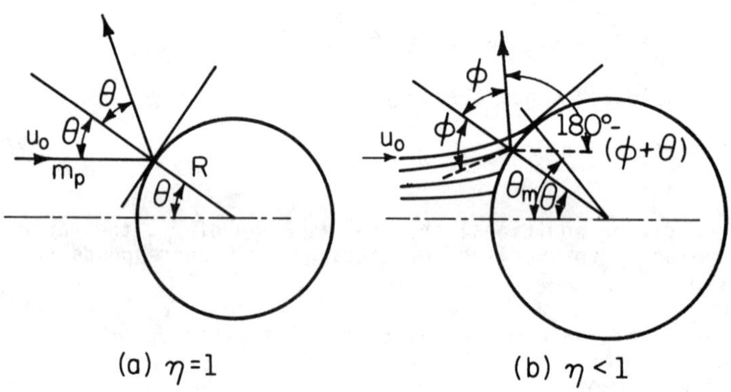

Figure 2 Collision and specular reflection of particles from a sphere

Since the area of contact A_{12} is given by πa_c^2, and the product $A_{12}\ \Delta t_{12}$ is a useful quantity in computing several transport processes, we denote

$$A_{12}\ \Delta t_{12} = \pi\ 2^{-8/5}(2.94)(\cos\ \theta)^{-1/5}\ \frac{(a_1\ a_2)^{3/2}}{\Delta u}\ (\frac{a_1}{a_2})^{6/5}(\frac{\bar{\rho}_{p1}}{\bar{\rho}_{p2}})^{2/5}$$

$$\times\ [(m_1/m_2)^{1/2} + (m_2/m_1)^{1/2}]^{-4/5}\ [(a_1/a_2)^{1/2} + (a_2/a_1)^{1/2}]^{-3/5}(N_{Im})^{4/5}$$

$$= \pi\ 2^{-8/5}(2/94)(\cos\ \theta)^{-1/5}(1 + \frac{m_2}{m_1})^{-4/5}(1 + \frac{a_2}{a_1})^{-3/5}(\frac{\bar{\rho}_{p2}}{\bar{\rho}_{p1}})^{2/5}\ \frac{N_{Im}}{\Delta u}^{4/5}\ a_2^3$$

$$\equiv (A_{12}\ \Delta t_{12})_o\ (\cos\ \theta)^{-1/5} \tag{3.8}$$

The above product gives a characteristic measure of area of contact A_{12} over time Δt_{12}.

4. TRANSFER OF MOMENTUM

Take the case of collision of a cloud of particles of radius a, mass m, number density n_p, and at velocity u_o with a target sphere of radius R, with simple scattering. With specular reflection, the rate of change of axial component of momentum of a particle approaching at a point of contact represented by the geometry in Fig. 2 is given by

$$m\ u_o - m\ u_o\ r^*\ \cos[\pi - (\theta + \phi)] = m\ u_o[1 + r^*\ \cos(\theta + \phi)] \tag{4.1}$$

For limiting contact angle θ_m for the tangential particle path ($\theta_m = \pi/2$ for $\eta = 1$) the force acting on the target sphere is given by

$$\mathcal{F} = n_p\ m\ u_o^2 \int_0^{\theta_m} [1 + r^*\ \cos(\theta + \phi)(R\ d\theta\ \cos\ \theta)(2\pi\ R\ \sin\ \theta) \tag{4.2}$$

We note that for $\eta \simeq 1$, $\theta_m \simeq \pi/2$, $\phi \sim \theta$, Eq. (4.2) integrates to

$$\mathcal{F} \simeq \eta\ \pi\ R^2\ \rho_p\ u_o^2 \tag{4.3}$$

The nature of scattering with specular reflection, regardless of the value of r* (including the case where all particles are stuck on impact), is such that the larger portion of the particles is not reversed in direction. The drag coefficient of a sphere due to particle impaction is given by:

$$C_{D_p} = \frac{\mathcal{F}}{(1/2)\ \rho_p\ u_o^2\ \pi\ R^2} \tilde{<} 2\eta(\Psi,\ \Phi) \tag{4.4}$$

Transfer of momentum from the fluid and from species 2 particles to species 1 particles is given by

$$du_{p1}/dt = F_{1f}(u - u_{p1k}) + F_{12}(u_{p2} - u_{p1}) \tag{4.5}$$

where F_{12} is the inverse relaxation time for momentum transfer from particles 2 to particles 1. The fraction impacted of particles 1 by particles 2 is given by the parameters based on reduced mass relation via Eqs. (2.8) and (2.9):

$$\Psi_{12}(du^*_{p2}/dt^*) = F^*_{2f}(u^* - u^*_{p2}) \tag{4.6}$$

with u^*2 $= u/\Delta u_{12}$, $u^*_{p2} = u_{p2}/\Delta u_{12}$, $u^*_{p2} = u_{p2}/\Delta u_{12}$, $t^* = t\ \Delta u_{12}/(a_1 + a_2)$. The force acting on a particle 1 due to collision by particles 2 is:

$$\mathcal{F}_{12} \approx \eta_{12}\ n_{p2}\ \pi(a_1 + a_2)\frac{m_1\ m_2}{m_1 + m_2}\ \Delta u_{12}(u_{p2} - u_{p1}) \tag{4.7}$$

and

$$F_{12} = \frac{3}{4}\ \eta_{12}\ \frac{(a_1 + a_2)^2\ \Delta u_{12}\ \rho_{p2}}{a_1^3\ \bar{\rho}_{p1}[1 + (m_2/m_1)]}$$

$$= \frac{3}{4}\ \eta_{12}\ \frac{[(a_1/a_2)^{1/2} + (a_2/a_1)^{1/2}]^2\ \Delta u_{12}\ \rho_{p2}}{(a_1/a_2)^{1/2}\ (\bar{\rho}_{p1}\ \bar{\rho}_{p2})^{1/2}[m_2/m_1)^{1/2} + (m_1/m_2)^{1/2}]} \tag{4.8}$$

For two clouds of interacting particles in a given volume, equality of action and reaction leads to

$$\rho_{p1}\ F_{12} = \rho_{p2}\ F_{21} \tag{4.9}$$

and

$$\eta_{21} = \eta_{21}(\Psi_{21}, \Phi_{21}) \tag{4.10}$$

Equation (4.8) has been validated experimentally [7] using large particles (species 1) of 5 to 8 mm diameter of glass, ceramics, and steel and small particles of 30 µm silica sand (species 2) at relative velocities of 2 to 10 m/s. Here $m_1 >> m_2$, $a_1 >> a_2$ Eq. (4.8) takes the form

$$F_{12} = (3/4)(\rho_2/\bar{\rho}_1)\ C_{D_{12}}|u_2 - u_1|\ (2a_1)^{-1} \tag{4.11}$$

For $C_D \sim 2\eta_{12}$, experimental results show that $\eta_{12} \sim 0.80$. Correlation of data was made via a quantity

$$\beta = (314)\ C_{D_{12}}\ \rho_2/(2a_1) \tag{4.12}$$

with

$$C_{D_{12}} \sim 0.7\ (2a_1)^{-0.146} \tag{4.13}$$

for $2a_1$ in m as an approximation.

When one of the solid phases becomes dense such that multiple scattering occurs and self-collision (1 to 1 collision) may become significant an apparent viscosity of that phase arises and analogy to kinetic theory of gases can be applied [2]. It is known [2], however, the viscosity due to particle-particle collision is non-Newtonian. An empirical correlation was postulated by Schugel [8] for the viscosity of a gas-solid show an increase in viscosity by the addition of solid particles. Schugel [8] postulated a correlation of experimental data obtained by shear viscous measurements to give for the mixture:

$$\mu_m = [4\ a^2\ \bar{\rho}_p/2.65 \times 10^{-2}]\ [4.35 + 0.13(\phi/\phi_o - \phi)]\ \mathrm{kg/m{\cdot}s}$$
$$= [4\ a^2\ \bar{\rho}_p/2.65 \times 10^{-2}]\ \mu_s^* \tag{4.14}$$

where ϕ is the volume fraction of solid and ϕ_o is that at minimum fluidization.

Schugel's data cover only moderate bed expansion $2 < \mu/(\phi_o - \phi) < 25$. He also recognized that for $\phi/(\phi_o - \phi) > 100$, non-Newtonian behavior is seen. Blake [9] accounted for the difficulty that μ_m does not reduce the μ_g of the gas phase alone by postulating for $0 < \phi/(\phi_o - \phi) < 2$

$$\mu_s^* = 0.5\ \mu_{s2}^*(\phi/\phi_o - \phi) \tag{4.15}$$

where μ_{s2}^* is the value of μ_s^* at $\phi/(\phi_o - \phi) = 2$. By postulating in this manner, the dependence of μ_m on $(\phi/(\phi_o - \phi)^2$ is accounted for, or

$$\mu_s^* = (4.61/4)(\phi/\phi_o - \phi)^2 \tag{4.16}$$

and

$$\mu_m = \mu_g + (4\ a^2\ \bar{\rho}_p/2.65 \times 10^{-2})\ (\mu_s^*), \tag{4.17}$$

In general, a polynomial expression for μ_s^* is useful to account for higher order collisions:

$$\mu_s^* = a_2(\phi/\phi_o - \phi)^2 + a(\phi/\phi_o - \phi)^3 + a_4(\phi/\phi_o - \phi)^4 \tag{4.18}$$

Fitting between $2 < \phi/(\phi_o - \phi) < 14$ gives, for $\phi_o = 0.52$,

$$a_2 = 1.650;\ a_3 = -0.2709;\ a_4 = 0.01114. \tag{4.19}$$

The non-Newtonian nature of the suspension has been treated in a preliminary way [10]. The result is in agreement with particle trajection obtained from motion pictures of tracer particles.

When applied to the situation where there is no lag of particle motion from fluid motion, Einstein [11] showed that the viscosity μ_m of an incompressible fluid containing solid spheres is given by

$$\mu_m = \bar{\mu}(1 + 2.5\ \phi) \tag{4.20}$$

for small ϕ where ϕ is the proportion of the total volume occupied by the particles. Further details are furnished by Happel [12].

5. HEAT TRANSFER

For the collision of a particle 1 by a cloud of particles 2 in a suspension, the temperature of particle 1 can be assumed to change uniformly. The time constant for transient heat conduction is to the order of α_1/a_1^2 (s^{-1}) where α_1 is the thermal diffusivity of material constituting particle 1 [13]. The energy equation of particle 1 can be written as:

$$d\ T_{p1}/dt = G_{1f}(T - T_{p1}) + G_{12}(T_{p2} - T_{p1}) \tag{5.1}$$

where G_{1f} is the time constant for heat transfer from the fluid to particle 1 [2] and G_{12} is that for heat transfer from particles 2 to particle 1 by collision. Heat transfer by 2 to 1 collision occurs according to the mechanism depicted in Fig. 1 with area and duration by contact given in Section 3. For characteristic thicknesses d_1' and d_2' (of order a_1 and a_2) over which the temperature gradient for heat conduction occur, an energy balance gives

$$\bar{\kappa}_{p1}/d_1'(T_1 - T_c) = \bar{\kappa}_{p2}/d_2'(T_c - T_2) \tag{5.2}$$

where T_c is the temperature at the contacting surface and $\bar{\kappa}_{p1}$, $\bar{\kappa}_{p2}$ are thermal conductivities of materials constituting particles 1 and 2. The time rate of change of temperature of each particle is given by

$$\begin{aligned} c_{p1}\ m_1\ d\ T_{p1}/dt &= -(\bar{\kappa}_{p1}/d_1')\ A_{12}(T_1 - T_c) \\ &= -(\bar{\kappa}_{p2}/d_2')\ A_{12}(T_c - T_2) = -C_{p2}\ m_2\ d\ T_{p2}/dt \end{aligned} \tag{5.3}$$

where c_{p1} and c_{p2} are specific beds of particles of 1 and 2, respectively. By eliminating T_c, Eq. (5.3) can be integrated to give

$$T_{p1} \sim T_{p1i} + (\bar{T}_{p2} - \bar{T}_{p1})[1 - \exp(-\frac{(\bar{K}_{p1}\ \bar{K}_{p2})^{1/2}\ A_{12}t}{m_1\ c_{p1}(d_1'\ d_2')^{1/2}})] \tag{5.4}$$

where $\overline{T}_p$'s are mean particle temperatures over time t. Over time Δt, the duration of contact and since the exponential term in Eq. (5.4) is usually small, the amount of heat transferred is given by:

$$q \sim (\overline{T}_{p2} - \overline{T}_{p1}) \frac{(\overline{\kappa}_{p1}\ \overline{\kappa}_{p2})\ A_{12}\ \Delta t}{(d_1'\ d_2')^{1/2}} \tag{5.5}$$

with A_{12} Δt given by Eq. (3.8). The resultant time constant for heat transfer is thus:

$$G_{12} = h_{12}\ 4\pi\ a_1^2/m_1\ C_{p1}$$

$$\sim [n_{12}\ \Delta u(\overline{\kappa}_{p1}\ \overline{\kappa}_{p2})^{1/2}/m_1\ C_{p1}\ 2(a_1\ a_2)^{1/2}]$$

$$\times \int_0^{\pi/2} A_{12}\ \Delta t(a_1 + a_2)\ d\theta\ 2\pi(a_1 + a_2)\sin\theta$$

$$\sim \frac{n_{12}\ n_{p2}(\overline{\kappa}_{p1}\ \overline{\kappa}_{p2})^{1/2}}{m_1\ C_{p1}\ 2(a_1\ a_2)^{1/2}}\ 2\pi(a_1 + a_2)^2\ A_o\ \Delta t(5/4)\ \Delta u \tag{5.6}$$

or

$$G_{12} = C_{12}(\rho_2/C_{p1})[(\overline{\kappa}_{p1}\ \overline{\kappa}_{p2})^{1/2}/\overline{\rho}_{p1}\ \overline{\rho}_{p2}]\ (1/a_1\ a_2) \tag{5.7}$$

with the collision coefficient C_{12} given by

$$C_{12} = 2.94\ (45/2^{3/5}\ 64)\ n_{12}[(m_1/m_2)^{1/2} + (m_2/m_1)^{1/2}]^{-4/5}$$

$$\cdot\ [(a_1/a_2)^{1/2} + (a_2/a_1)^{1/2}]^{7/5}\ N_{Im}^{4/5} \tag{5.8}$$

and G_{12} and G_{21} are related according to:

$$C_{p1}\ \rho_{p1}\ G_{12} = C_{p2}\ \rho_{p2}\ G_{21} \tag{5.10}$$

Note that in Eq. (5.7) $\overline{\kappa}_{p1}/C_{p1}\ \overline{\rho}_{p1}$ is the thermal diffusivity of the material of particle 1. The quantity $[\overline{\kappa}_{p1}/C_{p1}\ \overline{\rho}_{p}\ a_1\ a_2]\ (\overline{\kappa}_{p2}/\overline{\kappa}_{p1})^{1/2}$ correlates thermal diffusion in particle 1 with the characteristic diffusion length $(a_1\ a_2)^{1/2}$. When $a_1 \gg a_2$, $m_1 \gg m_2$, the above relations can be applied to collision with a wall.

Collision with a Wall To learn something about the order of magnitude of heat transfer by impaction, we now determine the heat transfer coefficient at a flat surface by impaction of solid particles with density ρ_p and intensity of turbulence $\langle u_p^2 \rangle^{1/2}$. We denote the wall material as (1) with constant wall temperature t_w. The incoming particle temperature is denoted as T_{pi}. The heat transfer coefficient h_w is given by

$$h_w = n_{21}\, n_{p2}(\bar{\kappa}_{p2}/a_2)\, \langle u_{p2}^2 \rangle^{1/2} \cos\theta\, A_o\, \Delta t(\cos\theta)^{-1/3}$$

$$= (2.94)([3/4)(2)^{-8/5}]\, \rho_{p2}(\bar{\kappa}_{p2}/\bar{\rho}_{p2}/a_2)\, (N_{Im})^{4/5}\, (\cos\theta)^{4/5} \qquad (5.11)$$

and $n_{21} = 1$,

$$N_{Im} = 5\pi^2 \langle u_{p2}^2 \rangle\, \bar{\rho}_{p2}[1 + (r^*/2)]\, (k_2 + k_w). \qquad (5.12)$$

It is further noted that since the temperature of the particle phase is that of the body temperature of the particles, heat conduction by surface contact must be accounted for. For elastic collision of particles with a wall, it is readily shown that an accommodation coefficient, α, is given by:

$$\alpha \sim \frac{3(2.94)}{8(2)^{3/5}} \left[\frac{(\bar{\kappa}_p\, \bar{\kappa}_w)^{1/2}}{c_p\, \bar{\rho}_p\, a(\Delta u_w)}\right] (N_{Im})^{4/5} \qquad (5.13)$$

The temperatures of an incoming particle T_{pi}, and a reflected particle T_{pr} are related by

$$(T_{pr} - T_{pi})/(T_w - T_{pi}) = \alpha \qquad (5.14)$$

Since $T_{pw} \sim (T_{pi} + T_{pr})/2$, the heat flux J_{qp}, due to impaction of particles at the wall is given by

$$J_{qp}/c_p\, \dot{m}_{pw} = (T_{pr} - T_{pi}) = (-2\alpha/2 - \alpha)(T_{pw} - T_w) \qquad (5.15)$$

where $\dot{m}_{pw}$ is the mass flux of impact at the wall. Equation (5.15) accounts for both the effect of conduction by surface contact at the wall and the effect of convection heat transfer with the fluid.

Measurement of heat transfer coefficient between a moving surface in a bed of slag particles was made by Dunsky, Zabrodsky, and Tamarin [14]. For relative velocities from 0.01 to 0.1 m/s, their system appears to conform to the above relations when extended to the multiple scattering model with local volume fraction solids from 0.4 to 0.1 and an "accommodation" coefficient of 0.8 between the particles and the wall in laminar flow [15]. Below 0.01 m/s, the effect of conductivity of the porous bed sliding past the surface appears to be more significant. Conceivably their experimental system may be used to check the single scattering model of particle-wall heat transfer at still higher relative velocities. Experimental data on thermal conductivity of particle clouds due to individual interactions are meager. The average thermal conductivity was correlated by Gelperin and Einstein [16] and Gorring and Churchill [17]. Refinement is needed.

Comparison of Thermal and Momentum Response The relative importance of momentum and thermal interaction is given by the ratio G_{12}/F_{12} of

$$G_{12}/F_{12} = [(2.94)(15)/32\ \pi 2^{3/5}][(m_1/m_2)^{1/2} + (m_2/m_1)^{1/2}]^{1/5}\ [(a_1/a_2)^{1/2} + (a_2/a_1)^{1/2}] \cdot \frac{[N_{Im}]^{3/5}[(\bar{\kappa}_{p1}\ \bar{\kappa}_{p2})/[(\bar{\rho}_{p1}\ \bar{\rho}_{p2}\ C_{p1}\ C_{p2})]}{\Delta u_{p2}(a_1\ a_2)^{1/2}} \tag{5.21}$$

which shows that $G_{12} \ll F_{12}$ for most granular materials.

6. EROSION AND ATTRITION

In general, erosion and attrition occur on contacting surfaces of piping components and granular material in various degrees depending on their properties and the dynamics of impact. The properties include hardness, yield strength, and surface roughness of each. The dynamics of impact are related to the kinetic energy of the granule, angle of impact, and friction coefficient of the surfaces. Further, attrition is usually a short term concern of the condition of the granules after handling while erosion is a long term concern relating to the maintenance and replacement costs of equipment. Erosion is a major concern of, for instance, a coal handling system. Attrition, however, is a major concern in grain handling.

The above treatment of momentum and heat transfer between particulate phases and particulate phase with a wall is readily extended to treating erosion of a wall by collision of particles and breakage of solid particles by collision with a solid surface.

The energies to remove a given volume of material are denoted as [21]: ε_d, related to ultimate yield stress for ductile wear σ_d and ε_b, related to ultimate impact resistance for brittle wear σ_b.

To relate the mechanisms of erosion to basic material properties, we need to identify the normal compressive force P_b and the tangential machining force P_d, area of contact A, the duration of contact t, and the ratio r* of the reflection speed to incoming speed V of a particle of mass m colliding with a surface at an angle of incidence θ with the plane of the surface. When the deviation from the elastic state of the surfaces of materials is small, the formulation in the above sections for contacting spheres consisting of spherical particle of radius a and a plane of infinite radius gives:

$$P_b t = (2.94)(5/16)\ mV(1 + r^*) \tag{6.1}$$

$$P_d t = f\ P_b t \tag{6.2}$$

$$At = A_o\ t_o (\sin\theta)^{-1/5} \tag{6.3}$$

and

$$A_o\ t_o = (2.94)(15\ \pi^2/8)\ 2^{-2/5}\ mV(k_1\ k_2)^{1/2}\ 1 + r^*)^{4/5}\ N_{Im}'^{1/5} \tag{6.4}$$

where $0 \leq r^* \leq 1$, $r^* \to 1$ as $\theta \to 0$, and f is the coefficient of friction; subscripts 1 and 2 denote each of the impacting pair, the dust particle and the plane surface; N_{Im}' is the impact number now redefined as:

$$N_{Im}' = (5\ \pi^2/2)\ \bar{\rho}_1\ V^2 (k_1\ k_2)^{1/2}\ [(k_1/k_2)^{1/2} + (k_2/k_1)^{1/2}]^4 \tag{6.5}$$

$\bar{\rho}_p$ is the density of the material forming the dust particle. These basic relations are applied to the two different modes of erosion [21]: the "ductile" mode (subscript d) and the "brittle" mode (subscript b). Erosion via tangential machining tends to be more prominent for ductile materials (metals) than for brittle materials (ceramics). While the normal compressive force has a reverse influence, erosion of a real material is a result of the combined effect of these modes. Again, for small deviation from the elastic states, a linear combination is seen to be valid.

Ductile Mode Experimental results show that the ductile mode which is typical of metal targets is characterized by a maximum erosion occurring at some intermediate incidence angle between 0 and 90 degrees, usually at 20 to 30 degrees [22]. The mechanism appears to be one of cutting or machining sustained by force P_d of Eq. (6.2). The mechanism is produced by a "friction" force and erosion occurs only when this force exceeds the yield strength σ_d of the surface being eroded. The erosion in volume loss w_d per impact is given by

$$w_d = \eta_d [C_d (P_d/A) - \sigma_d]\ V \cos\theta\ At/\varepsilon_d \tag{6.6}$$

where C_d corrects the non-sphericity of the actual particles, stress concentration at sharp edges, roughness of the surface impacted, and non-specular reflection from an actual surface $C_d \gtrsim 1$, usually for non-spherical particles. η_d is a mechanical efficiency of impact, including the effects of gliding, scattering, lifting of the particles by the gas stream; $\eta_d \ll 1$ usually. In terms of a volumetric erosion rate per unit area W_d produced by a dust stream of concentration ρ_1, we can express this rate as a dimensionless erosion energy parameter:

$$E_d^* \equiv W_d \, \varepsilon_d/\rho_1 \, V^3 \, C_d \quad f(1 + r^*)(2.94)(5/16) \, \eta_d$$
$$= \cos\theta[1 - K_d^*(\sin\theta)^{-1/5} \equiv F_d(\theta, K_d^*) \qquad (6.7)$$

and the resistance parameter K_d^* is given by

$$K_d^* = 6\pi^2 \, \sigma_d(k_1 \, k_2)^{1/2}/2^{2/5} \, f \, C_d(1 + r^*)^{1/5} \, (N'_{Im})^{1/5} \qquad (6.8)$$

Since $(1 + r^*)^{1/5}$ is nearly 1 in general, K_d^* can be treated as a parametric constant and the function $F_d(\theta, K_d^*)$ characterizes the ductile wear at various angles θ. The maximum value of function F_d or $F_{D/m}$ occurs at an angle θ_m given by:

$$K_d^* = 5(\sin\theta_m)^{11/5} \, (4 \sin^2\theta_m + 1)^{-1} \qquad (6.9)$$

It is seen that even with the uncertainty of the parameters C_d, σ_d, f and ε_d, the above relation providesfor the correlation of experimental results or to determine the design condition where erosion rate can be controlled. It is further noted that f is, in general, related to θ and r^*.

Brittle Mode The brittle mode which is typical of glass and ceramics is characterized by an erosion rate increasing with impingement angle up to, but below, 90 degrees [29]. The mechanism is one of cracking of the target surface by fatigue and brittle failure for stress above an impact yield stress σ_b [27]. The erosion in terms of volume loss per impact w_b is given by

$$w_b = \eta_b[C_b(P_b/A) - \sigma_b] \, V \sin\theta \, At/\varepsilon_b \qquad (6.10)$$

where C_b has an analogous meaning as C_d given before, although it may differ in magnitude even for similar particles and surfaces. The volume erosion rate per unit area W_b produced by a dust stream of concentration ρ_1 can be expressed in the form of a dimensionless erosion energy parameter E_b^*:

$$E_b^* \equiv W_b \, \varepsilon_b/\rho_1 \, V^3 \, C_b(1 + r^*)(2.94)(5/16) \, \eta_b$$
$$= \sin\theta[1 - K_b^*(\sin\theta)^{-1/5} \equiv F_b(\theta, K_b^*) \qquad (6.11)$$

where K_b^* has the same form as K_d^* with quantities of subscript b substituted by subscript d.

Equation (6.10) shows that maximum wear occurs at θ = 90 degrees. Experimentally, at θ = 90 degrees, interference of rebound particles with incoming particles is problemmatic (close to the condition of r^* = 0 plus sliding) and maximum wear rate is more demonstrable at θ < 90degrees [22]. The combined effect of ductile and brittle wear is verified experimentally [23,24] although uncertainty exists in the basic parameters of material properties. An example is given in Fig. 3 [23].

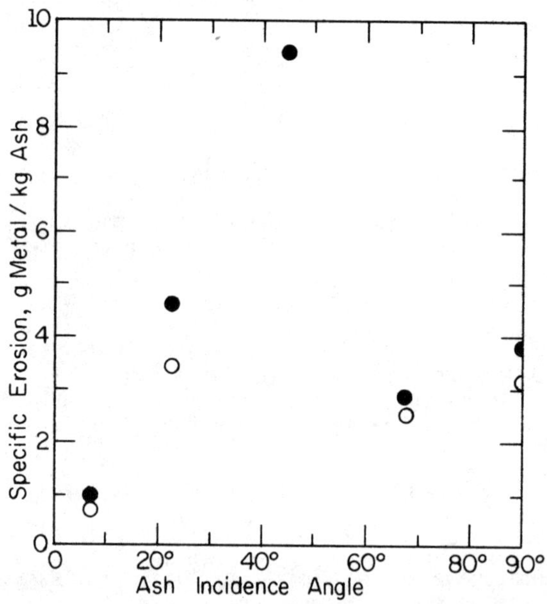

Figure 3 Specific erosion ($W \bar{\rho}_2/\rho_1$ V) versus ash incidence angle

(Stagnation temperature--------------------760 C
Gas velocity carrying particles-----------285 m/s
Specimen material-------------------------S-816
● From estimated dust loading-calculated
o From measured dust loading of 25 g/hr) [23]
$K_D^* \simeq 0.80$

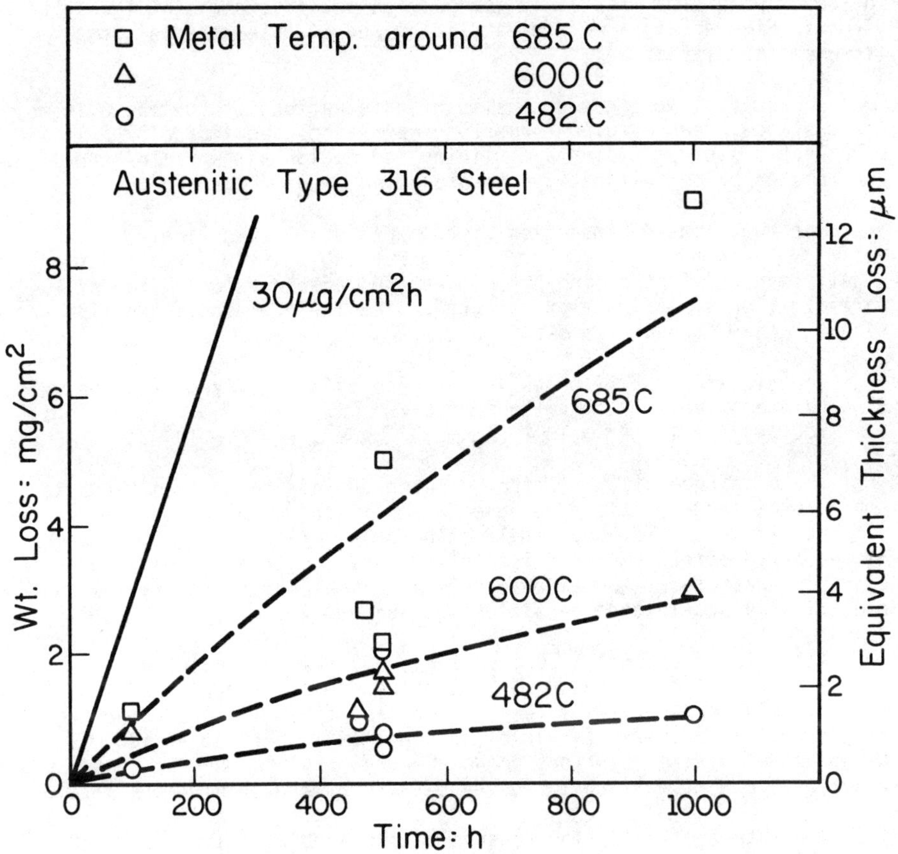

Figure 4 Weight loss of metallic surfaces embedded in fluidized bed combustor [25]

Similar considerations have been extended to treating wear by random motion of particles in a fluidized bed. A set of experimental results is given in Fig. 4 [25]. Wear by a sliding bed of bulk materials can be likewise formulated.

In reality, most cases consist of combinations of the above items. For instance, wear at pipe turns in general includes impact over a range of directions of particles and sliding bed motion as particles are accumulated by centrifugal forces along one side of the pipe.

For the cases of impaction of tempered steel by silica for $\varepsilon_d \simeq 10^9$ n/m^2, $\varepsilon_b \simeq 10^{10}$ n/m^2, $f \simeq 0.1$, $\eta_d C_d \simeq 10^{-3}$ showing the significance of lift action by the gas [20] or the velocity lag of particles, or small fraction impacted. The angle of incidence also spread over a range for each jet setting.

The situation still holds in the case of a fluidized bed--the wear of 316 stainless steel by dolomite particles (700 μm) for $1 - K_d^* \simeq 10^{-2}$, $\varepsilon_d \simeq 10^{10}$ n/m^2, $\eta_d C_d \simeq 10^{-4}$.

The magnitude of C_d is often greater or near to one and tends to be small for a large particle because of small contact area for a given mass of particle; it tends to be small also for a very small particle because of the large effect of air friction. η is, in general, small for gas-conveyed particles. A desirable reference test condition will be that of electrically accelerated particles in a vacuum.

Attrition of material by impact has not been extensively documented. Mills and Mason [34] reported degradation of quartz particles in various pipe materials (steel, nylon, and fiberglass) and the degree of fragmentation of quartz particles on steel of 10 percent at 100 m/s. Although definitive relations cannot be drawn in the general sense, their results show what designers should be aware of and should look into.

The extent of attrition depends on the number of impacts which a particle undergoes with the wall as it proceeds through the pipe system and the compression force acting on a particle during each impact. Hence, the major cause of attrition is the impact at pipe bends. The pressure at the point of contact during a collision may crush the particle locally. The maximum local compressive stress at the point of impact can be derived from the above relation:

$$P = K_A E_p (\bar{\rho}_p U_p^2 \sin\theta / E_p)^{1/5} (1 - \nu_p)^{-4/5} \qquad (6.12)$$

for the case where the modulus of elasticity of the wall E_w is much greater than that of the particle material E_p and ν_p is the Poisson ratio and θ is the angle of impact between the velocity vector and the wall. The above proportionality constant K_A is around 0.17 for a sphere and depends on details of shape, air resistance, etc. The amount of dust generated is seen to be proportional to the difference between P and the fracture strength of the particle material times the number of particles colliding with a surface per unit time. For particles in micrometer size range, the impact velocity is greatly reduced in the fluid boundary layer.

The reverse of attrition is agglomeration [13]. Cases include collection of dust by using water droplets [27].

Relative motion of particles of different or similar sizes can be produced acoustically. The collection efficiency varies with frequency and size of particles. Particle diffusivity plays an important part here, besides Brownian motion of small particles. Recent experiments by Scott [28] showed that relatively high efficiency can be achieved by using pulses of 250 to 400 Hz to agglomerate 1 μm ZnO_2 particles to 8 μm in a gaseous suspension of mass ratio of 10^{-3} to 10^{-2}. The power consumption was nearly 8 kW per m^3/s flow (or 5 W/cfm). This, however, is higher than the power consumption of an electrostatic precipitator. Study by Shaw [29] using longitudinal waves showed useful agglomeration characteristics at 3 to 10 kHz and 145 db of sound intensity. These interactions only concern dust generated from handling of granules.

7. CHARGE TRANSFER

When applied to charge transfer by surface contact and separation of two spheres, a direct analogy is seen to the heat transfer mechanism in Section 5.

The current density through the contract area A_{21} is given by

$$J_{21} = h_{21}(V_2 - V_1) = (\sigma_1/d_1')(V_c - V_1) = (\sigma_2/d_2')(V_2 - V_c) \qquad (7.1)$$

where h_{21} is the charge transfer coefficient through the contact surfaces, V_c is the potential at the contact area, d_1' and d_2' are not known in detail but are expected to be each of the order of a_1, a_2, respectively. The time rate of change of potential for each particle is

$$C_1(dV_1/dt) = -C_2(dV_2/dt) = J_{21} A_{21} = A_{21} h_{21}(V_2 - V_1) \qquad (7.2)$$

after eliminating V_c, where C_1 and C_2 are the capacitances of the particles, respectively, $C = 4\pi \varepsilon a$, ε is the permittivity, and

$$h_{21} = [\frac{\sigma_1}{d_1'}\frac{\sigma_2}{d_2'}]^{1/2} [(\frac{\sigma_1}{\sigma_1}\frac{d_2'}{d_2'})^{1/2} + (\frac{\sigma_1}{\sigma_2}\frac{d_2'}{d_1'})^{1/2}]^{-1} \qquad (7.3)$$

For free particles, having charges q_1 and q_2, and work functions ϕ_1 and ϕ_2

$$V_2 - V_1 = (\phi_2 + \frac{q_2}{4\pi \varepsilon_2 a_2}) - (\phi_1 + \frac{q_1}{4\pi \varepsilon_1 a_1}) \qquad (7.4)$$

For this case, integration of Eq. (5.4) for variables q's gives the net charge transfer per impact:

$$Q_{21} = \frac{C_1 \; C_2}{C_1 + C_2} (V_2 - V_1)(1 - e^{-\alpha \Delta t}) = -Q_{12} \sim h_{21} \; A_{21} \; \Delta t (V_2 - V_1) \quad (7.5)$$

where

$$\alpha = A_{21} \; h_{21} (C_1 + C_2)/C_1 \; C_2. \quad (7.6)$$

The charge transfer coefficient h_{21} given by Eq. (7.3) is based on clean surfaces. In reality, both h_{21} and ϕ's are influenced by adsorbed surface layer and humidity. In the case of mineral particles, the conductivity in logarithm scale increases almost linearly with an increase in relative humidity [6,7]. For collision of 16 μm coal dust on a 12.7 mm steel sphere at 18.3 to 36.6 m/s h_{21} ranges from 2 to 5 x 10^{-8} mho/cm^2, which is much lower than that based on material properties in Eq. (7.3). This magnitude was measured on coal dust having a saturation charge to mass ratio of 10^{-3} coul/kg in the flow system according to

$$\phi + q/4\pi \; \varepsilon a = \text{constant} \quad (7.7)$$

by collision with a brass surface (pipe). The average charge transfer per impact versus the impact number is given by Q = 1.5 x 10^{-18} N_{Im}. The charge transfer process involves an inverse relaxation time H_{21}, with

$$dV_2/dt = -H_{21}(V_2 - V_1) \quad (7.8)$$

$$H_{21} = K_{21}(a_2^2/\varepsilon_2)(\rho_1/\bar{\rho}_1)[h_{21}/(a_2 \; a_1)^{1/2}] \quad (7.9)$$

where K_{21} is a dimensionless constant given by

$$K_{21} = \frac{15}{32} (0.97) \; \eta_{21} [(m_2/m_1)^{1/2} + (m_1/m_2)^{1/2}]^{-4/5}$$

$$\cdot \; [(a_2/a_1)^{1/2} + (a_1/a_2)^{1/2}]^{7/5} \; [(N_{Im})_{21}]^{4/5} \quad (7.10)$$

Note that here η_{21} is affected by the electric field of interacting particles and may exceed 1 via mutual attraction [2]. Further, for charge transfer between two clouds of particles of species 1 and 2, we have the relation

$$C_2 \; n_2 \; H_{21} = C_1 \; n_1 \; H_{12} \quad (7.11)$$

based conservation of charges. It is seen that the asymptotic relation of Eq. (7.7) is for each species of particles in a cloud and for a cloud of particles of similar materials to acquire the condition of:

$$q/a = \text{constant} \quad (7.12)$$

by mutual collision. The above results in charge transfer is also the basis of electrostatic probes for a particle suspension. This phenomenon influences atmospheric sampling of aerosols and ice crystals.

We consider the case of sphere 2 impacted by a cloud of spheres 1 (see Section5.1). From Eq. (7.2), the change of potential of sphere 2 is given by [30]

$$C_2\,(dV_2/dt) = \eta_{21}\; n_1\; \Delta u_{21} \int_0^{\pi/2} 2\pi\, Q_{21}(a_1 + a_2)^2 \sin\theta\, d\theta, \tag{7.13}$$

where η_{21} is the fraction impacted, defined as the ratio of the cross-sectional area of stream of particles 1 to the projected area of particle 2 normal to the direction of flow of particles 1 and n_1 is the number density of particles 1. Since $C = dQ/dV$, the total electric current due to successive charge transfer is

$$i_2 = (5/4)\; 2\pi(a_2 + a_1)^2\; \eta_{21}\; n_1\; \Delta u_{21}(V_2 - V_1)\; h_{21}(A_{21}\; \Delta t)_0 \tag{7.14}$$

with mean value of $(V_2 - V_1)$ over Δt.

The relation given by Eq. (7.14) gives rise to an electrostatic probe theory for a suspension. When an impact probe is inserted into a flow stream of particles, charge transfer ocurs by impact. For probe dimensions smaller than the mean free path of particles, single scattering consideration would be adequate. The potential of each particle is given by

$$V_1 = \phi_1 + q_1/4\pi\, \varepsilon_1\, a \tag{7.15}$$

for adjustable probe voltage V_2.

For a stream of particles (species 1) impacting on a stationary probe in the form of a metal sphere, the current reading from its lead wire is given by Eq. (7.14). For $a_2 \gg a_1$:

$$\begin{aligned} i_2 &= \eta_{21}\; A\; \rho_1\; u_1\; Q_{21}/m_2 \\ &= (129.6)^1\, \eta_{21}^1\; h_{21}^{21}\; a_2^2(\rho_1/\bar{\rho}_1)[(\bar{\rho}_1\; u_1^2)(k_1 + k_2)(1+r^*/2)]^{4/5}(V_2 - V_1) \end{aligned} \tag{7.16}$$

where A is the projected area of the probe and Q_{21} is the charge transferred per impact. Equation (7.16) has been validated experimentally [30,31]. Extension to charge transfer in turbulent pipeflow of a suspension has been given; the results show a significant bearing on electrostatic safety in pneumatic transport [32].

8. CONCLUDING REMARKS

The above survey pertains to granular materials of, say, below 5 mm in size. In the course of handling, especially by pneumatic means, attrition of granules and erosion of conduits are often the design concerns. They are related to momentum and material transfers. Heat transfer is not effective unless high concentration of particles exists and depending on its application, granular insulators have been used. While agglomeration and hence deposition and electrostatic charging might not be important among the granules themselves, they affect significantly the influence of dust generated by handling and are major factors influencing safety from explosion and fire.

REFERENCES

1. S. L. Soo, Physics of Fluids 19, p. 757 (1976).

2. S. L. Soo, Fluid Dynamics of Multphase Systems, Blaisdell, 1967, Chaps. 5 and 10.

3. I. Langmuir and K. Blodgett "Mathematical Investigation of Water Droplet Trajectories," General Electric Research Lab. Report RL-225, 1944-5; J. Meteorology 5, 175 (1948).

4. S. L. Soo, Powder Technology, 7 267 (1973).

5. S. K. Friedlander, ARS J. 29 296 (1959).

6. S. Timoshenko, "Theory of Elasticity," McGraw-Hill, New York (1934) p. 339.

7. C. H. Wang, H. Arastoopour, and S. A. Weil, "Particle-Particle Interaction Force in a Dilute Gas-Solid System," Chem. Eng. Sci., 1982.

8. K. Schugel, Fluidization (eds., J. F. Davidson and D. Harrison), Academic Press, New York (1971), 261-292.

9. T. R. Blake, S. K. Garg, H. B. Levine, and J. W. Pritchett, "Computer Modeling of Coal Gasification Reaction," FE-1770-15, Annual Progress Report U.S. DOE Contract No. E(49-18)-1770, July 1976.

10. J. D. Gabor, Chem. Eng. J., 4 (1972) 118.

11. A. Einstein, Ann. Phys. Leipzig 19 289 (1960).

12. J. Happel, J. Appl. Phys. 28 (11) 1288 (1957).

13. S. L. Soo, "Flows of Suspension," von Karman Inst. for Fluid Dynamics, Lecture Series 51 (1973).

14. V. D. Dunsky, S. S. Zabrodsky, and A. I. Tamarin, Proc. 3rd Int. Heat Transfer Conf. 4 293 (1966).

15. E. R. G. Eckert and R. M. Drake, Jr., "Heat and Mass Transfer," McGraw-Hill, New York (1959).

16. N. I. Gelperin and V. G. Einstein, Fluidiztaion (eds., J. F. Davidson and D. Harison) Academic Press, New York (1971) 471-540.

17. R. L. Gorring and S. W. Churchill, Chem. Eng. Progress 57 (7) 53 (1961).

18. P. Michael, Physics of Fluids 8 1263 (1968).

19. I. Finnie, J. W. Dlak, and Y. H. Kabil, ASTM J. Materials 2 (1967) 682.

20. G. I. Sheldon, ASME Paper No. 89-WA/Met-7 (1969).

21. J. H. Neilson and A. G. Gilchrist, Wear 11 (1968) 111.

22. C. E. Smeltzer, M. E. Gulden, and W. A. Compton, Trans. ASME, J. Basic Eng. 92D (1970) 639.

23. W. M. Goldberger and H. Nack, Battelle Tech. Rev. 13 (11) (1964).

24. A. P. Fraas, ASME Paper No. 75-GT-123 (1975).

25. National Research Development Corp. (London) Report No. 85, Interim No. 1 (1971) NTIS PB-210673.

26. S. L. Soo and S. K. Tung, Powder Tech. 6 (1972) 283.

27. L. Cheng, I/EC Process Design and Rev. 12 (1973) 221.

28. D. S. Scott, J. Sound and Vibr. 43 (1975) 607.

29. D. T. Shaw, Recent Development in Aerosol Sciences, John Wiley (1979) Ch. 13.

30. L. Cheng and S. L. Soo, J. Appl. Phys. 41 (2) (1970) 585.

31. L. Cheng, S. K. Tung, and S. L. Soo, Trans. ASME, J. Eng. for Power 92 (A-2) (1970) 135.

32. S. L. Soo, J. Pipelines 1 (1) (1981) 57.

DEBRIS FLOW AND DEBRIS FLOW DEPOSITION

T. Takahashi

Disaster Prevention Research Institute, Kyoto University,
Kyoto, Japan

1. INTRODUCTION

Debris flow is a rapid flowage of debris with sufficient water to disperse grains throughout the whole depth. When a severe rainstorm attacks it often occurs on a steep mountain ravine carrying away all the accumulated debris along its course as deep as to the bedrock. It sometimes flows several kilometers and suddenly halts forming, on one occasion, a so-called debris cone and on another occasion a lobe. The front of the flow proceeds, taking the form of a positive step-like bore. It is in this part that huge stones are jostling with each other and which has an enormous destructive power incurring a great hazard.

In order to understand the phenomena and to develop some effective countermeasures, the writer has carried some fundamental studies on the mechanics of debris flow. This article is a rough review of some findings through his investigations.

2. CRITERIA FOR OCCURRENCE OF VARIOUS TYPES OF SEDIMENT TRANSPORT

When it rains, some of the rainfall will run off through as well as over the surface of the debris bed of a gully. Imagine a thick uniform layer of loosely packed non-cohesive grains, whose slope angle is θ . It is assumed that below the surface of the water flow, the pore spaces among grains are saturated and a parallel seepage flow without any excessive pore pressure occurs. The characteristic distribution of shear stress in the bed should be like one of those shown in Fig.1, in which τ is the applied tangential stress and τ_L the internal resistive stress, and these values at depth a are written for respective cases as:

Case 1:

$$\tau = g \sin\theta \{c_* \sigma a + (1-c_*)\rho(a-D+h)\} \tag{1}$$

$$\tau_L = g \cos\theta \{c_* \sigma (D-h) + c_*(\sigma-\rho)(a-D+h)\} \tan\phi \tag{2}$$

Case 2:

$$\tau = g \sin\theta \{c_*(\sigma-\rho)a + \rho(a+h) + c_b(\sigma-\rho)h_b\} \tag{3}$$

$$\tau_L = g \cos\theta \{c_*(\sigma-\rho)a + c_b(\sigma-\rho)h_b\} \tan\phi \tag{4}$$

in which σ and ρ are the densities of grains and fluid respectively; g the acceleration due to gravity; c_* the grain concentration in volume in the static bed; D the thickness of the debris bed;

c_b the concentration of bed load transport; h_b the thickness of the bed load transport layer; and ϕ is the internal friction angle.

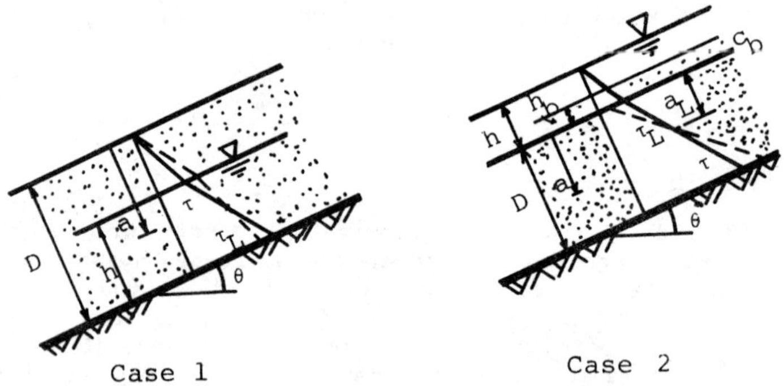

Fig.1 Characteristic shear-stress distributions.

Case 1 occurs when the bed slope satisfies

$$\tan\theta \geqq \frac{c_*(\sigma-\rho)}{c_*(\sigma-\rho)+\rho}\tan\phi \tag{5}$$

In this case the debris bed will slide as the level of seepage flow rises, but it will not turn into debris flow because the quantity of water will be insufficient.

For a slope flatter than that of Case 1, the debris bed will not slide before the seepage flow reaches the surface of the bed. With an increment of the runoff, however, the depth of the stream on the surface will increase. The result is that the tractive force of the flow will exceed the critical one and bed load transport will begin. By further increment of the surface flow the acting shearing stress due to the static weight of the surface flow, seepage water and bed load will exceed the resistance in the debris layer near the surface as shown in Case 2 in Fig.1. The thickness of this layer is shown as a_L in the figure. Accordingly, if the value of a_L is larger than the representative diameter of the debris bed, d, this layer will come loose and start moving as a so-called sediment gravity flow. Because there is enough water to fill the increased pore spaces, the flow is able to continue as long as the stream channel has an adequate slope.

The condition for the occurrence of sediment gravity flow, i.e., $a_L \geqq d$, is written from Eqs.3 and 4 as:

$$\frac{h}{d} \geqq \frac{c_*}{c_{d\infty}} - 1 + \frac{c_b}{c_{d\infty}}\frac{h_b}{d} \tag{6}$$

in which

$$c_{d\infty} = \frac{\rho\tan\theta}{(\sigma-\rho)(\tan\phi - \tan\theta)} \tag{7}$$

Although a sediment gravity flow will occur when a_L is larger than d, if a_L is considerably less than h, the particles in motion can not uniformly disperse throughout the depth, so the sediment flow is covered by a less densely concentrated water flow. These features of flow resemble the phenomenon of very active bed load transport rather than the debris flow. Therefore, if we define the debris flow as a kind of sediment gravity flow in which the grains are dispersed throughout its depth, the additional condition that $a_L \geq \kappa' h$ (κ'=0.7~1) is necessary for its occurrence. This latter condition can be rewritten by using Eqs.3 and 4, and the fact that $c_{d\infty} h \gg c_b h_b$, as

$$\tan\theta \geq \frac{c_*(\sigma-\rho)}{c_*(\sigma-\rho)+\rho(1+\kappa'^{-1})} \tan\phi \tag{8}$$

Eqs.5, 6, 8 and the equation for the critical tractive force [1] and the condition for the existence of a stable debris bed

$$\theta \leq \phi \tag{9}$$

define the criteria for the occurrence of various types of sediment movement as shown in Fig.2. For the calculation of Eq.6 in Fig.2, a suitable bed load equation is used[2].

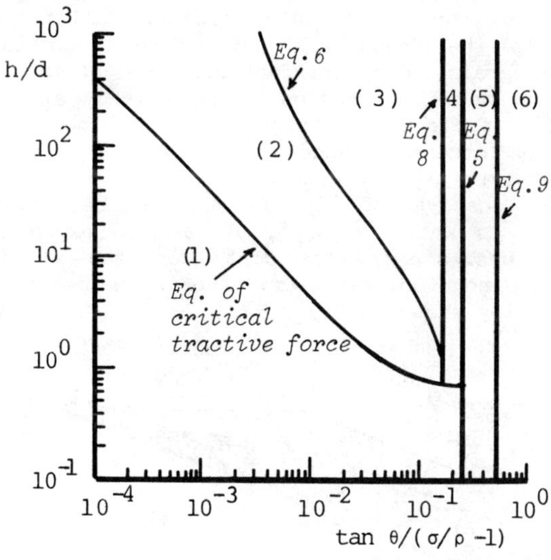

Fig.2 Criteria for the occurrence of various types of sediment transportation. The curves are obtained under the condition that c_*=0.7, σ=2.65g/cm³, ρ=1.0g/cm³, κ'=0.7, and tan φ=0.8.

The domain labeled 1 is that of no particle movement; 2 is the domain of individual particle movement due to dynamic force of the fluid flow, i.e., bed load transport; 3 is the domain of sediment gravity flow, in which the effect of dynamic force of fluid flow coexists and the flow is covered by a less densely concentrated water flow; 4 is the domain of debris flow in which the grains are dispersed in the whole layer; 5 is the domain of occurrence of both landslide and debris flow; and in 6 the sediment bed is unstable under no fluid flow.

Extension of the theory to a cohesive sediment bed has been also done [3,4].

The discussion herein mainly is concerned with in domain 4 and partly in domain 3.

3. MECHANICS OF QUASI-STEADY DEBRIS FLOW

3.1 Process of Formation

As soon as surface flow appears at the upstream end of the bed, whose slope angle is in domain 4 in Fig.2, debris above the depth a_L will begin to move and these grains mingled with water will form a bore. In the next instant, the bore front will tumble over the surface of the downstream bed. Then if the concentration of the bore front is thinner than the full freighting value, the bed beneath the front will be scoured, and the front will swell and its grain concentration increase. Thus, the front of a bore will grow until it reaches a quasi-steady state. The equilibrium or fully freighting concentration is obtained through the consideration of stress balance between the bed and moving layer with the result being $c_{d\infty}$ as shown in Eq.7.

After the passage of the front, the bed continues to be scoured as long as water is supplied, and consequently the length of asymptotic quasi-steady flow part elongates as it flows down. Therefore, the longitudinal profile of the debris flow should be as schematically shown in Fig.3[5].

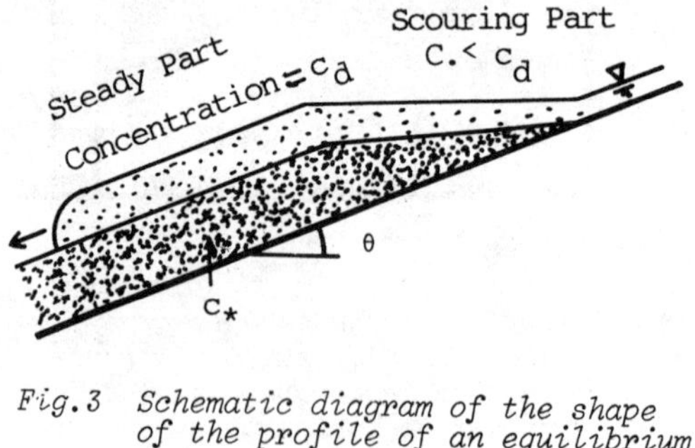

Fig.3 Schematic diagram of the shape of the profile of an equilibrium debris flow.

3.2 Velocity Distribution and the Front Translation

A flow of very high grain concentration necessarily induces frequent collision of particles. Consequently, there exists momentum transfer between grains of the two neighboring shear layers. This causes the dispersive stress that was first introduced by Bagnold[6]. Neglecting the shear due to distorsion of the inter-granular fluid, the equilibrium of shearing stress on an open channel in the fully inertial range (this condition should be easily met in a real debris flow) would be, according to Bagnold,

$$a_i \sigma \{ (\frac{c_*}{c_d})^{1/3} - 1 \}^{-2} d^2 (\frac{du}{dy})^2 \sin\alpha = \{ c_d (\sigma - \rho) + \rho \} g (h-y) \sin\theta \quad (10)$$

in which c_d is the volume concentration in debris flow; (du/dy) is the vertical velocity gradient; a_i is the experimental coefficient; α is the dynamic internal angle of friction.

Integration of Eq.10 gives

$$u = \frac{2}{3d} [\frac{g \sin\theta}{a_i \sin\alpha} \{ c_d + (1-c_d)\frac{\rho}{\sigma} \}]^{1/2} \{ (\frac{c_*}{c_d})^{1/3} - 1 \} \{ h^{3/2} - (h-y)^{3/2} \} \quad (11)$$

The cross-sectional mean velocity is

$$U = \frac{2}{5d} [\frac{g \sin\theta}{a_i \sin\alpha} \{ c_d + (1-c_d)\frac{\rho}{\sigma} \}]^{1/2} \{ (\frac{c_*}{c_d})^{1/3} - 1 \} h^{3/2} \quad (12)$$

Since no bed erosion occurs beneath the front part, there exists no overrun, so that U is equal to the translation velocity of the front.

Fig.4a shows an experimental particle velocity distribution on a rigid bed. In this case a_i value is approximately the same as that of Bagnold's experiments. While, as shown in Fig.4b, very large a_i values should be sometimes adopted to explain the velocities on the sediment beds. Although this may be mainly attributable to the effect of infiltration, more precise discussion would be necessary to explain the characteristics of velocity distribution, particularly on the existence of a secondary creeping layer beneath the virtual bed surface, which was neglected in the above discussion. Other remaining important problems should be as follows: As it is well known the velocity distribution in clay slurry follows Bingham fluid velocity equation, whereas the debris flow follows the Bagnold's dilatant fluid model. Which diameter will be the transition from either fluid model? In case of mixtures of very wide range of particle sizes, what role do the very fine particles play in the mechanics of flow?

3.3 Mechanics of Accumulation of Boulders at Front of Debris Flow

As soon as debris flow occurs on a well-graded sediment,

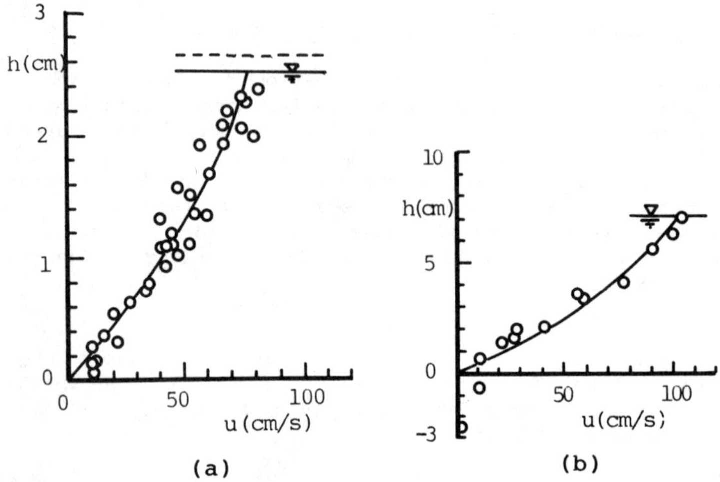

Fig.4 Experimental and theoretical velocity distributions.
(a) Velocity profile on a rigid bed. The broken line is the experimental water surface. The theoretical velocity distribution(full line) is obtained by a_i*=0.041.*
(b) Velocity profile on a sediment bed. The theoretical curve is obtained by a_i*=0.35.*

larger particles move upwards. Velocities in upper layers are faster than lower layers, and the larger particles are moved ahead as though they were on top of the tread of a caterpillar tractor. The particle reaching the front of the flow tumbles down to the bottom of the flow and is buried in the flow. But soon after that, if the buried particle is larger than the surrounding particles, it appears again on top of the flow and moves ahead. The biggest boulders accumulate at the front by the repetition of such processes along the distance of travel. As it is understood from the preceding description of the process of accumulation of boulders at the front, the mechanics of vertical segregation of grains are essential.

Bagnold's dispersive pressure concept claims that the resultant dispersive force acting on a particle is propotional to the difference between the square of velocity gradient, $(du/dy)^2$, at the top and the bottom of that particle. As it is clear from Eq.10, in the steady uniform flow, $(du/dy)^2$ value at the upper layer is smaller than that at the lower layer, and the resulting uplift force acting on a particle whose diameter is the same as the surrounding particles should be in equilibrium with the submerged weight of that particle. However, the uplift force acting on a larger particle than the surrounding ones exceeds the submerged weight of that particle; thence, the larger particle moves upwards, and vice-versa for the smaller particle.

Mathematical analysis of this process shows the quantitative characteristics of the upward or downward velocity of a particular particle in the uniform surrounding particles as in Fig.5. In this figure the ordinate K indicates the non-dimensional velocity

$$K = \frac{v^2}{gd_m} \frac{3}{4} C_D \frac{\{(\sigma-\rho)c_d+\rho\}}{(\sigma-\rho)\cos\theta}$$

and the abscissa r is the ratio of diameter, $r=(d_1/d_m)^3$, in which d_1 is the particular particle's diameter whose motion is being observed and d_m is that of surrounding ones, and C_D is the drag coefficient for the motion of d_1 particle. The curves in the figure satisfy the following relationship

$$K = r^{1/3}\{\frac{2}{1+r} \frac{(0.5r^{1/3}+0.5+\lambda^{-1})(r^{1/3}+\lambda^{-1})}{(1+\lambda^{-1})^5} - 1\} \qquad (13)$$

in which λ is the linear concentration which is defined as follows:

$$\lambda = \{(\frac{c_*}{c_d})^{1/3} - 1\}^{-1} \qquad (14)$$

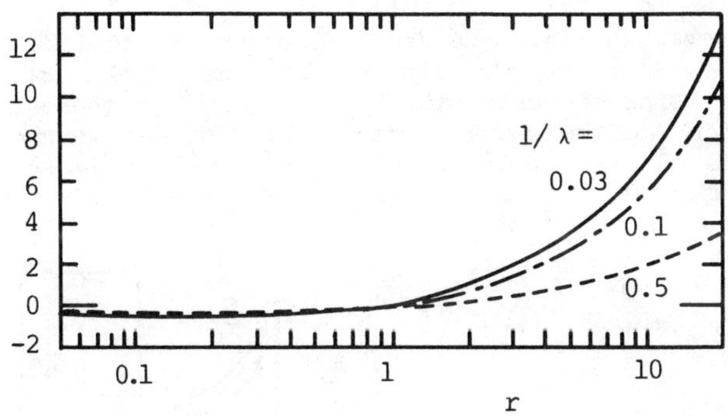

Fig.5 Nondimensional vertical velocity versus r.

Note that the larger the d_1 particle is than d_m, the faster the upward velocity. However, in the case when d_1 is smaller than d_m, its downward velocity is small. The results of laboratory experiments for two sets of r values verified this theory[7].

Although the extent of r for which the theory is valid is not known yet, very small particles will easily fall into the gap between larger particles, and hence they will not follow the theory.

3.4 Mechanics of Transportation of Huge Stones

How debris flow charged with coarse granular material can transport large blocks, yet flow on gentle slopes has been a cause of various arguments; some attribute to the strength of the slurry and the other to the buoyancy, and so on. The discussion in the preceding section on the upward motion of a larger particle may add a probable reasoning. Imagine a big boulder moving downstream at the surface of debris flow sticking a part of its body out of the flow as shown in Fig.6. The forces acting on the boulder perpendicular to the main flow direction is in this case its own weight, the buoyancy and the dispersive upward pressure due to collision with many other small particles. If we assume the effective bottom for collision of the boulder is the square area of $(B+s)^2$, $(=\{D\sqrt{1-(1-l/D)^2}+s\}^2)$, and the velocity of the boulder is equal to that of debris flow at the middle height of the draught, l, the condition of equilibrium in the forces is written as

$$\frac{2r}{1+r}c_d(\sigma-\rho)d^3\left(\frac{l}{2d}+\frac{1}{\lambda}+\frac{1}{2}\right)^2\frac{\{(D/d)\sqrt{1-(1-l/D)^2}+1/\lambda\}^2}{1+1/\lambda}g\cos\theta$$

$$=\frac{\pi}{6}\sigma D^3 g\cos\theta-\frac{\pi}{8}\rho D^3 g\cos\theta\left[\frac{2l}{D}\ \frac{1}{3}\left\{1-\left(1-\frac{2l}{D}\right)^3\right\}\right] \tag{15}$$

The curves in Fig.7 are depicted under the condition $\sigma=2.6\text{g/cm}^3$, $\rho=1\text{g/cm}^3$, $c_*=\pi/6$, and the circles are the results of laboratory experiments using the material $d=0.38$cm. Although there is considerable scatter in the experimental value, the tendency is rather well explained [8] . This result suggests that a boulder whose diameter is nearly twice as big as the depth of the flow can be transported as if it were floating.

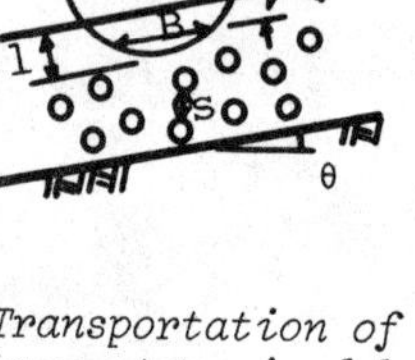

Fig.6 Transportation of a huge stone in debris flow.

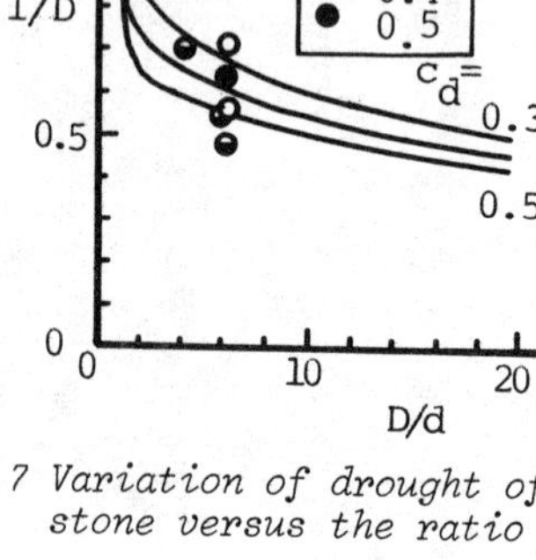

Fig.7 Variation of drought of the stone versus the ratio of diameters.

3.5 Roll Waves

Similar to the rapid water flow, the roll waves occur in debris flow. Fig.8 shows the non-dimensional celerity, period, wave height, and the ratio of the maximum height to the normal depth versus Froude number. The curves in the figure are theoretical ones which have been obtained for water flow [9], in which q_o is the discharge in unit width; h_m the maximum wave height; T the period; h_o the normal depth; c the wave celerity; and η is the momentum coefficient due to the velocity distribution. The experiments were done in a flume 20cm in width, and 18° and 15° in slope. Two kinds of materials were used, the diameters of which are 1.2mm and 4mm; the open circle and the closed circle in the figure, respectively, correspond to each material. The volume concentration of material in the debris flow was from

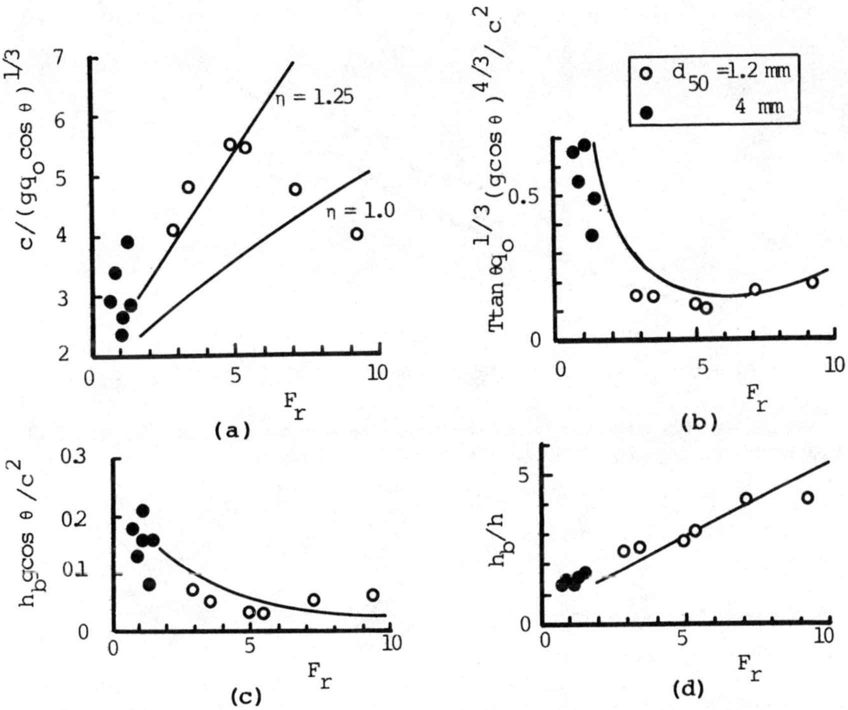

Fig.8 Characteristics of roll waves in debris flow.
(a) Celerity, (b) Period, (c) Maximum wave height,
(d) Ratio of the maximum wave height to the normal depth.

0.18 to 0.45.

As is clear from the figure, the characteristics of roll waves are similar to those of water flow, only the coefficient η is appropriately changed to be suitable to dilatant fluid, which is equal to 1.25. For thin concentrations, however, η value should be nearly equal to that of water flow.

4. DEPOSITION OF DEBRIS FLOW

4.1 Stoppage Mechanism

When a debris flow arrives at a place where the slope of a channel abruptly levels out without an expansion in width, it decelerates and thickens and finally stops. The profile of such a debris flow at time t and $(t+\Delta t)$ is modeled by the trapezoidal shape in Fig.9. It is assumed that all parts of the flow proceed at the same velocity v and have the same concentration equal to c_{du}.

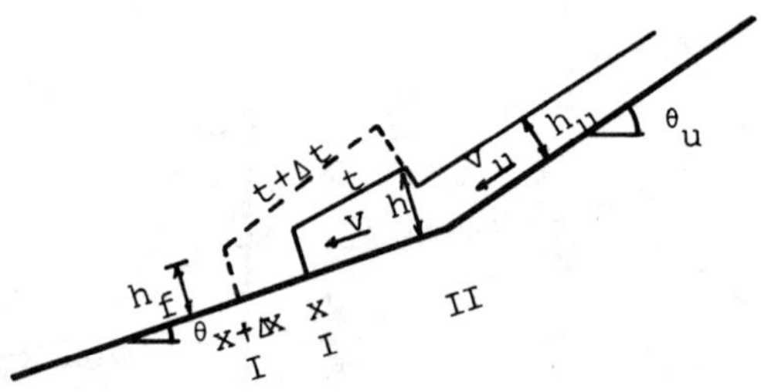

Fig.9 Process of the stoppage of forefront of a debris flow.

Thus, the momentum conservation between Section I and II (at the change in slope) is written as:

$$\frac{d}{dt}\left[\frac{1}{2}(h+h_f)\{(\sigma-\rho)c_{du}+\rho\}vx\right] = \frac{1}{2}(h+h_f)x\{(\sigma-\rho)c_{du}+\rho\}g\sin\theta$$

$$+\{(\sigma-\rho)c_{du}+\rho\}h_u v_u^2\cos(\theta_u-\theta)+\frac{1}{2}gh_u^2\cos\theta_u\cos(\theta_u-\theta)\{(\sigma-n\rho)c_{du}\kappa_a+n\rho\}$$

$$-\frac{1}{2}(\sigma-n\rho)gc_{du}(h+h_f)x\cos\theta\tan\alpha \tag{16}$$

where the left-hand side expresses temporal momentum change, the first term of the right-hand side is the force due to gravity, the second is the input of momentum from upstream of the change in slope, the third is the sum of hydrostatic and earth pressure at the slope change, and the fourth is the friction at the bottom. Tan α is the kinematic friction factor, κ_a is the coefficient similar the active earth pressure coefficient, n is the coefficient of the pressure rise due to settling of finer grains through the interstitial fluid, and the subscript u indicates a value upstream from the change in slope.

The equation of continuity is

$$\frac{1}{2}(h+h_f)x = h_u v_u t \tag{17}$$

From Eqs.16 and 17 one obtains the distance x_L within which the debris flow stops:

$$x_L = \frac{V^2}{G} \tag{18}$$

where

$$V = v_u\cos(\theta_u-\theta)\left[1+\frac{\{(\sigma-n\rho)c_{du}\kappa_a+n\rho\}\cos\theta_u}{2\{(\sigma-\rho)c_{du}+\rho\}}\frac{gh_u}{v_u^2}\right]$$

$$G = \frac{(\sigma-n\rho)gc_{du}\cos\theta\tan\alpha}{(\sigma-\rho)c_{du}+\rho} - g\sin\theta$$

Eq.18 means that if $G \leq 0$, the debris flow does not stop. This corresponds to

$$\tan\theta \gtrsim \frac{(\sigma-n\rho)c_{du}}{(\sigma-\rho)c_{du}+\rho}\tan\alpha \tag{19}$$

or if the debris flow stops, on the other hand,

$$\tan\theta < \frac{(\sigma-n\rho)c_{du}}{(\sigma-\rho)c_{du}+\rho}\tan\alpha \tag{20}$$

This equation implies that the denser the interstitial fluid and the less the concentration of the debris flow, the flatter the slope of the channel must be for the debris flow to stop. In addition, though the characteristics of n are not clear yet, it should be noted that the value of n for poorly graded mixture should essentially be unity, but for a well graded dense mixture it would become considerably large, say, 1.5.

As the debris flow continues to run down from upstream, it flows over already halted portion, and the deposit grows in both length and height. An early stage of this process can be depicted as in Fig.10.

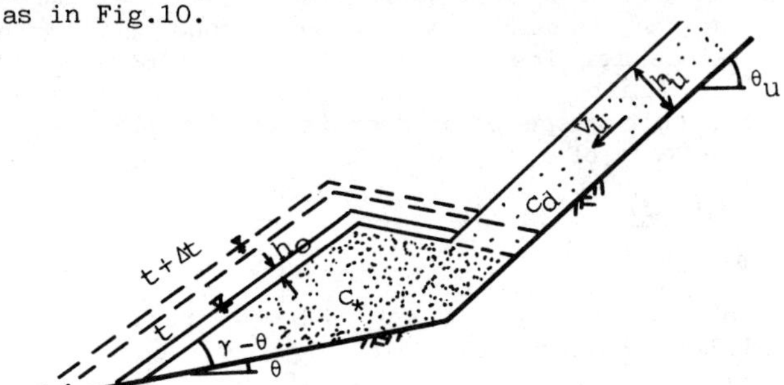

Fig.10 The early stage of deposition.

In this model, as long as the debris flow from upstream continues downward, the height of the deposit at the change in slope increases. But after it exceeds a certain critical height, the flow can no longer override it and some of the debris begins to be deposited upstream. Namely, in time the deposition moves upstream. This stage of the depositing process could be modeled as shown in Fig.11. Some experiments have proven that there is negligible energy loss in the process of flowing over the already halted portion. Therefore, the height of the rebound H' can be obtained by the following equation:

$$H' = \alpha' \frac{v_u^2}{2g} \cos\theta_u - \frac{v_o^2}{2g} \cos(\theta_u - \gamma) \tag{21}$$

where α' is the energy coefficient equal to 1.7 for the dilatant fluid flow, v_o the velocity of the surface water flow over the deposit, and γ is the surface slope of the deposit.

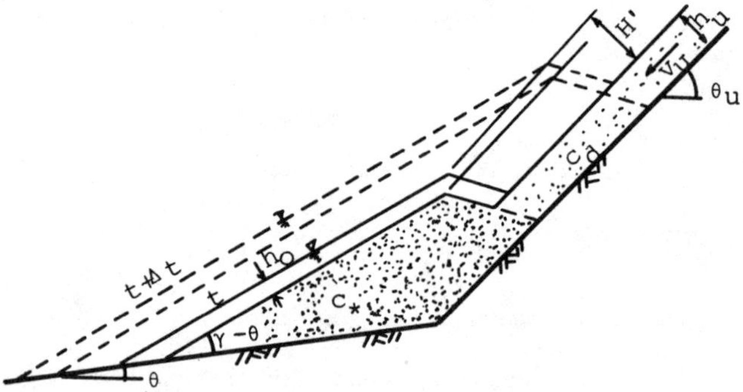

Fig.11 Upstream movement of the deposition.

The value of γ can be estimated by the following consideration; When the flow stops, the concentration of grains should change from c_{du} to c_*, meaning that the excess pore water will squeezed out of the deposit and flow onto its surface. If the slope of the surface of the deposit just before stoppage is greater than the critical slope for the occurrence of a sediment gravity flow, its slope will flatten out until it is equal to the critical one. Thus, a stable slope of a deposit can be obtained from Eq.6 neglecting the effect of c_b:

$$\tan\gamma = \frac{c_*(\sigma-\rho)}{c_*(\sigma-\rho)+\rho(1+h_o/d)} \tan\phi \tag{22}$$

The profile of the deposit at any time can be calculated by the model introduced provided the discharge of the debris flow is given. Fig.12 is one example of the calculation [10].

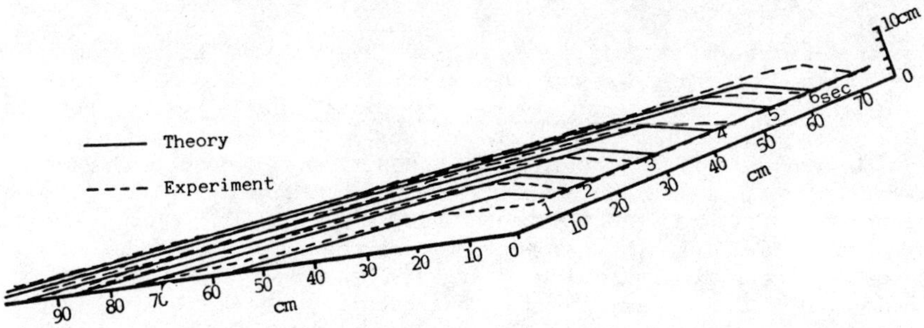

Fig.12 Comparison of the surface profiles in a process of deposition.

4.2 The Debris Cone Formation Process

When the width of the plane downstream from the mouth of a torrent is extremely wide, the debris flow forms a so-called "debris cone". Survey of micro-reliefs on debris cones so far reveal that they are constructed of many "lobes", each of them consisting an episode in a series of debris flows. Fig.13a,b are the examples of the results of experiments which show temporal variations in the shape of a debris flow. The number on each curve indicates the time in seconds which had elapsed from the moment the forefront of the debris flow arrived at the mouth.

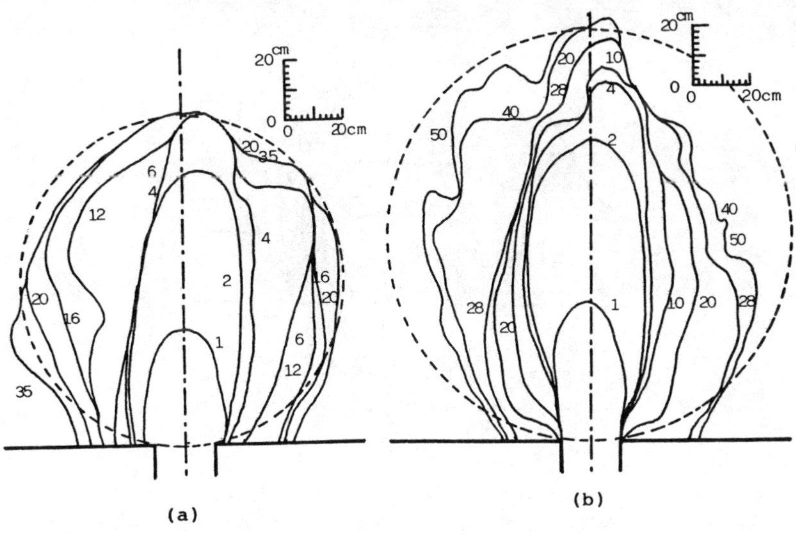

Fig.13 Temporal variation in the edges of a debris flow.

In the earliest stage, the path of the flowage downstream from the mouth is straight, and the width of it, i.e., the width of a lobe, is at most about twice the width of the mouth, but the lobe quickly reaches its maximum length (4 seconds in the case of Fig.13a). Then it begins to deposit the debris accompanying a hydraulic jump at the boundary of deposit downstream and flowage upstream. After a short time, when the hydraulic jump arrives at the outlet of the torrent (approximately 5 seconds in the case of Fig.13a), the flow changes its direction to the right, then to the left, and so on. Consequently, at the end of rather long wagging process as in the case of a large scale debris flow, the deposit is circular with a diameter approximately equaling the distance from the outlet to the distal end of the first lobe.

Fig.14 is the contour map of the debris cone formed in Fig.13a. The deposit shows a plane along the center line, which has a width about twice that of the outlet. The distribution of particle diameters in the debris cone is such that the biggest gravel accumulates near the toe and diminishes upstream.

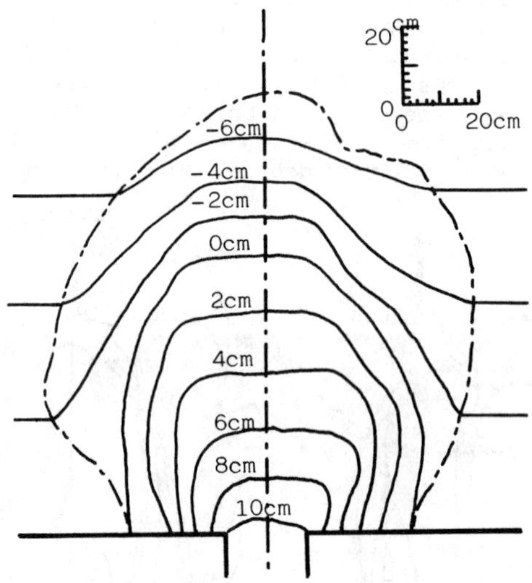

Fig.14 The contour map of a debris cone. The base level of the contour(0cm) is set to the height of the bottom of the outlet.

The experiments have shown that the diameter of the debris cone can be calculated by Eq.18 within ±8% error, and also that the longitudinal surface slope can be estimated by Eq.22 using the width of the plane along the center line of the deposit as the width of surface flow. In these experiments the debris flows stopped before the inside of the circular area(Fig.13) was completely filled with sediment. Provided the debris flow continued after this area was completely filled, the diameter of the cone would increase in order to accomodate the additional sediment while maintaining the same longitudinal surface slope.

5. SEDIMENT GRAVITY FLOW ON FLATTER SLOPE AND THE EROSION OF DEBRIS CONE

5.1 Mechanics of Sediment Gravity Flow

In domain 3 in Fig.2, the layer of sediment movement is covered by a less densely concentrated water flow. The ratio of the thickness of the sediment movement to the whole depth, however, thickens as large as one when the bed slope steepens to the extent to which defined by Eq.8. The schematic profile of the sediment gravity flow is therefore as shown in Fig.15.

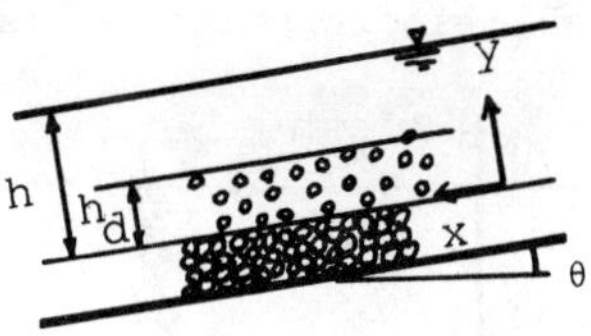

Fig.15 Sediment gravity flow.

Referring to the figure one obtains the equation of the ratio of the thickness of the sediment flow layer to the whole depth through the consideration of stress equilibrium condition at the surface of stable sediment bed:

$$\frac{h_d}{h} = \frac{c_{d\infty}}{c_{dc}}\left(1-\frac{\tau_{*c}}{\tau_*}\right) \tag{23}$$

in which c_{dc} is the concentration of sediment in the layer of sediment motion and it is nearly equal to $c_*/(1+\kappa'^{-1})$; τ_* is the non-dimensional tractive force, $h\sin\theta/\{(\sigma/\rho-1)d\}$; and τ_{*c} is the critical tractive force of the particle which is nearly equal to 0.05.

Application of the dilatant fluid model and the usual Newtonian turbulent flow model to the sediment moving layer and to the water flow layer respectively gives the following two equations:

$$a_i\sigma(\lambda d)^2\left(\frac{du}{dy}\right)^2\sin\alpha+\rho j^2\left(\frac{du}{dy}\right)^2 = \{(\sigma-\rho)c_{dc}(h_d-y)+\rho(h-y)\}g\sin\theta \tag{24}$$

$$\rho j'^2\left(\frac{du}{dy}\right)^2 = \rho g(h-y)\sin\theta \tag{25}$$

in which j is the mixing length in interstitial fluid; j' is the mixing length in covering water flow.

Fig.16 is an example of comparison of theoretical velocity distribution to the experimental result. The closed and open circles in the figure respectively represent the velocity of sediment particles and that of neutrally buoyant particles in the water flow. In the calculation of theoretical curve, the relationship $j = 3d/\lambda$ and $j' = 3d/\lambda + 0.4(y - h_d)$ are assumed.

The formula of sediment discharge and resistance to flow are obtained by integration of velocity distribution and through some appropriate abbreviations. The results are as follows [11].

$$q_{s*} = \frac{2}{3}\frac{(4.2-0.75c_{dc})}{\cos^2\theta(\tan\phi-\tan\theta)}\tau_*^2(1-\frac{\tau_{*c}}{\tau_*})^2 \tag{26}$$

$$\frac{U}{u_*} = (0.23+0.65\tan\theta)\frac{h}{d} \tag{27}$$

where $q_{s*} = q_s / u_* d$; q_s the sediment discharge per unit width; $u_*(=\sqrt{gh\sin\theta})$ the shear velocity; and U is the cross-sectional mean velocity.

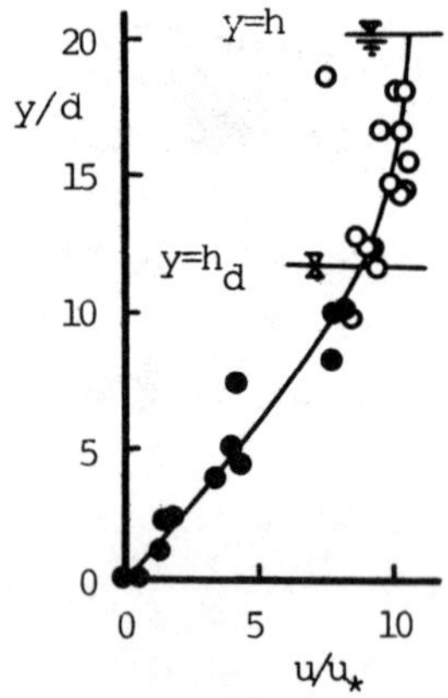

Fig.16 Velocity distribution.

5.2 Erosion of Debris Cone by a Flood Flow

The debris cones formed at the outlet of ravines are sometimes eroded by following flood flows. A narrow gorge is incised from the top of the cone and downstream of the so-called "intersection point" it spread out forming a new aluvial fan. This process in a laboratory model experiment is shown in Fig.17. One can see the deformation of contour lines of the debris cone and the formation of an aluvial fan by a steady flow of 0.5 l/s. The median diameter of the sediment is 2.03mm.

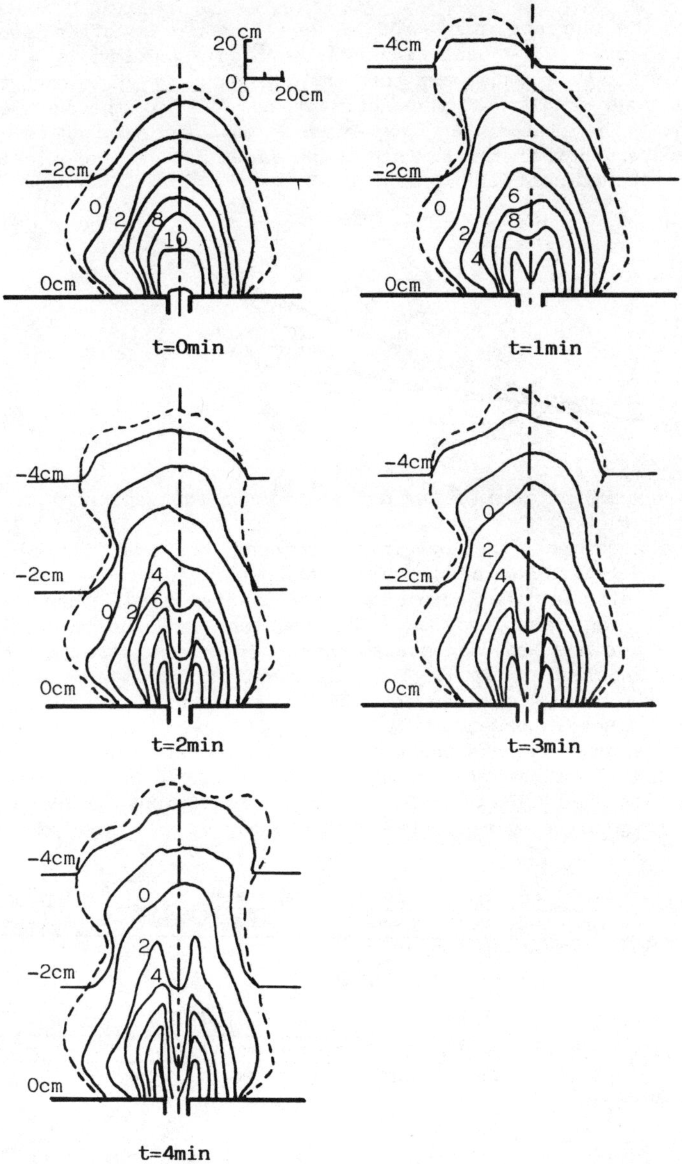

Fig.17 Process of erosion and redeposition of a debris fan.

Because the surface slope of a debris cone is often steep enough to be eroded by occurrence of sediment gravity flow, it is essential for the analysis of the process to disclose the mechanics of degradation of the sediment bed due to the sediment gravity flow. The laboratory experiments on the mechanics of degradation revealed that the surface slope variation is approximately linear as schematically shown in Fig.18.

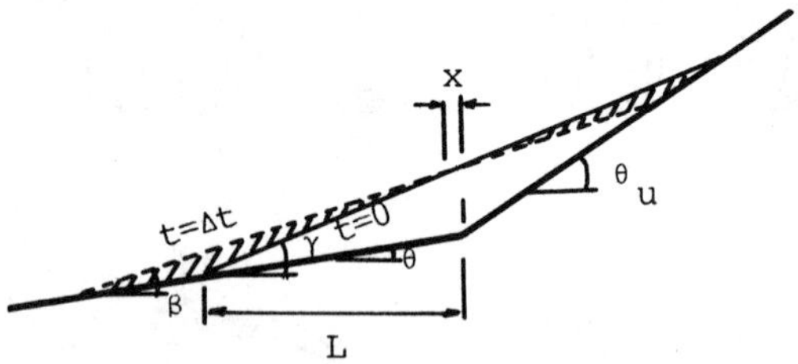

Fig.18 Conceptual diagram of the deformation of the debris pile.

In the experiment the original sediment bed was formed by deposition of debris flow at the slope change of the rigid uniform width flume. This is the straight line labeled $t=0$. Then some rate of water was supplied and the sediment bed upstream was eroded and it deposited downstream forming nearly straight bed (broken line labeled $t=\Delta t$). The intersection point is at the distance x from the slope change. The hatched area upstream and downstream of the intersection point should be equal to each other and this area corresponds to the total sediment discharge passed through the intersection point in time Δt. From the geometrical consideration one obtains the following equations which make it possible to analyse the process of deformation of the sediment bed;

$$\frac{x}{L}=\frac{\dfrac{\sin(\gamma-\theta)}{\sin(\beta-\theta)}+\dfrac{\sin(\theta_u-\gamma)}{\sin(\theta_u-\beta)}\,\dfrac{\tan\gamma-\tan\theta}{\tan\theta_u-\tan\gamma}-\dfrac{\tan\theta_u-\tan\theta}{\tan\theta_u-\tan\gamma}\sqrt{\dfrac{\sin(\gamma-\theta)\sin(\theta_u-\gamma)}{\sin(\beta-\theta)\sin(\theta_u-\beta)}}}{\dfrac{\sin(\gamma-\theta)}{\sin(\beta-\theta)}-\dfrac{\sin(\theta_u-\gamma)}{\sin(\theta_u-\beta)}} \tag{28}$$

$$\frac{q_s\Delta t}{1-n_o}=\frac{1}{2}\,\frac{\sin(\gamma-\beta)}{\cos^2\gamma}\,\frac{\sin(\gamma-\theta)}{\sin(\beta-\theta)}\,\frac{\tan\theta_u-\tan\theta}{\tan\theta_u-\tan\gamma}\,\frac{\sqrt{\dfrac{\sin(\gamma-\theta)\sin(\theta_u-\gamma)}{\sin(\beta-\theta)\sin(\theta_u-\beta)}}-\dfrac{\sin(\theta_u-\gamma)}{\sin(\theta_u-\beta)}}{\dfrac{\sin(\gamma-\theta)}{\sin(\beta-\theta)}-\dfrac{\sin(\theta_u-\gamma)}{\sin(\theta_u-\beta)}} \tag{29}$$

By this simple model the process was not only qualitatively but also quantitatively well explained[11].

For the application of this model to debris cone deformation process as shown in Fig.17, some more detailed observation on the spreading of sediment is necessary.

6. CONCLUDING REMARKS

The description herein explained various aspects of debris flow as well as sediment gravity flow from onset to end by applying the simple concept which Bagnold has exquisitely proposed. Though admitting that the dispersive stresses due to the collision of grains play the leading part, some dispute with Bagnold's original concept and therefore with the writer's application of it to the debris flow on account of the simplification such as assumption of perfect elasticity, neglecting of the effect of rotation and friction of grains, assumption of homogeneous vertical concentration of grains, etc. Refinement of the basic concept in the mechanics of flow of mixture of granular materials and fluid will modify the mechanical reasonings of the debris flow and the sediment gravity flow. It is to this end the writer hopes the progress in the mechanics of flow of granular materials will continue.

REFERENCES:

1. Ashida,K, Takahashi,T. and Mizuyama,T.,"Study on the initiation of the motion of sand mixtures in a steep slope channel," Shin-Sabo, 103, pp.6-13 (1977) [Japanese].

2. Ashida,K., Takahashi,T. and Mizuyama,T.,"Study on bed load equations for mountain streams," Shin-Sabo, 107, pp.9-17 (1978) [Japanese].

3. Takahashi,T.,"The occurrence and flow mechanism of debris flow," Tsuchi to Kiso, 26, pp.45-50 (1978)[Japanese].

4. Takahashi,T.,"Evaluation of the factors relevant to the initiation of debris flow," Proc. of International Symposium on Landslide, New Delhi, 3, pp136-140 (1980).

5. Takahashi,T., "Mechanical characteristics of debris flow," J. Hydraulics Div., ASCE, 104, pp.136-140 (1978).

6. Bagnold,R.A.,"Experiments on a gravity-free dispersion of large solid spheres in a Newtonian fluid under shear," Proc. R. Soc. London, Ser.A, 225, pp49-63 (1954).

7. Takahashi,T.,"Debris flow on a prismatic open channel," J. Hydraulics Div., ASCE, 106, pp381-96 (1980).

8. Yagi,H. and Takahashi,T.,"Mechanics of transportation of huge stones in debris flow," Proc. National Cong. of JSCE, 2, (1982)[Japanese].

9. Iwagaki,Y.,"Study on the mechanics of soil erosion," pp1-215 (1955)[Japanese].

10. Takahashi,T. and Yoshida,H.,"Study on the deposition of debris flow(1)-deposition due to an abrupt change of bed slope-," Annuals, Disaster Prevention Research Institute, Kyoto Univ., 22B-2, pp315-28 (1979)[Japanese].

11. Takahashi,T.,"Study of the deposition of debris flow (3)-erosion of debris fan-, Annuals, DPRI, Kyoto Univ.,25B-2, (Printing) [Japanese].

RHEOLOGY OF POWDER SYSTEMS

K. Umeya

Applied Chemistry Department, Faculty of Engineering,
Tohoku University,
Sendai, 980, Japan

ABSTRACT

Using the packing character shown in Table I, some systematic interpretations concerning to the rheological behavior of powder-liquid systems has been developed. As the results of this consideration, three sorts of packing stages can be distinguished as

(1) the slurry stage system in which a free medium phase has a continuous structure,
(2) the powder stage system in which particle phase constructs a continuous structure through the mutial contact points between particles,
(3) the mud stage system which has a continuous structure of a liquid film.

Rheological behavior, therefore, may be described as either the flow property in a slurry stage, the plastic deformable property of a packed bed in a powder stage and a flow-deformation property in a mud stage, respectively, because it is quite reasonable to consider that the continuous phase determines the mechanics at every these three stages. Unified interpretation through the packing character shown in Table I is desirable but can not be attainable now on account of the lack of sufficient rheological data over a whole range. The author is of opinion that a radial stress originating from an axial stress is a valid key to solve this problem. That is, the normal stress difference effect predominates with the solid concentration of suspension in the slurry stage and this effect produces a decay of the Rankine's coefficient at the extreme slurry concentration. Thus the results well known as the liquid content abnormality in the mud technology are obtainable as shown in Fig. 24. Finally, a solidlike character which can be represented by the lower value of the Rankine's coefficient arises in the powder stage. This interpretation may be considered as to construct one unified explanation.

A generalized explanation as to the whole range of the powder-liquid system may be reconsidered more precisely in this paper. The interpretation developed here may not be complete, however, it will be acceptable as a model for developing further experimental programs to provide a more complete explanation.

1. PACKING CHARACTER OF POWDER LIQUID SYSTEMS

The packing character of powder dispersing systems may be described as shown in Table I [1].

Table I. Packing character of powder-liquid systems.

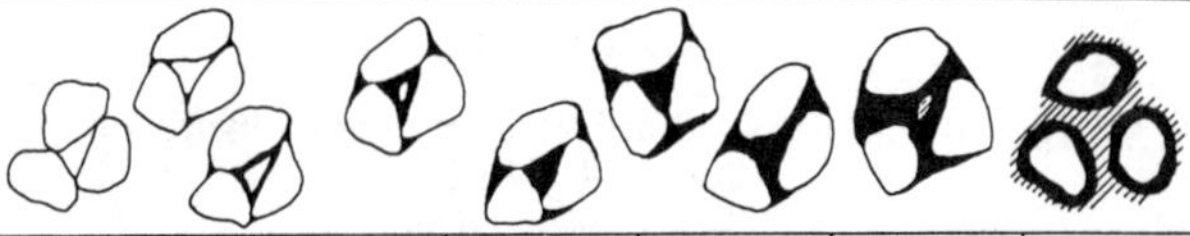

particle (solid phase)	continuous	continuous	continuous	dis-continuous	dis-continuous	dis-continuous
air (gaseous phase)	continuous	continuous	dis-continuous	—	—	—
liquid hull (liquid phase I)	dis-continuous	continuous	continuous	continuous	continuous	dis-continuous
free liquid (liquid phase II)				dis-continuous	continuous	continuous
stage	Pendular P	Funicular F-I	F-II	Capillary C-I	C-II	Slurry S
critical point	C-1	C-2	C-3	C-4	C-5	
in soil physics			P.L (plastic limit)		L.L (liquidus limit)	
rheological system	Powder system		mud system			slurry system

(1) Particles in a perfectly dry system exhibit a continuous structure throughout the powder bed due to the mutual interaction of the particles. In this case, the gaseous component, commonly air, is also continuous through the powder bed constructing a stereoscopic network structure.

(2) When a small quantity of liquid, e.g. water, is added to the above-mentioned perfectly dry powder system, the liquid forms a discontinuous structure in an ideal dispersed state because the fluid exists in this ideal dispersed state forming many ringed structures at the contact points between particles. This pattern, involving a concept mentioned in (1), is called the pendular stage structure in the field of the micromeritics.

(3) If yet more liquid is added, the individual liquid rings grow their size replacing the air phase forming a continuous structure. In this case, the solid phase (powder phase) also still has a continuous structure, because particles contact with each other at their contact points, and the gaseous phase (air phase) may also still have a continuous stereoscopic network structure. In micromeritics this pattern is called the funicular-I stage structure.

(4) When much more liquid is added to this system, the liquid displaces the air phase still leaving a continuous stereoscopic network structure behind, and at the critical stage the stereoscopic network structure of the air phase becomes to be constracted leaving many separate closed pores behind. Therefore, the gaseous phase becomes discontinuous. This pattern is called the funicular-II stage structure. In the extreme case of this stage, the system has no gaseous phase. This is called the Plastic Limit (P.L.) in the field of soil mechanics. If the powder surface is inactive enough, the packing pattern is completed at this stage. However, if active, three more

packing stages may follow, as indicated in Table I. In this case, however, it is most convenient to analyze this sequence from the view point of major liquid phase.

(5) When a particle is put into a liquid medium (typically water), the adsorbed layer may be constructed over the particle surface having a thickness (ℓ_0) determined by the surface activity of the powder. Such particles, coated with the surface layer of ℓ_0, are dispersed into the residual free liquid medium.

Therefore, at the begining, there is a continuous free medium phase, a discontinuous adsorbed liquid phase, and then a discontinuous powder phase. This pattern may be termed the slurry stage in the micromeritics.

(6) When a sufficient liquid phase, is removed from this slurry stage, the particles surrounded by the absorbed film interact with each other at their contact points producing a new critical state called the Liquidus Limit (L.L.) in the field of soil mechanics.

In such a way, the capillary-II stage is obtained having a continuous water film phase and a continuous free water phase as well as a discontinuous powder phase.

(7) When much more liqud is removed, the stereoscopic network structure of the free water phase becomes constracted leaving many separate enclosed free liquid drops in the system. This pattern is termed the capillary-I stage in the micromeritics.

Generally speaking, therefore, three stages of the packing structure can be distinguished in the case of the inactive powder system. On the other hand, three more stages may be added in the case of active powder system as shown in Table I. It can be seen from this table that the so-called powder system contains the pendular and funicular-I and the mud system includes the funicular-II and capillary stages, and finally, in the so-called slurry system the slurry stage is comprized.

From these classifications, it can be recognized as follows:

(1) the slurry system can be characterized by flow properties since only the free liquid medium is continuous,

(2) the mud system can be characterized by plastic properties since only an adsorbed liquid phase is continuous, and

(3) the powder system can be characterized by plastic and elastic deformation of the packed bed, not by the individual constituent particles themselves, because the solid phase can be said to be continuous through the contact points. Hence, the essential mechanical behavior of these three individual powder-liquid systems can be reasonably interpreted in terms of the rheological properties of the continuous phase, respentively.

Many prior rheological investigations have contributed to an understanding of powder-liquid systems. However, these were conducted as merely flow studies of the slurry system, or only deformation studies of the powder system, or merely deformation-flow evaluation of the mud system. In this paper, therefore, an attempt is provided towards an unified interpretation over an entire range shown in Table I using experimental data extracted mainly from the works obtained by the author [1]~[7].

2. CONSTITUTION EQUATIONS AND RHEOLOGICAL MODELS CORRESPONDING TO THE EVERY STAGE OF TABLE I

The mix which belongs to the P, F-1, F-2 stages is called a visco-plastico-elastic solid and the one belongs to the S stage is practically named a visco-plastico-elastic liquid, in the field of a powder-liquid dispersing rheology. In order to manifest a rheological behavior of such a visco-plastico-elastic solid or a visco-plastico-elastic liquid, equation (1) and (2) were proposed by Sawaragi [17], and, at the same time, some rheological models cited in Fig. 1 were introduced by him.

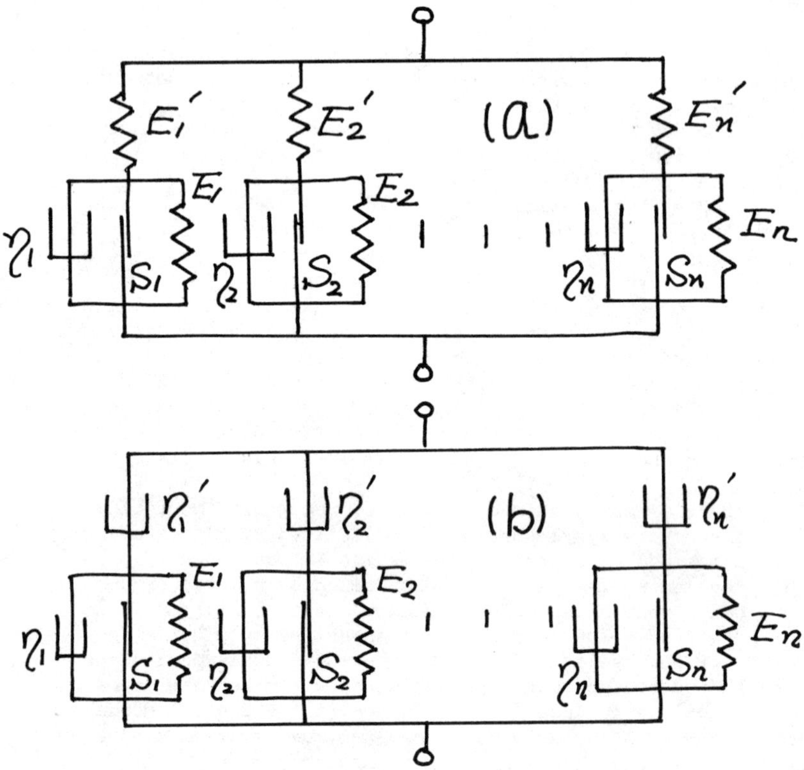

Fig. 1. Rheology Model for a visco-plastico-elastic solid (a) and for a visco-plastico-elastic liquid (b)

$$\sigma(t) = E\cdot\varepsilon(t) - \int_0^\infty d\mu \int_0^\varepsilon d\mu \frac{G(\mu\cdot\chi)}{\mu} \{\varepsilon(t) - \chi\}d\chi + \int_0^\infty d\mu \int_0^\varepsilon \frac{G(\mu\cdot\chi)}{\mu} \int_{t(x)}^t e^{-\mu(t-2)} \frac{d\varepsilon(2)}{dz} dz d\chi \tag{1}$$

$$\sigma(t) = \eta\cdot\dot{\varepsilon}(t) - \int_{9}^{\infty} d\mu \int_{o}^{\dot{\varepsilon}} \frac{H(\mu\cdot D)}{\mu} \{\dot{\varepsilon}(t) - D\} dD$$

$$+ \int_{o}^{\infty} d\mu \int_{o}^{\dot{\varepsilon}} H(\mu\cdot D) dD \int_{\varepsilon(D)^{-1}}^{t} \qquad (2)$$

$$\{\frac{d(t)}{dt} - D\} e^{-\mu(t-2)} dz$$

Here; σ, ε, t, τ mean stress, strain, elasped time and strain originating time. G(μ), H(μ) indicate distribution functions related to a relaxation time (1/μ = τ). G(χ), H(D) mean newly developed distribution functions dicided from the base of a yielding property.

Though some elaborate analysis is needed, one can decide the distribution functions G(μ) and H(μ) from some stress relaxation tests as seen in the latter section, and the another distribution functions G(χ) and H(D) from some constant straining tests which is also reconsidered in the latter section.

In the residual stage leaked in the above mentioned argument, that is, in C-1, C-2 stages, the mechanical behavior is more complicated which can be treated as a visco-plastico-elastic solid in some cases and a visco-plastico-elastic liquid in other cases.

A constitution equation for C stage, therefore, is thought to have combined formular of equation (1) and (2). (Manifestation of a delivered equation is omitted here.)

A rheological model indicating this case, however, is shown in Fig. 2, which might be called "a visco-plastico-elastic solid-like liquid or a visco-plastico-elastic liquid-like solid."

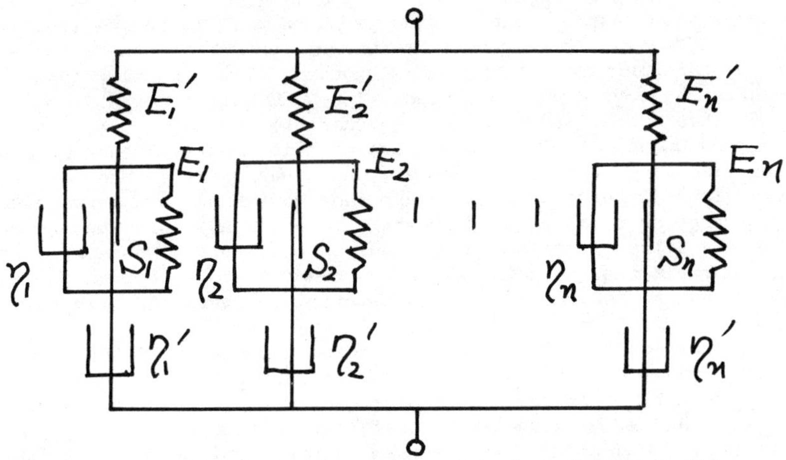

Fig. 2. Rheology Model for a visco-plastico-elastic solid-like liquid or liquid-like solid.

3. RHEOLOGICAL PROPERTIES IN SLURRY SYSTEMS

3.1. Flow Pattern in Steady and Transient States.

The flow properties of some suspended systems have been extensively investigated by the author [1]~[7], [18] and, in a steady state, most of the suspensions are elucidated to obey an "extended Ostwald flow pattern" illustrated in Fig. 3 when observed as an equilibrium response. The entire equilibrium flow curve is composed of many

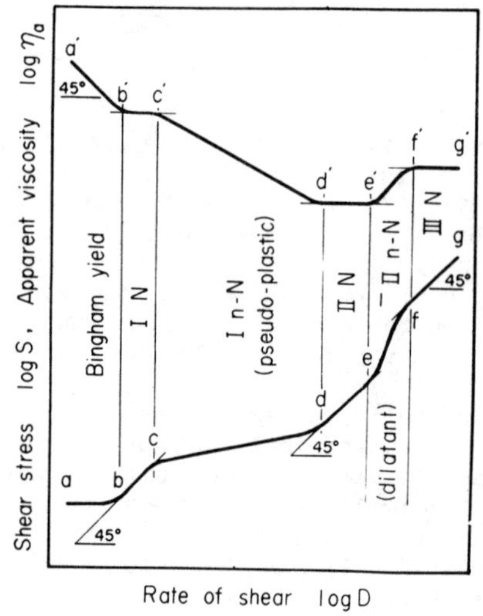

Fig. 3. Equilibrium flow pattern over a wide shear rate range.

essential regions such as the Bingham yielding (B.Y.), the first Newtonian (I N, exactly, apparent Newtonian), the first non-Nestonian (I n-N, pseudo-plastic), the second Newtonian (II N, exactly, apparent Newtonian), the second non-Newtonian (II n-N, dilatant, exactly speaking, rheologic dilatant), and the third Newtonian (III N, also apparent Neutonian). In the upper part of this figure, the calculated apparent viscosity curve is indicated, which has three constant viscosity regions corresponding to an apparent Newtonian, two decreasing viscosity regions corresponsing to the B.Y. and I n-N, and one increasing viscosity region corresponding to the II n-N. Therefore, the extended Ostwald flow pattern can be described as to have two softening and one hardening regions, each are essentially reversible with the shear rate [2, 3].

The extended Ostwald flow pattern is not completed if the time-dependent regions are not suplimented. Remarkable time-dependent parts are detectable in the middle part of the I n-N part as recognized by the relationship between loosely and tightly flocculated suspensions shown in Fig. 4. Another time-dependent character is obtained in the

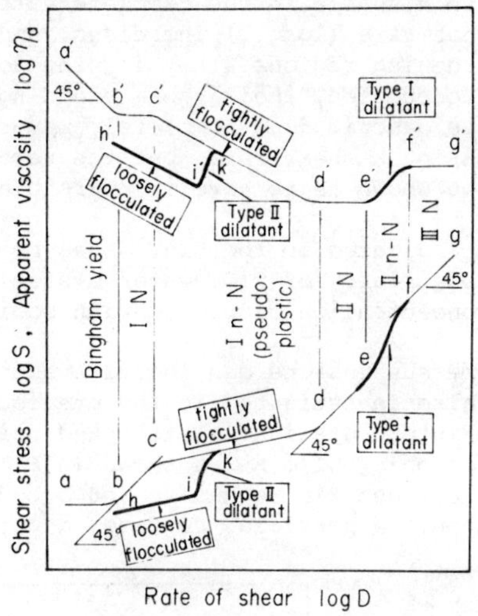

Fig. 4. Flow pattern over a wide shear rate range containing time-dependent parts.

dilatant (precisely rheologic dilatant, II n-N) region which is indicated as to have a remarkable L/R dependence having distiguished time dependent character. Here, L is the length and R is the radius of the measuring capillary [6]. As recognized above, the flow pattern for the loosely flocculated suspension is situated in the middle part of the extended Ostwald flow pattern as seen in Fig. 4, which shows softening character with increasing shear rate as shown by h'i' part in the figure, and is altered to harden toward the tightly flocculated part as shown by i'k' part. After a k' point is reached, the viscosity shows another softening tendency as shown by k'd' part [4]. Therefore, it may be recognized that some suspensions have two sorts of flocculated structures corresponding to the loosely and tightly flocculated structures and each of them can indicate the unique pseudo-plastic (I n-N) softening and hardening properties, respectively, and the transition region between these two pseudo-plastic structures can be indicated as Type II dilatancy (or may be called a structual dilatancy) having irreversible nature with shearing [3]. On the other hand, before cited dilatancy part detected at the end of the II N may be called the Type I dilatancy (or may be called a rheologic dilatancy). These two sorts of dilatancies has been distinguished for the first time in the previous works of the author [5, 6].

Therefore, as a whole, three sorts of the decreasing viscosity ranges and two sorts of the increasing viscosity ranges with increasing shear rate can be distinguishable in the complete extended Ostwald flow pattern as the shear rate (long D) increases, and the four among these softening and hardening regions (that is, the loosely flocculated, the tightly flocculated, the Bingham yielding and the Type I dilatant regions) can be described as completely reversible in nature by increasing or decreasing a shear rate, but the remaining one (the Type II dilatant) can be shown as to have irreversible character [4, 5].

The full relation indicated in the Fig. 4 has been acertained by the author using only one slurry of TiO_2-water system having 30 volume percentages of solid concentration dispersed with sodium pirophosphate ($Na_4P_2O_7 \cdot 10H_2O$).

The reason why some suspensions can indicate such two sorts of flocculated states is also investigated in the previous works of the author [6, 7], and the interpretation is attempted using the theory of the stability of lyophobic colloids as shown in Fig. 5(a) and (b). As the results of an electrostatic repulsive and a London-van der Waals attractive force acted on the particles surface, a potential relation

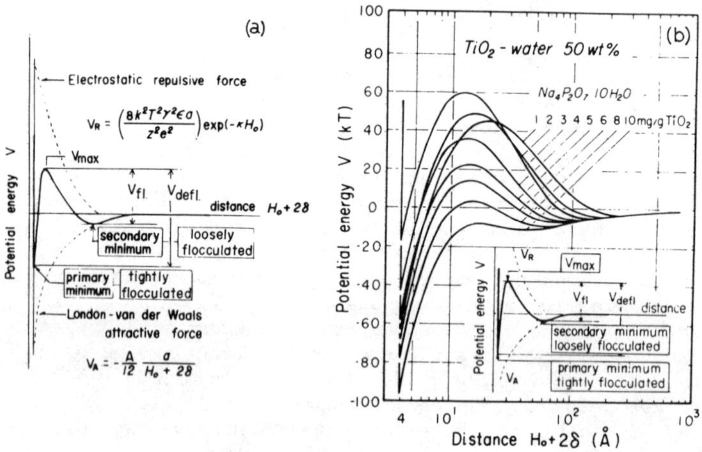

Fig. 5(a) Schematic representation of the potential energy curve as a function of the distance between two particles. (b) Potential energy curves calculated for TiO_2-water 50wt% suspensions stabilized with various amounts of $Na_4P_2O_7 \cdot H_2O$.

can be calculated as shown in Fig. 5(a), shematically, detailed analysis was provided in author's another works [5], which has a primary minimum having a deeper well in the vicinity of particle surface and a secondary minimum having a shallower well far from the particle surface. Therefore, two kinds of flocculated structures may be introduced as shown in Fig. 6 shematically, which shows primary flocculated units (m=6 or 7) can form a tight floc and loosely flocculated unit (units number is indicated by n) can be constructed

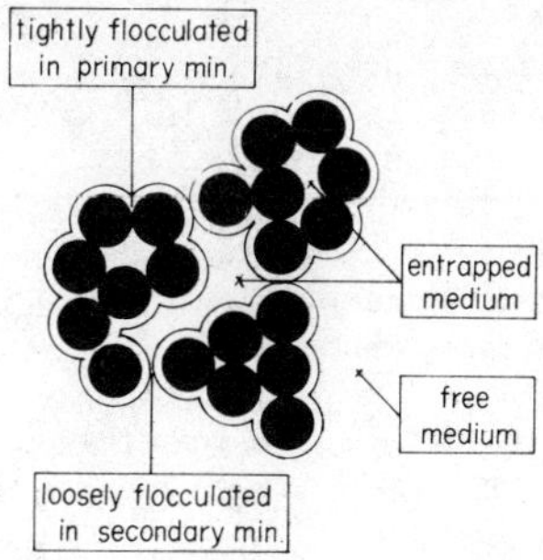

Fig. 6. Schematic illustration of loosely and tightly flocculated structures.

between them.

Though, in this instance, m=6 or 7 units are alowed to construct the tightly flocculated structure, the value of m and n may not be

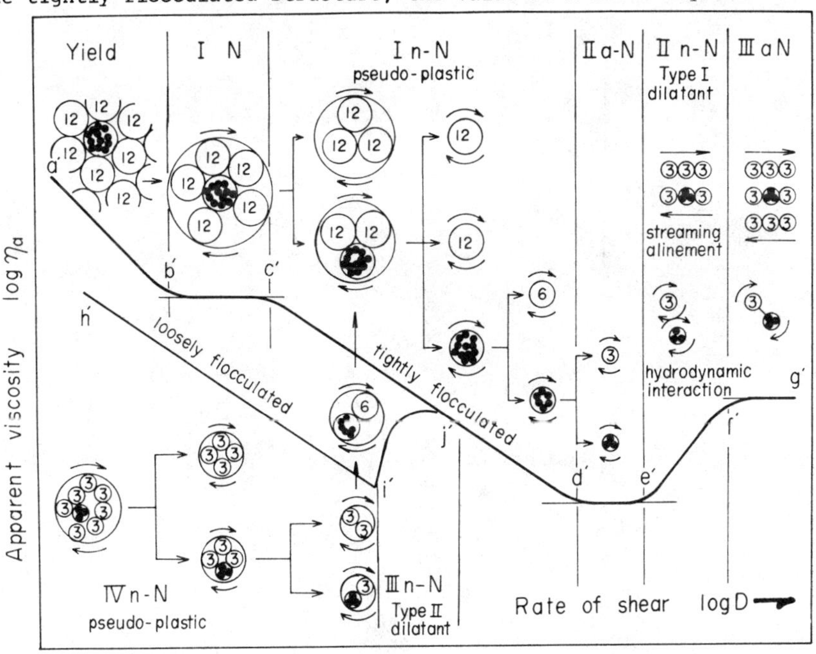

Fig. 7. Schematic illustration of the structural change in suspension corresponding to the extended Ostwald flow curve including an instantaneous flow curve.

restricted to this amount essentially. For instance, the structure having m=12 and n=1 initially is shown in Fig. 7 which can decrease or increase m and n value with the shearing history when breakdown or buildup mechanisems took place in the flocculated structure. It can be concluded that the whole pattern of the extended Ostwald flow curve is thus certificated under the conception of the micro rheology.

It can be recognized therefore that if the particles flocculated in to the secondary minimum in the potential energy curve, the loose structure will be resulted and on the other hand, when flocculated in to the primary minimum, the tight structure may be constructed [7].

3.2. Normal Stress Effect of Suspensions

In some concentrated suspensions, the normal stress difference effect cannot be ignored, which is usually measured by the PGB method [8, 9], the Jet method [10, 11] amd a Weisenberg effect independently. Some results obtained by the author using the zinc oxide suspensions of 73.6 and 70.4 wt% of solid concentrations are illustrated in Fig. 8. The normal stress difference effect confirmed by the PGB method is larger by about two orders, and the one obtained by a Jet method is

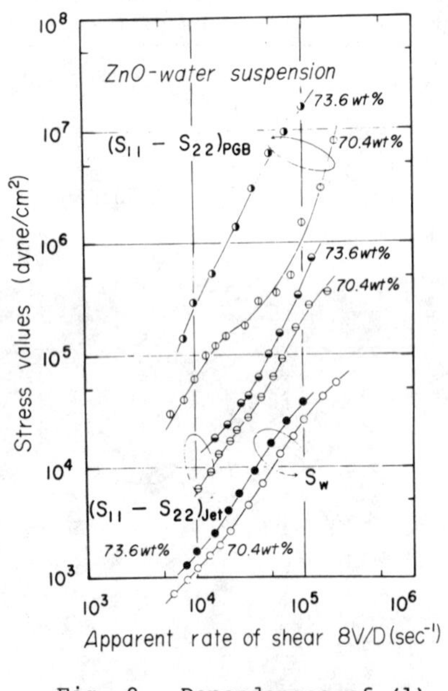

Fig. 8. Dependences of (1) shearing stress (Sw) and (2) normal stress difference $(S_{11}-S_{22})_{PGB}$, $(S_{11}-S_{22})_{Jet}$ on the apparent rate of shear (8V/D).

also larger by about one order than the value obtained in a normal shear flow character obtained in II n-N range [12]. This means that these suspensions possess the distinguished elastic properties besides the viscous effect as considered in the previous sections, and it is noticeable that this tendecy becomes remarkable with increasing the solid concentration as seen in the figure and as written in the later section.

4. SOME RHEOLOGICAL PROBLEMS IN THE POWDER SYSTEM

Putting off the problem in a mud stage, let us consider here rheological problems in the powder system beforehand, because the mechanical behavior of the mud system can be described as the emulative contribution of the viscous flow character in the slurry system and the plastic deformation character in the powder system.

The powder stage is composed of two kinds of continuous structure as written in the previous section: one is a continuous structure formed by particles interacting mutually through their contact points, and the other is a stereoscopic network structure of an air phase. Therefore, the mechanical behavior in this stage may be controlled by these two continuous phases; however, the viscous contribution by this air phase to a total viscous flow is so small that only frictional traction between particles must be considered. Therefore, if the particles are stiff enough, the deformability of a packed bed is decided by the transferability of particles in the piled bed, but not by the deformation character of the constituent particles tjemselves; however, if the particles are soft enough, the deformability of a packed bed is governed mainly by the deformation character of the constituent particles. To make this point clearer, some measurements are attempted using plastic beads (polystyrene powder) whose characteristics are given in Table II.

Table II. Characteristics of polystyrene beads

Shape	Spherical
Mean diameter	300 μ
Softening point	~ 100°C
Density	1.087 g/cm^3 (25°C)

We made use of this sample because the rheological characteristics of polystyrene beads (deformation properties) have been extensively studied in the field of the polymer rheology. These materials can behave as stiff powder at lower temperatures or under higher shear rates; on the other hand, they behave as soft powder at higher temperatures or under lower shear rates as seen by Fig. 9 or 10.

4.1. Deformation pattern (stress-strain diagram)

In an ordinary compaction procedure, a compaction schedule indicated by the insertion figure (A) in Fig. 9 is applied; that is, after the prearranged compaction period has been elapsed, the compressive stress (σ_1) is quickly removed to take the compact from a mold. When these loading and unloading rates are assumed to be constant as shown in the insertion figure, deformation character (a stress-strain σ_1-ε_1 diagram) of the packed bed can be obtained under various testing temperatures as illustrated in the main figure of Fig. 9. According

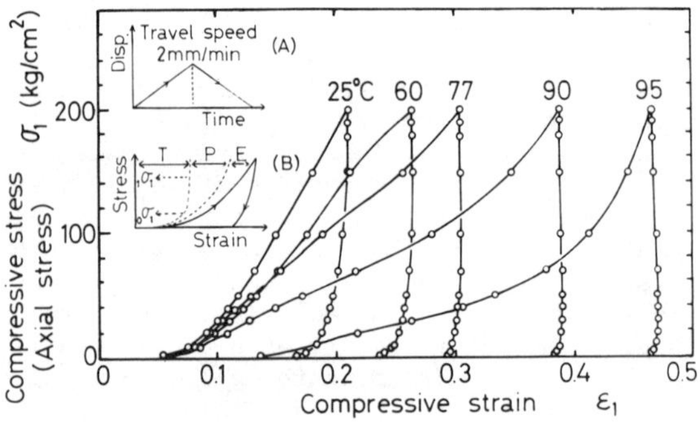

Fig. 9. Stress (σ_1) - strain (ε_1) diagrams during the loading and the unloading procedures.

to the conclusion hitherto obtained in the field of the micromeritics [13, 14], a strain value in a stress-strain diagram is composed of an elastic part (E), a plastic part (P) of constituent particles and a translational part (T) raised from a sliding movement between particles, which also contributes to a plastic deformation of a packed bed, as indicated in the insertion figure. That is, the total strain (ε_1) is expressed as

$$\varepsilon_1 = T + P + E \tag{3}$$

under the stress of ${}_0\sigma_1$, but the translational part is saturated at a higher stress (${}_1\sigma_1$). When the stress is removed, only an elastic part (E) can be restored and the residual strains are not recovered. With increasing temperature, T and P parts increase, but on the other hand, E part decrease; therefore, the hysteresis area which indicates the energy consumption during the compaction procedure increases with the testing temperature.

The radial stress effect is most pronounced in the powder system in the whole powder-liquid systems in Table I, because stress transmission is achieved only through the contact points between particles without interference of adsorbed liquid films. To investigate this problems more precisely, some stress path diagrams constructed between an axial stress (σ_1) and a radial stress (σ_2) are investigated and described in Fig. 10, which indicates that a radial stress (σ_2) increases approximately in proportion to an axial stress (σ_1) during a loading path. In an unloading path, however, this proportionallity is

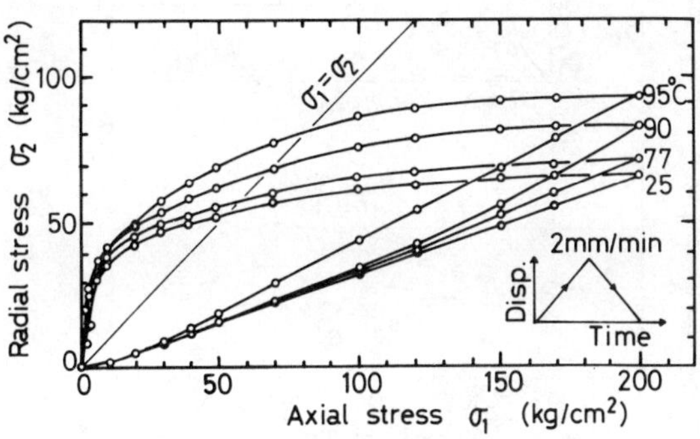

Fig. 10. Stress path diagrams during the loading and the unloading procedures.

not secured and σ_2 even exceeds σ_1 after the unloading stress path has crossed the line of $\sigma_1=\sigma_2$.

4.2. Determination of Rheological Properties of a Packed Bed.

(1) Stress relaxation of a packed bed

As one of the most effective means to ascertain this problem more precisely, the stress relaxation properties of the piled bed in the die are investigated employing the measuring program shown in the insertion figure in Fig. 11. In an usual case, relaxation properties are treated essentially as shearing character merely, not as dilatometric one. However, in such a powder compaction system treated here, both relaxation properties must be considered because an appreciable volumetric strain is governed during compaction process. Therefore, an exial stress (σ_1) and a radial stress (σ_2) are converted to a normal stress (σ_{smp}) and a shear stress (τ_{smp}) which stands for volumetric and shear contributions, respectively, acting upon spatial mobilized plane imaginally constructed within the powder bed. Equations (4) and (5) in Fig. 13 were utilized to calculate these convensions. Obtained results are indicated in Figs. 11 and 12. In an ordinary relaxation

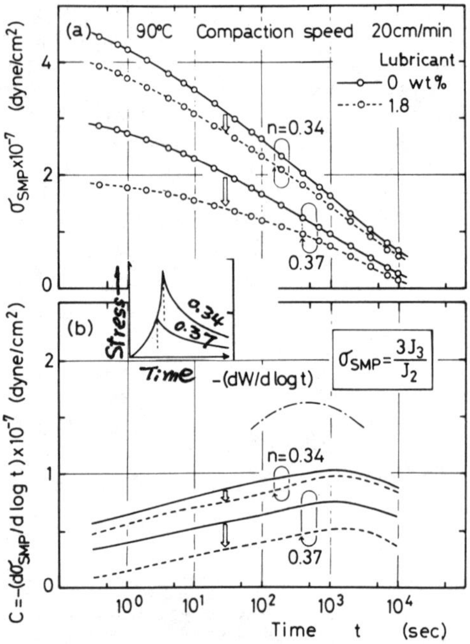

Fig. 11. Relaxation of normal stress on SMP σ_{SMP} under constant void fraction n.

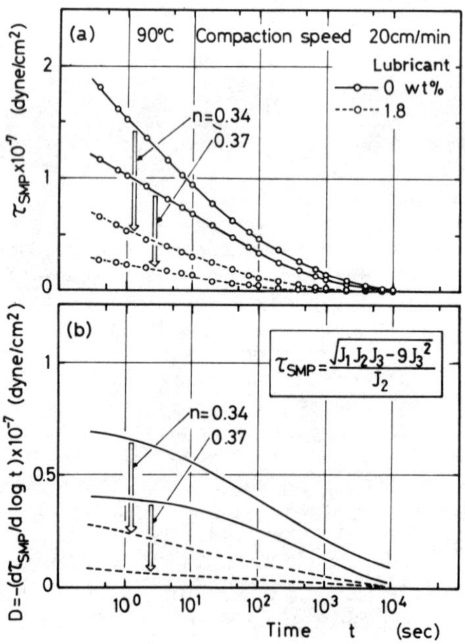

Fig. 12. Relaxation of shear stress on SMP τ_{SMP} under constant void fraction n.

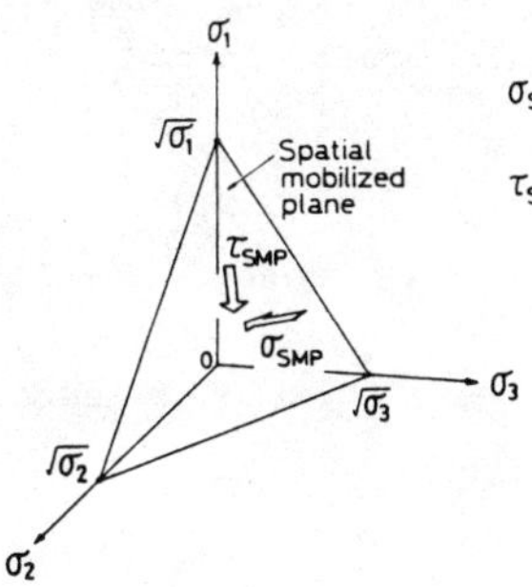

$$\sigma_{SMP} = \frac{3J_3}{J_2} \qquad (4)$$

$$\tau_{SMP} = \frac{\sqrt{J_1 J_2 J_3 - 9 J_3^2}}{J_2} \qquad (5)$$

$$J_1 = \sigma_1 + \sigma_2 + \sigma_3$$
$$J_2 = \sigma_1\sigma_2 + \sigma_2\sigma_3 + \sigma_3\sigma_1$$
$$J_3 = \sigma_1 \sigma_2 \sigma_3$$
$$\left(\begin{array}{l}\sigma_2 = \sigma_3 \\ \text{for axial symmetry}\end{array}\right)$$

Fig. 13. Projection to a Spatial Mobilized Plane.

phenomenon, a relaxation modulus G(t) obtained for an infinitesinal strain gives a spectrum H(log λ) concerning to the relaxation time λ which is calculated as $C = -(d\sigma_{SMP}/d\log t)$ and $D = -(d\tau_{SMP}/d\log t)$. These calculated C and D values will provide an effective key to evaluate the relaxation phenomenon of the packed structure. Calculated results are indicated in the lower portion of the same figures, from which it can be seen that the spatial mobilized shear stress (τ_{SMP}) reduces more rapidly than the spacial mobilized normal stress (σ_{SMP}), and that the maximum value of the C-spectrum is found at the time scale of 10^3 - 10^4 sec for every packed compact ranging from tightly (n=0.34) to loosely n=0.41) compacted beds. On the other hand, only decaying trend is observed in the D-spectrum for every compact. It can be concluded that, in the relaxation of a compacted powder bed in the die, the reduction by shearing which means a dilatometric character originated by a sliding motion between particles is governed initially, and the relaxation by deformation character of individual particles is followed, having the owen relaxation time respectively as mentioned above.

(2) Relaxation Spectra G(μ) and G(χ)

Thus, one could determine the shape about the first type distribution function described as G(μ) in the equation (1) which manifests the stress relaxation after effect.

In the second step, let us try to decide about the second type distribution function describe as $G(\chi)$ in the equation (1) which manifests the stress hereditary effect.

For this purpose, the second derivative of the equation (1) is calculated after treating in the limiting condition of

$$\tau < 0, \quad \varepsilon(\tau) = 0$$

$$\tau > 0, \quad \dot{\varepsilon}(\tau) = V_{\tau}, \quad t(\chi) = \chi/V, \quad V = \text{const.} \tag{3}$$

As the results of this treatments, the next equation is established

$$\frac{d^2(\sigma)V = 0}{d\varepsilon^2} = G(\chi) \tag{4}$$

The second type distribution function $G(\chi)$ is thus determined here.

Calculated results are shown in Fig. 14. By the aid of this figure, it is clear that the second type distribution function has its peaks at about ε=0.05 in the lower temperature (25°C), at about ε= 0.3 in the higher temperature (95°C) and in the middle temperature between them (60°C) two distinguished peaks are observed.

Adding the result obtained in the previous section to this conclusion, it can be concluded that the first peak is corresponded to the transrational mechanism and the second peak to the deformation mechanism of constituent particles themselves.

The same treatment obtained by a visco-plastico-elastic solid can be applied to a visco-plastico-elastic liquid or a visco-plastic-elastic solid-like or liquid-like materials using equation (2), however, descriptions about them are omitted here.

(3) Shearing properties of a packed bed.

As another means to evaluated the rheological properties of a packed bed, the shearing tests under a constant volume and constant load are employed using a shearing box. The dependency of a shear stress (τ) and volume dilation (ΔV) on a shearing displacement (d) (τ-d, ΔV-d relations) obtained under a constant load, and the dependence of a shear stress (τ) and a compaction stress (σ) on a shearing displacement (d) (τ-d, σ-d relations) obtained under a constant volume are investigated and the results are shown in Figs. 15 and 16 using the same polystyrene powder employed in the previous section. As known in the micromeritics, τ indicates its maximum value (τ_m) during the volume dilation takes place (dotted curves in Fig. 15) or the compaction stress (σ) has an increasing tendency (dotted curves in Fig. 16) and indicates a monotonously increasing trend to the equilibrium value (τ_m) when the volume shrinkage takes place (solid curves

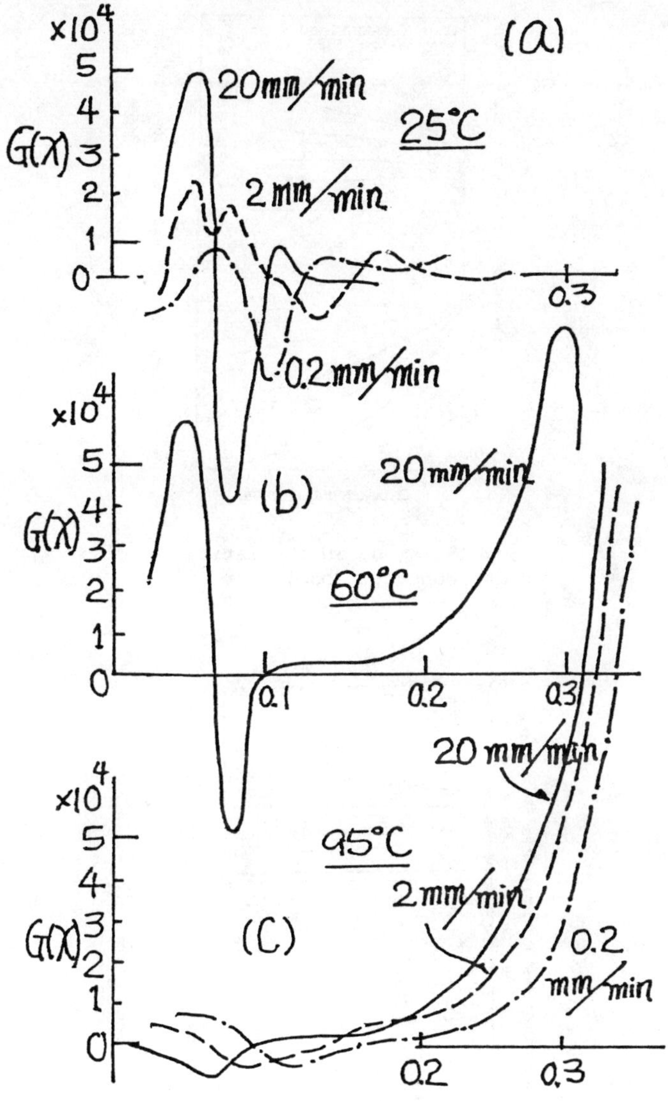

Fig. 14. The second type distribution functions obtained at various testing temperatures.

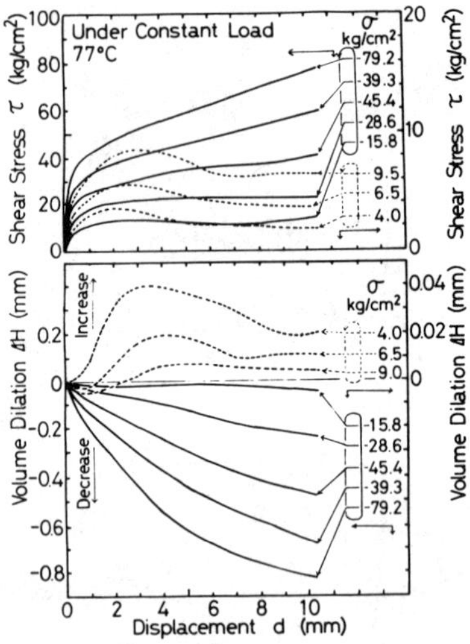

Fig. 15. τ-d, ΔH-d relation ships under constant load.

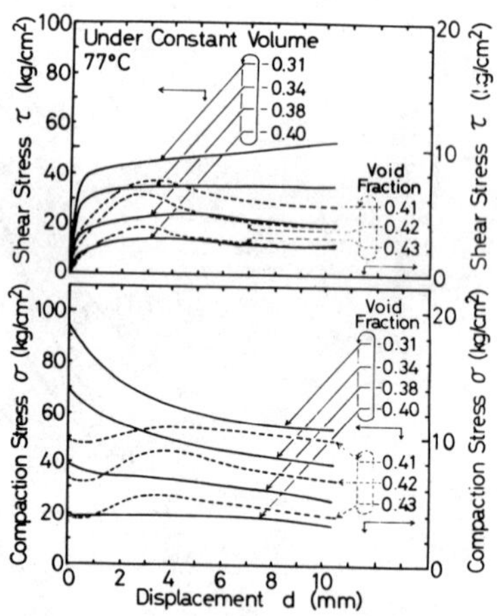

Fig. 16. τ-d, σ-d relation ships under constant volume.

in Fig. 15) or the compaction stress has a decreasing tendency (solid curves in Fig. 16). From these results, the relation of a shear stress (τ) vs a compaction stress (σ) can be estimated as shwon in Fig. 17(a) under a constant load and in Fig. 17(b) under a constant volume, in which a parameter means a shearing displacement (d). Three τ-σ relations selected at different displacements (0.2, 3 and 9mm) are indicated in Fig. 18, from which it can be concluded that τ-σ relations

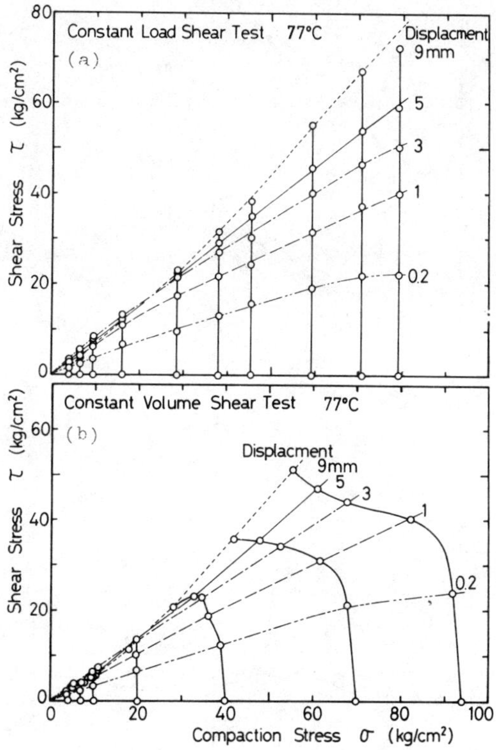

Fig. 17. τ-σ relations obtained from shear tests, (a) under constant load, (b) under constant volume.

under a constant load and under a constant volume show a good agreement in the higher stress range, but the results under constant load show always a larger value in the lower stress range. The difference observed between these two types of τ-σ relations was once interpreted by the author as that the translational movement of individual particles

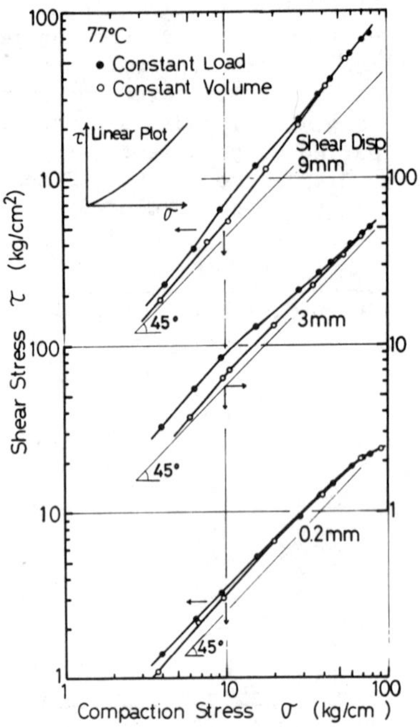

Fig. 18. A cross reference of τ-σ relations obtained from shear tests under constant load and under constant volume.

was governed mainly in the lower stress range but the deformational contribution of the constituent particles predominated in the higher shearing range [15]. However, accompanyed supplemental explanation must be added here to interpret this phenimenon more sufficiently, which is ascertained also by the author using a two-dimensional shearing box and described as the bridging effect due to the dilatant behavior or the particles near the sharing plane in the shear box [16]. These results obtained from this direct observation by naked eyes also show that the transmitting movement of the constituent particles is predominated at the loosely packed stage but the deformation behavior of constituent particles is pronounced at tightly packed stage, which supports the conclusions mentioned in the previous section (3.2(1)).

5. SOME RHEOLOGICAL PROBLEMS IN THE MUD SYSTEM

The rheological properties of the mud system may be supposed to have both viscous character of slurry stage and plastic character of powder stage. Therefore, the mechanical behavior of the mud system may be quite complicated. The most striking phenomenon in the mud region is the so-called liquid content abnormality as can be recognized from the Fig. 19 obtained for the clay powder-vaseline (vaseline is treated as liquid in this case) system investigated in the author's laboratory. As seen in the figure, pressure gradient in the die indicats the maximum

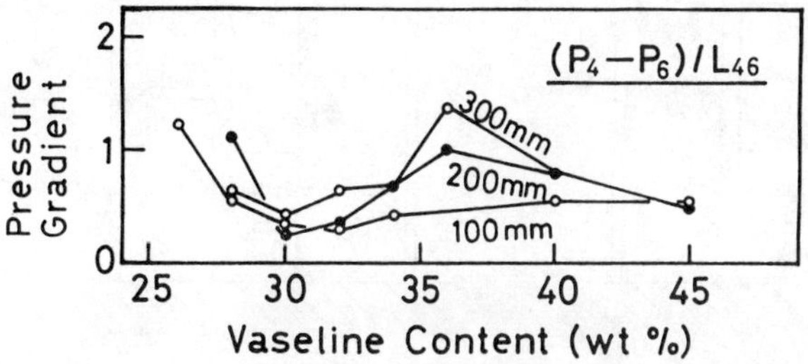

Fig. 19. Liquid content abnormality obtained clay powder - vaseline system.

value at about 36wt% of vaseline content and the minimum value at about 30wt%.

5.1. Relation between an Axial and a Radial Pressure in the extruding Procedure.

In order to investigate a consolidation properties within the die, exhausting properties in the die and extruding properties from the die, an apparatus shown schematically in Fig. 20 was constructed. The mud specimen held in the die is consolidated initially by the upper piston measuring an axial upper consolidation pressure ($P_{axi.U}$), an axial lower consolidation pressure ($P_{axi.L}$) and a radial pressure (P_{rad}).

The measuring procedure is as follows; after this consolidation process, an exhausting process is replaced, fixing the upper piston (Fig. 20(b)) and pulling up the container (the die) upward, which means the downward movement of the mud through the die, measuring also $P_{axi.U}$, $P_{axi.L}$ and P_{rad} values. When the mud specimen becomes naked from the lowest part of the die during this exhausting process, an extruding process takes place, measuring also $P_{axi.U}$, $P_{axi.L}$ and P_{rad} values. The measured results are shown in Fig. 21. In the consolidation range, the similar tendency mentioned in the powder stage is observed. In the exhausting range, however, the variation tendency of $P_{axi.U}$, $P_{axi.L}$ and especially P_{rad} are quite striking. It can be concluded that, during the exhausting movement, the axial stress ($P_{ax.U}$) is replaced by the radial stress (P_{rad}), but, after entired into the extruding range, all stresses including the radial stress show the

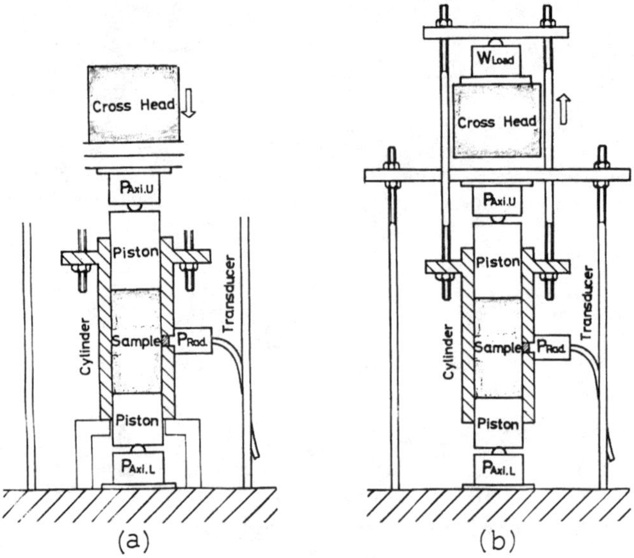

Fig. 20. Employed apparatus to measure the axial and radial stresses in a consolidation, an exhausting and an extruding procedures.

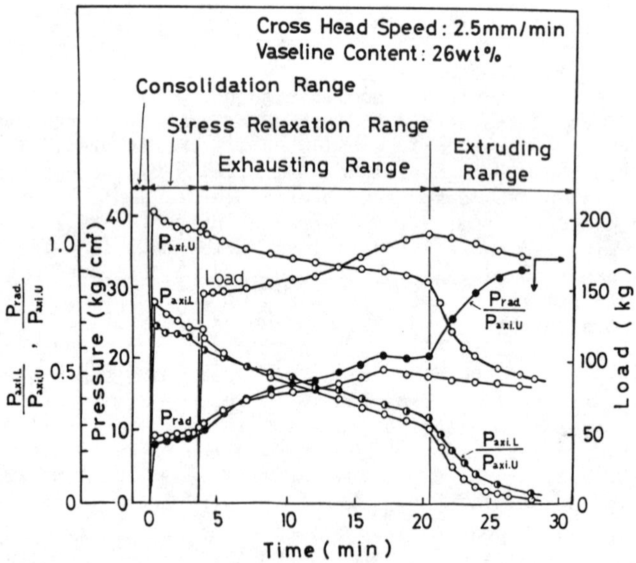

Fig. 21. $P_{axi.U}$, $P_{axi.L}$, P_{rad}, $P_{rad}/P_{axi.U}$ and $P_{axi.L}/P_{axi.U}$ values obtained in the consolidation, exhausting and extruding procedures.

decreasing tendency.

5.2. Consideration of Liquid Content Abnormality from the Concept of Rankine's Coeffecient.

A Rankine's coefficient ($P_{rad}/P_{axi.U}$) and a so-called Shaxby coefficient ($P_{axi.L}/P_{axi.U}$) which mean transmitting properties of $P_{axi.U}$ to the radial and axial directions respectively, are calculated and shown in the same figure. It can be seen that the transmitting pro-

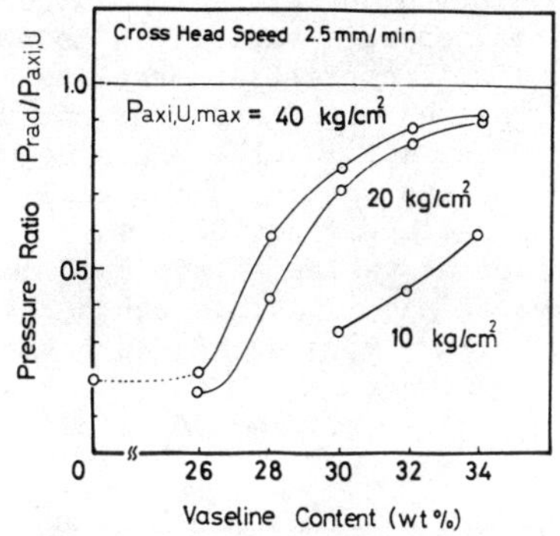

Fig. 22. Vaseline content dependence of Rankine's coefficient ($P_{rad}/P_{axi.U}$).

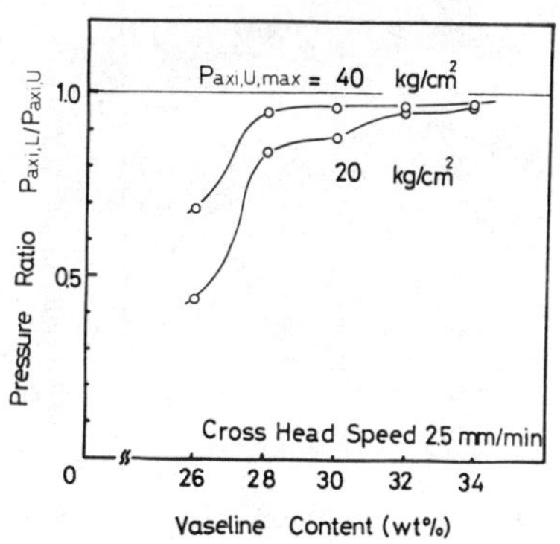

Fig. 23. Vaseline content dependence of Sahxby value ($P_{axi.U}/P_{axi.U}$).

perties to the radial dirction are improved remarkably sacrificing the transmitting properties to the downward die with the testing time, in the exhausting and especially in the extruding period, respectively. A Rankine's coefficient and a Shaxby value obtained at the entrance to the exhausting period are plotted against the vaseline content (liquid content) in Figs. 22 and 23 as a typical example, which show an increasing tendency from lower transmitting character ($P_{rad}/P_{axi.L} = 0.2$, $P_{axi.L}/P_{axi.U} = 0.45$) at vaseline contents of 26wt% with a saturating tendency into the uniform stress field ($P_{rad}/P_{axi.L} = 0.9$, $P_{axi.L}/P_{axi.U} = 0.97$) at the vaseline content of 34wt%. The similar results are also observed in any exhausting or extruding periods investigated by another mud tested in auther's laboratory but they are not indicated here to avoid complexity. It may be said that this result obtained from Fig. 22 is to provide an effective key to solve the so-called liquid content abnormality in the mud range shown in Fig. 19. Because the radial stress gradient valueds are introduced as an ordinate measure in Fig. 17, the pressure gradient $(P_4-P_6)/L_{46}$ value can be influenced by the Rankine's coefficient as it means the transmitting capasity to the radial direction. Therefore it can be said that if the Rankine's coefficient varies with liquid content as shown in Fig. 24, the flow properties of mud materials can be altered from the flow curve corresponding to R. C. = 0.9 in the 34wt% vaseline content to the other flow curve corresponding to C = 0.2 in the 24% vaseline content gradually as indicated in Fig. 24; here, two dotted decreasing curves in

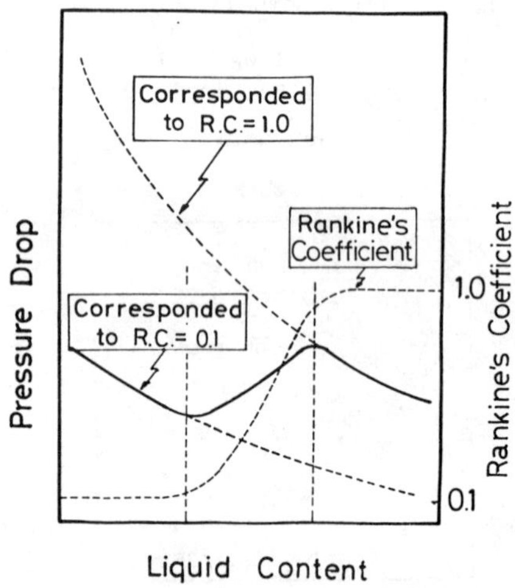

Fig. 24. Relation between pressure drop and Rankine's coefficient obtained in the mud range of clay powder - vaseline system.

the figure show the pressure droping trend when it is assumed that Rankine's coefficients are not affected by the liquid content. Eventually, it can be said that the mud stage is described as the transition range from the liquidlike region having a large Rankine's coefficient (R.C. = 0.9) to the solidlike region having a smaller Rankine's coefficient (R.C. = 0.2), and this transition phenomenon can be evaluated as the liquid content abnormality in the mud stage.

6. CONCLUSIONS

Using the packing character scheme shown in Table I, a systematic interpretation of rheological behavior for powder-liquid systems has been developed. From these considerations, three stages of packing can be defined.

They are;

(1) the slurry stage in which a free medium phase is continuous,

(2) the powder stage in which a particle phase is essentially continuous, and

(3) a mud stage which exhibits a continuous film of liquid.

Therefore, rheological behavior may be described as either the flow of a slurry stage, the plastic deformation of a packed bed in a powder stage, and the flow-deformation property of a mud stage. It is thus reasonable to consider that the continuous phase determines its own mechanics at each stage. A complete interpretation influenced through the packing character, schematically shown in Table I, is desirable but not yet attainable due to insufficient rheological data over the full composition range. The author is of opinion that a radial stress originating from an axial stress is a key to resolve this problem. That is, the normal stress defference effect predominates with the concentration of suspension in the slurry stage and this effect predominates with the concentration of suspension in the slurry stage and this effect induces a decay of the Rankine's coefficient at extreme slurry concentration. Thus, the phenomena proposed as the liquid content abnormality in the mud stage can be introduced as shown in Fig. 24. Finally, a solidlike character which can be represented by the lower value of the Rankine's coefficient arises in the powder stage. In this interpretation, for instance, the corelation between the normal stress difference effect in the slurry stage and the radially transmitted stress in the powder stage is not elucidated sufficiently. Moreover, the relation between viscous drag in the slurry stage and the yielding character of the mud stage comprehended as a frictional drag is not yet well understood.

A generalized explanation for the whole range of the powder-liquid compositions may be reconsidered after these unknown factors are elucidated. However, the interpretation developed in this paper may be useful as one model for developing further experimental programs to provide a more complete explanation.

REFERENCES

1. Umeya, K., "Rheology for Chemical Investigaters", ed. by S. Onogi, p.175 (1959).

2. Umeya, K., T. Isoda, T. Ishii, and K. Sawamura, Powder Technol., 3, 259 (1969/70).

3. Umeya, K., "Proc. 5th Int. Cong. Rheol.", vol.2, p.295 (1970) ed. by S. Onogi, University of Tokyo Press.

4. Umeya, K., T. Kanno, M. Wagatsuma, and H. Igarashi, Preprint presented in "Joint Meeting U. S. - Japan Societies of Rheology", Honolulu, Hawaii, June 3, 1975.

5. Umeya, K., T. Kanno, and M. Wagatsuma, J. Soc. Rheol., Japan, 3, 108 (1975).

6. Umeya, K., T. Kanno, M. Wagatsuma, and H. Igarashi, J. Soc. Rheol., Japan, 3, 145 (1975).

7. Umeya, K., T. Kanno, and K. Kato, J. Soc. Rheol., Japan, 2, 17 (1974).

8. Philippoff, W., and F. H. Gaskins, Trans. Soc. Rheol., 2, 263 (1958).

9. Bagley, E. B., Trans. Soc. Rheol., 5, 355 (1961).

10. Metzner, A. B., W. T. Houghton, R. A. Sailor, and J. L. White, Trans. Soc. Rheol., 5, 133 (1961).

11. Bogue, D. C., and J. L. White, "Engineering Analysis of Non-Newtonian Fluid, Section 4", AGARDograph 144 (1971).

12. Umeya, K., T. Kanno, and A. Toyooka, J. Soc. Mater. Sci., Japan, 22, 410 (1973).

13. Cooper, A. R., and L. E. Eaton, J. Amer. Ceram. Soc., 45, 97 (1962).

14. Umeya, K., R. Hara, J. Kikuta, and T. Isoda, J. Soc. Mater. Sci., Japan, 23, 549 (1974).

15. Umeya, K., R. Hara, T. Yoshida, and S. Narihiro, J. Soc. Mater. Sci. Japan, 19, 521 (1970).

16. Umeya, K., R. Hara, and J. Kikuta, J. Chem. Eng., Japan, 8, 56 (1975).

17. Sawaragi, G., Zairyo-Shiken (Japan), 3, 317 (1953), etc.

18. Umeya, K., Continum Mech. and Stast. Approaches in the Mechanics of Granular Mater. p.222 (1978).

SECTION 6

MECHANICS OF GRANULAR MATERIALS AND SOILS

AN ELASTIC-PLASTIC ISOTROPIC CONSTITUTIVE MODEL FOR SAND

George Y. Baladi

Research Civil Engineer, Geomechanics Division, Structures Laboratory,
U.S. Army Engineers Waterways Experiment Station
Vicksburg, MS, USA

ABSTRACT

A three-dimensional elastic-plastic isotropic constitutive model has been developed for simulating the stress-strain response of Ottawa sand. The constitutive model contains 20 material constants that can be determined from standard laboratory tests such as isotropic consolidation and triaxial tests. The model is capable of reproducing the hysteretic behavior of the material under both hydrostatic and deviatoric states of stress, and it accounts for the effect of superimposed hydrostatic stress on the shearing response. In addition, the model accounts for shear-induced volume change.

Two sets of data were used in developing the model. The first set was used in formulating and fitting the model; the second in analyzing the model's predictive capability. It can be concluded that the model does a very good job both qualitatively and quantitatively of representing the behavior of Ottawa sand.

1. INTRODUCTION

The constitutive model described in this paper is of the "cap" type [1-5]. It is used to give a mathematical description of the behavior of geological materials which is based on the classical theory of plasticity. The yield surface for the model consists of a failure envelope, which is usually fixed in stress space (ideally plastic), and a moveable cap whose position is a function of the volumetric strain history of the material. The model uses an associated flow rule to satisfy theoretical continuity and uniqueness requirements. It controls dilatancy while it allows hysteresis and apparent Bauschinger effects, in a load-unload cycle. With a hardening failure envelope, the model can represent cyclic hysteresis. Also, material anisotropy, rate-dependency, and the effects of pore fluid can all be included in the model.

The model is the natural development of earlier work, starting with the Drucker-Prager elastic-plastic models [6]. The idea of adding a moveable cap is due to Drucker, Gibson, and Henkel [7] and Jenike and Shield [8]. Finally, Roscoe and his co-workers [9] at

Cambridge University have formulated a cohesionless model which involves a strain-hardening yield function of volumetric compressive strain, but has doubtful uniqueness and continuity properties.

The cap model has been used to simulate the behavior of a wide range of geological and man-made materials such as soil, rock, concrete, etc. This paper describes the development of a three-dimensional elastic-plastic isotropic cap model for Ottawa sand. The constitutive relationship is capable of reproducing the hysteretic behavior of the material under both hydrostatic and deviatoric states of stress; it also accounts for shear-induced volume change and the effect of superimposed hydrostatic stress on shearing response. The capability of the constitutive relationship for simulating the observed stress-strain behavior of Ottawa sand is examined.

2. CONSTITUTIVE MODEL

The basic premise of elastic-plastic constitutive models is the assumption that certain materials are capable of undergoing small plastic (permanent) as well as elastic (recoverable) strain at each loading increment. Mathematically, the total strain increment, $d\varepsilon_{ij}$, is assumed to be the sum of the elastic strain increment, $d\varepsilon^E_{ij}$, and the plastic strain increment, $d\varepsilon^P_{ij}$, i.e.,

$$d\varepsilon_{ij} = d\varepsilon^E_{ij} + d\varepsilon^P_{ij} \quad . \tag{1}$$

2.1 Elastic Behavior

The behavior of the model in the elastic range is governed by the following incremental equations:

$$dJ_1 = 3K\, d\varepsilon^E_{kk} \tag{2}$$

$$dS_{ij} = 2G\, de^E_{ij} \quad , \tag{3}$$

in which dJ_1 = the first invariant of the stress increment tensor; $d\varepsilon^E_{kk}$ = the elastic volumetric strain increment; dS_{ij} and de_{ij} = the stress and strain deviation increment tensors, respectively; and K and G = the elastic bulk and shear moduli, respectively.

The elastic bulk modulus describes the unloading portion of the pressure-volumetric strain response of the material. The following expression is proposed for the elastic bulk modulus:

$$K = \frac{K_i}{1 - K_1}\,[1 - K_1 \exp(-K_2 J_1)] \quad , \tag{4}$$

where K_i = initial elastic bulk modulus and K_1 and K_2 = material constants.

The elastic shear modulus describes the unloading portion of the deviatoric stress-strain response of the material. The following expression is proposed for the elastic shear modulus:

$$G = G_i + G_1[1 - \exp(-G_2\, W^{Pf})] \, , \tag{5}$$

where G_i = initial elastic shear modulus; G_1 and G_2 = material constants; and W^{Pf} = the plastic work associated with the yield envelope.

2.2 Plastic Behavior

The behavior of the model in the plastic range is governed by the flow rule

$$d\varepsilon_{ij}^{P} = d\lambda \frac{\partial \mathfrak{f}}{\partial \sigma_{ij}} \, , \tag{6}$$

in which $d\varepsilon_{ij}^{P}$ and σ_{ij} = the plastic strain increment and the stress tensor, respectively; $\mathfrak{f}$ = the loading function; and $d\lambda$ = a positive scalar factor of proportionality, which is nonzero only when plastic deformation occurs and is dependent on the particular form of the loading function $\mathfrak{f}$.

The loading function $\mathfrak{f}$ is assumed to be isotropic and to consist of two parts: an elliptically shaped strain-hardening yield surface described by

$$F(J_1 , \sqrt{\bar{J}_2} , \kappa) = (J_1 + C)^2 + 6\bar{J}_2 - [X(\kappa) + C]^2 = 0 \, , \tag{7}$$

and a Prager-Drucker type yield envelope defined by

$$f(J_1 , J_2 , \alpha) = \frac{\sqrt{\bar{J}_2}}{J_1 + C} - \alpha = 0 \, , \tag{8}$$

where $\bar{J}_2$ = the second invariant of the stress deviation tensor; C = a material constant representing the location of the center of the strain-hardening yield surface F and coinciding with the intersection of the yield envelope f and the J_1 axis; κ = a hardening parameter which is generally a function of the history of plastic volumetric strain; $X(\kappa)$ = the intersection of the strain-hardening surface with the J_1 axis; and α = a work-hardening and work-softening parameter which allows the yield envelope f to expand and contract isotropically (Fig. 1).

The hardening parameter κ in this model was chosen to be

$$\kappa = \varepsilon_{kk}^{PH} = A[1 - \exp(-DX)] \,, \tag{9}$$

where D and A = material constants and ε_{kk}^{PH} = the plastic volumetric strain associated with the strain-hardening yield surface F. The parameter A defines the maximum plastic volumetric compaction that the material can experience under hydrostatic loading. Note that the strain-hardening surface F is chosen so that the plastic strain increment vector associated with it is always parallel to the yield envelope f.

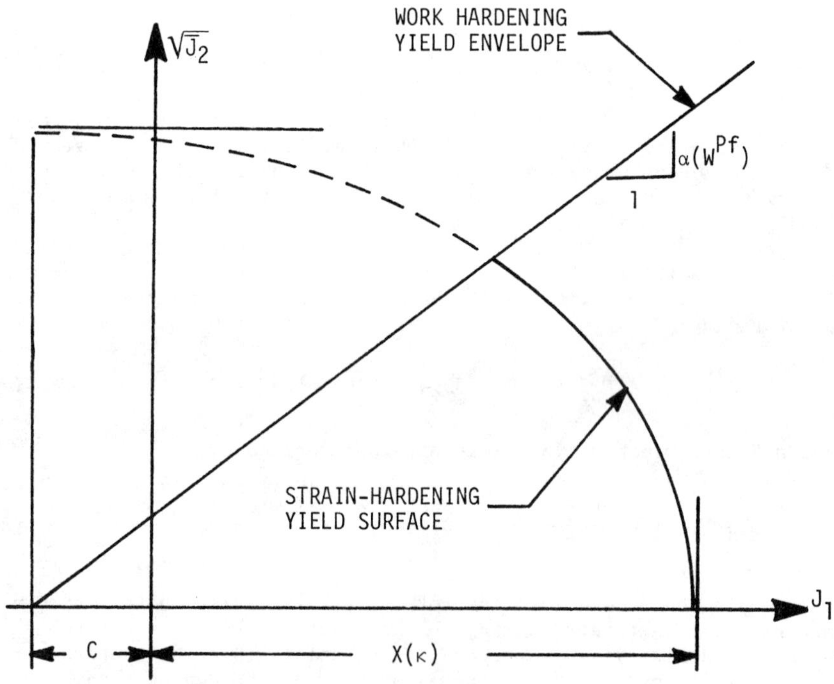

FIG. 1.--Proposed Yield Surfaces for Ottawa Sand

The parameter α in Eq. 8 is assumed to be a function of the plastic work W^{Pf} associated with the yield envelope f through the following expression:

$$\alpha = C_1[1 - \exp(-C_2 W^{Pf})] + C_3[1 - \exp(-C_4 W^{Pf})]$$

$$+ C_5[1 - \exp(-C_6 W^{Pf})] + \alpha_1 , \quad (10)$$

where C_2, C_4, and C_6 = material constants, and α_1 = initial value of α, and C_1, C_3, and C_5 = functions of the stress ratio determined through the following expressions:

$$C_1 = (1 - m) C_{1max} + mC_{1min} \quad (11)$$

$$C_3 = (1 - m) C_{3max} + mC_{3min} \quad (12)$$

$$C_5 = (1 - m) C_{5max} + mC_{5min} \quad (13)$$

$$\alpha_1 = (1 - m) \alpha_{1max} + m\alpha_{1min} , \quad (14)$$

where C_{1max}, C_{1min}, C_{3max}, C_{3min}, C_{5max}, C_{5min}, α_{1max}, and α_{1min} = material constants. The proportionality factor $d\lambda$ in Eq. 6 can be shown to be (Baladi and Rohani [3])

$$d\lambda = \frac{3K \frac{\partial F}{\partial J_1}\left(d\varepsilon_{kk} - d\varepsilon_{kk}^{Pf}\right) + \frac{G}{\sqrt{J_2}} \frac{F}{\partial\sqrt{J_2}} S_{ij}\left(de_{ij} - de_{ij}^{Pf}\right)}{9K\left(\frac{\partial F}{\partial J_1}\right)^2 + G\left(\frac{\partial F}{\partial\sqrt{J_2}}\right)^2 - 3\frac{\partial F}{\partial J_1}\frac{\partial F}{\partial\varepsilon_{kk}^{PH}}} \quad (15)$$

for the strain-hardening surface. In the case of the yield envelope f (Eq. 8), due to the hardening parameter α (Eq. 10), it is more convenient to express $d\lambda$ in terms of plastic work. Thus,

$$d\lambda = \frac{dW^{Pf}}{\frac{\partial f}{\partial\sigma_{ij}}\sigma_{ij}} \quad (16)$$

for the yield envelope f, where $d\varepsilon_{kk}^{Pf}$ and de_{ij}^{Pf} = the volumetric and deviatoric components, respectively, of the plastic strain increment tensor associated with f. The total strain increment tensor given by Eq. 1 can, therefore, be written as

$$d\varepsilon_{ij} = d\varepsilon^{E}_{ij} + d\varepsilon^{PH}_{ij} + d\varepsilon^{Pf}_{ij} \quad , \tag{17}$$

where $d\varepsilon_{ij}$, $d\varepsilon^{E}_{ij}$, $d\varepsilon^{PH}_{ij}$, and $d\varepsilon^{Pf}_{ij}$ = the total strain increments, the elastic strain increments, the plastic strain increments associated with the hardening function, and the plastic strain increments associated with the yield condition, respectively.

In summary, there are 20 material constants that must be determined experimentally. The material constants K_i, K_1, K_2, A, and D describe the complete hydrostatic response of the material (loading, unloading, and reloading). The material constants G_i, G_1, and G_2 describe the shearing response of the material in the elastic range (i.e., during unloading). The material constants C_{1max}, C_{1min}, C_2, C_{3max}, C_{3min}, C_4, C_{5max}, C_{5min}, C_6, α_{1max}, α_{1min}, and C describe the yield condition of the material and the rate of its expansion with plastic work. These constants can be readily determined from standard triaxial compression (with radial strain measurements) and isotropic consolidation tests.

3. COMPARISONS OF LABORATORY TEST DATA WITH MODEL BEHAVIOR

3.1 Experimental Program

The material used in the experimental program was an Ottawa sand whose gradation is given in Fig. 2. The sand was compacted by aerial pluviation to 1.73 gm/cc (relative density of 87 percent) and tested in the dry state three-dimensionally for its stress-strain properties. The stress paths shown in Fig. 3 were used to generate the data base for formulating and fitting the constitutive model. Then, the model was used to predict the stress-strain response of the same material when tested in the same test equipment along stress paths different from those used in the model formulation. The stress paths used for these predictions are shown in Fig. 4.

The test equipment used is a flexible, fluid cushion, cubical device with stress control. The vertical axis, z, and the horizontal axes, x and y, are principal stress axes.

3.2 Material Constants

As indicated above, there are 20 material constants in the present model that must be determined experimentally. The material constants K_i, K_1, and K_2 were determined from the unloading portion of the isotropic consolidation test. The material constants A and D

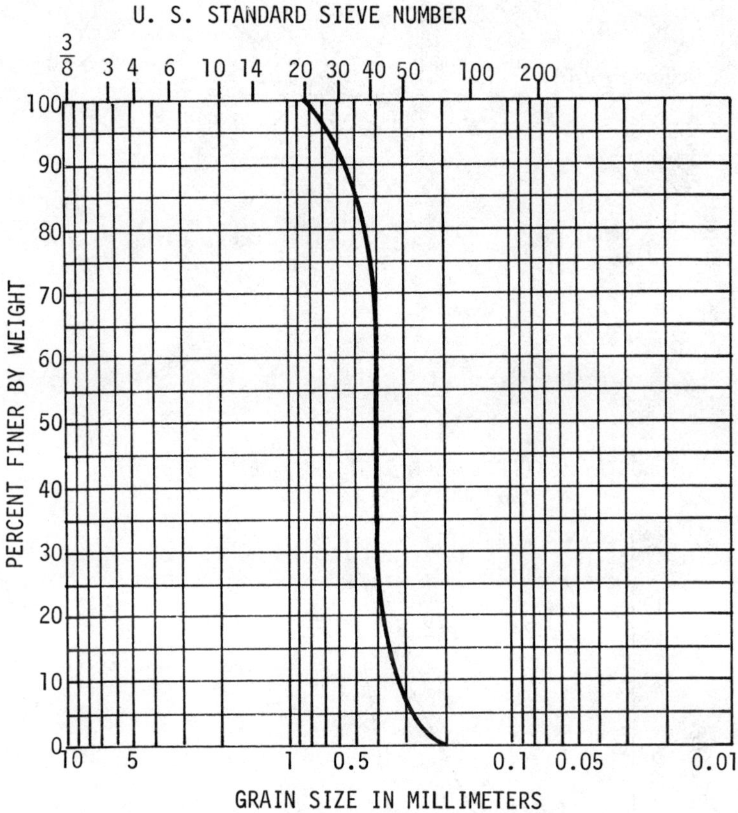

GRAVEL	SAND			SILT OR CLAY
FINE	COARSE	MEDIUM	FINE	

FIG. 2.--Grain Size Distribution for Ottawa Sand

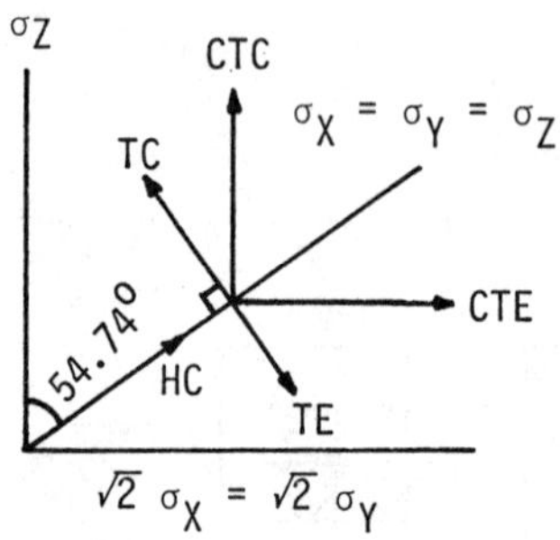

FIG. 3.--Stress Paths Used to Generate Data Base for Modeling

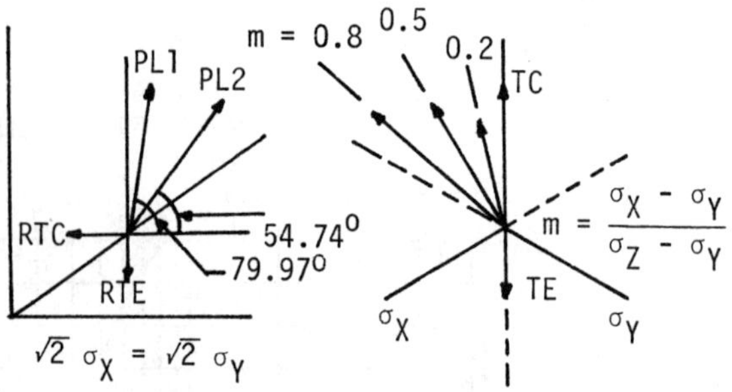

FIG. 4.--Stress Paths Used for Predictions

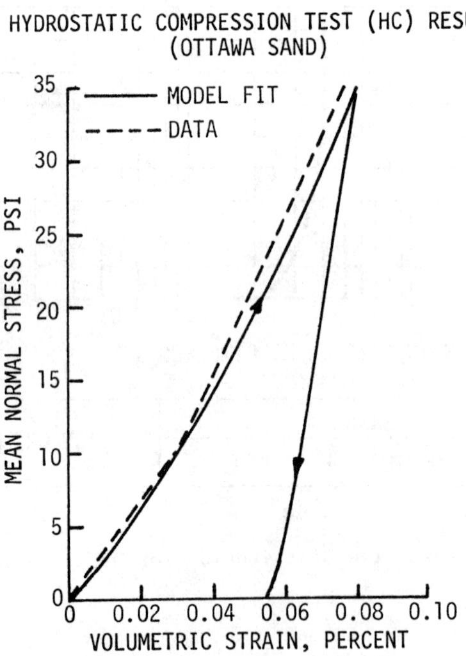

FIG. 5.--Isotropic Consolidation Response, Laboratory Measurements Versus Model Behavior

were determined from the loading portion of the pressure-volumetric strain response of the material using K_i, K_1, and K_2. The material constants G_i, G_1, and G_2 were determined from the unloading portion of the deviatoric stress-strain response of the material. The remaining constants were determined from the yield condition of the material and the rate of its expansion. The numerical values of these material constants are given in Table 1.

3.3 Correlation with Test Results

Actual test results for Ottawa sand are compared with model fits in Figs. 5-14. The comparison for effective mean normal stress versus volumetric strain for the isotropic compression test is shown in Fig. 5. Conventional triaxial compression test results are shown in Figs. 6 and 7 for constant lateral stresses of 10 and 5 psi and a stress ratio of 0. Model behavior versus test results for a conventional triaxial extension test is shown in Fig. 8. Triaxial compression test results for constant mean normal stresses of 5, 10, and 20 psi and for stress ratios of 0 and 1 are shown in Figs. 9-14.

4. MODEL PREDICTIONS

The following presents comparisons of the model predictions with the second set of data (see Fig. 4). The set of data was not used to fit the model. The predictions are for several simple shear tests with constant mean normal stresses, a reduced triaxial compression test with a stress ratio of 0, a reduced triaxial extension test with a stress ratio of 1, and two proportional loading tests with a stress ratio of 0.

The stress-strain behavior predicted by the model is compared to that measured for the Ottawa sand in Figs. 15-21. The format of these figures was chosen so that the ability of the model to predict the shear behavior and the shear-induced volume change of Ottawa sand could be easily examined. It is clear from these figures that the model does a very good job both qualitatively and quantitatively of representing the behavior of Ottawa sand, especially in the octahedral shear stress-octahedral shear strain space. However, the prediction in volumetric strain-octahedral shear strain space could be improved if unloading isotropic consolidation test results were available.

Table 1.--Numerical Values of Material Constants for Ottawa Sand

Material Constant			
Name	Notation	Unit	Numerical Value
Hardening failure envelope parameter	C	psi	2
	C_{max}	---	0.075
	C_{1min}	---	0.05
	C_2	lb-in.$^{-1}$	4,000
	C_{3max}	---	0.1
	C_{3min}	---	0.0667
	C_4	lb-in.$^{-1}$	800
	C_{5max}	---	0.085
	C_{5min}	---	0.0567
	C_6	lb-in.$^{-1}$	125
	α_{1max}	---	0.1
	α_{1min}	---	0.0667
Hardening surface parameter	A	---	0.0024
	D	psi^{-1}	0.002
Bulk modulus function	K_i	psi	37,500
	K_1	---	0.75
	K_2	psi^{-1}	0.05
Shear modulus function	G_i	psi	5,000
	G_1	psi	-3,000
	G_2	lb-in.$^{-1}$	250

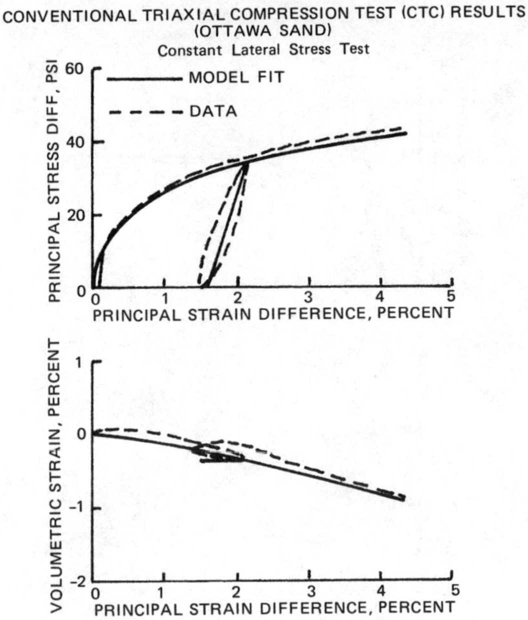

FIG. 6. - Principal Stress Difference and Volumetric Strain Versus Principal Strain Difference, Laboratory Measurements Versus Model Behavior (Lateral Stress = 10 psi; Stress Ratio, m = 0)

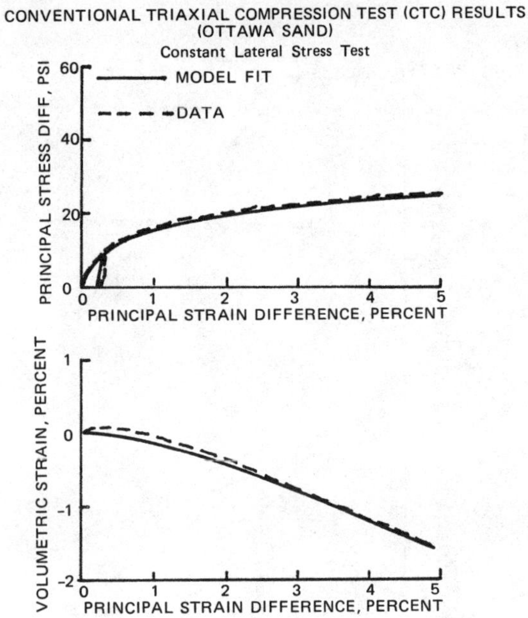

FIG. 7. - Principal Stress Difference and Volumetric Strain Versus Principal Strain Difference, Laboratory Measurements Versus Model Behavior (Lateral Stress = 5 psi; Stress Ratio, m = 0)

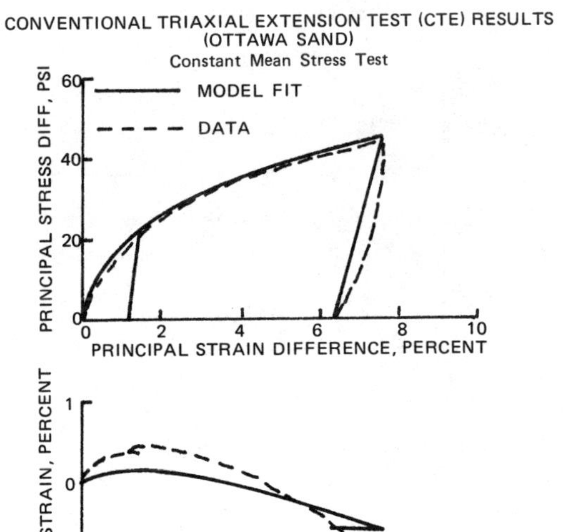

FIG. 8. - Principal Stress Difference and Volumetric Strain Versus Principal Strain Difference, Laboratory Measurements Versus Model Behavior (Stress Ratio, m = 1)

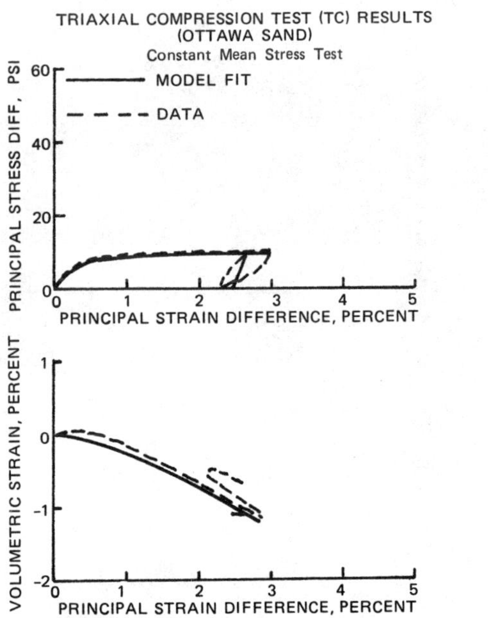

FIG. 9. - Principal Stress Difference and Volumetric Strain Versus Principal Strain Difference, Laboratory Measurements Versus Model Behavior (Mean Stress = 5 psi; Stress Ratio, m = 0)

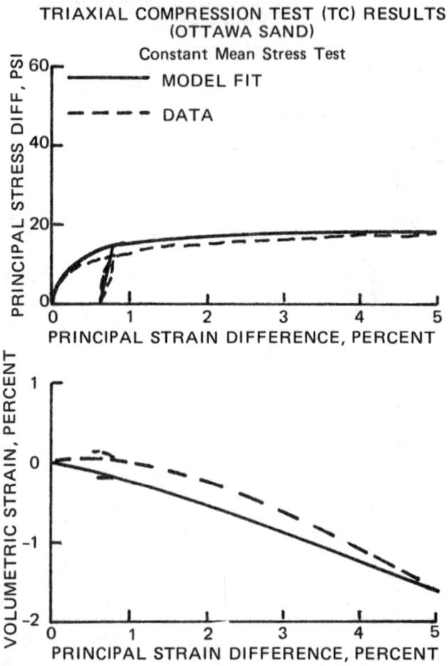

FIG. 10. - Principal Stress Difference and Volumetric Strain Versus Principal Strain Difference, Laboratory Measurements Versus Model Behavior (Mean Stress = 10 psi; Stress Ratio, m = 0)

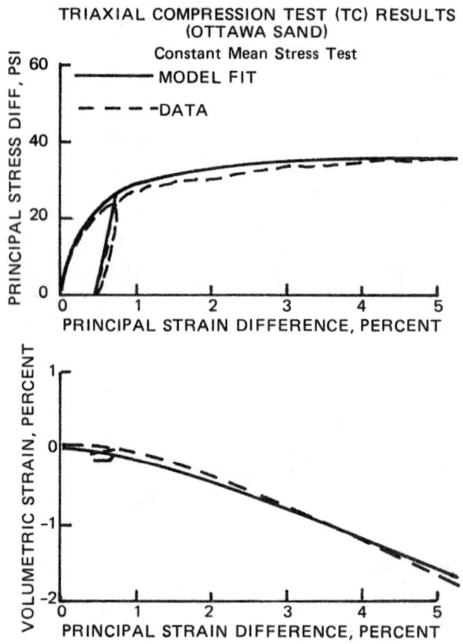

FIG. 11. - Principal Stress Difference and Volumetric Strain Versus Principal Strain Difference, Laboratory Measurements Versus Model Behavior (Mean Stress = 20 psi; Stress Ratio, m = 0)

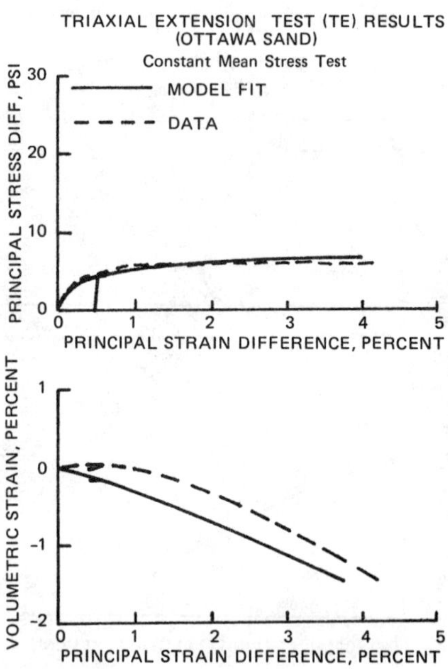

FIG. 12. - Principal Stress Difference and Volumetric Strain Versus Principal Strain Difference, Laboratory Measurements Versus Model Behavior (Mean Stress = 5 psi; Stress Ratio, m = 1)

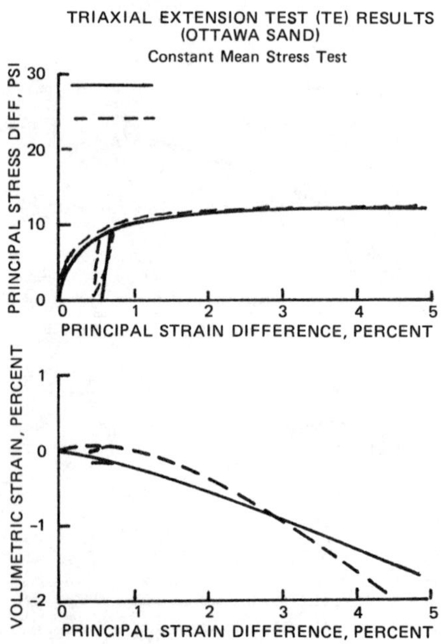

FIG. 13. - Principal Stress Difference and Volumetric Strain Versus Principal Strain Difference, Laboratory Measurements Versus Model Behavior (Mean Stress = 10 psi; Stress Ratio, m = 1)

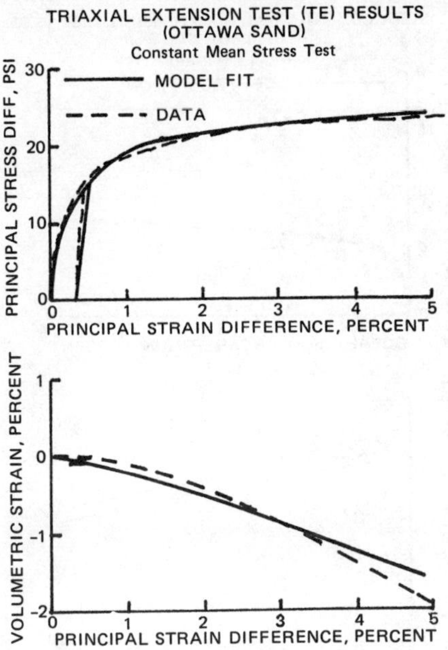

FIG. 14. - Principal Stress Difference and Volumetric Strain Versus Principal Strain Difference, Laboratory Measurements Versus Model Behavior (Mean Stress = 20 psi; Stress Ratio, m = 1)

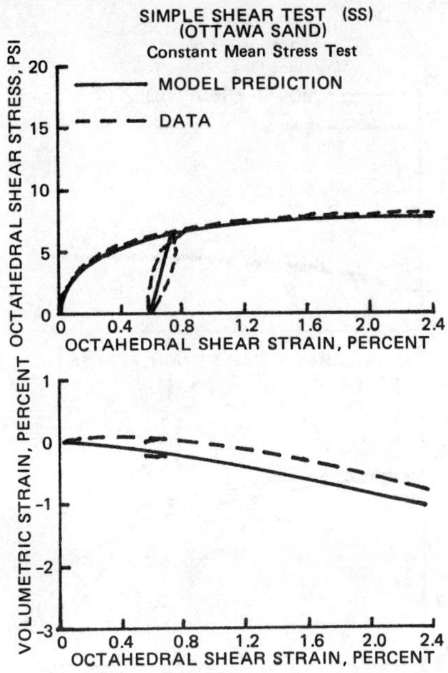

FIG. 15. - Octahedral Shear Stress and Volumetric Strain Versus Octahedral Shear Strain, Model Prediction Versus Laboratory Measurements (Mean Stress = 10 psi; Stress Ratio, m = 0.2)

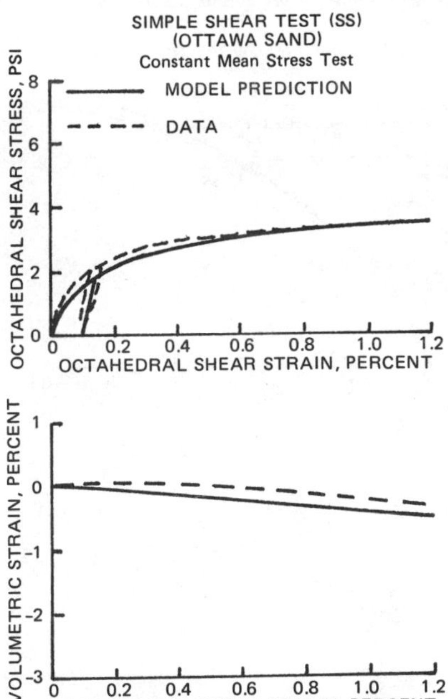

FIG. 16. - Octahedral Shear Stress and Volumetric Strain Versus Octahedral Shear Strain, Model Prediction Versus Laboratory Measurements (Mean Stress = 5 psi; Stress Ratio, m = 0.5)

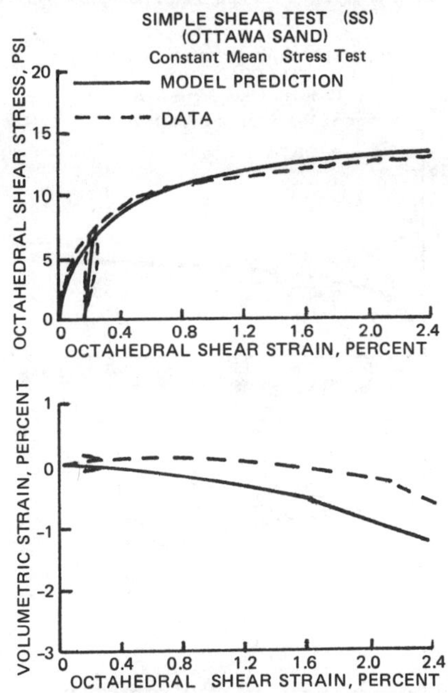

FIG. 17. - Octahedral Shear Stress and Volumetric Strain Versus Octahedral Shear Strain, Model Prediction Versus Laboratory Measurements (Mean Stress = 20 psi; Stress Ratio, m = 0.5)

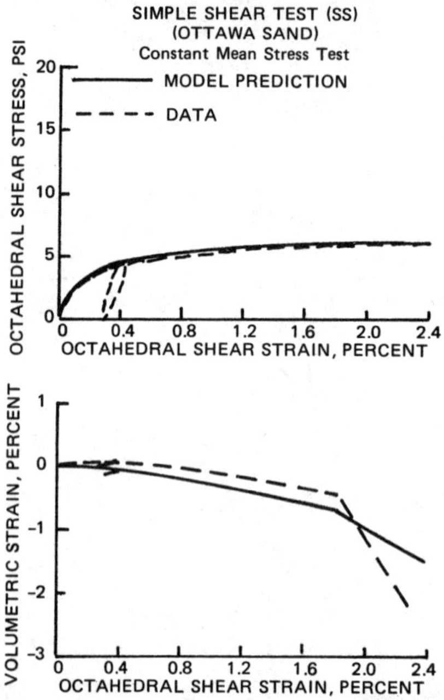

FIG. 18. - Octahedral Shear Stress and Volumetric Strain Versus Octahedral Shear Strain, Model Prediction Versus Laboratory Measurements (Mean Stress = 10 psi; Stress Ratio, m = 0.8)

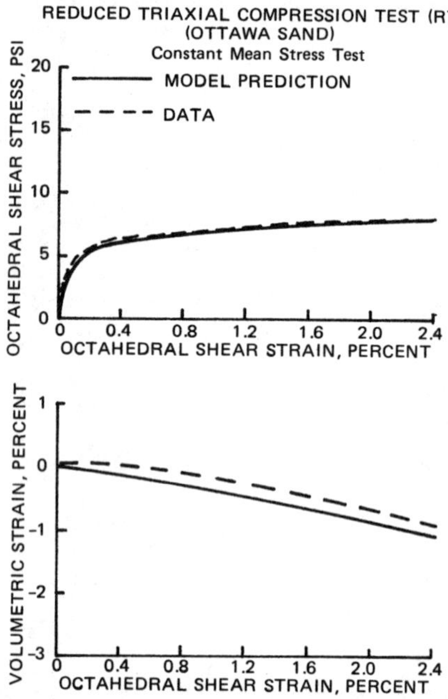

FIG. 19. - Octahedral Shear Stress and Volumetric Strain Versus Octahedral Shear Strain, Model Prediction Versus Laboratory Measurements (Stress Ratio, m = 0)

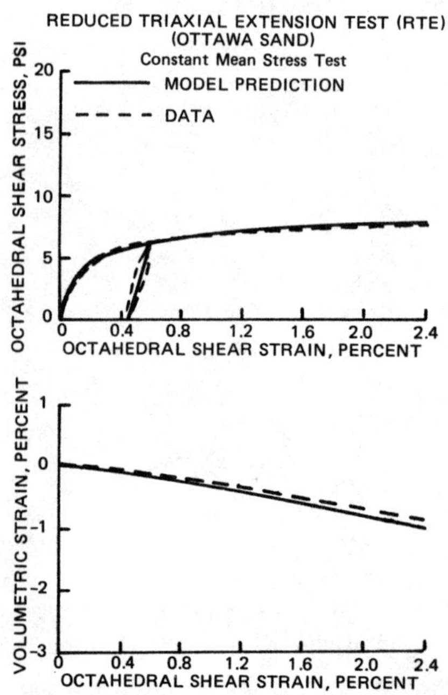

FIG. 20. - Octahedral Shear Stress and Volumetric Strain Versus Octahedral Shear Strain, Model Prediction Versus Laboratory Measurements (Stress Ratio, m = 1)

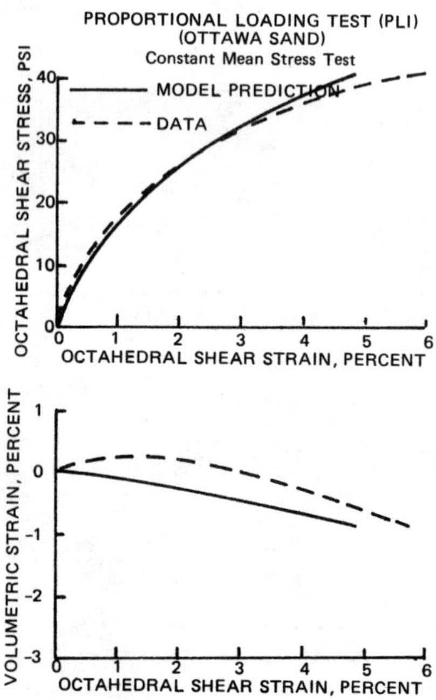

FIG. 21. - Octahedral Shear Stress and Volumetric Strain Versus Octahedral Shear Strain, Model Prediction Versus Laboratory Measurements (Stress Ratio, m = 0)

REFERENCES

1. Baladi, G. Y., "The Latest Development in the Nonlinear Elastic-Nonideally Plastic Work Hardening Cap Model," Symposium on Plasticity and Soil Mechanics, Cambridge, England, pp. 51-55 (1973).

2. Sandler, I. S., DiMaggio, F. L., and Baladi, G. Y., "Generalized Cap Model for Geological Materials," Journal of the Geotechnical Engineering Division, ASCE, Vol. 102, No. GT7, Proc. Paper 12243, pp. 683-699 (1976).

3. Baladi, G. Y., "Numerical Implementation of a Transverse-Isotropic Inelastic, Work-Hardening Constitutive Model," Transactions of the 4th International Conference on Structural Mechanics in Reactor Technology, Vol. M, Methods for Structural Analysis, San Francisco, Calif. (1977).

4. Baladi, G. Y. and Rohani, B., "Liquefaction Potential of Dams and Foundations; Development of an Elastic-Plastic Constitutive Relationship for Saturated Sand," Research Report S-76-2, Report 3, U. S. Army Engineer Waterways Experiment Station, CE, Vicksburg, Miss. (1977).

5. Baladi, G. Y. and Rohani, B., "Liquefaction Potential of Dams and Foundations; Development of a Constitutive Relation for Simulating the Response of Saturated Cohesionless Soil," Research Report S-76-2, Report 5, U. S. Army Engineer Waterways Experiment Station, CE, Vicksburg, Miss. (1978).

6. Drucker, D. C. and Prager, W., "Soil Mechanics and Plastic Analysis or Limit Design," Quarterly of Applied Mathematics, Vol. 14 (1956).

7. Drucker, D. C., Gibson, R. E., and Henkel, D. J., "Soil Mechanics and Work-Hardening Theories of Plasticity," Transactions, American Society of Civil Engineers, Vol. 122 (1957).

8. Jenike, A. W. and Shield, R. T., "On the Plastic Flow of Coulomb Solids Beyond Original Failure," Journal of Applied Mechanics (1959).

9. Schofield, A. N. and Wroth, P., Critical State Soil Mechanics, McGraw-Hill, New York (1968).

MECHANICAL BEHAVIOR OF SOILS DESCRIBED BY HYPOELASTICITY EQUATIONS

Lucia Drăguşin

Catedra Matematici II
Institutul Politehnic Bucuresti
Bucarest 77266, Romania

1. INTRODUCTION

Soil structure around the infrastructure of building works includes: mineral grains, constituent water, adsorbed water, pore-liquid, gas.

Fairly large sized mineral grains are somewhat spherical in shape and they are typical of cohesionless soils. The smaller sized ones are disc shaped and they are characteristic of cohesive soils.

The constituent water incorporates the hydroxyls to be found in mineral grains.

The adsorbed water covers the solid grains with a liquid film of a certain rigidity and pseudocrystalline structure. It infiltrates the foils of unstable clays.

The pore liquid can freely circulate through the voids.

Hence soil is a mixture of three phases: solid, fluid and gaseous (for details, see [1]).

The solid phase implies therefore the mixture of mineral grains, constituent water and adsorbed water whereas the fluid phase refers to the pore liquid.

With a view to developing a mathematical model, the soil will not be regarded as a three phased material but as a continuum provided with the properties characteristic of soils.

2. SOME BASIC CONCEPTS OF SOIL MECHANICS

The transition from a three phase body to a continuum requires the introduction of some new notions unused in the ordinary continuum mechanics.

We shall replace a soil sample with an ideal continuous body, B, which occupies at time t an region b_t of volume $v(b_t)$ within the three-dimensional Euclidean space E^3.

A particle χ from the ideal body B has the position $X(X_1,X_2,X_3)$ at $t=t_o$ and $x(x_1,x_2,x_3)$ at t, relative to a given system of rectangular Cartesian coordinates.

We shall assume the three phases to be continuously distributed within the region b_t and the solid, fluid or gaseous phases to be sharing in their one contribution at every point x.

The apparent mass density of solid phase ρ_s will be the ratio of the solid mass phase m_s lying in the region b_t to the total volume $v(b_t)$ and the effective mass density of solid phase ρ^s will be the ratio of the solid phase mass m_s to the solid phase volume $v(b_t^s)$ to be found in region b_t.

In other words, if we assume that b_t is a set in E^3, Lebesgue measurable with strictly positive measure $v(b_t)$ and if we assume the existence of the function ρ_s: $b_t \to R^+$, measurable and Lebesgue integrable on b_t, we shall define a measure m_s: $b_t \to R^+$, $m_s(b_t) = \int_{b_t} \rho_s(x,t)\, dv$.

It is shown [2] that the measure m_s is absolutely continuous, i.e. for any sequence of measurable sets $(b_n)_{n\in N}$, $b_n \subset b_t$, $v(b_n) > 0$ and $\lim_{n\to\infty} v(b_n) = 0$, we have $\lim_{n\to\infty} m_s(b_n) = 0$.

Then, for any sequence of measurable sets $(b_n)_{n\in N}$, $b_n \subset b_t$, $v(b_n) > 0$, $\bigcap_{n\in N} b_n = \{x\}$ and $\lim_{n\to\infty} v(b_n) = 0$, we have

$$\rho_s(x,t) = \lim_{n\to\infty} \frac{m_s(b_n)}{v(b_n)} = \lim_{n\to\infty} \frac{m_s(b_n)}{v(b_n^s)} \frac{v(b_n^s)}{v(b_n)} = \rho^s(x,t) \lim_{n\to\infty} \frac{v(b_n^s)}{v(b_n)} .$$

Similarly, we can define the apparent mass density of fluid phase ρ_f, the effective mass density of fluid phase ρ^f, the apparent mass density of gaseous phase ρ_g and the effective mass density of gaseous phase ρ^g (see [3], part III).

The total mass m of body B is: $m = m_s+m_f+m_g$, where m_f, m_g are the mass of the fluid and gaseous phases in the moment t; then the total mass density $\rho(x,t)$ in point x at moment t is defined by the sum $\rho(x,t) = \rho_s(x,t) + \rho_f(x,t) + \rho_g(x,t)$.

If we assume ρ^s = const. then the variation of porosity n(x,t), $n(x,t) = 1 - \frac{\rho_s(x,t)}{\rho^s}$, or that of the void ratio e(x,t), e(x,t) =

$\frac{\rho^s}{\rho_s(x,t)} - 1$ may be replaced by the apparent mass density of the solid phase $\rho_s(x,t)$ variation.

Let P be an arbitrary part of an ideal body B, respectively P_t is closed subset of set b_t. In any point x belonging to the boundary of P_t we shall define three vector fields $t_s(x,P_t)$, $t_f(x,P_t)$, $t_g(x,P_t)$ determined by the action of part B - P of part P, respectively, upon the solid, fluid and gaseous phase to be found in the part at moment t.

We assume the existence of three vector-valued functions $t_s(x,n_t)$, $t_f(x,n_t)$, $t_g(x,n_t)$ defined for all points $x \varepsilon b_t$ and for all unit vectors n_t, n_t being the internal noraml of the boundary of P_t in point x at moment t, so that:

$$t_s(x,P_t) = t_s(x,n_t),\ t_f(x,P_t) = t_f(x,n_t),\ t_g(x,P_t) = t_g(x,n_t).$$

The resultant of all actions of part B - P upon part P in x at t will be $t(x,n_t)$, where $t(x,n_t) = t_s(x,n_t) + t_f(x,n_t) + t_g(x,n_t)$.

We assume these functions to be continuous and we define on b_t the tensor fields $T_s(x,t)$, $T_f(x,t)$, $T_g(x,t)$, $T(x,t)$ so that

$$T_s(x,t)n_t = t_s(x,n_t) \qquad T_f(x,t)n_t = t_f(x,n_t)$$

$$T_g(x,t)n_t = t_g(x,n_t) \qquad T(x,t)n_t = t(x,n_t)\ .$$

In the mechanics of soils, tensor $T_f(x,t) = u_f(x,t)I$ is called the <u>pore-fluid pressure tensor</u>, whereas $T_g(x,t) = u_g(x,t)I$ is called the <u>pore-gas pressure tensor</u> (I is unit tensor), u_f and u_g being two scalar valued functions.

In the solid phase various forces are at work such as Van der Waals forces, Coulomb's attraction forces, attraction forces between the grains caused by the water film adsorbed, attraction forces between the grains and the water adsorbed, which have been influenced by the water polar moment (accounts for the thixotropy phenomenon). The existence of these forces leads to the necessity of introducing the <u>initial internal stress tensor T^i</u>. It depends on:

- the grain size,
- the degree of soil loosening,
- the degree of soil stability in the presence of water,
- the environment humidity (by trapping in some water molecules,

the menisci formed between two grains decrease their curvature).

Tensor T^i is an index of solid phase strength to destruction and it does not depend on the external load. (For details on the concept see the discrete theory of solids [4], ch. I).

In Coulomb's theory the cohesion and the internal friction are taken to represent the degree of this strength.

Tensor T_s results by adding the stress tensor determined by external loads T_s^{ext} to the initial stress tensor T^i

$$T_s = T^i + T_s^{ext} .$$

Assuming the soil to be in a state of disagregation if $T_s = o$, knowing the external load having caused this phenomenon, one may find tensor $T^i = -T_s^{ext}$. In soils mechanics, tensor T_s is called the effective stress tensor and it is usually considered to be equal to the solid phase stress tensor created by external loads T_s^{ext}.

The mechanical behavior of soils may be also dependent on the initial stress tensor T^o determined by the load to which the sample has been extracted in situ. It depends on:

- the depth of sample location, the sample being subjected to the weight of upper layers,
- the degree of soil consolidation.

3. MOTION EQUATIONS

We shall say that a body B is a stable system if the solid phase mass is invariant in the presence of water (for ex. sand, kaolinite) and an unstable system if it is influenced by water presence (for ex. montmorillonite).

a) Ideal motion of a stable open system

Definition. We shall say that a body B is a stable open (drained system if the solid phase mass is invariant in time, the total mass being variable, that is $\dot{m}_s(b_t) = o$, $\dot{m}(b_t) \neq o$, where $\dot{m}_s$, $\dot{m}$ are the material time derivatives of a m_s and m.

Experimentally, a soil sample is subjected to a load at t_0. Then the sample is left to consolidate for a time interval d whereafter another load is applied, etc.

Body B occupies at t_o a region b_{t_o} which undergoes a change until t+d. The consolidation is over when the position of all particles χ becomes constant in time. The duration d depends on soil permeability.

Definition. Let B be a stable open system and M a function

$$M: \; b_{t_o} \times R \to b_t \; .$$

If: a_1) $M(.,t)$ is a diffeomorphism of class C^n $(n \geq 2)$,

a_2) there exists a real positive number d_M, so that

$$\frac{\partial M}{\partial t} = 0 \quad , \quad \forall t \geq t_o + d_M$$

then partial function of class $C^n (n \geq 2)$

$$M(x,.): \; [t_o, t_o + d_M] \to E^3, \; \forall X \varepsilon B \; .$$

is called a process of consolidation of initial moment t_o and duration d_M. For facility, we note this process by $M(b_{t_o}, t)$.

Definition. Let $M_1, M_2, \ldots, M_n$ be a succesion of consolidation processes of duration $d_{M_1}, d_{M_2}, \ldots, d_{M_3}$, so that the initial configuration of a process should coincide with the final configuration of the interior process, that is

$$M_1(b_o, d_{M_1}) = M_2(b_{d_{M_1}}, d_{M_1})$$

$$M_2(b_{d_{M_1}}, d_{M_1} + d_{M_2}) = M_3(b_{d_{M_1} + d_{M_2}}, d_{M_1} + d_{M_2})$$

. . . .

$$M_{n-1}(b_{d_{M_1} + \ldots + d_{M_{n-2}}}, d_{M_1} + \ldots + d_{M_{n-1}}$$

$$= M_n(b_{d_{M_1} + \ldots + d_{M_{n-1}}}, d_{M_1} + \ldots, d_{M_{n-1}}) \; .$$

We shall say that $x = \chi(X,.)$ describes an ideal motion of body B if det (Grad χ) > 0 and

$$\chi(X, d_{M_1}) = M_1(b_o, d_{M_1})$$

$$\chi(X, d_{M_1}+d_{M_2}) = M_2(b_{d_{M_1}}, d_{M_1}+d_{M_2})$$

. . . .

$$\chi(X, d_{M_1}+\ldots+d_{M_n}) = M_n(b_{d_{M_1}+\ldots+d_{M_{n-1}}}, d_{M_1}+\ldots+d_{M_n}) .$$

In other words, the ideal motion of a stable open system is defined by means of final configurations of the process of consolidation (see [3], part III).

Since, in this case, only the solid phase mass is preserved, the equation of continuity is $\dot{\rho}_s + \rho_s \operatorname{div} \dot{x} = o$, where $\dot{x}$ is the velocity of particle $\chi \varepsilon B$.

b) Motion of a stable closed undrained system

Definition. We shall say that a body B is a stable closed undrained system if the solid, fluid and gaseous phase masses are invarient in time, that is: $\dot{m}_s(b_t) = o$, $\dot{m}_f(b_t) = o$, $\dot{m}_g(b_t) = o$.

Since the total mass is conserved, the equation of continuity is

$$\dot{\rho} + \rho \operatorname{div} \dot{x} = o$$

if the system is compressible and $\operatorname{div} \dot{x} = o$ if the system is incompressible.

In the case of stable closed undrained systems, the real motion coincides with the ideal motion.

c) Motion of an unstable closed undrained system

Definition. We shall say that a body B is an unstable closed undrained system if total mass is invariant in time, the solid phase mass, the fluid phase mass, the gaseous phase mass being variable, that is $\dot{m}_s \neq o$, $\dot{m}_f \neq o$, $\dot{m}_g \neq o$, $\dot{m} = o$.

If $\dot{m}_g = o$, since the total mass is conserved, we have $\dot{m}_s = -\dot{m}_f$ and the equations of continuity are:

$$\frac{\partial \rho_s}{\partial t} + \operatorname{div}(\rho_s \dot{x}_s) = \rho \alpha_s , \quad \frac{\partial \rho_f}{\partial t} + \operatorname{div}(\rho_f \dot{x}_f) = -\rho \alpha_s ,$$

$$\frac{\partial \rho_g}{\partial t} + \operatorname{div}(\rho_g \dot{x}_g) = o ,$$

where $\dot{x}_s, \dot{x}_f, \dot{x}_g$ are the velocities corresponding to the solid, fluid and gaseous phase, respectively, and

$$\rho\alpha_s(t) = \lim_{t_1 \to t} \frac{m_s(t_1-t)-m_s(t)}{t_1-t}$$

is the index of soil swelling.

4. Constitutive equations

A sample of soil is actually taken out from a certain depth, the natural load causing the initial stress tensor $T^o = T(o)$.

Bringing it to soil level partially discharges the sample. Let t_o be the beginning moment of the experiment. The sample is brought back to its natural state by motion.

$$x = \chi^e(X,t) \qquad t\varepsilon[t_o, t^3) \ . \tag{1}$$

The upper coefficient "e" from (1) shows that the motion achieved during the interval $[t_o, t^e)$ may be considered as a reversible deformation if the stress path upon loading has been the same with the one of unloading. If before extraction, the sample occupies the region b_o at t=o, when motion (1) is finished, the sample will occupy the same region b_o at t, excepting an arbitrary rigid motion.

If the material is further stressed so that the irreversible deformation occurs, then the body motion, up to the moment t_D, when dilatancy occurs, is described by equation

$$x = \chi^{P_1}(X,t) \ , \quad t\varepsilon[t^3, t^D) \ . \tag{2}$$

It is a continuation of previous motion (1), hence

$$\chi^{P_1}(X,t^e) = \lim_{t \to t^e} \chi^e(X,t) \ .$$

If the loading is continued up to the moment of failure t^f, the motion equation is:

$$x = \chi^{P_2}(X,t) \quad , \ t\varepsilon[t^D, t^f) \tag{3}$$

where $\chi^{P_2}(X,t^D) = \lim_{t \to t^D} \chi^{P_1}(X,t)$.

Therefore, the motion from the beginning of the experiment up to the failure is described by three equations: one in the domain of reversible deformations; another in the domain of dilatancy-free irreversible deformations and then in the domain of irreversible deformations with dilatancy.

Observation work on soils led to the following conclusions:

1) soils are cohesive or cohesionless materials,
2) soils are materials having memory, their behavior depending on the initial stress T^o and the stress path,
3) soils are work hardening or softening materials,
4) soils are reversibly and irreversibly deformable.

A hypoelastic material was defined by Truesdell [5] and Bernstein [6] (see also Truesdell & Noll [7]) through

a_1) a constitutive equation $\overset{o}{T} = H(T)[D]$,

where: $D = \frac{1}{2}(L+L^T)$ is the deformation rate tensor ($L=-\text{grad}\,\dot{x}$ denotes the spatial gradient of velocity and L^T-transpose of L), $\overset{o}{T} = \dot{T}+WT+TW$ - the Jaumann-Noll stress rate tensor ($W=\frac{1}{2}(L-L^T)$ is the spin tensor) and H(T) an isotropic function in T of class C^1 in D from the vector space $(D_{11},D_{22},D_{33},D_{12}D_{13},D_{23})$ to the vector $\overset{o}{T}_{11},\overset{o}{T}_{22},\overset{o}{T}_{33},\overset{o}{T}_{12},\overset{o}{T}_{13},\overset{o}{T}_{23})$. D is the domain in R^6 where det $H(T) \neq o$.

a_2) a corresponding equivalence class [T,F], determined by the initial conditions $(T^o,F^o]$ (F=-Grad X is the deformation gradient).

Let the hypoelastic material have a constitutive equation

$$\overset{o}{T} = -3apI_DI + 9apD + 3bpI_DT + 3(b-c)p\,tr(TD)I - 3c\,tr(TD)T \qquad (4)$$

where I_D=tr D; $p=\frac{1}{3}$tr T is the mean stress; a,b,c-the strictly positive material constants, b-c were assumed to be strictly positive as well.

It may be shown that these materials have the following properties (see [3] part I, II, for details):

1) There are two classes: N-type (if b-2c>o) and C-type (if b-2c<o). The N-type materials behavior resembles that of the cohesionless soils and whereas the C-type materials behavior resembles that of the cohesive soils.

2) Let $T^* = T - pI$ be the stress deviator and $q^2 = \frac{3}{2}tr(T^*)^2$. There are two potentials ϕ,ψ, determined apart from the arbitrary additives constants $k_1, k_2 \varepsilon R$

$$\phi(p,q) = \frac{c}{6b(b-c)} \ln[p^2|9ap-bq^2|^{(b-2c)/c}] + k_1 ,$$

$$\psi(p,q) = \frac{1}{6(b-c)} \ln \frac{p^2}{|9ap-bq^2|} + k_2 ,$$

defined on $D = \{T\varepsilon R^2 | \; p\neq o \wedge 9ap-bq^2\neq o\} = \Delta_1 U \Delta_2 U \Delta_3$ (see Fig. 1)

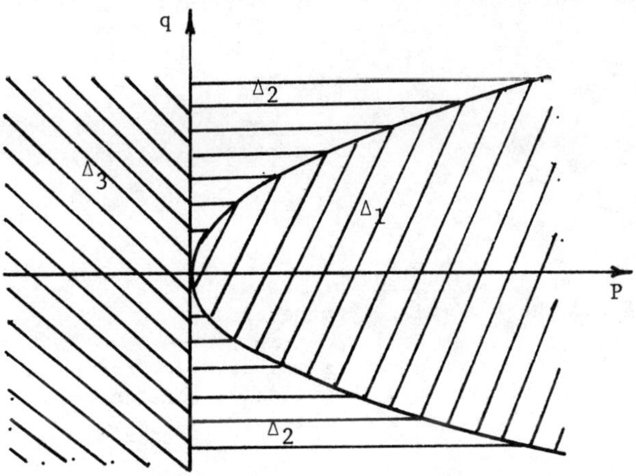

Fig. 1. Definition Domain

Its properties are: $\dot{\phi} = pI_D$, $\dot{\psi} = tr(TD)$. The materials having the constitutive equation (4) are divided in three classes depending on whether the initial condition (p_o,q_o) belongs to any of the subsets $\Delta_1,\Delta_2,\Delta_3$. Materials for which $(p_o,q_o)\varepsilon\Delta_1$ are similar in behavior with the normally consolidated ones, whereas materials for which $(p_o,q_o)\varepsilon\Delta_2$ are similar in behavior with overconsolidated soils.

3) The behavior of these materials depends on the stress history $T = T(\tau), \tau\varepsilon[t_o,t]$.

4) The stress work in the deformation of a body B made from such a materials is:

$$[W(T(.))](t) = \int_{t_o}^{t} \left(\int_{v(b_t)} tr(TD)dv \right) d\tau = \int_{v(b_{t_o})} \left(\int_{t_o}^{t} \frac{\rho_o}{\rho} tr(TD) d\tau \right) dv$$

$$= \int_{v(b_{t_o})} \left(\int_{t_o}^{t} \frac{\rho_o}{\rho} \dot{\psi}\, d\tau \right) dv$$

and it depends on stress history.

In other words, although the curvilinear integral $\int_{t_o}^{t} \dot{\psi}(T(\tau))d\tau = \psi(p,q) - \psi(p_o,q_o)$ does not depend on the path within the stress space, the product of $\dot{\psi}$ and the density ratio $\frac{\rho_o}{\rho}$ is not fully integrable (see [3], part I).

Since $I_D = \frac{\dot{\rho}}{\rho}$, the same thing will hold for $p\frac{\dot{\rho}}{\rho} = \dot{\phi}$.

Reversing the eqn. (4) results into constitutive equation

$$D = \frac{1}{9ap} \{\overset{o}{T} + 3(aI-bT)\dot{\phi} - 3[(b-c)pI-cT]\dot{\psi}\} \quad . \tag{5}$$

The mathematical model suggested for the mechanical behavior of isotropic soils has the following hypotheses:

1) the mechanical behavior of soils may be described by means of the constitutive eqn. (5) wherein the material constants are a, b,c in the motion (1), a_1,b_1,c_1 in the motion (2), a_2,b_2,c_2 in the motion (3) and

$$\frac{a}{b} = \frac{a_1}{b_1} = \frac{a_2}{b_2} , \quad \frac{c}{b} = \frac{c_1}{b_1} = \frac{c_2}{b_2} \quad .$$

This hypothesis assumes that the material changes it mechanical properties, hence its material constants, for a given stress path when passing from the domain of reversible deformations to that of irreversible deformations.

2) The cohesionless soils are of the N-type materials ($2\frac{c}{b} < 1$) and the cohesive soils are of the C-type materials ($2\frac{c}{b} > 1$).

3) The consolidation degree may be described by means of the initial stress invariants (p_o,q_o). A normally consolidated soil has $q_o^2 < 9\frac{a}{b}p_o$ and a overconsolidated soil has $p_o > 0$, $q_o^2 > 9\frac{a}{b}p_o$.

4) Let

$$\dot{E} = b\dot{\psi}^{e} = b_1\dot{\psi}^{P_1} = b_2\dot{\psi}^{P_2} = \frac{(9\frac{a}{b}p-2q^2)\dot{p} + 2pq\dot{q}}{6(1-\frac{c}{b})p(9\frac{a}{b}p-q^2)} .$$

Then $E(p,q) = E(p_o,q_o)$ may be considered as a yield condition with the equation

$$\frac{p^2}{9\frac{a}{b}p-q^2} = \frac{p_o^2}{9\frac{a}{b}p_o-q_o^2} .$$

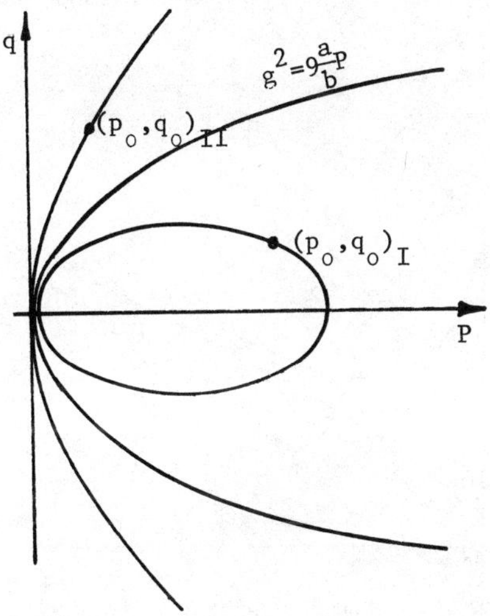

Fig. 2. Yield Condition

It is represented in Fig. 2. The shape of these curves have been proved experimentally [8] (see also [9]).

The occurrence of irreversible deformations is determined only by the stress invariants position in the (p,q) plane namely by the intersection of the stress path with the yield condition (fixed by the initial stress (p_o,q_o)).

5) The loading and the neutral loading are described by:

$\dot{E}(p(t),q(t)) \geq o$,

and the unloading by $\dot{E}(p(t),q(t)) < o$.

6) Let

$$\dot{E}_v = b\dot{\phi}^e = b_1\dot{\phi}^{P_1} = b_2\dot{\phi}^{P_2} = \frac{9(\frac{a}{b}p-2\frac{c}{b}q^2)\dot{p} - 2(1-2\frac{c}{b})pq\dot{q}}{6(1-\frac{c}{b})p(9\frac{a}{b}p-q^2)}$$

and p_M, q_M the stress invariants in the moment when the mass density of the solid phase ρ_s reaches its maximum or minimum value. Then, the relation $E_v(p,q) = E_v(p_M,q_M)$ may be regarded as a dilatancy condition. It has the equation

$$p^2|9\frac{a}{b}p-q^2|^{(b-2c)/c} = p_M^2|9\frac{a}{b}p_M-q_M^2|^{(b-2c)/c}$$

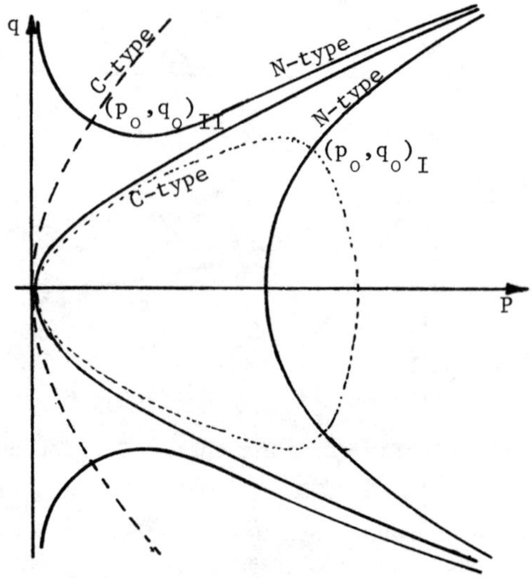

Fig. 3. Dilatancy Condition

It is represented in Fig. 3 and it was experimentally proved in [10,11] for sand. Hence the occurrence of dilatancy is also determined only by the position of stress invariants on (p,q) plane. On these curves $I_D = (\dot{\rho}/\rho) = o$.

7) In a corresponding equivalence class determined by the initial state $[T_s^o, F^o]$, the constitutive equations of a stable open (drained) system are (see [3], part IV)

$$D = \begin{cases} \frac{1}{a}\tau_s , & \text{if } E(p,q) < E(p_o,q_o) \text{ or } \begin{cases} E(p,q) \geq E(p_o,q_o) \\ \dot{E} < o \end{cases} \\ \frac{1}{a_1}\tau_s, & \text{if } E(p,q) \geq E(p_o,q_o) \text{ and } \dot{E} \geq o , \dot{E}_v \geq o \\ \frac{1}{a_2}\tau_s, & \text{if } E(p,q) \geq E(p_o,q_o) \text{ and } \dot{E} \geq o , \dot{E}_v < o \end{cases}$$

where $\tau_s = \frac{1}{9p} \{\overset{o}{T}_s + 3(\frac{a}{b}I - T_s)\dot{E}_v - 3[(1-\frac{c}{b})pI - \frac{c}{b}T_s]\dot{E}\}$.

$p = \frac{1}{3}\mathrm{tr}\, T_s , \quad q^2 = \frac{3}{2}\mathrm{tr}(T_s^*)^2$.

8) In a corresponding equivalence class determined by the initial state $[T^o, F^o]$, the constitutive equations of a stable closed undrained compressible system are the same, the only difference being that the effective stress tensor T_s will be replaced by the total stress tensor $T = T_s + u_f I + u_g I$.

9) In a corresponding equivalence class determined by the initial state $[T^o + u^o I, F^o]$, the constitutive equations of a stable closed undrained incompressible system are

$$D = \begin{cases} \frac{1}{a}\tau , & \text{if } E(p+u,q) < E(p_o+u^o,q_o) \text{ or } \begin{cases} E(p+u,q) \geq E(p_o+u^o,q_o) \\ \dot{E} < o \end{cases} \\ \frac{1}{a_1}\tau , & \text{if } E(p+u,q) \geq E(p_o+u^o,q_o) \text{ and } \dot{E} \geq o \end{cases}$$

where u is an arbitrary hydrostatic pressure

$$\tau = \frac{1}{9(p+u)} \{\overset{o}{T} + \dot{u}I - 3](1-\frac{c}{b})(p+u)I - \frac{c}{b}(T+uI)]\dot{E}\} ,$$

$$p = \frac{1}{3}\,\mathrm{tr}\, T , \quad q^2 = \frac{3}{2}\,\mathrm{tr}\,(T^*)^2 , \quad T = T_s + u_f I + u_g I .$$

Since in this case $I_D = o$, the only loading path imposed by this constitutive equations is

$$q^2 = (p+u)^{-2c/(b-2c)}\{9\frac{a}{b}[(p+u)^{b/(b-2c)}-(p_o+u^o)^{b/(b-2c)}]$$

$$+ q_o^2(p_o+u^o)^{2c/(b-2c)}\} \quad .$$

It is represented in Fig. 3 where p will be replaced by p+u.

5. Comparison between theoretical results and experimental data

The comparison of theoretical results with the experimental data in "triaxial" was made (see [3], part II, IV)

$$D(t) = \begin{bmatrix} -d_1(t) & o & o \\ o & -d_1(t) & o \\ o & o & d_3(t) \end{bmatrix}, \quad T(t) = \begin{bmatrix} T_1(t) & o & o \\ o & T_1(t) & o \\ o & o & T_3(t) \end{bmatrix}$$

$$\varepsilon_r = \frac{X_1-x_1}{X_1} = \frac{X_2-x_2}{X_2} = -d_1(o)t \quad , \quad \varepsilon_a = \frac{X_3-x_3}{X_3} = d_3(o)t \ .$$

Then: $$d_1(t) = \frac{d_1(o)}{1+d_1(o)t} = -\frac{\dot{\varepsilon}_r}{1-\varepsilon_r} \ , \ d_3(t) = \frac{d_3(o)}{1-d_3(o)t} = \frac{\dot{\varepsilon}_a}{1-\varepsilon_a} \ ,$$

$$p = \frac{1}{3}(2T_1+T_3) \ , \quad q = T_3-T_1 \quad .$$

Since $\frac{\dot{\rho}}{\rho} = 2D_{11}+D_{33} = -2d_1(t)+d_3(t)$,

$$\dot{\varepsilon} = \frac{2}{3}(D_{33}-D_{11}) = \frac{2}{3}(d_1(t)+d_3(t)) \ ,$$

there results

$$\frac{\rho}{\rho_o} = \frac{(1-\varepsilon_r^o)^2(1-\varepsilon_a^o)}{(1-\varepsilon_r)^2(1-\varepsilon_a)} \ , \quad \varepsilon-\varepsilon_o = \frac{2}{3}\ln\left|\frac{1-\varepsilon_r}{1-\varepsilon_a}\frac{1-\varepsilon_a^o}{1-\varepsilon_r^o}\right| \quad .$$

In drained tests there are usually measured the axial strain ε_a and the volume strain $\frac{\Delta v}{v}$ or the void ratio e. Then

$$\frac{\rho_s}{\rho_s^o} = \frac{1}{1-\frac{\Delta v}{v}} = \frac{1+e_o}{1+e} \quad , \quad \varepsilon-\varepsilon_o = -\ell n\left[\frac{1-\varepsilon_a}{1-\varepsilon_a^o}\left(\frac{\rho_s}{\rho_s^o}\right)^{1/3}\right] .$$

In Fig. 4 one may notice good concordance existing between the experimental data [10] acquired with drained sand in a test wherein the radial stress has been kept constant $(T_1)_s^c = 414.233$ kPa and the theoretical results computed for an N-type material where:

$$\frac{a}{b} = \frac{q_{cr}^2}{3\, q_{cr}+3(T_1)_s^c} = 238.667 \text{ kPa} \ ,$$

$$\frac{c}{b} = \frac{2q_M^2 + 3(2(T_1)_s^c - \frac{a}{b})q_M - 9\frac{a}{b}(T_1^2)_s^c}{2q_M(q_M+6(T_1)_s^c} = 0.1165$$

$$b_1 = 0.081(\text{kPa})^{-1} \ , \quad b_2 = 0.0246\ (\text{kPa})^{-1} \ ,$$

$$T^i = o \ , \quad p_o = 414.233 \text{ kPa} \ , \quad q_o = o$$

$$q_M = 448.7523 \text{ kPa},$$

$$q_{cr} = q_{failure} = 1366.93 \text{ kPa}$$

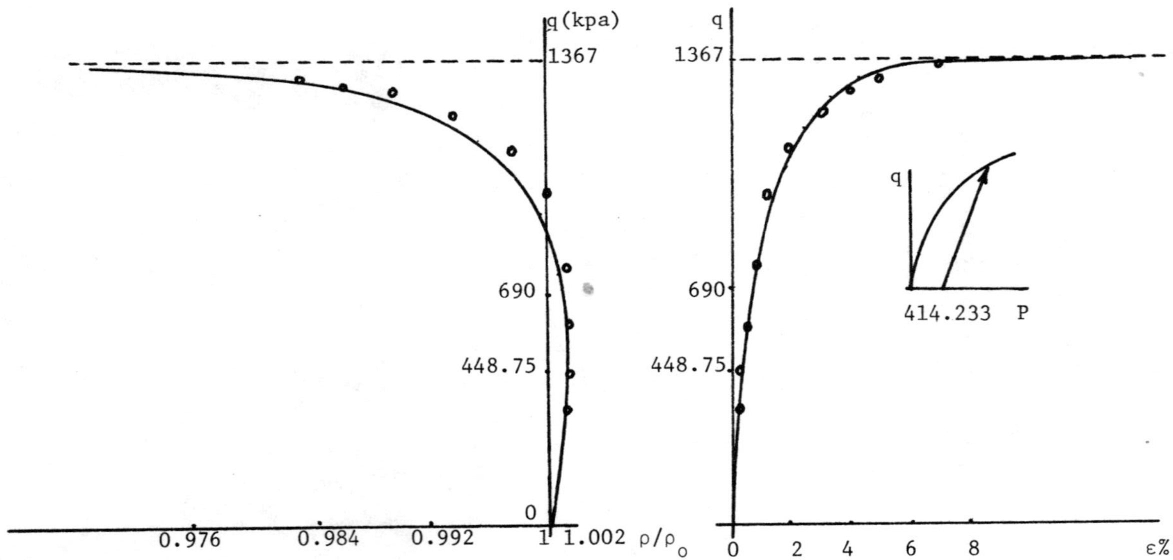

Fig. 4. Conventional drained test for saturated sand ___________, theoretical line; o, o, o, experimental data [10, Fix. 2].

REFERENCES

1. Caquot, A., Kerisel, J., Traité de Mécanique du Sols, Gauthier-Villars Ed, Paris (1966).

2. Nicolescu, M., Mathematical Analysis-III (in Romanian), Tech. Ed., Burcarest (1960).

3. Dragusin, L., A hypoelastic model for soils I-IV, Int. J. Engng. Sci. 19, 511-552 (1981).

4. Soos, E., Discrete and continuous models of the solids (in Romanian) Sci. Ed., Burcarest (1974).

5. Truesdell, C., Hypo-elasticity, J. Rational Mech. Anal. 4, 83-133 (1955).

6. Berstein, Hypo-elasticity and elasticity, Arch. Rational Mech. Anal. 6, 89-104 (1960).

7. Truesdell, C., Noll, W., Non-linear field theories of mechanics, Handbuch der Phÿsik, Vol. III/3, Springer-Verlag, Berlin (1965).

8. Schofield, A., Wroth, P., Critical state soil mechanics, McGraw-Hill Book Company (1968).

9. Roscoe, K.H., Bruland, J.B., On the generalized stress-strain behavior of "wet" clays, Cambridge Univ. Press, Massachussetts (1968).

10. Thurairajah, A., Shear behavior of sand under stress reversal, Proc. 8th Int. Conf. Soil. Mech. Found. Engng. 1.2, 439-445 (1973).

11. Arnold, M., Mitchell, P.W., Proc. 8th Int. Conf. Soil. Mech. Found. Engng. 1.1, 11-18 (1973).

A DISCRETE LIMIT ANALYSIS OF FOUNDATION STRUCTURES

T. Kawai[1] and N. Takeuchi[2]

[1]Professor, Institute of Industrial Science, University of Tokyo,
Tokyo, Japan
[2]Manager of the Systems Research Laboratory,
Kyowa Engineering Consultants Co., Ltd.
Tokyo, Japan

ABSTRACT

Basing on the experimental evidence of solids under the ultimate state of loading, the first author proposed a family of new discrete models in 1976. These models consist of rigid bodies and two types of connection springs, one of which resists the dilatational deformation, while the other the shear deformation. Theoretical basis of the models are described first and then application of these models is proposed to soil and rock mechanics. Numerical examples are shown in the limit analysis of the ultimate bearing capacity of the sandy soil foundation and top of the embankment slope. Stability analysis of the fill-up ground and the flexible sheet piles are also given in this paper.

NOMENCLATURE

C: cohesion
$\mathbf{f}$: element force vector
$\mathbf{k}$: element stiffness matrices
k_n: normal spring constant
k_s: tangential spring constant
$\mathbf{u}$: displacement vector
$\mathbf{u}^{(1)}$: linear component of $\mathbf{u}$
$\mathbf{u}^{(2)}$: nonlinear component of $\mathbf{u}$
$\boldsymbol{\delta}$: relative displacement vector
$\boldsymbol{\delta}^{(1)}$: linear component of $\boldsymbol{\delta}$
$\boldsymbol{\delta}^{(2)}$: nonlinear component of $\boldsymbol{\delta}$
$\mathbf{D}$: stress-strain matrix
$\mathbf{B}$: shape function of an element
σ_n: normal stress
τ_s: tangential stress
$\mathbf{k}_{ij}$: stiffness matrices of spring system connecting ⓘth and ⓙth elements
$\boldsymbol{\varepsilon}$: strain vector
ϕ: angle of internal friction
$\bar{\mathbf{F}}$: external force

INTRODUCTION

It is extremely important from the design point of view to conduct stability analysis of foundation structures, and it has been carried out by basing on the theory of elasticity and rigid plastic bodies.
These type of problems, however, constitute extremely difficult nonlinear boundary value problems. Therefore, analytical solutions can be obtained only in a few cases, and it is the present day practice to obtain more or less approximate solutions by using numerical methods.
The finite element method which was developed by the structural engineers in aerospace industries and made remarkable progress with advancement of computer industry. It has become one of the most useful tools of structural analysis. Power of this method, however, is well recognized only in the linear elasticity problems or elasto-plastic problems with small plastic strains and in the strong nonlinear problems such as the plastic collapse problem where effects of material nonlinearity are profound, accuracy of the numerical solutions can hardly be guaranteed and computing time necessary to carry out these analysis is usually prohibitive [1],[2],[3].
And therefore the practical application of the finite element nonlinear analysis is still not well established. To consider the soil and rock materials as continua they are generally too nonuniform, inhomogenous and easy to slip internally under applied loading. Basing on such considerations, the method of limit analysis was introduced many years ago in the field of soil mechanics, and naturally it has been extensively used. This simplified method, however is a practical but rather crude method which has been developed by ingenious use of elasticity and plasticity theories because the stress field is determined without consideration on the displacement and strain hysteresis.
Furthermore construction of the plastic flow field in the rigid-plastic bodies which is compatible to the velocity boundary conditions is extremely laborious task. Consequently it is almost impossible to find the admissible velocity field for actual foundation structures and its application has been quite limited. Considering such a status of existing methods the first author proposed a family of new discrete models in 1976 by basing on the experimental evidence on the flow and fracture of solids. He considered that the essential feature of flow and fracture of solids is slip or cracking of particles or elements which compose the solids, and therefore it is strongly needed to develop a new method of analysis in which effects of such discontinuous characteristic is fully taken into account.
In these models actual structures or solids are idealized as a set of rigid elements interconnected by two types of spring system, one of which resists the dilatational, the other shearing deformation. Therefore sliding or separation of two adjacent elements can be made easily.
As nodal parameters, the rigid body displacement is taken at the centroid of a given element. Consequently the strain energy of a given element is evaluated by that of the spring system lumped on the interelement boundaries and incremental analysis can determine automatically the slip surfaces. Although the stress distribution in the elements is not considered, the solution obtained always satisfies the mechanism and equilibrium of surface tractions on the interelement boundaries.

Therefore it gives clearly one upper bound solution.
In other words, these models are discrete models suitable for generalized limit analysis. In early stage of development these models were mainly applied to limit analysis of the metallic structures and their validity has been almost verified by a series of numerical studies.
There must be big difference in structural behavior between the polycrystalline structures like metallic materials and granular materials such as the clay and sand. Undoubtedly the plastic flow or fracture is represented by the discontinuous displacement between elements or particles which is regarded as slip movements in either materials.
In this paper application of these models to analysis of foundation structures will be attempted and the general method of discrete limit analysis will be described with some verification examples.

THEORETICAL BASIS OF NEW DISCRETE MODELS

Generally speaking all materials are composed of very large number of particles and therefore it is obvious that their deformation may be controlled not only by the strength of the particle itself but also by the intergranular strength. From the standpoint of the solid mechanics, solids are considered as continuum and generally the intergranular strength is not taken into account. The present authors believe that at this point there exists a limitation of the plasticity theory in its application to collapse or failure analysis of solids and structures where slip movements play an important role.
Experimental evidence of the above discussion is shown in Fig. 1.
It can be seen from this figure that the essential feature of plastic deformation is the slip movement in view of Fig. 2 and at the limiting state solids or structures may move like a set of rigid blocks.
Basing on such consideration Kawai proposed a new discrete model entitled "Rigid Body-Spring Model as shown in Fig. 3 [4],[5],[6].
They are assumed to be in equilibrium with external loads, and reaction forces are produced by the spring system which is distributed over the contact surface of two adjacent bodies. For further development of new element models, it will be assumed that the contact area is known and fixed. It should be mentioned here that in actual contact problem, the contact surface are not known a priori, and therefore it can be determined only in iterative way. Taking two such rigid bodies under contact, an infinitesimal deformation of the spring system is considered (Fig. 4).
The displacement $\mathbf{u}$ of an arbitrary point in a rigid body can be given by the following vector equation:

$$\mathbf{u} = \mathbf{u}_G + \mathbf{\Theta} \times (\mathbf{r}-\mathbf{r}_G) \qquad (1)$$

where $\mathbf{u}_G$ is the displacement vector of the centroid, $\mathbf{\Theta}$ is the infinitesimal rotation vector and $(\mathbf{r}-\mathbf{r}_G)$ is the position vector of an arbitrary point with respect to the centroid before deformation.

$$\mathbf{u}_G = (u_g, v_g, w_g), \qquad \mathbf{\Theta} = (\theta, \phi, \chi) \qquad (2)$$

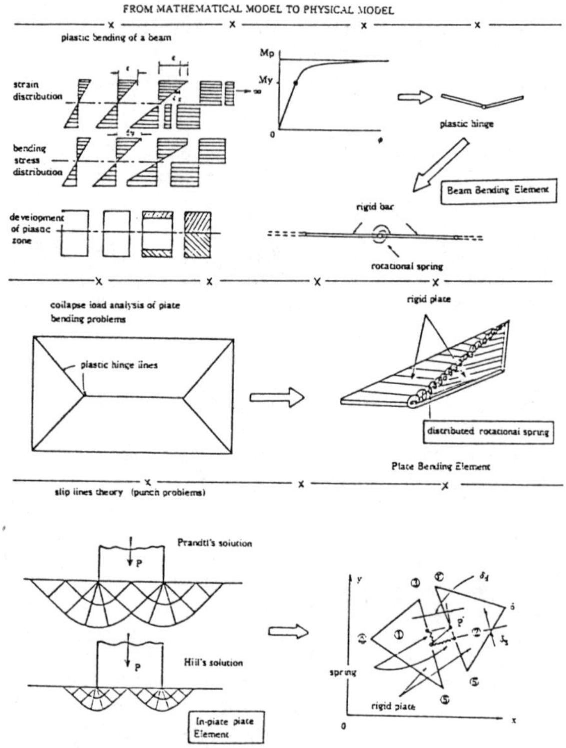

Fig. 1 Concept of the rigid body-spring model

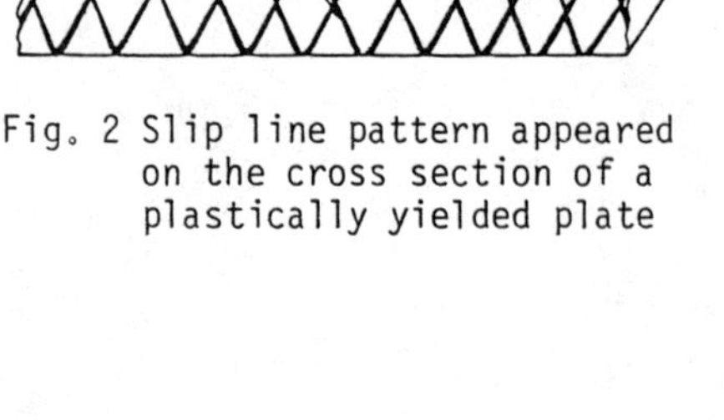

Fig. 2 Slip line pattern appeared on the cross section of a plastically yielded plate

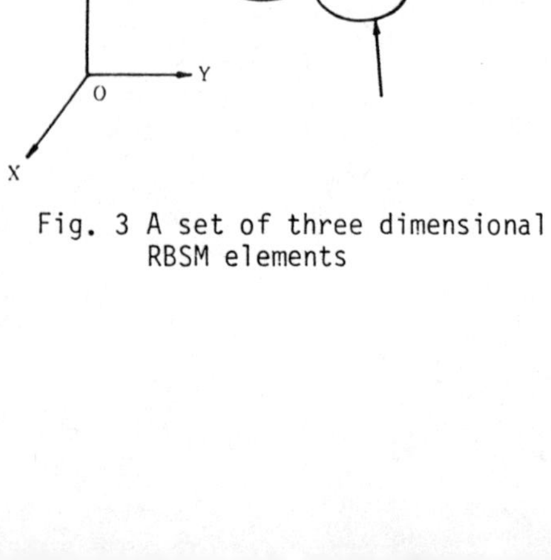

Fig. 3 A set of three dimensional RBSM elements

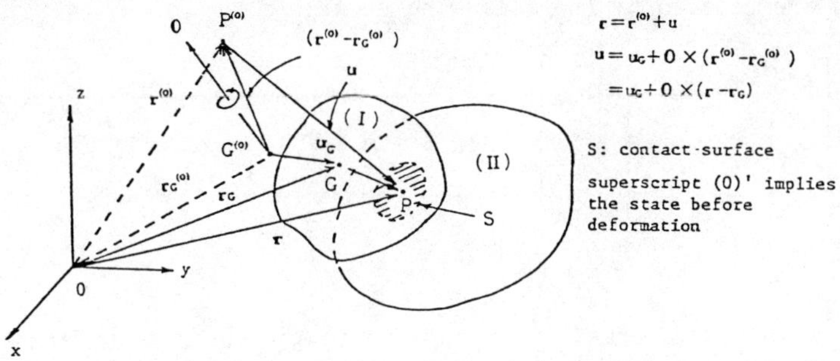

Fig. 4 Definition of rigid bodies-spring system

Denoting the displacement vectors of an arbitrary point P(x,y,z) in bodies (I) or (II) by $\mathbf{u}'$, $\mathbf{u}''$, respectively, the relative displacement vector of the point P can be defined as follows:

$$\overrightarrow{P'P''} = \mathbf{u}''-\mathbf{u}' = \boldsymbol{\delta} = (\delta_x, \delta_y, \delta_z) \tag{3}$$

This relative displacment is also represented by a vector $(\delta_t, \delta_b, \delta_n)$ defined with respect to the local coordinates as shown in Fig. 5.

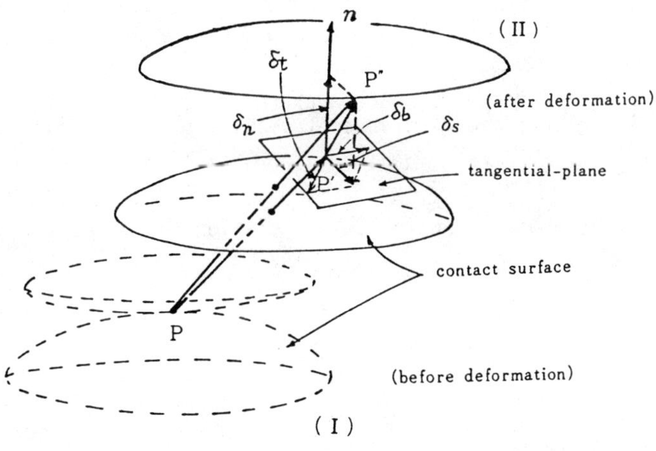

Fig. 5 Relative displacements of P on the contact surfaces in bodies (I) and (II)

And they are related by the following equation:

$$\left\{\begin{matrix}\delta_x\\ \delta_y\\ \delta_z\end{matrix}\right\} = \begin{bmatrix}\ell_1 & m_1 & n_1\\ \ell_2 & m_2 & n_2\\ \ell_3 & m_3 & m_3\end{bmatrix}\left\{\begin{matrix}\delta_t\\ \delta_b\\ \delta_n\end{matrix}\right\} \quad \text{or } \boldsymbol{\delta} = \mathbf{A}\,\boldsymbol{\delta}_0 \tag{4}$$

$$\begin{matrix}\| & & \| & & \|\\ \boldsymbol{\delta} & & \mathbf{A} & & \boldsymbol{\delta}_0\end{matrix}$$

Where $\mathbf{n}$ is the unit normal drawn outward to the contact surface at the point P, $\mathbf{t}$ and $\mathbf{b}$ are unit orthogonal vectors to compose the local coordinates as shown in Fig. 5.
With respect to this local coordinates, the stress-strain matrix of spring system is defined by the following equation:

$$\left\{\begin{matrix}f_t\\ f_b\\ f_n\end{matrix}\right\} = \begin{bmatrix}k_{11} & k_{12} & k_{13}\\ k_{21} & k_{22} & k_{23}\\ k_{31} & k_{32} & k_{33}\end{bmatrix}\left\{\begin{matrix}\delta_t\\ \delta_b\\ \delta_n\end{matrix}\right\} \tag{5}$$

$$\begin{matrix}\| & & \| & & \|\\ \mathbf{f}_0 & & \mathbf{D}_0 & & \boldsymbol{\delta}_0\end{matrix}$$

or

$$\mathbf{f}_0 = \mathbf{D}_0\ \boldsymbol{\delta}_0 \tag{6}$$

This reaction force vector $\mathbf{f}_0$ can be expressed by a force vector $\mathbf{f}$ with respect to the global coordinates as follows:

$$\mathbf{f} = \mathbf{A}\ \mathbf{f}_0 \tag{7}$$

From eqs.(4),(6) and (7) the following relation can be easily derived:

$$\mathbf{f} = \mathbf{D}\ \boldsymbol{\delta} \tag{8}$$

where

$$\mathbf{D} = \mathbf{A}^T\ \mathbf{D}_0\ \mathbf{A} \tag{9}$$

Basing on the above preliminaries, the strain energy V can be expressed by the following equation:

$$V = \frac{1}{2}\iint \boldsymbol{\delta}^T\mathbf{f}ds = \frac{1}{2}\iint \boldsymbol{\delta}^T\mathbf{D}\boldsymbol{\delta}ds \tag{10}$$

In case where the infinitesimal rigid body displacement field is assumed for the element displacement field, the following relation is easily obtained:

$$\underbrace{\left\{\begin{array}{c}\delta_x\\ \hline \delta_y\\ \hline \delta_z\end{array}\right\}}_{\{\delta\}} = \underbrace{\left[\begin{array}{c:c:c:c:c:c:c:c:c:c:c:c} -1 & 0 & 0 & 0 & -(z-z_1) & (y-y_1) & 1 & 0 & 0 & 0 & (z-z_2) & -(y-y_2) \\ \hdashline 0 & -1 & 0 & -(z-z_1) & 0 & -(x-x_1) & 0 & 1 & 0 & (z-z_1) & 0 & (x-x_2) \\ \hdashline 0 & 0 & -1 & -(y-y_1) & (x-x_1) & 0 & 0 & 0 & 1 & (y-y_2) & -(x-x_2) & 0 \end{array}\right]}_{[\mathbf{B}]} \underbrace{\left\{\begin{array}{c} u_1\\ v_1\\ w_1\\ \theta_1\\ \phi_1\\ \chi_1\\ u_2\\ v_2\\ w_2\\ \theta_2\\ \phi_2\\ \chi_2 \end{array}\right\}}_{\{\mathbf{u}\}} \qquad (11\text{-a})$$

or

$$\{\delta\} = [\mathbf{B}]\{\mathbf{u}\} \qquad (11\text{-b})$$

And therefore the following well-known expression for V can be derived:

$$V = \frac{1}{2}\,\mathbf{u}^T\,\mathbf{k}\,\mathbf{u} \qquad (12)$$

where

$$\mathbf{k} = \iint \mathbf{B}^T\,\mathbf{D}\,\mathbf{B}\,ds \qquad (13)$$

It may not be necessary to mention that $\mathbf{k}$ is the (12 x 12) stiffness matrix of the RBSM 3D elements from which the beam, plate and shell elements can be derived without any difficulty.
Details of the derivation can be seen in the literature listed at the end of this paper [7],[8].
Spring constants defined in eq.(5) can be determined in the following way.
In case of the isotropic elastic materials the stress-strain matrix $\mathbf{D}$ is given by

$$\mathbf{D} = \begin{bmatrix} K_n & 0 & 0 \\ 0 & K_s & 0 \\ 0 & 0 & K_t \end{bmatrix} \quad \text{and} \tag{14}$$

$$\left. \begin{aligned} K_n &= \frac{1}{h} \cdot \frac{E(1-\nu)}{(1+\nu)(1-2\nu)} \\ K_s &= K_t = \frac{1}{h}\left(\frac{E}{1+\nu}\right) \end{aligned} \right\} \tag{15}$$

h is the projceted length of a vector connecting centroids along the normal drawn on the tangential plane of two adjacent elements.
For determination of spring constants in plastic range, Yamada's method is adopted [9].
First it is assumed that plastic yielding will occur if stresses in these spring system satisfy the following condition:

$$f(\boldsymbol{\sigma}) = 0 \; (\boldsymbol{\sigma}^t = \lfloor \sigma_n, \tau_s, \tau_b \rfloor) \tag{16}$$

where $f(\boldsymbol{\sigma})$ is the plastic potential in the flow theory of plasticity.
Based on the plastic flow rule, the relation between stress increments $\Delta\boldsymbol{\sigma}$ and strain increments $\Delta\boldsymbol{\varepsilon}$ can be finally obtained in the following form:

$$\Delta\boldsymbol{\sigma} = \left[\mathbf{K}^e - \frac{\left(\frac{\partial f}{\partial \boldsymbol{\sigma}}\right)^t \mathbf{K}^{(e)} \cdot \mathbf{K}^{(e)} \frac{\partial f}{\partial \boldsymbol{\sigma}}}{\left(\frac{\partial f}{\partial \boldsymbol{\sigma}}\right)^t \mathbf{K}^{(e)} \frac{\partial f}{\partial \boldsymbol{\sigma}}} \right] \Delta\boldsymbol{\varepsilon} \tag{17}$$

It should be mentioned here that the method described was proposed for the rigid-plastic materials and therefore the effect of strain hardening is not taken into account. (See Fig. 6)

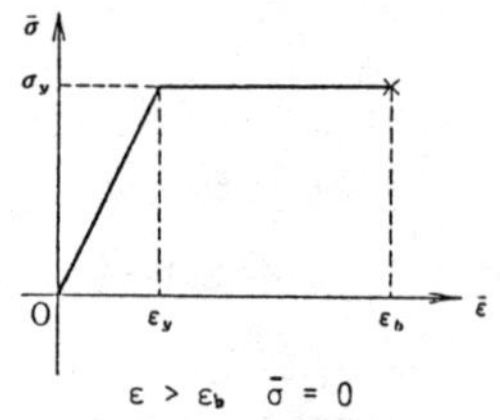

Fig. 6 A condition of the crack initiation of the interelement boundary

It is needless to say that a pertinent yield condition should be selected depending upon the material to be used.
Contact problems can be also handled easily by using the spring characteristics as shown in Fig. 7.

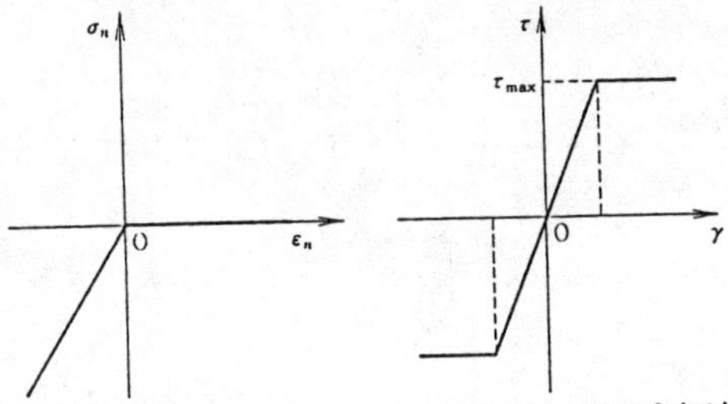

$\tau_{max} = \mu\sigma_n$: maximum frictional force(x area)
μ: static coefficient of friction
$|\tau| < \tau_{max}$ no slip
$\tau \geqq \tau_{max}$ slip

Fig. 7 Characteristics of the interelement springs in contact problems

However, in case of actual soil or rock materials, one single yield criterion does not always fit to their structural behavior and therefore a reliable semi-empirical method should be established for determination of the plastic potential $f(\boldsymbol{\sigma})$.
In general analysis of the soil and rock materials, it is often important to consider the coupled dynamic problems of structures and foundations due to earthquake motion or creep deformation problems under the gravitational loads and seepage effect.
For this purpose spring system connecting two adjacent rigid bodies are replaced by a rheological elasto-viscoplastic system composed of springs, sliders and dashpots as shown in Fig. 8.
General method of solution on the elasto-viscoplastic problems of solids has been proposed by Toi in 1979 [10],[11].
In case of the displacement with finite rotation, eq.(1) has to be replaced by the following equation:

$$\mathbf{u} = \mathbf{u}_G + (\mathbf{T} - \mathbf{I})(\mathbf{r} - \mathbf{r}_G) \tag{18}$$

where $\mathbf{T}$ is a coordinate transformation matrix of local coordinates attached to the centroid between before and after deformation, and $\mathbf{I}$ is an unit matrix [12].
In case where rotational displacements (θ,ϕ,χ) are sufficiently small that approximations such as $\cos\theta \fallingdotseq 1-\theta^2/2$, $\sin\theta \fallingdotseq \theta$ are allowed, the matrix $[\mathbf{T} - \mathbf{I}]$ can be given by the following equation:

$$[\mathbf{T}-\mathbf{I}] = \begin{bmatrix} -\frac{1}{2}(\phi^2+\chi^2) & -\chi + \frac{1}{2}\phi\theta & \phi + \frac{1}{2}\chi\theta \\ \chi + \frac{1}{2}\phi\theta & -\frac{1}{2}(\chi^2+\theta^2) & -\theta + \frac{1}{2}\chi\phi \\ -\phi + \frac{1}{2}\chi\theta & \theta + \frac{1}{2}\chi\phi & -\frac{1}{2}(\theta^2+\phi^2) \end{bmatrix} \tag{19}$$

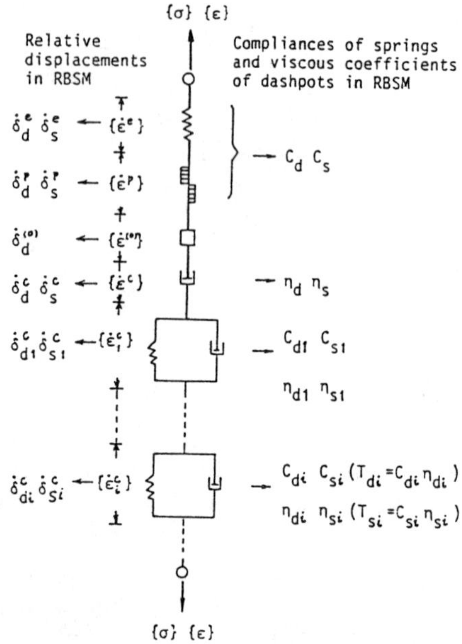

$\{\varepsilon^e\}$: elastic strain $\{\varepsilon^p\}$: plastic strain
$\{\varepsilon^{(o)}\}$: initial strain (thermal strain)
$\{\varepsilon^c\}$: stationary creep strain
$\{\varepsilon^c_i\}$: transient creep strain

Fig. 8 Mechanical model of viscoelastic-plastic constitutive equation

With such preliminaries, the following incremental form of the virtual work equation can be derived:

$$\left.\begin{aligned}
&\Sigma\iint\delta(\Delta\delta^{(1)})^t\,(\mathbf{f}^{(0)}+\Delta\mathbf{f})ds+\Sigma\iint\delta(\Delta\boldsymbol{\delta}^{(2)})^t\mathbf{f}^{(0)}ds\\
&-\Sigma\iiint\delta(\Delta\mathbf{u}^{(1)})^t(\bar{\mathbf{P}}^{(0)}+\Delta\bar{\mathbf{P}})dv-\Sigma\iiint\delta(\Delta\mathbf{u}^{(2)})^t(\bar{\mathbf{P}}^{(0)}+\Delta\bar{\mathbf{P}})dv\\
&-\Sigma\iint\delta(\Delta\mathbf{u}^{(1)})^t(\bar{\mathbf{f}}^{(0)}+\Delta\bar{\mathbf{f}})ds-\Sigma\iint\delta(\Delta\mathbf{u}^{(2)})^t(\bar{\mathbf{f}}^{(0)}+\Delta\bar{\mathbf{f}})ds=0
\end{aligned}\right\}\qquad(20)$$

Where superscript(o) implies the initial state, (i) (i=1,2) represents the linear and nonlinear components respectively. Body forces and surface traction are represented by $\bar{\mathbf{P}}$ and $\bar{\mathbf{f}}$ respectively.
X components of non-linear as well as linear displacements, i.e., $\Delta U^{(1)}$ and $\Delta U^{(2)}$ are given by

(i) $\Delta U^{(1)}$ (linear component)

$$\left.\begin{aligned}\Delta U^{(1)} &= u_0 + (z-z_0)\Delta\phi - (y-y_0)\Delta\chi - (x-x_0)(\phi^{(0)}\Delta\phi + \chi^{(0)}\Delta\chi)\\ &\quad + \frac{1}{2}(y-y_0)(\phi^{(0)}\Delta\theta + \theta^{(0)}\Delta\phi) + \frac{1}{2}(z-z_0)(\chi^{(0)}\Delta\theta + \theta^{(0)}\Delta\chi)\end{aligned}\right\} \quad (21)$$

(ii) $\Delta U^{(2)}$ (nonlinear component)

$$\left.\begin{aligned}\Delta U^{(2)} &= -\frac{1}{2}(x-x_0)\{(\Delta\phi)^2 + (\Delta\chi)^2\} + \frac{1}{2}(y-y_0)(\Delta\phi)(\Delta\theta)\\ &\quad + \frac{1}{2}(z-z_0)(\Delta\chi)(\Delta\theta)\end{aligned}\right\} \quad (22)$$

Similary x components of $\Delta\delta^{(1)}$, and $\Delta\delta^{(2)}$ are given by the following equations:

$$\left.\begin{aligned}\Delta\delta_x^{(1)} &= \Delta U_2^{(1)} - \Delta U_1^{(1)}\\ &= (\Delta u_2 - \Delta u_1) + \{(z-z_2)\Delta\phi_2 - (z-z_1)\Delta\phi_1\} - \{(y-y_2)\Delta\chi_2\\ &\quad - (y-y_1)\Delta\chi_1\} - \{(x-x_2)(\phi_2{}^{(0)}\Delta\phi_2 + \chi_2{}^{(0)}\Delta\chi_2) - (x-x_1) \times\\ &\quad (\phi_1{}^{(0)}\Delta\phi_1 + \chi_1{}^{(0)}\Delta\chi_1)\} + \frac{1}{2}\{(y-y_2)(\phi_2{}^{(0)}\Delta\theta_2 + \theta_2{}^{(0)}\Delta\phi_2)\\ &\quad -(y-y_1)(\phi_1{}^{(0)}\Delta\theta_1 + \theta_1{}^{(0)}\Delta\phi_1)\} + \frac{1}{2}\{(z-z_2)(\chi_2{}^{(0)}\Delta\theta_2 +\\ &\quad \theta_2{}^{(0)}\Delta\chi_2) - (z-z_1)(\chi_1{}^{(0)}\Delta\theta_1 + \theta_1{}^{(0)}\Delta\chi_1)\}\end{aligned}\right\} \quad (23)$$

$$\left.\begin{aligned}\Delta\delta_x^{(2)} &= \Delta U_2^{(2)} - \Delta U_1^{(2)}\\ &= -\frac{1}{2}[(x-x_2)\{(\Delta\phi_2)^2 + (\Delta\chi_2)^2\} - (x-x_1)\{(\Delta\phi_1)^2 + (\Delta\chi_1)^2\}]\\ &\quad + \frac{1}{2}\{(y-y_2)(\Delta\phi_2)(\Delta\theta_2) - (y-y_1)(\Delta\phi_1)(\Delta\theta_1)\}\\ &\quad + \frac{1}{2}\{(z-z_2)(\Delta\chi_2)(\Delta\theta_2) - (z-z_1)(\Delta\chi_1)(\Delta\theta_1)\}\end{aligned}\right\} \quad (24)$$

In derivation of eq.(20), the terms of higher order products of the incremental displacements were neglected.
Eq.(20) is further transformed into the following equation:

$$\left.\begin{aligned}
&\Sigma\iint\delta(\Delta\boldsymbol{\delta}^{(1)})^t\,\mathbf{D}\Delta\delta^{(1)}ds + \Sigma\iint\delta(\Delta\boldsymbol{\delta}^{(2)})^t\mathbf{f}^{(0)}ds \\
&-\Sigma\iiint\delta(\Delta\mathbf{u}^{(1)})^t\Delta\bar{\mathbf{P}}\,dv - \Sigma\iiint\delta(\Delta\mathbf{u}^{(2)})^t(\bar{\mathbf{P}}^{(0)}+\Delta\bar{\mathbf{P}})dv \\
&-\Sigma\iint\delta(\Delta\mathbf{u}^{(1)})^t\Delta\bar{\mathbf{f}}\,ds - \Sigma\iint\delta(\Delta\mathbf{u}^{(2)})^t(\bar{\mathbf{f}}^{(0)}+\Delta\bar{\mathbf{f}})ds = \mathbf{R} \\
&\qquad\mathbf{R} = -\Sigma\iint\delta(\Delta\boldsymbol{\delta}^{(1)})^t\mathbf{f}^{(0)}ds - \Sigma\iiint\delta(\Delta\mathbf{u}^{(1)})^t\bar{\mathbf{P}}^{(0)}dv \\
&\qquad\quad -\Sigma\iint\delta(\Delta\mathbf{u}^{(1)})^t\bar{\mathbf{f}}^{(0)}ds
\end{aligned}\right\}\tag{25}$$

It may not be difficult to derive the following matrix equation in incremental form after some calculations.

$$\mathbf{K}_T\,\Delta\mathbf{u} = \Delta\bar{\mathbf{F}} - \bar{\mathbf{F}}_r \tag{26}$$

where $\mathbf{K}_T$ is the total tangential stiffness matrix given by

$$\mathbf{K}_T = \mathbf{K} + \mathbf{K}_0 + \mathbf{K}_G \tag{27}$$

$\mathbf{K}$ is the original stiffness matrix, $\mathbf{K}_0$ the initial displacement matrix and $\mathbf{K}_G$ corresponds to the initial stress (geometric) matrix in the well known finite element methods.
It should be mentioned here that $\bar{\mathbf{F}}_r$ is an unbalanced force due to manipulation error in previous stage of loading.

CONVERGENCY OF THE RBSM SOLUTIONS

Generally speaking the elastic deformation of solids is characterized by the deformation of the finite elements without slip on the interelement boundaries. The finite element displacement models can satisfy this displacement continuity on the interelement boundaries so that convergency of the FEM elastic solutions can be usually guaranteed. However, convergency of the FEM inelastic solutions up to the limit loading stage can hardly be guaranteed because the FEM analysis may always require such displacement continuity along the interelement boundaries even beyond the elastic limit. This might cause substantial error of the calculated results from the true solutions.
On the other hand, in the RBSM models the elements are assumed to be rigid and therefore the element deformations are not considered, but the interelement displacement can be taken into account by introducing two types of springs connecting two adjacent elements.

Consequently it is extremely difficult to represent the elastic deformation by using the RBSM models, and then convergency of the RBSM elastic solutions can not be guaranteed in general except in the cases of some elements such as beam element, ring element for the axisymmetric deformation of shells of revolution.
However, once the structural deformation reaches the stage of the large scale yielding, deformation due to slip on the intergranular boundaries may become predominant and consequently the elastic deformation of the particles may become negligibly small. And therefore if the real collapse mechanism is included in the assumed mesh pattern, the incremental RBSM solutions always yield the true collapse loads. Or even if this is not the case, it can be proved that such incremental method can always give the best upper bound solutions.
From the practical point of view, however, it is still not sufficient to guarantee accuracy of the RBSM solutions unless the corresponding lower bound solutions can be derived. Kawai has considered that a lower bound solution may be obtained from the RBSM solutions by following the diagram as shown in Fig. 9.

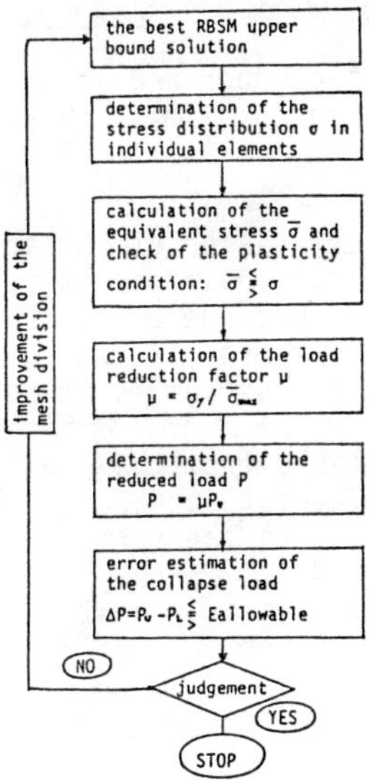

limit load	P_U	P_L
mechanism condition	OK	?
equilibrium condition	OK	OK
plasticity condition	?	OK

$\therefore \boxed{P_L \leq P_c \leq P_U}$

(P_c is the true collapse load)

E is the allowabl error

Fig. 9 Flow diagram for evaluation of the reliable limit loads

The practical program to obtain such lower bound solutions has been developed recently by Takeuchi and he verified the effectiveness of his method by solving several simple examples, one of which is shown in Fig. 10 [13].

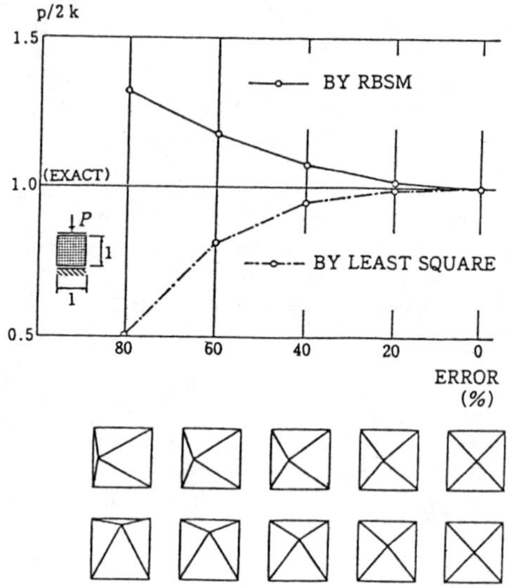

Fig. 10 Determination of the upper and lower bound solutions in the present limit analysis

In this way a reliable method for determination of collapse loads has been developed by using the RBSM models. The rigorous mathematical proof, however, has not been completed, because exact determination of the stress distribution in individual elements is extremely difficult so that approximate methods of solution such as the least square method are obliged to use.
Nevertheless this method may be acceptable from the limit analysis point of view. Another important factor related to the reliability of the RBSM solutions is rational determination of the spring constants and their constitutive law already mentioned.
Formula for determination of the elastic spring constants eq.(14) is generally not accurate enough because of the two following reasons;

(a) Exact formula for σ_n is given by the following equation:

$$\sigma_n = \frac{E}{(1+\nu)(1-2\nu)}\{(1-\nu)\varepsilon_n + \nu\varepsilon_s + \nu\varepsilon_b\} \tag{28}$$

In eq.(16), therefore, effects of two other strain components ε_b and ε_s are neglected.

(b) The following strain-displacement relations used for derivation of eq.(15) is not accurate.

$$\varepsilon_n = \frac{\delta_n}{h}, \qquad \gamma_s = \frac{\delta_s}{h} \tag{29}$$

The more accurate finite difference formula for the strain-displacement relation can be derived on the basis of the variational difference method. This leads to very laborious analysis and accuracy estimation of the proposed formula is extremely difficult, and therefore accuracy of the RBSM elastic solution can not be guaranteed.

It is opinion of the present authors that improvement of the RBSM elastic solutions may be possible, but it may not be economically feasible because any other numerical method is not considered to be superior to the finite element method as far as the elastic analysis is concerned, and then use of the RBSM models should be positively restricted to nonlinear analysis of solids.

NO-TENSION AND TENSION CRACK ANALYSIS OF SOIL AND ROCK FOUNDATIONS BY MEANS OF NEWLY PROPOSED DISCRETE ELEMENTS [14]

Zienkiewicz and others proposed a method of "No-Tension Analysis" to study behavior of the granular materials [2],[3].
Fig. 11 shows the flow diagram of their method.

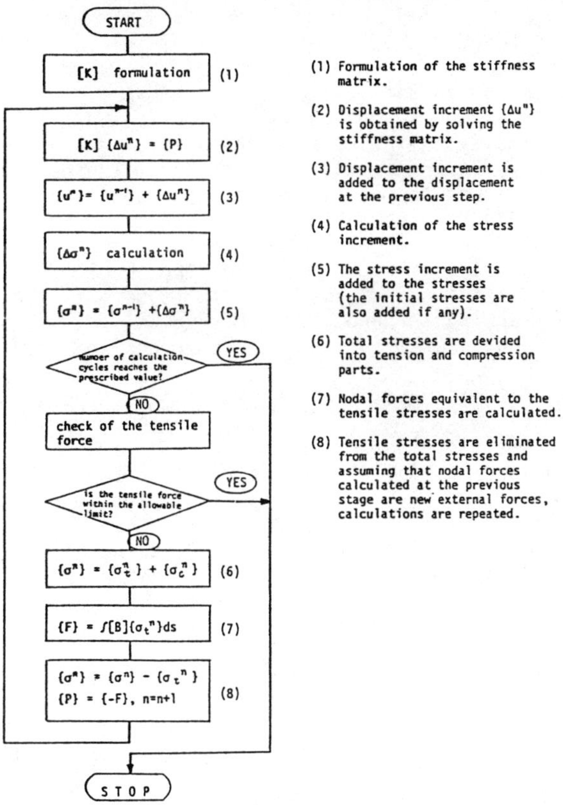

Fig. 11 Flow chart for no-tension analysis

This method consists of the following three steps:

(1) Assuming that a given structures is elastic, stress distribution is determined first. If there exists an initial stress distribution, total stresses at this step are determined by adding the initial stresses to the elastic stresses obtained.

(2) If the magnitude of the tensile stress exceeds the allowable limit, such stresses have to be eliminated without causing any additional structural deformation. For this purpose a system of restraining forces must be imposed in order to maintain the equilibrium.

(3) Since such restraining forces may not actually exist, a system of nodal forces of the same magnitude but opposite sign should be added to the structure. It should be mentioned here that the structure may behave elastically during the elimination process of the restraining forces.

The second and third steps should be repeated until the tensile stresses are eliminated.

In the above-mentioned method of analysis, the initial stiffness matrix can be used for the entire process of calculation, and therefore reduction of computing time may be possible, but the accuracy of displacement obtained by this method may not be guaranteed if the crack initiates somewhere. Especially in the case of analysis of granular materials, actual behavior may be completely different from those predicted by this no-tension analysis and to make it worse, the contact problem can not be handled easily by this method.
Takeuchi considered that a checking scheme for contact of two adjacement elements in the process of incremental loading should be incorported to the program of no-tension analysis in order to make it more realistic in analysis of soil and rock foundations.
Fig. 12 is the flow diagram for the tension crack analysis thus developed.

Outline of this method consists of the following steps:

(1) Elastic stress distribution is obtained for a given structure. If there exists initial stress distribution the total stress is obtained by adding it to the calculated elastic stresses.

(2) If there appear tensile stresses exceeding the allowable value in the stress distribution, portion of the maximum stresses is taken out and the spring in this portion is cut by applying fictitious forces to eliminate such exceeding tensile stresses.

(3) Elastic calculation is made based on the calculation of stress release at the second step, and possibility of re-contact is checked. If no contact is made calculation returns to the first step, while if re-contact is observed calculation goes to the 4th step.

(4) The minimum magnification factor r_{min} which makes the relative

displacement zero at the 3rd step is obtained by using the relative displacements at the (n-1) and (n+1) steps, and the stresses multiplied by r_{min} is considered as the stresses at the present step. Then the spring action is recovered and multiplying the released force by $(1-r_{min})$ elastic calculation is made. This step of calculation will be repeated until recontact is eliminated and calculation will return to the second step.

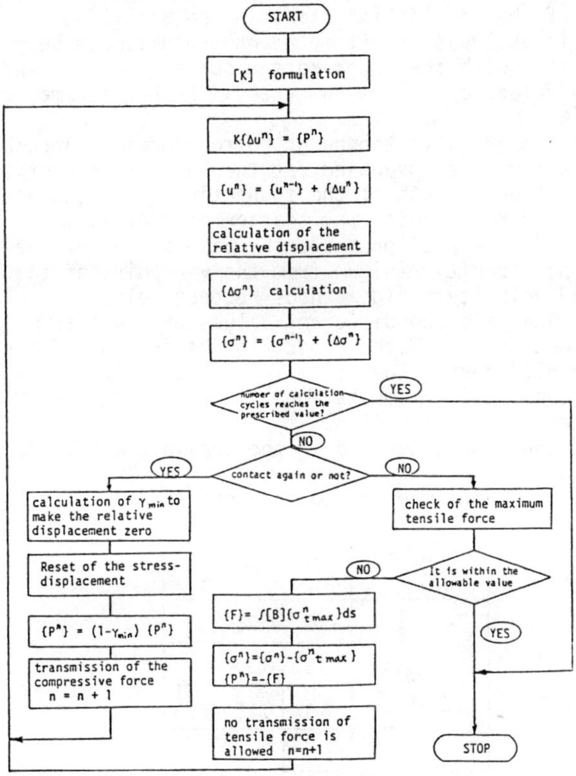

Fig. 12 Flow chart for tension-crack analysis

In the above mentioned method incremental calculation is made by releasing the unbalanced force. During the stress-releasing process possibility of the re-contact of adjacent elements is considered so that the algorism becomes somewhat complicated, and computing time may be increased due to various checking schemes.
In this computational scheme introduction of the crack initiation, consideration of initial stresses and application to the contact as well as elasto-plastic analyses may be possible and these advantages may fully compensate the above-mentioned drawback.

Some Typical Examples of Application

(i) Analysis on the ultimate bearing capacity of the foundation structures

In the analysis on the ultimate bearing capacity of the footings resting on the sandy soil foundation, laborious numerical analysis is usually inevitable and it has not been well studied because consideration of deadweight of the soil is important in the analysis.
Therefore, limit analysis is attempted on the ultimate bearing capacity of a strip footing with the rough bottom surface on the sandy soil and the calculated values of N_γ are compared with the recommended design values [15],[16].
In case of the footing with the rough bottom surface, a number of recommendations have been proposed for the base angle ψ of the wedge to be formed just underneath of the foundation. And consequently the estimated bearing capacity may scatter considerably.
Meyerhof [17] and Chen [18] proposed the base angles of the wedge which make the bearing capacity minimum from the viewpoint of the limit equilibrium and velocity field methods respectively.
In the present analysis two different values of ψ are assumed by basing on their recommendation. With respect to the internal friction angle of the soil, four different values of ϕ', i.e., $\phi'=30°$, 35°, 40° and 45° are assumed. And considering computational limitation a small value $C'=0.001$ tf/m^2 is adopted for the cohesion.
Cases analysed and constants used in the analyses are listed in Table 1.

Table 1 Cases considered in the present analysis

ψ (deg)		ϕ' (deg)	$2b/B$
Meyerhof ($1.2\phi'$)	36	30	0.28~.43
	42	35	0.3 ~ .6
	48	40	0.25~.55
	54	45	〃
Chen	45	30	0.3 ~ .45
	50	35	0.3 ~ .6
	54	40	〃
	58	45	〃

$C'=0.001 tf/m^2$, $E=1500 tf/m^2$, $\nu=0.45$

With respect to the mesh division in the transient range, four different mesh patterns which are based on the logarithmic spirals with poles $P_1 \sim P_4$ as shown in Fig. 13 are conveniently selected.

In these analyses, b values shown in Fig. 13 are taken as shown in Table 1 after preliminary calculations on the location of the pole in each case.
Fig. 14 shows the mesh divisions employed.
Coulomb's condition is adopted as the failure criterion.
One example of slip lines at the failure is shown in Fig. 15 where bold lines implies slip lines to form the collapse mechanism.
In the transient area of these slip lines there can be seen small gaps. This may be mainly due to approximation of transient area by straingt lines along the logarithmic spiral. Fig. 16 shows calculated values of N_γ corresponding to base angles ψ of the two different wedges.

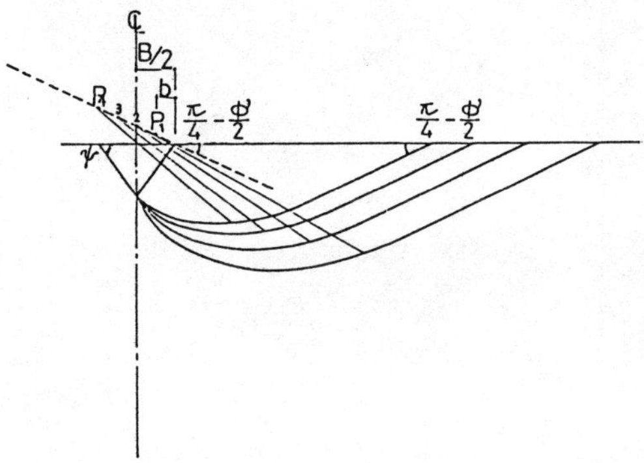

Fig. 13 Assumed slip line patterns in the transient area

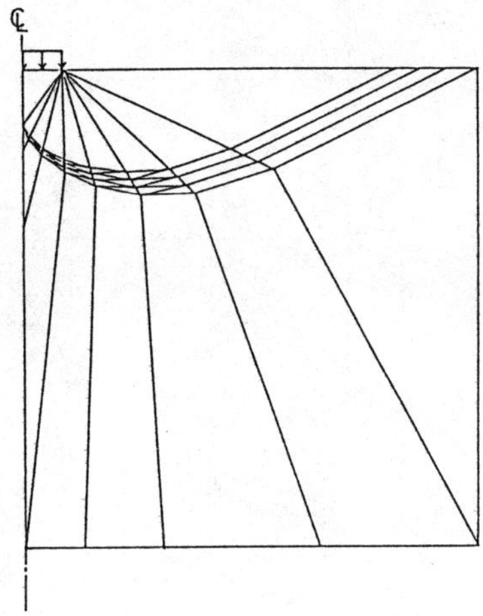

Fig. 14 Mesh division used

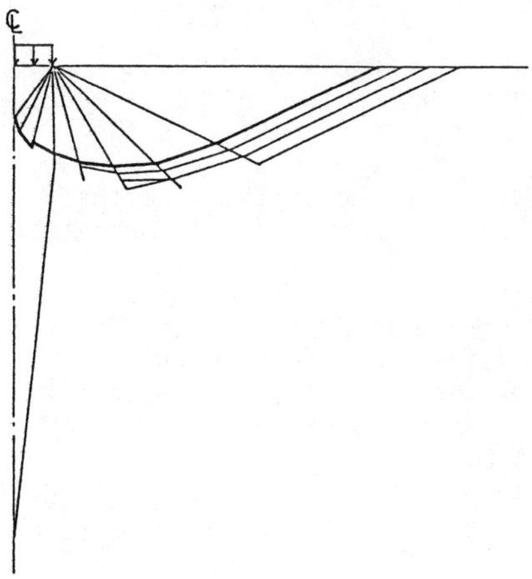

Fig. 15 Slip line pattern at the collapse stage

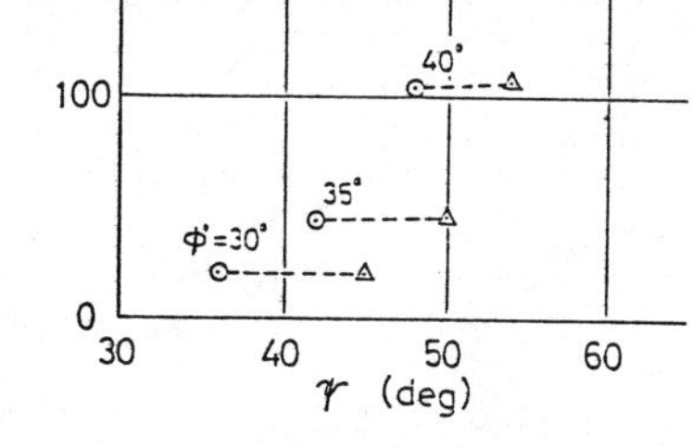

Fig. 16 Base angles of the wedge vs $N\gamma$ values calculated

Meyerhof reported that the minimum values of N_γ may fractuate 36%, for $\phi'=30°$ and 17% for $\phi'=45°$ in the domain $\phi' \leqq \psi \leqq \pi/4 + \phi'/2$ respectively. In the present analysis, however, values of N_γ corresponding to each ϕ' did not change.
Calculated values of N_γ and recommended values for $\phi'=30° \sim 45°$ are plotted in Fig. 17.

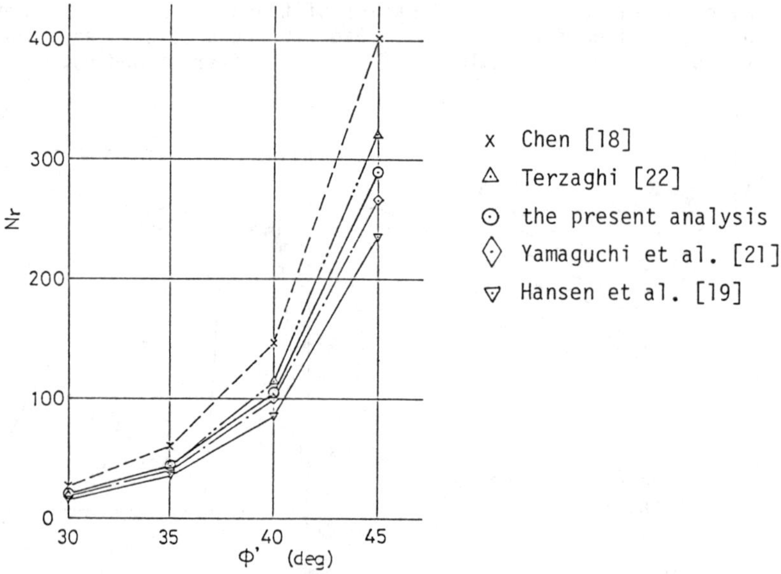

Fig. 17 Calculated values of N_γ and recommended values

Hansen's value [19] in this figure was obtained by the method of characteristic curves which constructs slip line net successively in the foundation by using Kötter's equation given by Lungren et al. [20].
A method proposed by Yamaguchi et al. [21] is based on the same idea, but their assumption of the wedge just underneath of a given foundation is different.
More precisely, in the former method a line which intersects the central axis of the foundation with the angle of $\pi/4 + \phi'/2$ among many slip lines passing at the edge of the footing is considered to form the side surface of the wedge, while in the latter method a rigid wedge of the equilateral triangle is considered just underneath of the foundation basing on the experimental evidence.
Terzaghi's value [22] as a typical solution of the approximate methods is 10 ∿ 35% greater than these rigorous values, while Chen's value obtained from the conventional velocity field is also 50 ∿ 80% greater than these values.
The calculated values by means of the present method coincide with values obtained by other methods for $\phi'=30°$ and 35°, but for $\phi'=40°$ and 45° they

are between Terzaghi's and Yamaguchi's solutions. It can be concluded that they give good upper bounds for comparatively simple mesh divisions.

(ii) Analysis of the ultimate bearing capacity of the slope [23]

The slip line and ultimate bearing capacity of top of the embankment slope varies considerabley depending on gradient of the slope and loading point.
Kagawa et al. [24] conducted an experiment on the ultimate bearing capacity of a model slope which is considered the sandy soil and he carried out a simplified analysis in which slip surface is assumed and the finite element analysis where the material nonlinearity is considered. In the former method, however, representation of the compound slide etc. is difficult while in the latter method the slip surface can not be determined clearly.
Furthermore the displacement field determined by the finite element method is essentially continuum approximation. Therefore the displacement just underneath of the loading plane may be different from the experimental result because the slip movement is predominant there.
In view of this, limit analysis of this experiment is made by using the present models to verify usefullness of the proposed method.
As for the material constants used in the analysis, the specific weight of the soil γ_d = 0.0016kgf/cm^3, the angle of internal friction ϕ = 25°, the cohesion $\bar{C}$ = 0.25kgf/cm^2. Poisson's ratio ν = 0.35 and E = 60 kgf/cm^2 is assumed for the coefficient of deformation from the result of triaxial compression test shown in Fig. 18.

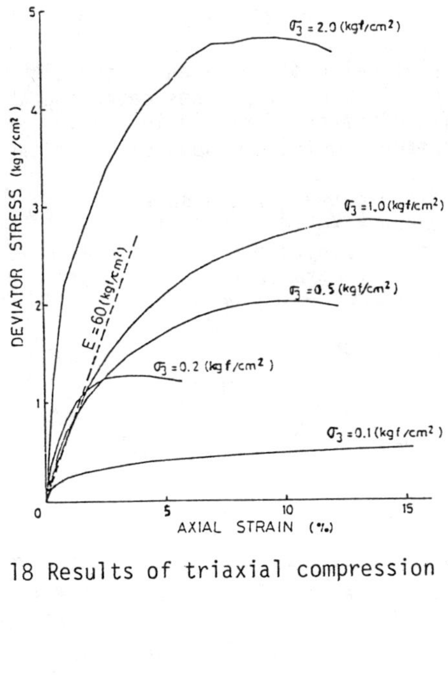

Fig. 18 Results of triaxial compression test

The structural model and mesh division used for the analysis are shown in Figs. 19 and 20 respectively.

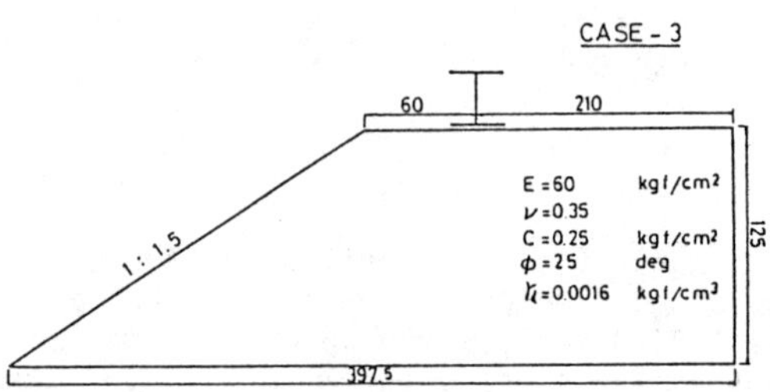

Fig. 19 Experimental model used for computation

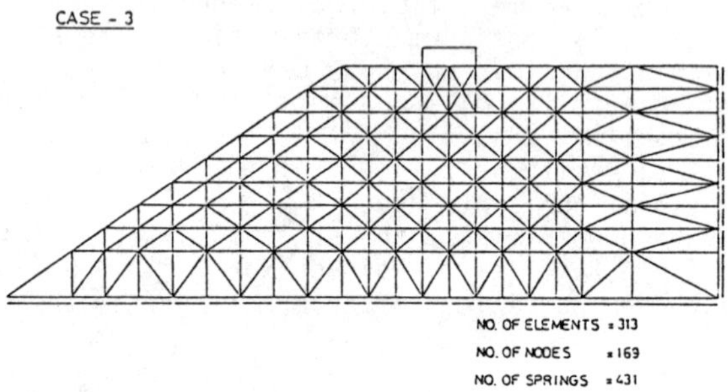

Fig. 20 Mesh division

In Fig. 21 results of calculation based on the circular arc method are compared with the slip lines determined by experiments, and naturally the failure patterns are considerably different. This shows difficulty in representing the compound slide in the circular arc method.

Fig. 22 show the distribution of the safety factor of the nonlinear finite element analysis where a solid line is the slip line determined expreimentally. In the finite element analysis compound slide can not be represented easily and the failure tends to spread downward of the slope like in case of the circular arc method.

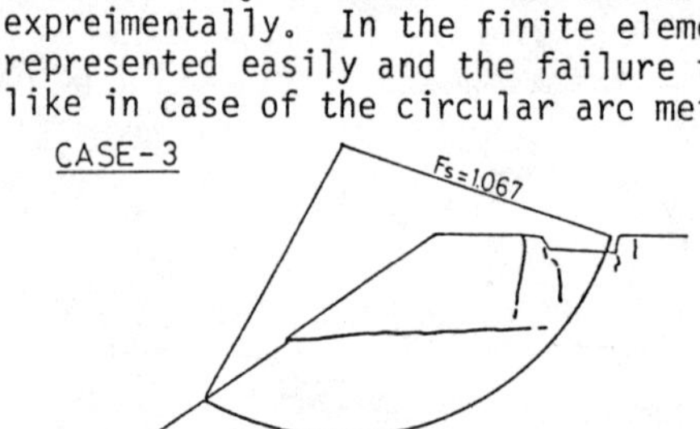

Fig. 21 Results of the circular arc method

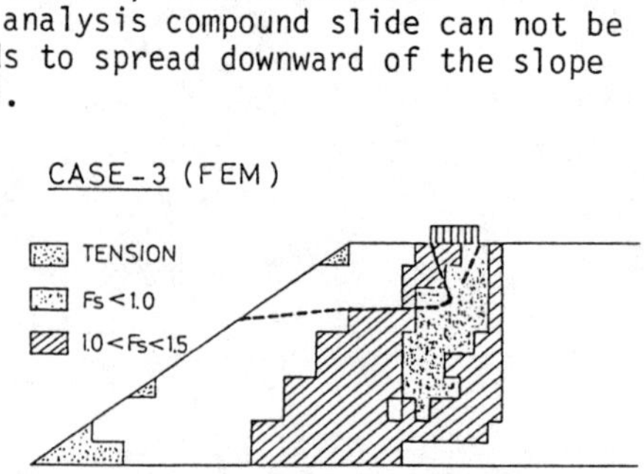

Fig. 22 Results of the finite element nonlinear analysis

On the other hand, results of the present analysis is shown in Fig. 23. and the solid line shows the slip line determined by calculation and the broken line represents experimentally determined slip lines.

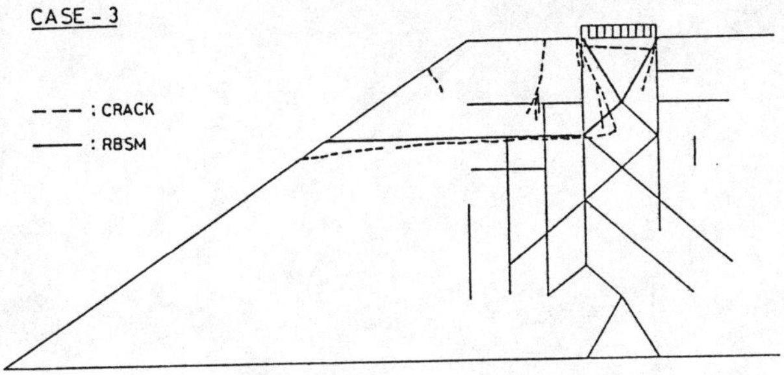

Fig. 23 Slip line pattern

The difference of calculation and experiment was found about 16%. Comparison of the measured and calculated earth pressure is shown in Fig. 24 where the solid line represents calculated and the broken line the measured results. It can be clearly seen rise of the earth pressure with increase of the applied load.

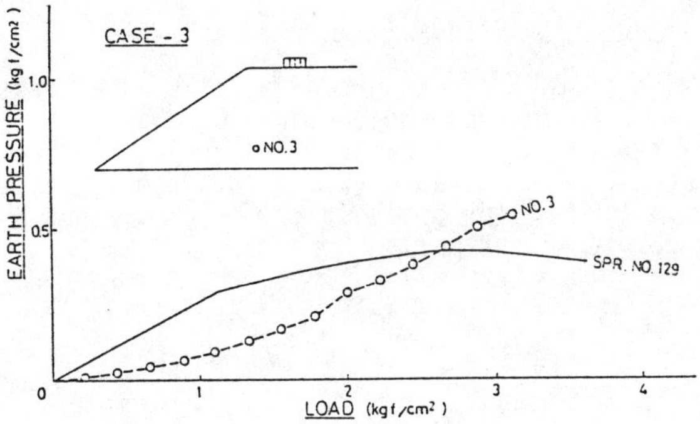

Fig. 24 Earth pressure distribution

The load-settlement curve is also plotted in Fig. 25. Good agreement of calculation and experiment can be seen. The calculated value of the ultimate bearing capacity was 3.611 kg/cm^2, and that of experiment was 3.11 kg/cm^2. The calculated collapse load is slightly greater than the experimental value, but the solution obtained must be good enough if uncertainty of the constitutive equation of a given sandy soil is considered.

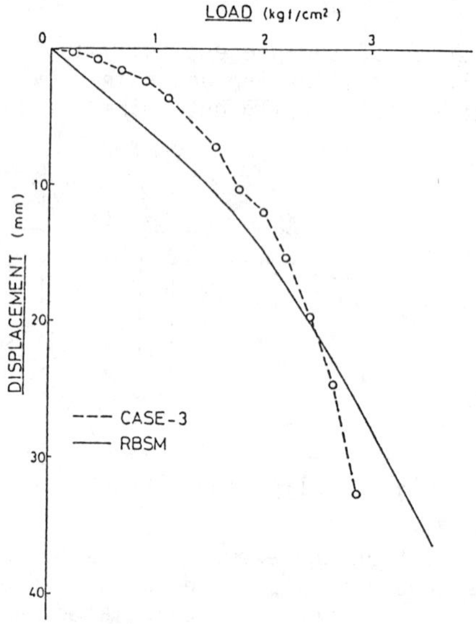

Fig. 25 Load-settlement curve

(iii) Stability analysis of the fill-up ground [25]

In the design of the fill-up ground effects of successive banking up to the finally planned slope height should be analysed.
In view of this, analysis was carried out by using the present models and supplementing an algorithm to add new elements due to successive banking. Fig. 26 is the model used for analysis of the embankment where Ⓑ is the bank and Ⓒ the counterweight fill.

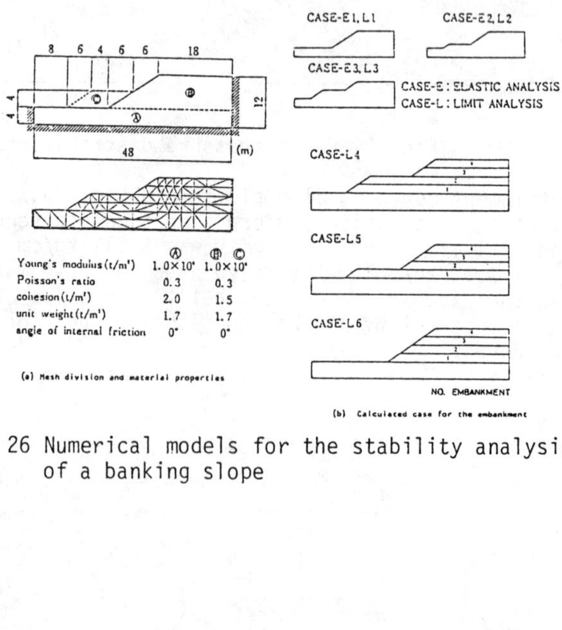

Fig. 26 Numerical models for the stability analysis of a banking slope

In this figure cases analysed are also shown.
Ⓛ implies a case where limit analysis was made and Ⓔ stands for cases where elastic analyses were made E1, L1 and L6 represents cases without counterweight fill and in the remaining cases counterweight fill of 2m ∼ 4m height was made.
Fig. 27 shows one of the calculated slip line and load-displacement curves. It can be seen that the slope instability may occur due to the bottom failure when the embankment is made at one time. On the other hand in case of the successive banking the slope lost the stability due to before the total deadweigh has loaded, development of the compound slip surface.

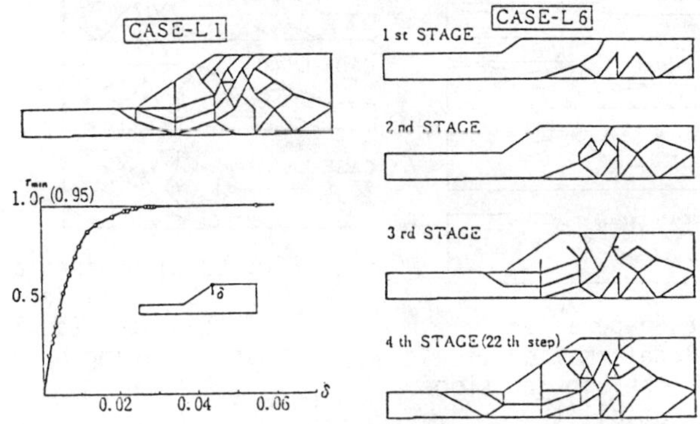

Fig. 27 Slip lines and load-displacement curve for the banking slope stability analysis without loading bank

This fact may be explained as follows:
In the former case the result of analysis is equivalent to the solution where 95% of the total weight of the embankment was loaded at one time, while in the latter case stress distribution may change with the progress of successive banking and collapse occurred due to the final banking. Therefore the stress distribution may be entirely different from that of the former case.
The total displacement was divided by the maximum displacement and such specific deformation mode is plotted in Fig. 28.
Form this figure it can be concluded that cases of embankment mentioned above are clearly different each other and the deformation pattern due to the successive banking may be more reasonable and realistic than the pattern due to the banking at one time. The result of calculation in case of counterweight fill of 2m height is shown in Fig. 29.
In this case the slope did not loose stability. The result of analysis in case of 4m height is shown in Fig. 30.
In cases where the slope did not collapse completely, results of calculations for two different cases of banking were found nearly the same. This may be due to the fact that the stress distribution in both cases have no appreciable difference unless the slope does collapse.

It can be seen that the structural behavior of two different cases of embankment may be naturally different if the time effect is taken into account in the analysis.

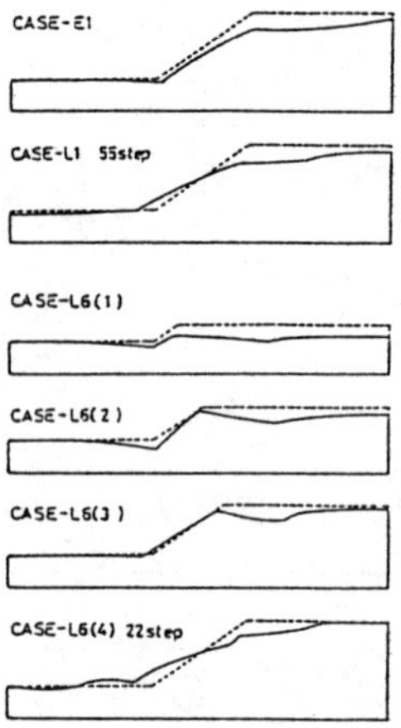

Fig. 28 Shape after deformation — the banking slope stability

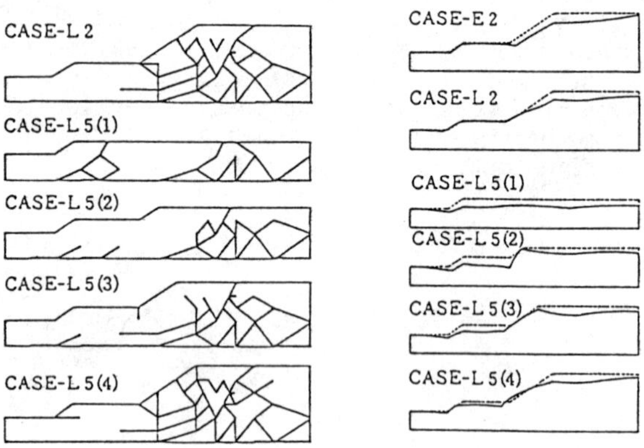

Fig. 29 Slip lines and shape after deformation — the banking slope stability analysis with loading bank (2m)

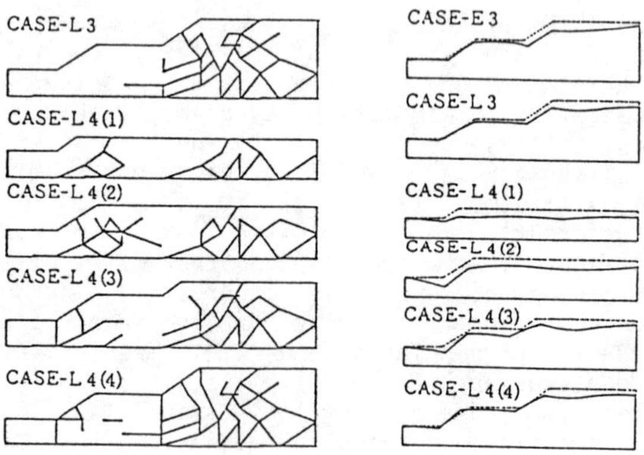

Fig. 30 Slip lines and shape after deformation — the banking slope stability analysis with loading bank (4m)

(iv) Analysis of the flexible sheet piles [26]

As can be seen in Fig. 31 the inplane RBSM element has three degrees of displacement freedom which are taken at the centroid, i.e., two translational displacements (u,v) and one rotational displacement (θ). And the strain energy of a given element is stored in the connection springs distributed on the interelement boundaries.

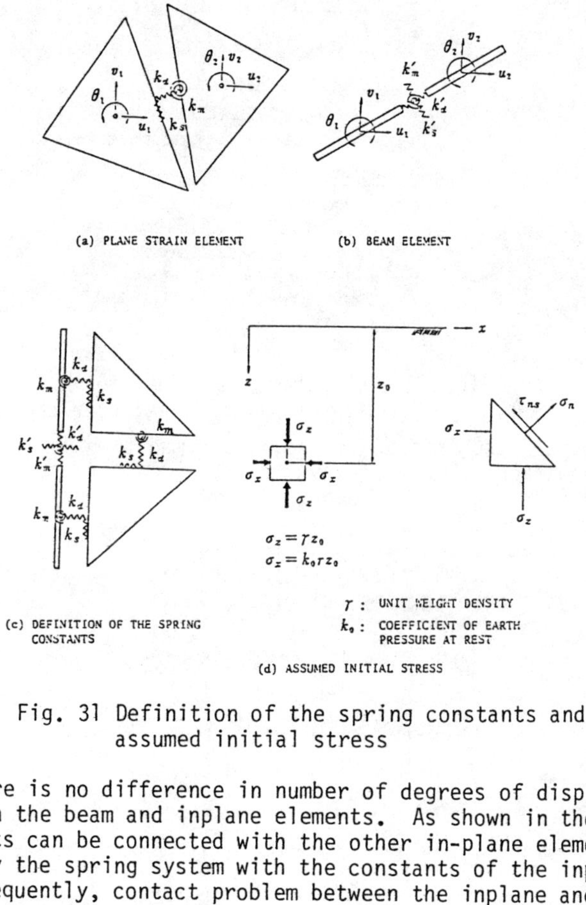

Fig. 31 Definition of the spring constants and assumed initial stress

Therefore, there is no difference in number of degrees of displacement freedom between the beam and inplane elements. As shown in the figure inplane elements can be connected with the other in-plane element or beam element by the spring system with the constants of the inplane element. Consequently, contact problem between the inplane and beam elements can be considered by adjusting the shear spring constant. As an example of such consideration mentioned above, the result of analysis on the simple flexible sheet piles shown in Fig. 32 is described in this section. The boundary conditions and material constants employed are shown in the figure and the number in ( ) is the coefficient of friction on the interface between the sheet pile and the foundations. It is also assumed that the failure occurs only in the foundations and the sheet pile is considered to behave elastically.

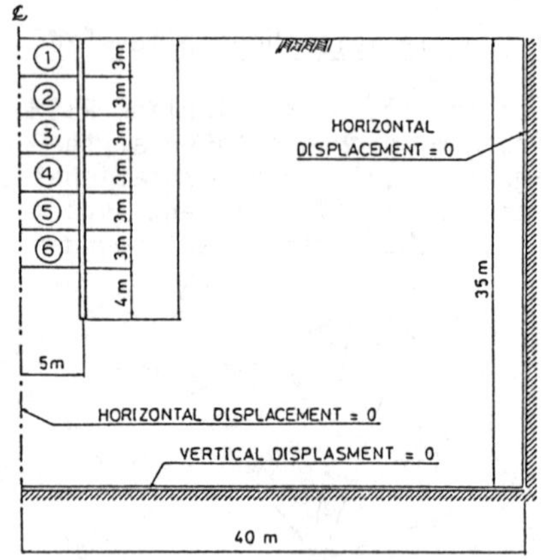

FOUNDATION		SHEET PILE	
E (t/m²)	5000	E (t/m²)	2.1×10^7
ν	0.3	ν	0.3
γ (t/m³)	1.7	A (cm²)	447
C (t/m²)	0.5 (0.1)	I (cm⁴)	180000
ϕ	30°		
K_0	0.4		

Fig. 32 A computational model for a limit analysis of the cantilever sheet pile

In this analysis the initial stress is assumed to be given by the following equation: $\sigma = k_0 \gamma z$
where k_0 is the static coefficient of earth pressure γz is the earth pressure due to deadweight.
Excavation is made successively in the sequence of the number encircled in the figure. Fig. 33 shows the failure patterns obtained by the plastic analysis at each step of excavation.

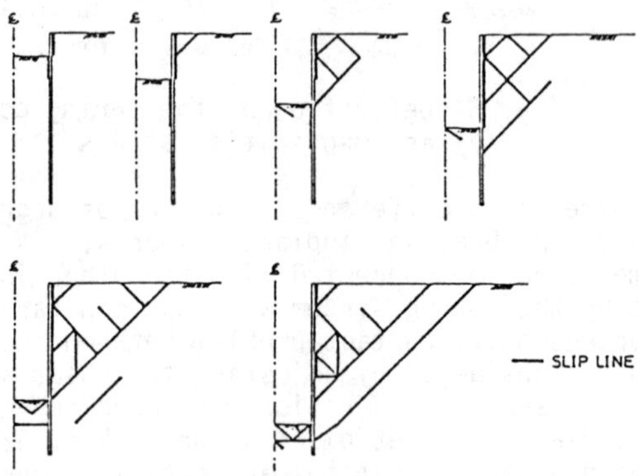

Fig. 33 Slip line patterns in the limit analysis of the cantilever sheet pile problem

It was found that at the early steps of excavation, slip tend to occur on the interface between the sheet pile and foundation rather than in the foundation itself, but failure in the foundation behind the sheet pile takes place as the excavation proceeds, and furthermore failure occurs on the excavated bottom surface. At the 6th step of excavation the bottom surface clearly failed and function of the embankment was not fulfilled.

Fig. 34 shows the displacement modes obtained from the results of plastic analysis at each step of excavation. In the early steps of excavation, the sheet pile was apt to fall forward, but it became apt to move forward as a whole after excavation proceeds to a certain extent. At the same time swelling of the excavated bottom surface became larger as the excavation proceed. Fig. 35 shows the bending moment diagrams at each step of excavation. It can be seen that large bending moments are produced not only in the portion already excavated but also even in the foundation.

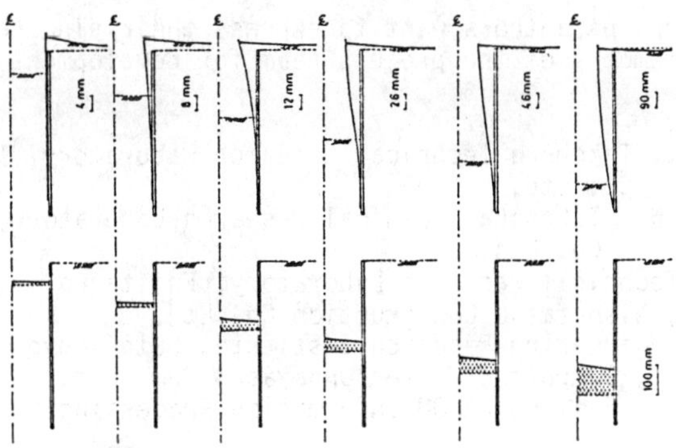

Fig. 34 Displacement modes of the limit analysis corresponding to the specific stage of excavation

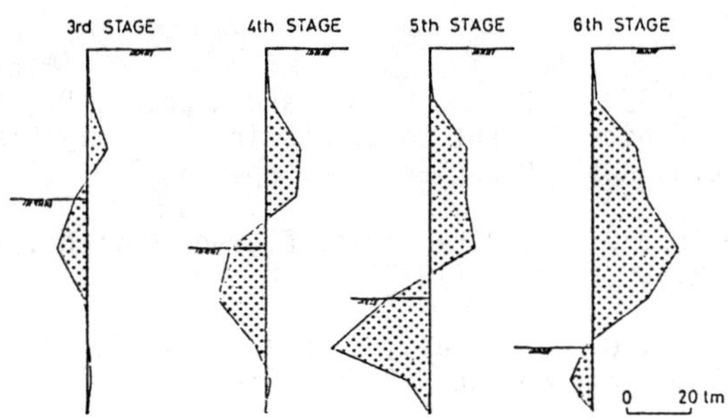

Fig. 35. Calculated bending moment diagram at each step of excavation

CONCLUSION

To attack the highly nonlinear structural problems a new discrete method of limit analysis is proposed by using the simplified elements which were developed by the first author Kawai six years ago.
The second author Takeuchi extended this method to the method of tension-crack analysis which is especially suitable for analysis of soil and rock mechanics problems.
Numerical examples shown in this paper are simple and may not be too realistic, but results obtained are proved to be consistent with experiments. Since these discrete models are ideally suitable to simulate granular materials, extensive application of these models to analysis of the actual geotechnical problems will be expected.

ACKNOWLEDGEMENTS

The present authors want to express their sincere thanks to the following members of the project team for development of the "EARTH-version I":

Kakurai, M., Takenaka Technical Research Laboratory, Takenaka Komuten Co. Ltd.
Yamashita, K., Takenaka Technical Research Laboratory, Takenaka Komuten Co. Ltd.
Hada, M., Technical Research Laboratory, Fujita Co.
Kumeta, T., Nishimatsu Construction Co. Ltd.
Yada, K., Engineering Research Institute, Sato Kogyo Co. Ltd.
Tokunaga, M., Computing Center, Ohbayashi Gumi Ltd.
Nakamura, S., Fujitsu FACOM Information Processing Ltd.

Without their enthusiastic cooperation and sacrificing efforts shown in the course of the program development, this paper could not be written. They are also grateful to Miss Sueko Suzuki for typing manuscript and drawing figures.

REFERENCES

1. Nagtegaal, J. C., Parks, D. M., and Rice, J. R., "On Numerically Accurate Finite Element Solutions in the Fully Plastic Range", Comp. Meth. Appl. Mech. Eng., 4, pp. 153-178, (1974).

2. Zienkiewicz, O. C., "The Finite Element Method", McGraw-Hill Book Co. (U. K.), 3rd ed., (1977).

3. Zienkiewicz, O. C., et al., "Stress Analysis of Rock as 'No-tension' Material", Geotechnique, 18, pp. 56-66, (1968).

4. Kawai, T., "New Discrete Structural Models and Generalization of the Method of Limit Analysis", Finite Elements in Nonlinear Mechanics, Vol. 2, Norwegian Institute of Technology, Trondheim, pp. 885-906, (1977).

5. Kawai, T., "New Discrete Models and Their Application to Seismic Response Analysis of Structures", Nuclear Engineering and Design, Vol. 48, pp. 207-229, (1978).

6. Kawai, T., and Toi, Y., "A New Discrete Analysis on Dynamic Collapse of Structures", Journal of the Society of Naval Architects of Japan, Vol. 143, pp. 275-281, (1978), Vol. 145, pp. 112-119, (1979).

7. Kawai T., Takeuchi, N., and Kamio, Y., "A New Discrete Limit Analysis of Underground Structures", Recent Advances in Lifeline Earthquake Engineering in Japan, ASME Publication PVP-43, August, (1980).

8. Kawai, T., "Some Considerations on the Finite Element Method", International Journal for Numerical Methods in Engineering, Vol. 16, pp. 81-120, (1980).

9. Yamada, Y., Yoshimura, N., and Sakurai, T., "Plastic Stress-Strain Matrix and its Application for the Solution of Elastic-Plastic Problems by the Finite Element Method", Int. J. Mech. Sci., 10, pp. 343-354, (1968).

10. Toi, Y., and Kawai, T., "Transient Response Analysis of Visco-Elastic Problems", Journal of the 'Seisan Kenkyu', Institute of Industrial Science, University of Tokyo, Vol. 29, No. 8, pp. 430-434. (1977).

11. Toi, Y., and Kawai, T., "Transient Response Analysis of an Elasto-Viscoplastic Beam Subjected to Transverse Impact", 'Seisan Kenkyu' Institute of Industrial Science, University of Tokyo, Vol. 29, No. 12, pp. 662-665, (1977).

12. Kawai, T., and Takeuchi, N., "New Discrete Models and Their Application to Mechanics of Granular Materials', proc. of the U.S.-Japan Seminar on New Models and Constitutive Relations in the Mechanics of Granular Materials, August 23-27, Cornell University, Ithaca, New York, (1982).

13. Takeuchi, N., and Kawai, T., "A Method for the Error Evaluation of the New Discrete Limit Analysis Solutions", Journal of the 'Seisan Kenkyu', Institute of Industrial Science, University of Tokyo, Vol. 33, No. 2, pp. 80-83, (1981), (in Japanese).

14. Takeuchi, N., Kawai, T., et al., "Limit Analysis of Soil and Rock Foundations by Means of New Discrete Models", (in Japanese), Journal of the 'Seisan Kenkyu', Institute of Industrial Science, University of Tokyo,

(i) "An Analysis of No-Tension Materials", Vol. 32, No. 6, pp. 301-304, (1980).
(ii) "Bearing Capacity Problem", Vol. 32, No. 8, pp. 372-375, (1980).
(iii) Slope Stability", Vol. 32, No. 8, pp. 376-379, (1980).

(iv) "Flexible Sheet Pile", Vol. 32, No. 10, pp. 460-463, (1980).
(v) "Tunnel Problems", Vol. 33, No. 5, pp. 182-185, (1981).
(vi) "Ultimate Bearing Capacity of Foundations on Slopes (1)", Vol. 33, No. 7, pp. 313-316, (1981).
(vii) "Ultimate Bearing Capacity of Foundations on Slopes (2)", Vol. 33, No. 7, pp. 317-319, (1981).
(viii) "Ultimate Bearing Capacity of Foundation on Sand Overlying Clay", Vol. 33, No. 8, pp. 345-348, (1981).

15. Takeuchi, N., and Kawai, T., "Limit Analysis of Soil and Rock Foundation by Means of New Discrete Models (2n Report) - Bearing Capacity Problem -, Journal of the 'Seisan Kenkyu', Institute of Industrial Science, University of Tokyo, Vol. 29, No. 4, pp. 204-207, (1977), (in Japanese).

16. Yamashita, K., Kakurai, M., and Tomono, M., "Discrete Limit Analysis of the Bearing Capacity of Footings on the Sand", (in Japanese), prep. of the 37th JSCE Annual Meeting, (1982).

17. Meyerhof, G. G., "Influence of Roughness of Base and Ground-Water Conditions on the Ultimate Bearing Capacity of Foundations", Geotechnique, Vol. 5, No. 3, pp. 227-242, (1955).

18. Chen, W. F., "Limit Analysis and Soil Plasticity", Elsevier Scientific Publishing Company, (1975).

19. Hansen, B., and Christensen, N. H., "Discussion of Theoretical Bearing Capacity of Very Shallow Footings", ASCE, SM6, Vol. 95, pp. 1568-1572, (1969).

20. Lundgren, H., and Mortensen, K., "Determination by the Theory of Plasticity of the Bearing Capacity of Continuous Footings of Sand", the 3rd ICSMFE, Vol. 1, pp. 409-412, (1953).

21. Yamaguchi, H., Kimura, T., Fujii, N., and Saito, K., "Bearing Capacity Test by means of a Centrifuge", (in Japanese), prep. of the 12th Japan National Conference on Soil Mechanics and Foundation Engineering, (1977).

22. Terzaghi, K., and Peck, R. B., "Soil Mechanics in Engineering Practice", Wiley, New York, 2nd edition, (1967).

23. Takeuchi, N., Hada, M., and Kawai, T., "Limit Analysis of Soil and Rock Foundations by Means of New Discrete Models (7th Report) - Ultimate Bearing Capacity of Foundations on Slopes (2) -", Journal of the 'Seisan Kenkyu', Institute of Industrial Science, University of Tokyo, Vol. 33, No. 7, pp. 317-319, (1981), (in Japanese).

24. Goto, T., Kagawa, Y., "Analysis of the Stresses and Deformation of the Slope under Strip Type Loading", (in Japanese), prep. of the 9th Japan National Conference on Soil Mechanics and Foundation Engineering, (1974).

25. Takeuchi, N., and Kawai, T., "Limit Analysis of Soil and Rock Foundation by Means of New Discrete Models (3rd Report) - Slope Stability - ", Journal of the 'Seisan Kenkyu', Institute of Industrial Science, University of Tokyo, Vol. 32, No. 8, pp. 376-379, (1980).

26. Takeuchi, N., and Kawai, T., "Limit Analysis of Soil and Rock Foundation by Means of New Discrete Models (4th Report) - Flexible Sheet Pile -", Journal of the 'Seisan Kenkyu', Institute of Industrial Science, University of Tokyo, Vol. 32, No. 10, pp. 460-463, (1980).

DEFORMATION AND STRENGTH OF GRANULAR MATERIALS BASED ON THE THEORY OF "COMPOUNDED MOBILIZED PLANES" (CMP) AND "SPATIAL MOBILIZED PLANE" (SMP)

H. Matsuoka

Department of Civil Engineering, Nagoya Institute of Technology,
Nagoya, Japan

ABSTRACT

A stress-strain model for granular materials is proposed by consideration of the mechanism of fabric change under shear. From paying attention to "two particles" across the potential sliding plane and considering the mechanism of disappearance and generation of interparticle contacts with the third particle, a stress-strain relationship is derived under two-dimensional stress conditions. The concept of "Compounded Mobilized Planes"(CMP) and "Spatial Mobilized Plane"(SMP) is then introduced to extend the stress-strain relationship to the three-dimensional one. The stress-strain relations based on the CMP and the SMP, and the failure criterion based on the SMP are formulated in the three-dimensional state and are compared with the various kinds of shear tests on sand. Deformation and strength of artificially deposited anisotropic sand are also analyzed by taking account of the angle between the bedding plane of the sample and the average sliding direction of soil particles on the basis of the above mentioned theory of "mobilized plane".

1. MICROSCOPIC ANALYSIS OF SHEAR DEFORMATION OF GRANULAR MATERIALS [1, 2, 3]

1.1. Introduction

It is one of the most interesting and difficult problems in the mechanics of granular materials to derive their stress-strain

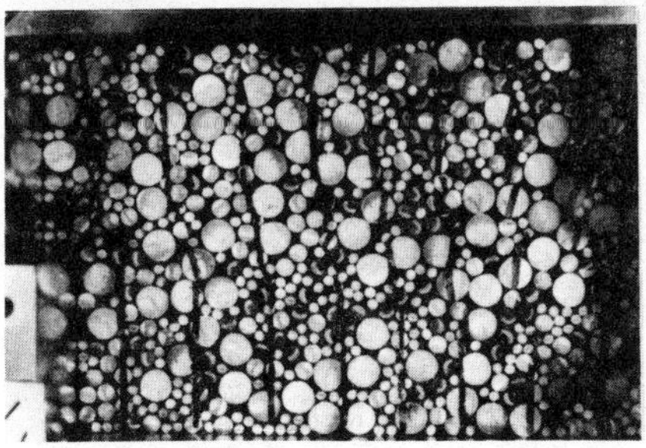

Fig. 1 Direct shear test on stack of aluminium rods

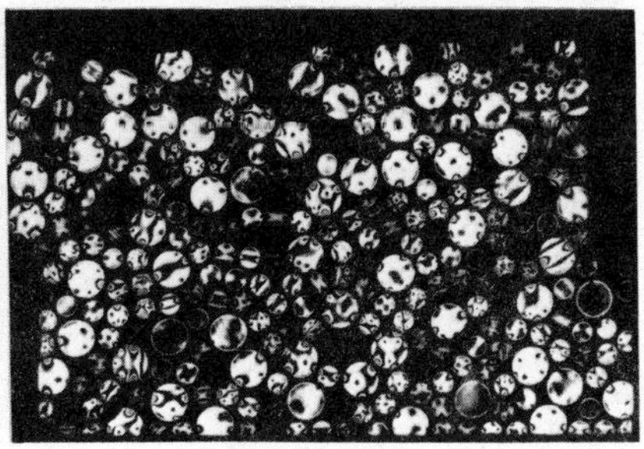

Fig. 2 Direct shear test on stack of photoelastic rods

relationship from the mechanism of fabric change in granular materials. Direct shear tests and simple shear tests have been performed using a stack of aluminium rods (ϕ 3, 5 and 9mm) or photoelastic rods (ϕ 6.2 and 10mm) as a two-dimensional model of granular materials as shown in Figs. 1 and 2. Based on the observation of these shear tests, the relationships among the shear-normal stress ratio (τ/σ_N), the shear strain (γ) and the normal strain (ε_N) on the potential sliding plane are derived by the medium of the interparticle contact angles (θ) along the potential sliding plane. The stress-strain behaviours under cyclic loading is also analyzed by the proposed model [3].

1.2 Relationship between Shear-Normal Stress Ratio (τ/σ_N) on Potential Sliding Plane and Average Interparticle Contact Angle ($\bar{\theta}$)

Paying attention to the interparticle contacts along the potential sliding plane as shown in Fig. 3 and denoting the interparticle contact angle (abbreviated "contact angle") by θ_i, the interparticle force by f_i and the average value of the mobilized interparticle friction angle by ϕ_μ, the following equation can be obtained from the equilibrium of interparticle forces on the potential sliding plane.

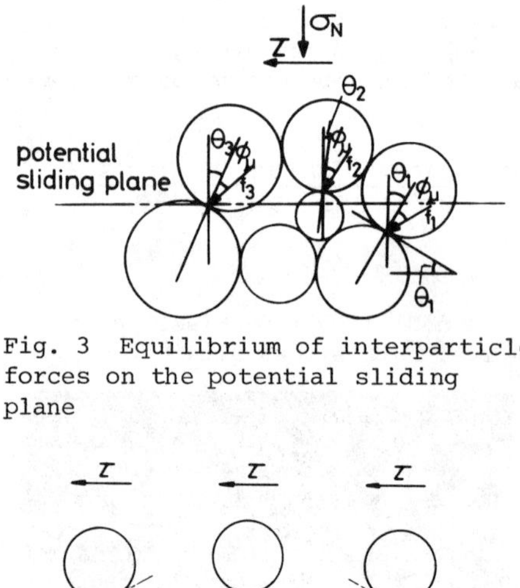

Fig. 3 Equilibrium of interparticle forces on the potential sliding plane

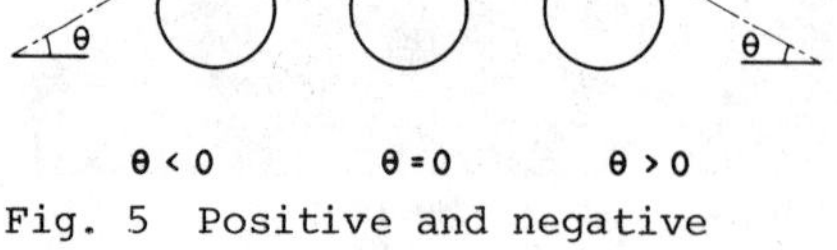

Fig. 5 Positive and negative signs for interparticle contact angle θ

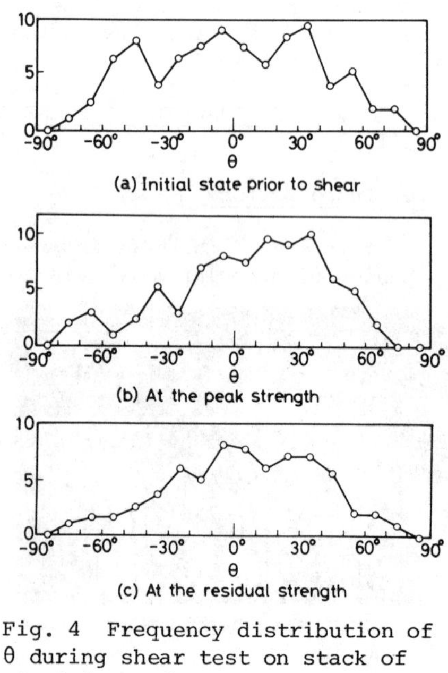

Fig. 4 Frequency distribution of θ during shear test on stack of aluminium rods

$$\frac{\tau}{\sigma_N} = \frac{\sum_{i=1}^{n} f_i \cdot \sin(\theta_i + \phi_\mu)}{\sum_{i=1}^{n} f_i \cdot \cos(\theta_i + \phi_\mu)} \tag{1}$$

where n means the number of the interparticle contacts along the potential sliding plane. It is assumed on the basis of the results of direct shear tests on a stack of photoelastic rods (see Fig. 2) that the interparticle forces f_i are the same at all of the interparticle contacts to first approximation. Then, if the frequency distribution of contact angles θ is denoted by N(θ), Eq. 1 can be transformed as follows:

$$\frac{\tau}{\sigma_N} = \frac{\int_{-\pi/2}^{\pi/2} N(\theta) \cdot \sin(\theta + \phi_\mu) \cdot d\theta}{\int_{-\pi/2}^{\pi/2} N(\theta) \cdot \cos(\theta + \phi_\mu) \cdot d\theta} \tag{2}$$

Figure 4 shows the variation of the frequency distribution N(θ) during a direct shear test on a stack of aluminium rods. Figures 4(a), (b) and (c) correspond to the frequency distributions at the initial state, at the peak strength and at the residual strength respectively. At the initial state, N(θ) tends to show a symmetrical distribution with respect to $\theta=0°$. The frequency is usually low at the both ends of θ (at $\theta \fallingdotseq \pm 90°$) because there is little probability that the center line of two particles concerned with the equilibrium of interparticle forces intersects the potential sliding plane (see Fig. 3). At the peak strength, N(θ) has the distribution one-sided to the positive zone of θ where the angle θ is effective to resist the applied shear stress (see Fig. 5). At the residual strength, N(θ) returns to an almost symmetrical distribution again.

Now, we consider the reason why N(θ) is one-sided to the positive zone of θ, though the particles on the potential sliding plane slide to the direction of the applied shear stress, namely, the negative direction of θ. Figure 6 expresses the frequency distribution of θ obtained by a simple shear test on a stack of aluminium rods, in which

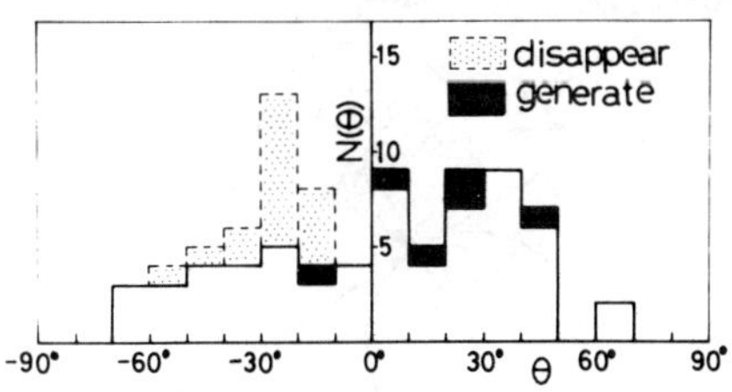

Fig. 6 Frequency distribution of θ showing disappearance and generation of contact angles during shear

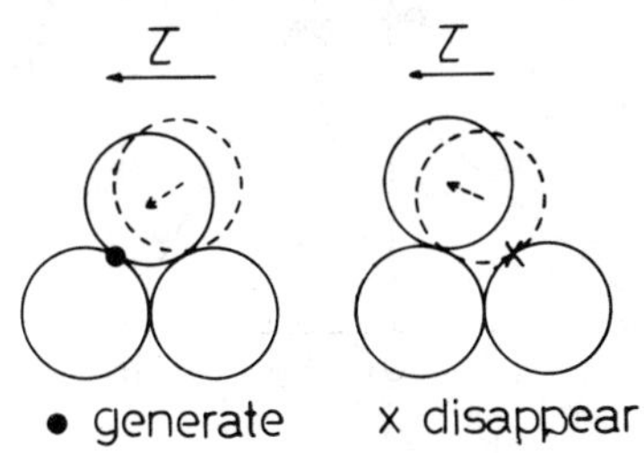

Fig. 7 Mechanism of disappearance and generation of interparticle contacts

disappearance and generation of contact angles during shear (shear strain γ=5 to 15%) are shown. It is seen from this figure that the bias of the frequency distribution of θ toward its positive side is caused by disappearance of the negative angles and generation of the positive angles of θ. The main mechanism of these disappearance and generation of θ may be visualized by Fig. 7.

If the frequency distribution of θ is approximated by a triangular distribution as shown in Fig. 8, $N(\theta)$ can be expressed by the following equations.

$$\left.\begin{aligned} N(\theta) &= h\cdot\frac{\theta+\pi/2}{\theta_p+\pi/2} \qquad (-\pi/2\leq\theta\leq\theta_p) \\ N(\theta) &= h\cdot\frac{\theta-\pi/2}{\theta_p-\pi/2} \qquad (\theta_p\leq\theta\leq\pi/2) \end{aligned}\right\} \tag{3}$$

The following equation is obtained by substituting Eq. 3 into Eq. 2.

$$\frac{\tau}{\sigma_N} = \frac{\pi\cdot\sin(3\bar{\theta}+\phi_\mu)-6\bar{\theta}\cdot\cos\phi_\mu}{\pi\cdot\cos(3\bar{\theta}+\phi_\mu)+6\bar{\theta}\cdot\sin\phi_\mu} \tag{4}$$

where $\bar{\theta}$ is the average value of θ ($\bar{\theta}=\int N(\theta)\cdot\theta\cdot d\theta/\int N(\theta)\cdot d\theta=\theta_p/3$). Equation 4 can be approximated within the practical range of $\bar{\theta}$ by the following formula [1].

$$\frac{\tau}{\sigma_N} \fallingdotseq \lambda\bar{\theta} + \mu \tag{5}$$

where λ and μ are parameters ($\mu=\tan\phi_\mu$).

1.3 Relationship between Shear Strain (γ) and Normal Strain (ε_N) on Potential Sliding Plane, and Average Interparticle Contact Angle ($\bar{\theta}$)

Considering these disappearance, generation and "steady" (interparticle contact is kept during shear) behaviour of the contact angles,

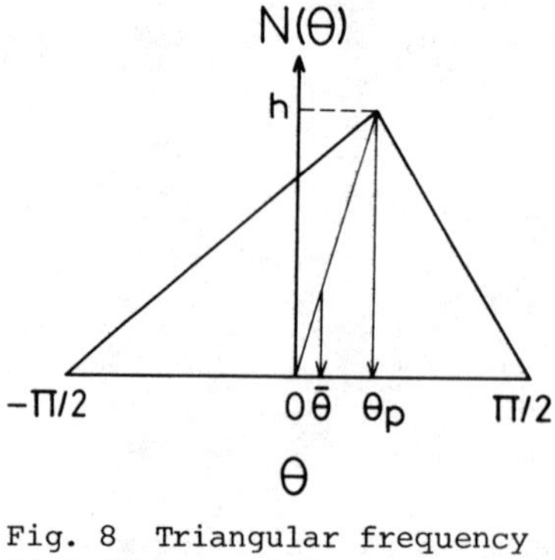

Fig. 8 Triangular frequency distribution of θ

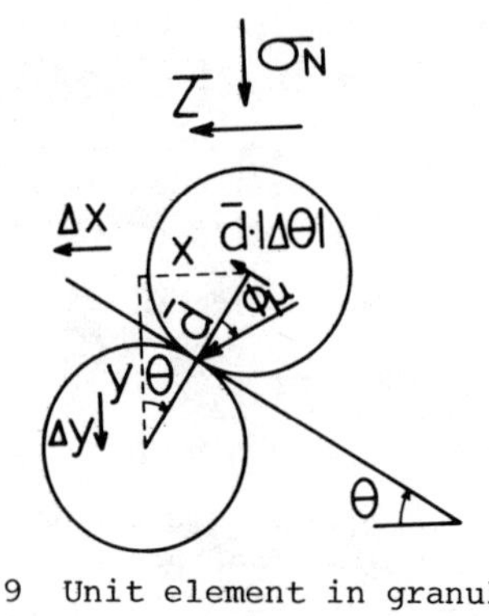

Fig. 9 Unit element in granular materials--"two particles model"

the sliding displacements and the sliding strains due to shear are formulated. When the particles on the potential sliding plane slide averagely to the amount of $\bar{d}\cdot|\Delta\theta|$ to the shearing direction, the horizontal and vertical sliding displacement increments at a contact point having a certain value of θ can be expressed as follows (see Fig.9):

$$\Delta\dot{x} = \bar{d}\cdot|\Delta\theta|\cdot N(\theta)\cdot\cos\theta \tag{6}$$

$$\Delta\dot{y} = -\bar{d}\cdot|\Delta\theta|\cdot N(\theta)\cdot\sin\theta \tag{7}$$

where $\bar{d}$ is the average distance between centers of two adjacent particles. Substituting Eq. 3 into Eqs. 6 and 7, and integrating them in the range of $-\pi/2\leq\theta\leq\pi/2$, the sliding displacement increments for a certain frequency distribution of $\bar{\theta}$ are obtained as follows:

$$\begin{aligned}\Delta x &= \int_{-\pi/2}^{\pi/2}\bar{d}\cdot|\Delta\theta|\cdot N(\theta)\cdot\cos\theta\cdot d\theta\\ &= \bar{d}\cdot h\cdot\frac{\pi\cdot\cos 3\bar{\theta}}{(\pi/2+3\bar{\theta})(\pi/2-3\bar{\theta})}\cdot|\Delta\theta|\end{aligned} \tag{8}$$

$$\begin{aligned}\Delta y &= -\int_{-\pi/2}^{\pi/2}\bar{d}\cdot|\Delta\theta|\cdot N(\theta)\cdot\sin\theta\cdot d\theta\\ &= \bar{d}\cdot h\cdot\frac{6\bar{\theta}-\pi\cdot\sin 3\bar{\theta}}{(\pi/2+3\bar{\theta})(\pi/2-3\bar{\theta})}\cdot|\Delta\theta|\end{aligned} \tag{9}$$

Equations 8 and 9 must be integrated with respect to $\bar{\theta}$ in order to obtain the sliding displacements, because θ_p $(=3\bar{\theta})$ in Fig. 8 is varied with shear as described in the previous section. For this integration, it is necessary to get the relationship between $|\Delta\theta|$ and $\Delta\bar{\theta}$.

If any change in the average contact angle $\Delta\bar{\theta}$ is produced by the disappearance, generation and "steady" behaviour of the contact angles, the following equation is obtained [2].

$$\Delta\bar{\theta}\cdot n = -\int_{-\pi/2}^{0} N_d(\theta)\cdot\theta\cdot d\theta+\int_{-\pi/2}^{0} N_g(\theta)\cdot(-\theta)\cdot d\theta-|\Delta\theta|\cdot\left(n-\int_{-\pi/2}^{0} N_d(\theta)\cdot d\theta\right) \tag{10}$$

where n is the total number of the contact angles $(n=\pi h/2)$. $N_d(\theta)$ and $N_g(\theta)$ are the frequencies of disappearance and generation of contact angles respectively. The left side of Eq. 10 corresponds to the total change in θ. The first term on the right side expresses the total increase in θ caused by disappearance of the negative contact angles. The second term expresses the total increase in θ caused by generation of the positive contact angles. The range of integration in this term is chosen $-\pi/2$ to 0 from the following assumption. The main mechanism of the generation of θ is assumed that a particle having one negative contact angle generates a new contact at the symmetrically positive angle with respect to $\theta=0°$ through movement of the particle to the shearing direction. Finally, the third term means the total decrease in θ caused by sliding of the "steady" contacts to the negative direction.

It is seen from the results of direct shear tests on a stack of photoelastic rods that the main mechanism of disappearance and generation of contact angles can be illustrated by Fig. 10. At the initial state, the interparticle contact points are stable under the normal stress σ_N on the potential sliding plane. If the shear stress τ is applied, the interparticle force at the positive contact angle becomes large and that at the negative contact angle becomes small generally. These phenomena may be considered a kind of generation and disappearance of contact angles. Therefore, the number of the generated and disappeared contact angles is estimated to be quite large at the beginning of shear. As it is seldom that the generated (or disappeared) contact angles are disappeared (or generated) during the same shearing process, the number of generation and disappearance of contact angles is considered to decrease with shear. Figure 11 shows the relationship between $\bar{\theta}$ and P, which is a ratio defined by the following equation [4].

$$P = \frac{1}{|\Delta\theta|} \cdot \frac{N_g(\theta)+N_d(\theta)}{N(\theta)} \tag{11}$$

P means the ratio of the number of generation and disappearance of contact angles ($N_g(\theta)+N_d(\theta)$) to the number of all contact angles ($N(\theta)$) per $|\Delta\theta|=1$ radian. The plots in this figure indicate the observed values from the direct shear tests on a stack of photoelastic rods. It is seen from this figure that P decreases extremely with shear (with increase in $\bar{\theta}$). Then, the relationship between P and $\bar{\theta}$ is assumed to be expressed by the following exponential form.

$$P = a \cdot \exp(-b \cdot \bar{\theta}) \tag{12}$$

where a and b are parameters to express the degree of fabric change with shear. The plots in Fig. 11 can be approximated by the parameters of a=100 and b=3. From Eqs. 11 and 12, the following relationship can be obtained.

$$N_g(\theta)+N_d(\theta) = a \cdot \exp(-b \cdot \bar{\theta}) \cdot N(\theta) \cdot |\Delta\theta| \tag{13}$$

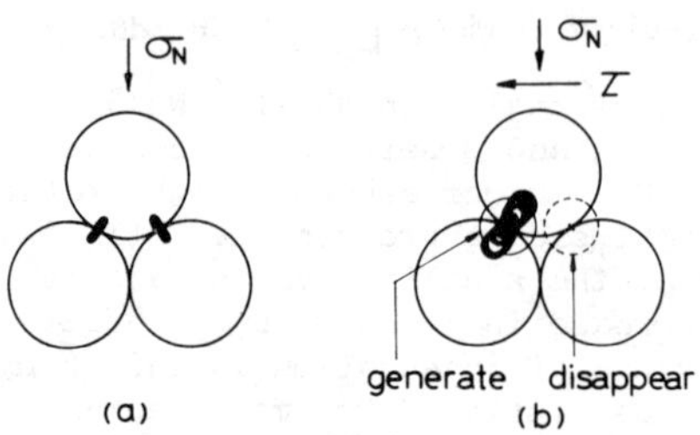

Fig. 10 Disappearance and generation of interparticle contacts observed by photoelastic rods

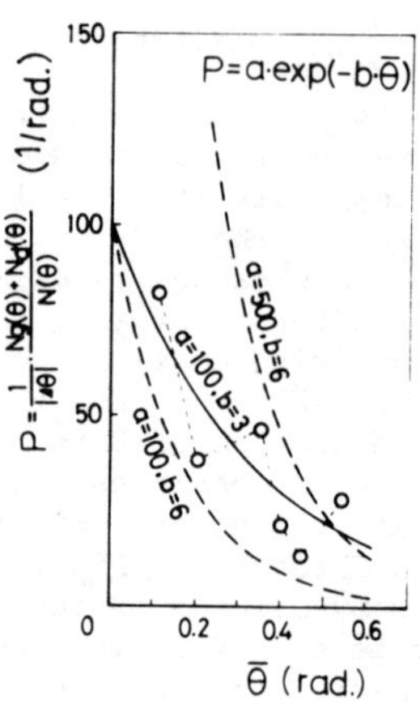

Fig. 11 Variation of disappearance and generation of interparticle contacts with shear

By substituting Eq. 13 into Eq. 10, the following relationship among $|\Delta\theta|$, $\bar{\theta}$ and $\Delta\bar{\theta}$ is obtained.

$$\left.\begin{aligned} |\Delta\theta| &= \frac{\pi/2+3\bar{\theta}}{(\pi^2/24)\,a\cdot\exp(-b\bar{\theta})-(\pi/2+3\bar{\theta})}\cdot\Delta\bar{\theta} \qquad (\bar{\theta}\geq 0) \\ |\Delta\theta| &= \frac{\pi/2-3\bar{\theta}}{a\cdot\exp(-b\bar{\theta})\cdot(3\bar{\theta}^2-(\pi/2)\cdot\bar{\theta}+(\pi^2/24))-(\pi/2-3\bar{\theta})}\cdot\Delta\bar{\theta} \qquad (\bar{\theta}<0) \end{aligned}\right\} \quad (14)$$

The sliding displacement increments per one contact point can be obtained by substituting Eq. 14 into Eqs. 8 and 9, and dividing them by the total number of the contact angles ($\pi h/2$). If these displacement increments are divided by the average width of one shear layer D, which corresponds to the average normal component of the distances between two adjacent sliding particles to the potential sliding plane, they may be converted to the strain increments. Thus, the shear strain increment $\Delta\gamma$ and the normal strain increment $\Delta\varepsilon_N$ on the potential sliding plane can be expressed as follows:

$$\left.\begin{aligned} \Delta\gamma &= \frac{\pi/2+3\bar{\theta}}{(\pi^2/24)\cdot a\cdot\exp(-b\bar{\theta})-(\pi/2+3\bar{\theta})}\cdot\Delta\bar{\theta} \qquad (\bar{\theta}\geq 0) \\ \Delta\gamma &= \frac{\pi/2-3\bar{\theta}}{a\cdot\exp(-b\bar{\theta})\cdot(3\bar{\theta}^2-(\pi/2)\cdot\bar{\theta}+(\pi^2/24))-(\pi/2-3\bar{\theta})}\cdot\Delta\bar{\theta} \qquad (\bar{\theta}<0) \end{aligned}\right\} \quad (15)$$

$$\Delta\varepsilon_N = \frac{6\bar{\theta}-\pi\cdot\sin 3\bar{\theta}}{\pi\cdot\cos 3\bar{\theta}}\cdot\Delta\gamma \quad (16)$$

where

$$D = \frac{\int_{-\pi/2}^{\pi/2} N(\theta)\cdot\bar{d}\cdot\cos\theta\cdot d\theta}{\int_{-\pi/2}^{\pi/2} N(\theta)\cdot d\theta} = \frac{2\bar{d}\cdot\cos 3\bar{\theta}}{(\pi/2+3\bar{\theta})(\pi/2-3\bar{\theta})} \quad (17)$$

From Eq. 4 or 5 and Eqs. 15 and 16, a stress-strain relationship for granular materials under shear can be obtained by the medium of the average contact angle $\bar{\theta}$ which is considered to be an index to express the fabric of granular materials.

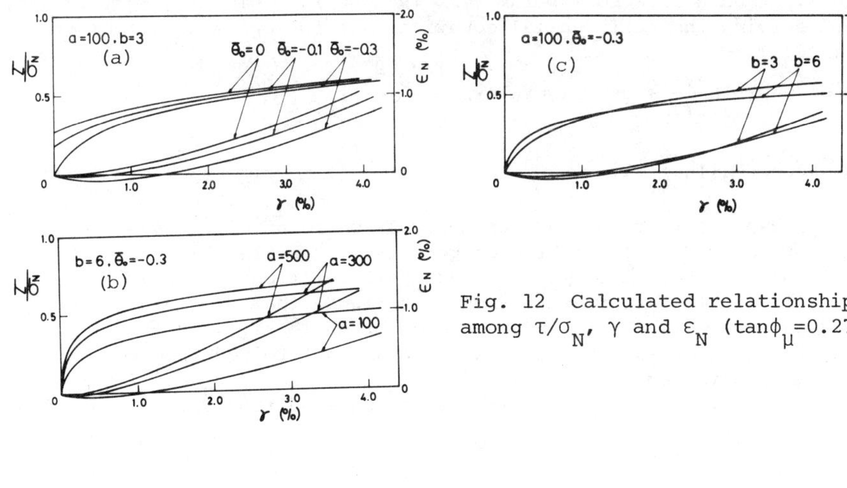

Fig. 12 Calculated relationships among τ/σ_N, γ and ε_N ($\tan\phi_\mu$=0.27)

Figures 12 (a), (b) and (c) show the calculated relationships among τ/σ_N, γ and ε_N ($\tan\phi_\mu=0.27$). The reason why τ/σ_N is not equal to 0 at $\gamma=0$ in Fig. 12 (a) is due to the assumption that the interparticle friction angle ϕ_μ is fully mobilized from $\gamma=0$. In Fig. 12 (a) $\bar{\theta}_0$ denotes the average contact angle at the start of shear. It can be seen from Figs. 12 (b) and (c) that the parameters a and b have close connection with strain hardening behaviour. The strain hardening is considered to be originally caused by generation and disappearance of the contact angles (see Fig. 11). If $\bar{\theta}$ at failure is denoted by $\bar{\theta}_f(>0)$, the following relationship can be obtained from the condition that the denominator in Eq. 14 becomes zero at failure ($\Delta\gamma\to\infty$).

$$(\pi^2/24)\cdot a\cdot \exp(-b\cdot\bar{\theta}_f)-(\pi/2+3\bar{\theta}_f) = 0 \tag{18}$$

Therefore, the parameters in the proposed stress-strain relationship are "ϕ_μ" to denote the interparticle friction angle, "$\bar{\theta}_0$ and $\bar{\theta}_f$" to express the fabric at the start of shear and at failure, and "a" (or b) to express the degree of fabric change during shear.

1.4 Basic Stress-Strain Relationship on Potential Sliding Plane

The author has already proposed the following shear-normal stress ratio (τ/σ_N) vs. normal-shear strain increment ratio ($-d\varepsilon_N/d\gamma$) relationship and shear-normal stress ratio (τ/σ_N) vs. normal-shear strain ratio ($-\varepsilon_N/\gamma$) relationship on the "mobilized plane" [1, 5]. The "mobilized plane" is such a plane on which the shear-normal stress ratio (τ/σ_N) is maximized and is considered to correspond to the potential sliding plane approximately.

$$\frac{\tau}{\sigma_N} = \lambda\cdot\left(-\frac{d\varepsilon_N}{d\gamma}\right)+\mu \tag{19}$$

$$\frac{\tau}{\sigma_N} = \lambda\cdot\left(-\frac{\varepsilon_N}{\gamma}\right)+\mu' \tag{20}$$

where λ, μ and μ' are soil parameters ($\mu=\tan\phi_\mu$). Combining Eqs. 19 and 20 and solving the differential equation, the following equations are obtained.

$$\frac{\tau}{\sigma_N} = (\mu'-\mu)\cdot\ln\frac{\gamma}{\gamma_0} + \mu \qquad \left(\gamma=\gamma_0\cdot\exp\left(\frac{\tau/\sigma_N-\mu}{\mu'-\mu}\right)\right) \tag{21}$$

$$\varepsilon_N = \frac{\mu-\mu'}{\lambda}\cdot\gamma\cdot\left\{\ln\frac{\gamma}{\gamma_0}-1\right\} \tag{22}$$

where γ_0 denotes γ at the maximum compression point of ε_N and is a parameter to express the fabric of granular materials [5].

Figure 13 shows the τ/σ_N vs. ($-d\varepsilon_N/d\gamma$) relationship calculated from Eqs. 5, 15 and 16. It is seen from this figure that Eq. 19 holds good ($\lambda=0.9$ and $\mu=0.27$). If the average contact angle at failure $\bar{\theta}_f$ is given, the sets of the values of a and b are determined from Eq. 18. The relationships among τ/σ_N, γ and ε_N, and between τ/σ_N and ($-\varepsilon_N/\gamma$) are shown in Figs. 14 and 15 which are calculated from Eqs. 5, 15 and

16, using the four sets of the values of a and b determined for the case of $\bar{\theta}_f=0.60$ ($(\tau/\sigma_N)_f=0.81$). It can be seen from these two figures that the four different stress-strain curves in Fig. 14 are arranged within a narrow zone by the τ/σ_N vs. $(-\varepsilon_N/\gamma)$ relationships in Fig. 15. The four τ/σ_N vs. γ curves in Fig. 14 coincide at a large value of γ because they have the same shear strength ($\bar{\theta}_f=0.60$). The schematic figure of the τ/σ_N vs. $(-d\varepsilon_N/d\gamma)$ relationship and the τ/σ_N vs. $(-\varepsilon_N/\gamma)$ relationships is shown in Fig. 16. It is seen from this figure that the τ/σ_N vs. $(-\varepsilon_N/\gamma)$ relationships become convex curves between $(-\varepsilon_N/\gamma)=\bar{\theta}_0$ and $\bar{\theta}_f$. Equation 20 shown by the broken lines in Fig. 16 can be explained to be an approximate linear relationship for the main parts of these convex curves. As these convex curves are considered to be similar figures approximately, the following equation is obtained [6].

$$\mu'-\mu = \kappa\cdot(\bar{\theta}_f-\bar{\theta}_0) \tag{23}$$

where the parameter κ is nearly constant ($\kappa\fallingdotseq0.2$) as can be seen from Fig. 15. Therefore, μ' is considered a parameter to reflect the fabric parameters $\bar{\theta}_0$ and $\bar{\theta}_f$.

Substituting Eqs. 5 and 23 into Eq. 21, and denoting $\bar{\theta}$ and γ at failure by $\bar{\theta}_f$ and γ_f respectively, the following equation is obtained.

$$\gamma_0 = \gamma_f\cdot\exp(-\frac{\lambda}{\kappa})\cdot\exp(\frac{\lambda\cdot(-\bar{\theta}_0)}{\kappa\cdot(\bar{\theta}_f-\bar{\theta}_0)}) \tag{24}$$

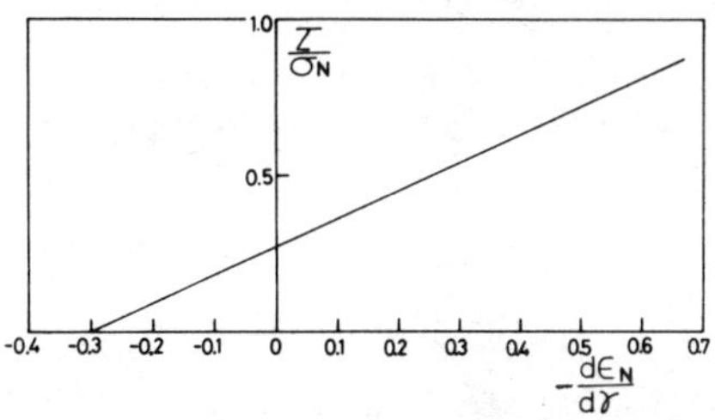

Fig. 13 Calculated relationships between τ/σ_N and $-d\varepsilon_N/d\gamma$ ($\lambda=0.9$, $\mu=0.27$)

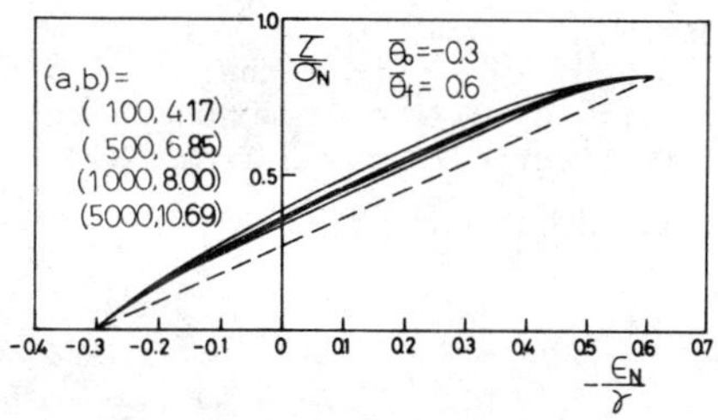

Fig. 15 Calculated relationships between τ/σ_N and $-\varepsilon_N/\gamma$ ($\lambda=0.9$, $\mu=0.27$)

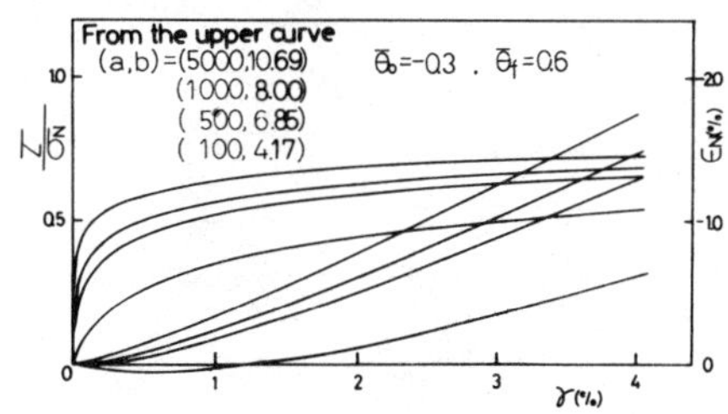

Fig. 14 Calculated relationships among τ/σ_N, γ and ε_N ($\lambda=0.9$, $\mu=0.27$)

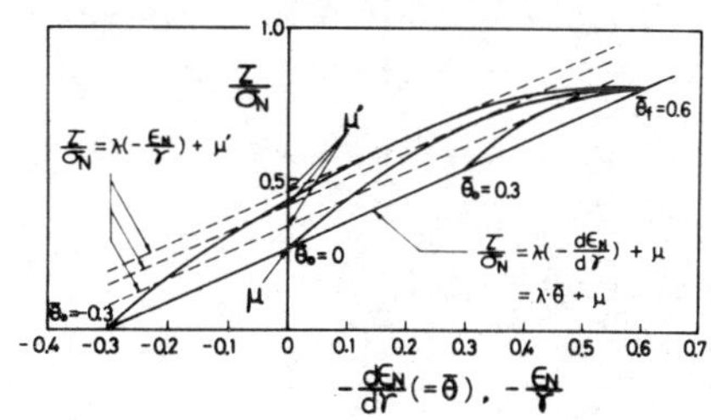

Fig. 16 Schematic relationships between τ/σ_N and $-d\varepsilon_N/d\gamma$, and between τ/σ_N and $-\varepsilon_N/\gamma$

If it is assumed that γ_f is the same no matter what the initial fabric ($\bar{\theta}_0$) is, the following relationship for γ_0 is derived from Eq. 24.

$$\gamma_0 = \gamma_{0(\bar{\theta}_0=0)} \cdot \exp\left(\frac{\lambda\cdot(-\bar{\theta}_0)}{\kappa\cdot(\bar{\theta}_f-\bar{\theta}_0)}\right) \qquad (25)$$

where $\gamma_{0(\bar{\theta}_0=0)}$ means γ_0 at $\bar{\theta}_0=0$. It is seen from Eq. 25 that γ_0 is also a parameter to reflect the fabric of granular materials.

As an example of shear stress history, the stress-strain behaviour under cyclic shear stress is analyzed here considering the mechanism of fabric change as mentioned before. Figure 17 shows the conceptional relationship between a stress-strain behaviour under cyclic loading and the variation of the triangular frequency distribution of contact angles. As the particles are rearranged so as to resist the shear stress τ_{1st} by the 1st loading, the frequency distribution is one-sided to the right side of θ. If the particles are then subject to τ_{2nd} in the reverse direction, they slide down easily as shown in Fig. 18, and thus the stress-strain curve under the 2nd loading becomes gentle comparing with that under the 1st loading. For example, $\bar{\theta}_0$ under the 2nd loading is considered to be $-\bar{\theta}_{1st}$ where $\bar{\theta}_{1st}$ is $\bar{\theta}$ at the end of the 1st loading (see Fig. 17). Generally $\bar{\theta}_0$ under cyclic loading is given from Eq. 5 by the following equation.

$$\bar{\theta}_0 = -(\tau/\sigma_N-\mu)/\lambda \qquad (26)$$

where τ/σ_N in Eq. 26 represents τ/σ_N at the end of the preceding loading.

Usually $\bar{\theta}_0$ under the 1st loading is considered to be nearly equal to zero ($\bar{\theta}_0 \fallingdotseq 0$), so if μ' under the 1st loading is denoted by μ'_{1st}, the coefficient κ is given from Eq. 23 as follows:

$$\kappa = (\mu'_{1st}-\mu)/\bar{\theta}_f \qquad (27)$$

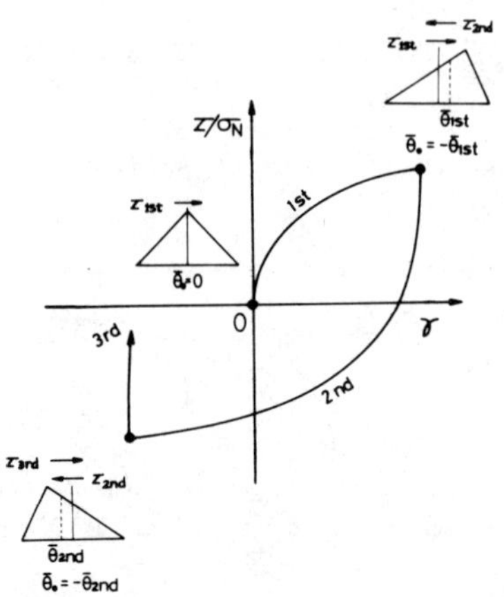

Fig. 17 Conceptional relationship between cyclic stress-strain behaviour and variation in frequency distribution of θ

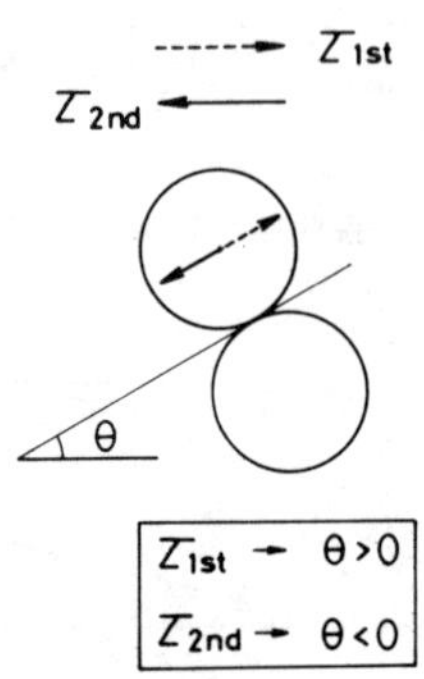

Fig. 18 Positive and negative signs of θ under cyclic shear stress

If γ_0 under the 1st loading is denoted by $\gamma_{0_{1st}}$, the parameter γ_0 is similarly expressed from Eq. 25 by the following equation.

$$\gamma_0 = \gamma_{0_{1st}} \cdot \exp\left(\frac{\lambda \cdot (-\bar{\theta}_0)}{\kappa \cdot (\bar{\theta}_f - \bar{\theta}_0)}\right) \tag{28}$$

The parameters λ and μ are considered to be constant and $\bar{\theta}_f$ can be assumed to be constant under cyclic loading. Therefore, the parameters μ' and γ_0 under cyclic loading can be estimated from Eqs. 23, 26, 27 and 28 using parameters such as λ, μ, μ'_{1st}, $\gamma_{0_{1st}}$ and $\bar{\theta}_f$ which are obtained from the 1st loading test (the conventional monotonous loading test).

2. CONCEPT OF "COMPOUNDED MOBILIZED PLANES"(CMP) AND "SPATIAL MOBILIZED PLANE"(SMP) [5, 7, 8, 9]

The behaviour of soil particles is considered to be governed by the frictional law, i.e., the shear-normal stress ratio. The plane on which the shear-normal stress ratio is maximized has been called "mobilized plane", and considered to be the plane where soil particles are most mobilized on the average [10]. In order to extend the two-dimensional stress-strain relationship derived from the microscopic consideration in the preceding section to the three-dimensional one, three two-dimensional mobilized planes of soil particles are supposed between the respective two principal stress directions. The stress states on the three mobilized planes correspond to the three points at which three straight lines from the origin are tangent to three Mohr's stress circles as shown in Fig. 19. When three different principal stresses σ_1, σ_2 and σ_3 are applied to the directions of I, II and III axes, three two-dimensional mobilized planes between the respective two principal stress directions are expressed in Fig. 20. These three mobilized planes have been named "Compounded Mobilized Planes"(CMP) [5]. The above mentioned two-dimensional stress-strain relationship is applicable on each of the three mobilized planes as described in the following section.

Paying further attention to a plane ABC in Fig. 21 with these three mobilized planes as the three sides, it has been defined as "Spatial Mobilized Plane"(SMP) [7, 8, 9]. It should be noted that the SMP is not a stress plane, but a geometrical plane. The values at points A, B and C at which the SMP intersects the three axes are proportional to the square roots of the respective principal stresses

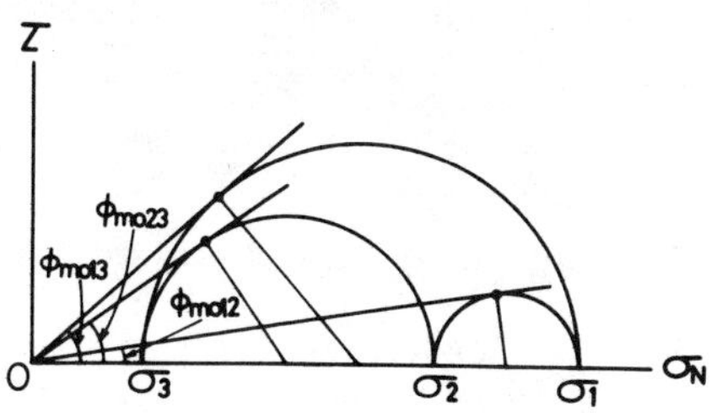

Fig. 19 Three mobilized planes where shear-normal stress ratio is maximized under respective two principal stresses

according to the following equation as shown in Fig. 21.

$$\tan(45°+\frac{\phi_{moij}}{2}) = \sqrt{\frac{1+\sin\phi_{moij}}{1-\sin\phi_{moij}}} = \sqrt{\frac{\sigma_i}{\sigma_j}} \quad (i,j=1,2,3;\ i<j) \tag{29}$$

Accordingly, the SMP coincides with the octahedral plane in isotropic stress conditions ($\sigma_1=\sigma_2=\sigma_3$) and is variable with possible change in the stresses. In its physical significance, it is considered to be the plane on which soil particles are most mobilized on the average in the three-dimensional stresses. The direction cosines (a_1, a_2, a_3) of the normal of the SMP are expressed as follows:

$$a_i = \sqrt{J_3/(\sigma_i J_2)} \qquad (i=1,2,3) \tag{30}$$

where J_1, J_2 and J_3 are the first, second and third effective stress invariants which are expressed by the following equations.

$$J_1=\sigma_1+\sigma_2+\sigma_3,\ J_2=\sigma_1\sigma_2+\sigma_2\sigma_3+\sigma_3\sigma_1,\ J_3=\sigma_1\sigma_2\sigma_3 \tag{31}$$

The stress-strain relationship derived from the microscopic consideration is also applicable on the SMP.

3. STRESS-STRAIN RELATIONSHIP BASED ON THE THEORY OF "COMPOUNDED MOBILIZED PLANES"(CMP) [5, 11, 12]

As mentioned in the preceding section, Eqs. 19-22, which are the stress-strain relations derived from the microscopic analysis of shear deformation in the two-dimensional state, are considered to hold good on the mobilized plane. Differentiating Eq. 21 with respect to $\tau/\sigma_N \equiv X$, the follwoing equation is obtained.

$$d\gamma = \frac{\gamma_0}{\mu'-\mu}\cdot\exp(\frac{X-\mu}{\mu'-\mu})\cdot dX \tag{32}$$

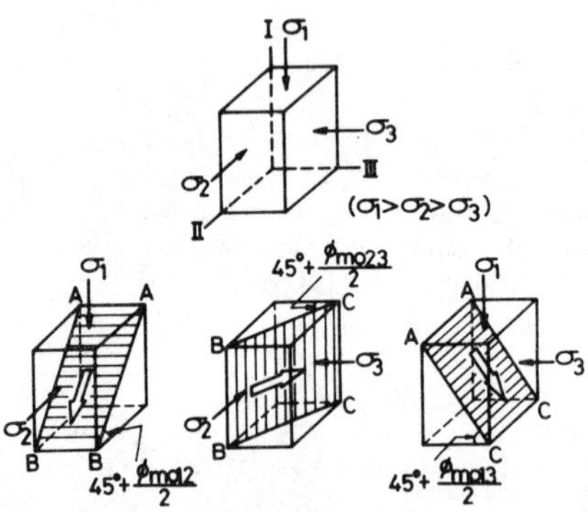

Fig. 20 Three two-dimensional mobilized planes (Compounded Mobilized Planes) under three different principal stresses

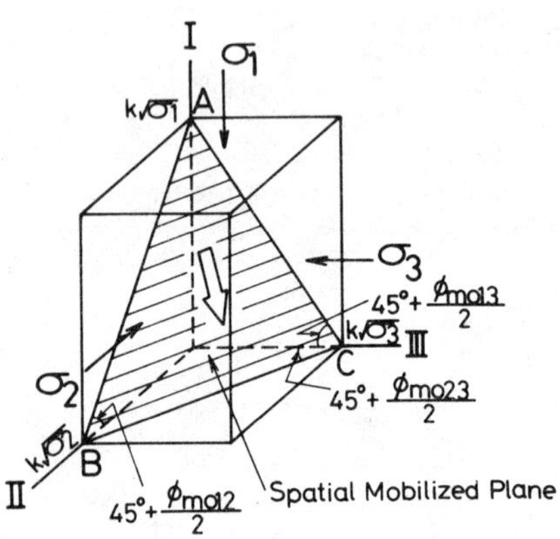

Fig. 21 Spatial Mobilized Plane under three principal stresses

The following is also obtained from Eq. 19.

$$d\varepsilon_N = \frac{\mu - X}{\lambda} \cdot d\gamma \tag{33}$$

The relationships among the shear strain increment ($d\gamma$) and the normal strain increment ($d\varepsilon_N$) on one mobilized plane, for example AC in Fig. 20, and the principal strain increments ($d\varepsilon_1$ and $d\varepsilon_3$) are expressed as follows, if it is assumed that the direction of the principal strain increments coincides with that of the principal stresses.

$$\frac{d\gamma}{2} = \frac{d\varepsilon_1 - d\varepsilon_3}{2} \cdot \sin 2(45° + \frac{\phi_{mo}}{2}) = \frac{d\varepsilon_1 - d\varepsilon_3}{2} \cdot \cos\phi_{mo} \tag{34}$$

$$\begin{aligned} d\varepsilon_N &= \frac{d\varepsilon_1 + d\varepsilon_3}{2} + \frac{d\varepsilon_1 - d\varepsilon_3}{2} \cdot \cos 2(45° + \frac{\phi_{mo}}{2}) \\ &= \frac{d\varepsilon_1 + d\varepsilon_3}{2} - \frac{d\varepsilon_1 - d\varepsilon_3}{2} \cdot \sin\phi_{mo} \end{aligned} \tag{35}$$

where $\sin\phi_{mo} = (\sigma_1 - \sigma_3)/(\sigma_1 + \sigma_3)$ and $\cos\phi_{mo} = 2\sqrt{\sigma_1\sigma_3}/(\sigma_1 + \sigma_3)$. The following equations are obtained from Eqs. 34 and 35.

$$d\varepsilon_1 = d\varepsilon_N + \sqrt{\frac{\sigma_1}{\sigma_3}} \cdot \frac{d\gamma}{2} \tag{36}$$

$$d\varepsilon_3 = d\varepsilon_N - \sqrt{\frac{\sigma_3}{\sigma_1}} \cdot \frac{d\gamma}{2} \tag{37}$$

Substituting Eqs. 32 and 33 into Eqs. 36 and 37, the following principal stress ratio vs. principal strain increments relationships are obtained.

$$d\varepsilon_1 = \frac{\gamma_0}{\mu' - \mu} \cdot \left(\frac{\mu - X}{\lambda} + \frac{1}{2}\sqrt{\frac{\sigma_1}{\sigma_3}} \right) \cdot \exp\left(\frac{X - \mu}{\mu' - \mu} \right) \cdot dX \equiv f'\left(\frac{\sigma_1}{\sigma_3}\right) \cdot dX \tag{38}$$

$$d\varepsilon_3 = \frac{\gamma_0}{\mu' - \mu} \cdot \left(\frac{\mu - X}{\lambda} - \frac{1}{2}\sqrt{\frac{\sigma_3}{\sigma_1}} \right) \cdot \exp\left(\frac{X - \mu}{\mu' - \mu} \right) \cdot dX \equiv g'\left(\frac{\sigma_1}{\sigma_3}\right) \cdot dX \tag{39}$$

where $X \equiv \tau/\sigma_N = \tan\phi_{mo} = (\sqrt{\sigma_1/\sigma_3} - \sqrt{\sigma_3/\sigma_1})/2$.

Under three different principal stresses, three two-dimensional mobilized planes are generally considered between the respective two principal stress directions as shown in Fig. 20. If each of three principal strain increments $d\varepsilon_1$, $d\varepsilon_2$ and $d\varepsilon_3$ is assumed to be expressed as linear summation of two components produced by the corresponding two mobilized planes, the following equations are obtained from Eqs. 38 and 39 [5, 11, 12].

$$d\varepsilon_1 = f'\left(\frac{\sigma_1}{\sigma_3}\right) \cdot dX_{13} + f'\left(\frac{\sigma_1}{\sigma_2}\right) \cdot dX_{12} \tag{40}$$

$$d\varepsilon_2 = f'\left(\frac{\sigma_2}{\sigma_3}\right) \cdot dX_{23} + g'\left(\frac{\sigma_1}{\sigma_2}\right) \cdot dX_{12} \tag{41}$$

$$d\varepsilon_3 = g'\left(\frac{\sigma_1}{\sigma_3}\right)\cdot dX_{13} + g'\left(\frac{\sigma_2}{\sigma_3}\right)\cdot dX_{23} \tag{42}$$

where

$$f'\left(\frac{\sigma_i}{\sigma_j}\right) = \frac{\gamma_0}{\mu'-\mu}\cdot\left(\frac{\mu-X_{ij}}{\lambda} + \frac{1}{2}\sqrt{\frac{\sigma_i}{\sigma_j}}\right)\cdot\exp\left(\frac{X_{ij}-\mu}{\mu'-\mu}\right) \tag{43}$$

$$g'\left(\frac{\sigma_i}{\sigma_j}\right) = \frac{\gamma_0}{\mu'-\mu}\cdot\left(\frac{\mu-X_{ij}}{\lambda} - \frac{1}{2}\sqrt{\frac{\sigma_j}{\sigma_i}}\right)\cdot\exp\left(\frac{X_{ij}-\mu}{\mu'-\mu}\right) \tag{44}$$

$$X_{ij} \equiv \tan\phi_{moij} = \left(\sqrt{\sigma_i/\sigma_j} - \sqrt{\sigma_j/\sigma_i}\right)/2 \qquad (i,j=1,2,3;\ i<j) \tag{45}$$

In the case of triaxial compression conditions ($\sigma_1>\sigma_2=\sigma_3$), the principal strain increments can be expressed from Eqs. 40-42 as follows:

$$d\varepsilon_1 = 2\cdot f'(\sigma_1/\sigma_3)\cdot dX_{13},\quad d\varepsilon_2 = d\varepsilon_3 = g'(\sigma_1/\sigma_3)\cdot dX_{13} \tag{46}$$

In the case of triaxial extension conditions ($\sigma_1=\sigma_2>\sigma_3$), the principal strain increments are expressed as follows:

$$d\varepsilon_1 = d\varepsilon_2 = f'(\sigma_1/\sigma_3)\cdot dX_{13},\quad d\varepsilon_3 = 2\cdot g'(\sigma_1/\sigma_3)\cdot dX_{13} \tag{47}$$

In the case of plane strain conditions ($d\varepsilon_2=0$), the principal strain increments may be approximately expressed as follows:

$$\left.\begin{aligned} d\varepsilon_1 &= f'(\sigma_1/\sigma_3)\cdot dX_{13} + f'(\sigma_1/\sigma_2)\cdot dX_{12} \fallingdotseq f'(\sigma_1/\sigma_3)\cdot dX_{13} \\ d\varepsilon_3 &= g'(\sigma_1/\sigma_3)\cdot dX_{13} + g'(\sigma_2/\sigma_3)\cdot dX_{23} \fallingdotseq g'(\sigma_1/\sigma_3)\cdot dX_{13} \end{aligned}\right\} \tag{48}$$

The approximation in Eq. 48 means that $f'(\sigma_1/\sigma_3)\cdot dX_{13}$ is much larger than $f'(\sigma_1/\sigma_2)\cdot dX_{12}$ and $g'(\sigma_1/\sigma_3)\cdot dX_{13}$ is also much larger than $g'(\sigma_2/\sigma_3)\cdot dX_{23}$ in most cases of plane strain conditions. In order to obtain the relationships on one mobilized plane from test results, it is clear from Eqs. 46 and 47 that the half of the measured value of ε_1 must be used in triaxial compression tests and the half of the measured value of ε_3 must be used in triaxial extension tests. It is also seen from Eqs. 46 and 47 that ε_1 in triaxial compression is twice the size of ε_1 in triaxial extension and ε_3 in triaxial extension is twice the size of ε_3 in triaxial compression, if the test specimens have the same initial granular structure.

Figure 22 shows the results of triaxial compression and triaxial extension tests under constant mean principal stresses σ_m on the Toyoura sand as arranged in terms of the τ/σ_N vs. $-d\varepsilon_N/d\gamma$ relationship on one mobilized plane. The Toyoura sand has a specific gravity of 2.65, a mean particle size of 0.2mm and a uniformity coefficient of 1.3. The maximum and minimum void ratios are 0.95 and 0.58 respectively. The initial void ratio prior to shearing is about 0.68. The effective mean principal stresses applied are σ_m=196kN/m^2 and σ_m=392kN/m^2 [14]. It is seen from Fig. 22 that Eq. 19 holds good under both triaxial compression and triaxial extension conditions. Figure 23 shows the same test results as in Fig. 22 as arranged in terms of the τ/σ_N vs. $-\varepsilon_N/\gamma$

relationship on one mobilized plane. It can be seen from Fig. 23 that Eq. 20 also holds under both triaxial compression and triaxial extension conditions. Figure 24 represents the same test results under σ_m =392kN/m^2 as in Fig. 22 as arranged in terms of the relation among τ/σ_N, γ and ε_N on one mobilized plane. This figure corresponds to Eqs. 21 and 22 or to Eqs. 32 and 33. The two solid curves in Fig. 24 are calculated from Eqs. 21 and 22 (or Eqs. 32 and 33) using parameters λ=1.2, μ=0.20, μ'=0.36 and γ_0=0.08(%). It should be noticed that the results of both triaxial compression and triaxial extension tests can be explained by the same parameters. Figure 25 shows the principal stress ratio vs. principal strains relationship of the same test results as in Fig. 24. The solid curves in this figure are calculated from Eqs. 46 and 47 using the above mentioned parameters. It is proved from Figs. 22-25 that the theory of "Compounded Mobilized Plane"(CMP) is applicable to such granular material as sands.

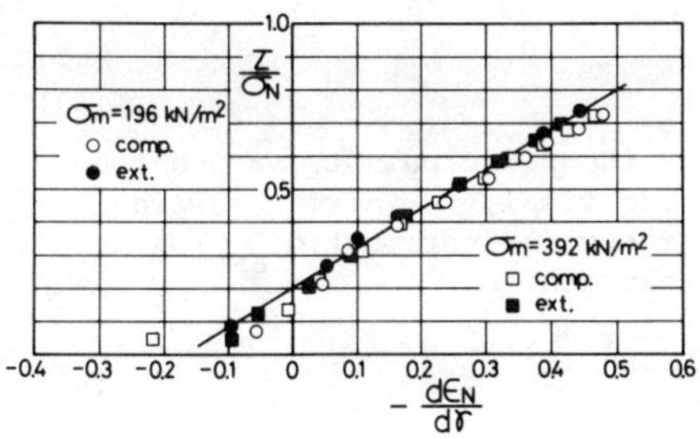

Fig. 22 Relationship between τ/σ_N and $-d\varepsilon_N/d\gamma$ on one mobilized plane obtained by triaxial compression and extension tests on Toyoura sand

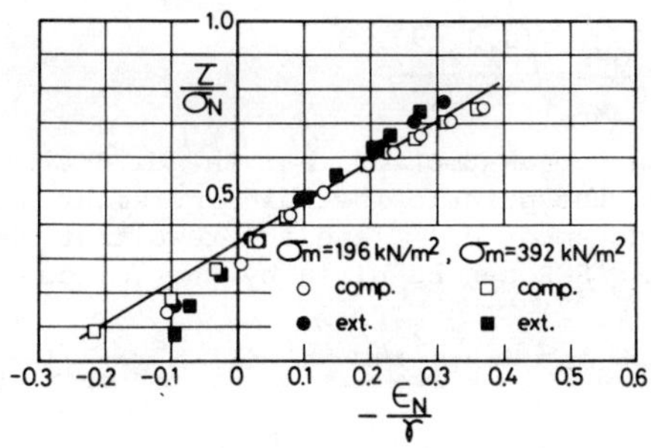

Fig. 23 Relationship between τ/σ_N and $-\varepsilon_N/\gamma$ on one mobilized plane obtained by triaxial compression and extension tests on Toyoura sand

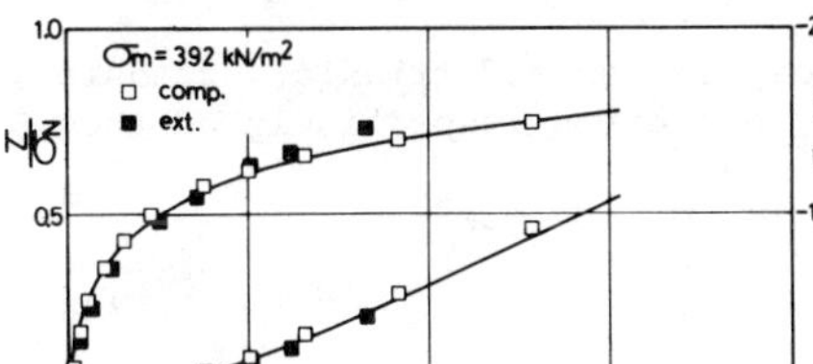

Fig. 24 Relationship among τ/σ_N, γ and ε_N on one mobilized plane obtained by triaxial compression and extension tests on Toyoura sand

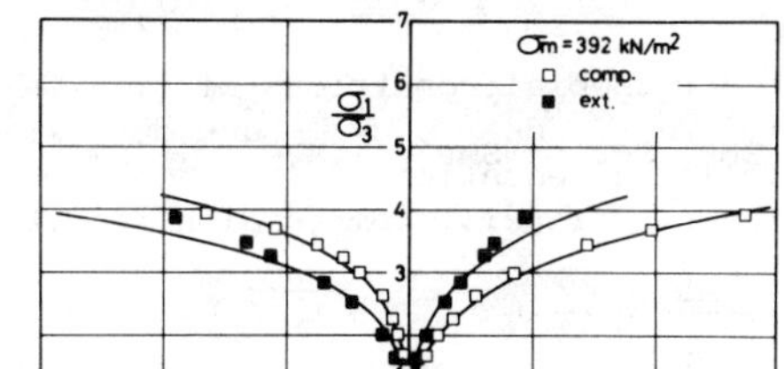

Fig. 25 Relationship among σ_1/σ_3, ε_1 and ε_3 obtained by triaxial compression and extension tests on Toyoura sand

4. STRESS-STRAIN RELATIONSHIP AND FAILURE CRITERION BASED ON THE THEORY OF "SPATIAL MOBILIZED PLANE"(SMP) [7, 8, 9, 13]

Since the direction cosines (a_1, a_2, a_3) of the normal of the SMP are given by Eq. 30, the normal stress σ_{SMP}, the shear stress τ_{SMP} and the shear-normal stress ratio τ_{SMP}/σ_{SMP} on the SMP can be expressed as follows:

$$\sigma_{SMP} = \sigma_1 a_1^2 + \sigma_2 a_2^2 + \sigma_3 a_3^2 = \frac{3J_3}{J_2} \tag{49}$$

$$\begin{aligned}\tau_{SMP} &= \sqrt{(\sigma_1-\sigma_2)^2 a_1^2 a_2^2 + (\sigma_2-\sigma_3)^2 a_2^2 a_3^2 + (\sigma_3-\sigma_1)^2 a_3^2 a_1^2} \\ &= \frac{\sqrt{J_1 J_2 J_3 - 9J_3^2}}{J_2}\end{aligned} \tag{50}$$

$$\frac{\tau_{SMP}}{\sigma_{SMP}} = \sqrt{\frac{J_1 J_2 - 9J_3}{9J_3}} \tag{51}$$

On the assumption that the direction of the principal stress and that of the principal strain increment are identical, the normal strain increment $d\varepsilon_{SMP}$ and the resultant shear strain increment $d\gamma_{SMP}$ on the SMP can be given by the following expressions.

$$d\varepsilon_{SMP} = d\varepsilon_1 \cdot a_1^2 + d\varepsilon_2 \cdot a_2^2 + d\varepsilon_3 \cdot a_3^2 = \frac{J_3}{J_2} \cdot \left(\frac{d\varepsilon_1}{\sigma_1} + \frac{d\varepsilon_2}{\sigma_2} + \frac{d\varepsilon_3}{\sigma_3} \right) \tag{52}$$

$$\begin{aligned}\frac{d\gamma_{SMP}}{2} &= \sqrt{(d\varepsilon_1-d\varepsilon_2)^2 a_1^2 a_2^2 + (d\varepsilon_2-d\varepsilon_3)^2 a_2^2 a_3^2 + (d\varepsilon_3-d\varepsilon_1)^2 a_3^2 a_1^2} \\ &= \frac{J_3}{J_2} \sqrt{\frac{(d\varepsilon_1-d\varepsilon_2)^2}{\sigma_1\sigma_2} + \frac{(d\varepsilon_2-d\varepsilon_3)^2}{\sigma_2\sigma_3} + \frac{(d\varepsilon_3-d\varepsilon_1)^2}{\sigma_3\sigma_1}}\end{aligned} \tag{53}$$

Under special conditions of triaxial compression and triaxial extension, τ_{SMP}/σ_{SMP}, $d\varepsilon_{SMP}$, $d\gamma_{SMP}$ and $d\varepsilon_{SMP}/d\gamma_{SMP}$ are given respectively by the following equations.

Under a triaxial compression condition ($\sigma_2=\sigma_3$, $d\varepsilon_2=d\varepsilon_3$):

$$\frac{\tau_{SMP}}{\sigma_{SMP}} = \frac{\sqrt{2}}{3}\left(\sqrt{\sigma_1/\sigma_3} - \sqrt{\sigma_3/\sigma_1}\right) \tag{54}$$

$$d\varepsilon_{SMP} = \frac{2\sigma_1 d\varepsilon_3 + \sigma_3 d\varepsilon_1}{2\sigma_1 + \sigma_3} \tag{55}$$

$$d\gamma_{SMP} = \frac{2\sqrt{2}\sqrt{\sigma_1\sigma_3}\,(d\varepsilon_1 - d\varepsilon_3)}{2\sigma_1 + \sigma_3} \tag{56}$$

$$\frac{d\varepsilon_{SMP}}{d\gamma_{SMP}}=\frac{2\sqrt{\sigma_1/\sigma_3}\cdot(d\varepsilon_3/d\varepsilon_1)+\sqrt{\sigma_3/\sigma_1}}{2\sqrt{2}\{1-(d\varepsilon_3/d\varepsilon_1)\}} \tag{57}$$

Under a triaxial extension condition ($\sigma_1=\sigma_2$, $d\varepsilon_1=d\varepsilon_2$):

$$\frac{\tau_{SMP}}{\sigma_{SMP}}=\frac{\sqrt{2}}{3}(\sqrt{\sigma_1/\sigma_3}-\sqrt{\sigma_3/\sigma_1}) \tag{58}$$

$$d\varepsilon_{SMP}=\frac{\sigma_1 d\varepsilon_3+2\sigma_3 d\varepsilon_1}{\sigma_1+2\sigma_3} \tag{59}$$

$$d\gamma_{SMP}=\frac{2\sqrt{2}\sqrt{\sigma_1\sigma_3}\ (d\varepsilon_1-d\varepsilon_3)}{\sigma_1+2\sigma_3} \tag{60}$$

$$\frac{d\varepsilon_{SMP}}{d\gamma_{SMP}}=\frac{\sqrt{\sigma_1/\sigma_3}\cdot(d\varepsilon_3/d\varepsilon_1)+2\sqrt{\sigma_3/\sigma_1}}{2\sqrt{2}\{1-(d\varepsilon_3/d\varepsilon_1)\}} \tag{61}$$

If the two basic stress-strain relationships on the mobilized plane (Eqs. 19 and 20) are applicable on the SMP, the following equations are obtained.

$$\frac{\tau_{SMP}}{\sigma_{SMP}}=\lambda\cdot(-\frac{d\varepsilon_{SMP}}{d\gamma_{SMP}})+\mu \tag{62}$$

$$\frac{\tau_{SMP}}{\sigma_{SMP}}=\lambda\cdot(-\frac{\varepsilon_{SMP}}{\gamma_{SMP}})+\mu' \tag{63}$$

The shear strain increment $d\gamma_{SMP}$ and the normal strain increment $d\varepsilon_{SMP}$ on the SMP can be obtained from Eqs. 62 and 63 as follows (see Eqs. 32 and 33):

$$d\gamma_{SMP}=\frac{\gamma_0}{\mu'-\mu}\cdot\exp(\frac{X-\mu}{\mu'-\mu})\cdot dX \tag{64}$$

$$d\varepsilon_{SMP}=\frac{\mu-X}{\lambda}\cdot d\gamma_{SMP} \qquad (X\equiv\tau_{SMP}/\sigma_{SMP}) \tag{65}$$

where λ, μ, μ' and γ_0 are stress-strain parameters. The parameters of λ, μ and μ' are almost constant for a given sample, and γ_0 is variable with the initial granular structure and the magnitude of the confining stress.

On the assumption that the principal stress and the principal strain increment coincide in direction, the direction cosines of $d\varepsilon_{SMP}$ are given by Eq. 30. If it is assumed that the direction of $d\gamma_{SMP}$ coincides with that of τ_{SMP}, the direction cosines of $d\gamma_{SMP}$ are given by the direction cosines (b_1, b_2, b_3) of τ_{SMP} as follows:

$$b_i = \frac{\sigma_i - \sigma_{SMP}}{\tau_{SMP}} \cdot a_i = \frac{\sigma_i J_2 - 3J_3}{\sqrt{\sigma_i J_2 (J_1 J_2 - 9J_3)}} \qquad (i=1,2,3) \tag{66}$$

Therefore, the strain increments on the SMP ($d\varepsilon_{SMP}$, $d\gamma_{SMP}$) can be transformed into the principal strain increments ($d\varepsilon_1$, $d\varepsilon_2$, $d\varepsilon_3$) by the following equation.

$$\begin{aligned} d\varepsilon_i &= d\varepsilon_{SMP} + \frac{b_i}{a_i} \cdot \frac{d\gamma_{SMP}}{2} \\ &= d\varepsilon_{SMP} + \frac{\sigma_i - \sigma_{SMP}}{\tau_{SMP}} \cdot \frac{d\gamma_{SMP}}{2} \qquad (i=1,2,3) \end{aligned} \tag{67}$$

Substituting Eqs. 64 and 65 into Eq. 67, the principal stress vs. principal strain relationship can be calculated.

A new failure criterion for soil has been proposed on the basis of the theory of the "Spatial Mobilized Plane"(SMP). The shear-normal stress ratio on the SMP τ_{SMP}/σ_{SMP} can be transformed from Eq. 51 as follows:

$$\begin{aligned} \frac{\tau_{SMP}}{\sigma_{SMP}} &= \sqrt{\frac{J_1 J_2 - 9J_3}{9J_3}} = \frac{2}{3}\sqrt{\frac{(\sigma_1-\sigma_2)^2}{4\sigma_1\sigma_2} + \frac{(\sigma_2-\sigma_3)^2}{4\sigma_2\sigma_3} + \frac{(\sigma_3-\sigma_1)^2}{4\sigma_3\sigma_1}} \\ &= \frac{2}{3}\sqrt{\tan^2\phi_{mo12} + \tan^2\phi_{mo23} + \tan^2\phi_{mo13}} \end{aligned} \tag{68}$$

If it is assumed that soil fails when this stress ratio τ_{SMP}/σ_{SMP} reaches a limiting value K, one can derive quite a simple form of failure criterion as follows:

$$\frac{\tau_{SMP}}{\sigma_{SMP}} = \sqrt{\frac{J_1 J_2 - 9J_3}{9J_3}} = K = \text{const.} \tag{69}$$

or

$$\tan^2\phi_{mo12} + \tan^2\phi_{mo23} + \tan^2\phi_{mo13} = \left(\frac{3}{2}K\right)^2 = \text{const.} \tag{70}$$

or

$$\frac{J_1 J_2}{J_3} = 9(K^2+1) = \text{const.} \tag{71}$$

If this criterion is expressed on the octahedral plane, one obtains a curve circumscribing the hexagon representing the Mohr-Coulomb criterion as shown in Fig. 26.

Figures 27 and 28 show the results of triaxial compression and triaxial extension tests on the Toyoura sand as arranged in terms of the τ_{SMP}/σ_{SMP} vs. $-d\varepsilon_{SMP}/d\gamma_{SMP}$ relationship and the τ_{SMP}/σ_{SMP} vs. $-\varepsilon_{SMP}/\gamma_{SMP}$ relationship on the SMP respectively. The test results are the same as in Figs. 22-25 [14]. It should be noted that the results of triaxial compression and triaxial extension tests are arranged almost on the same lines. Figure 29 represents the relationship among τ_{SMP}/σ_{SMP}, γ_{SMP} and

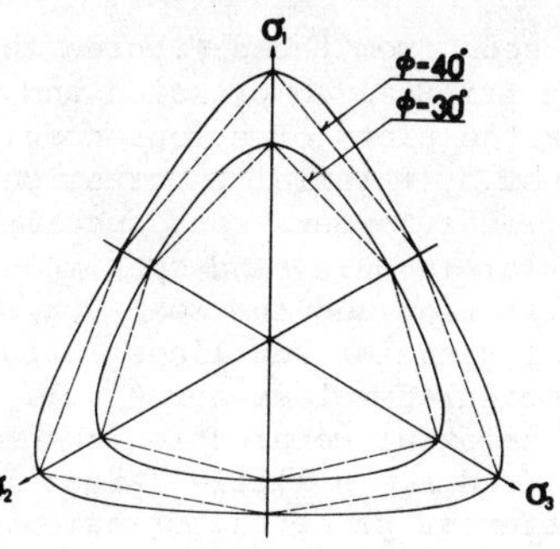

Fig. 26 Proposed SMP failure criterion and Mohr-Coulomb failure criterion on octahedral plane

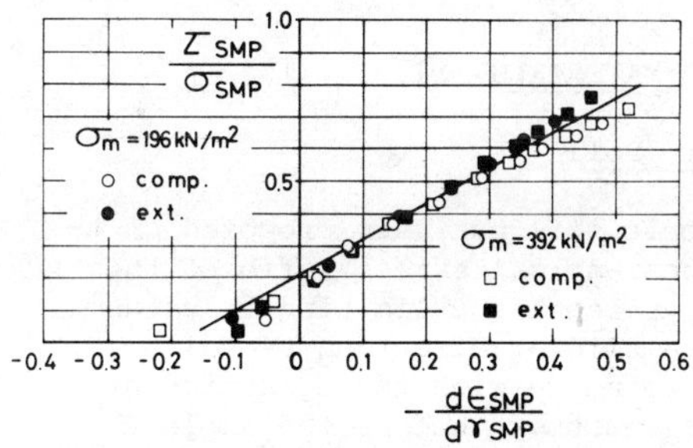

Fig. 27 Relationship between τ_{SMP}/σ_{SMP} and $-d\varepsilon_{SMP}/d\gamma_{SMP}$ obtained by triaxial compression and extension tests on Toyoura sand

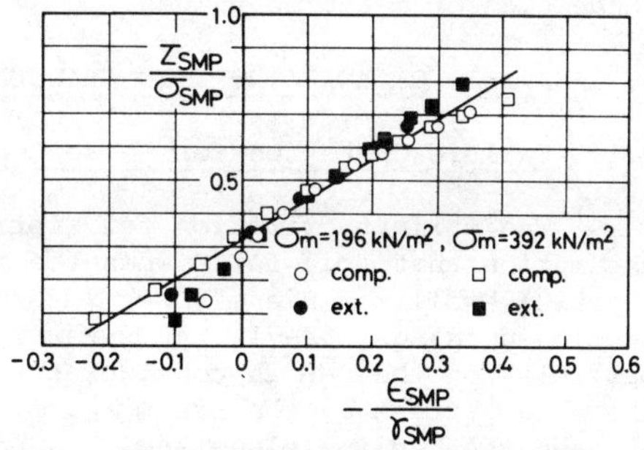

Fig. 28 Relationship between τ_{SMP}/σ_{SMP} and $-\varepsilon_{SMP}/\gamma_{SMP}$ obtained by triaxial compression and extension tests on Toyoura sand

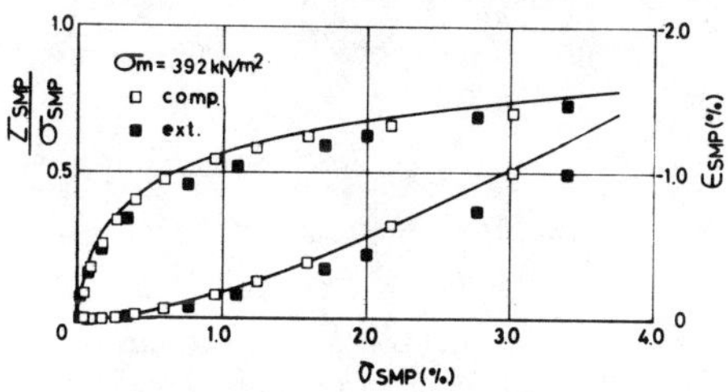

Fig. 29 Relationship among τ_{SMP}/σ_{SMP}, γ_{SMP} and ε_{SMP} obtained by triaxial compression and extension tests on Toyoura sand

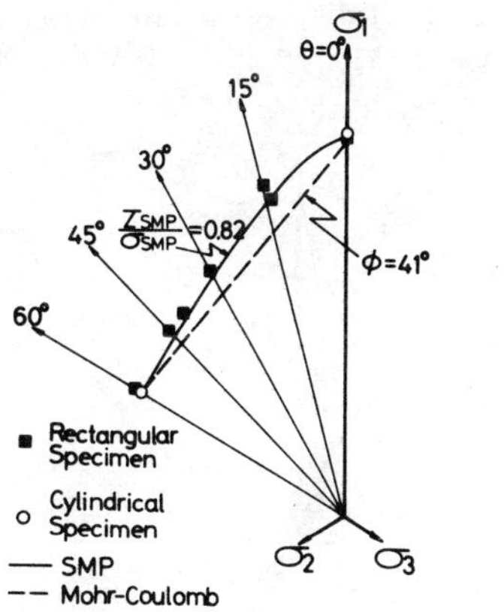

Fig. 30 Measured stress conditions at failure and SMP and Mohr-Coulomb criteria shown on octahedral plane

ε_{SMP} of the same test results. It is seen from these figures that Eqs. 62-65 hold pretty good under both triaxial compression and triaxial extension conditions. Figure 30 shows the plots of stress conditions at failure on the octahedral plane obtained by triaxial compression and triaxial extension tests on the cylindrical Toyoura sand specimens and by true triaxial tests on the rectangular Toyoura sand specimens [15]. In Fig. 30 the proposed SMP failure criterion and the Mohr-Coulomb failure criterion are also drawn by solid and broken lines respectively. It can be seen from Fig. 30 that the measured values agree well with the Mohr-Coulomb criterion under both triaxial compression (θ=0°) and triaxial extension (θ=60°) conditions, and are a little larger than the Mohr-Coulomb criterion under three different principal stresses. The SMP criterion shown by the solid line in Fig. 30 explains well such tendencies of measured values.

5. ANALYSIS OF ANISOTROPY IN GRANULAR MATERIALS [11, 12, 15]

5.1. Failure Criterion for Anisotropic Soil

The failure criterion for isotropic soil has been proposed on the assumption that soil fails when the shear-normal stress ratio on the "Spatial Mobilized Plane"(SMP) τ_{SMP}/σ_{SMP} reaches a limiting value as mentioned in Eqs. 69-71. As the average sliding direction of soil particles on the SMP is considered to coincide with the direction of the shear stress on the SMP τ_{SMP}, it is assumed that as the angle ζ between the bedding plane and τ_{SMP} becomes large, the strength defined by the SMP criterion has a large value. Based on this postulate, the failure criterion for isotropic soil is extended for anisotropic soil. For example, if the principal stress ratios at failure $(\sigma_1/\sigma_3)_f$ are 5 from a triaxial compression test (θ=0°) and 4 from a triaxial extension test (θ=180°) on a horizontally deposited sand sample, the relationship between τ_{SMP}/σ_{SMP} and ζ can be given at two points (see Fig. 31 for ζ

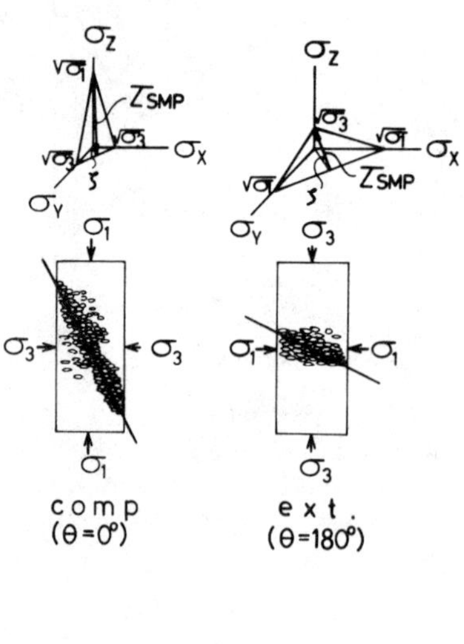

Fig. 31 Relationship between horizontal bedding plane and spatial mobilized plane(SMP) at θ=0° and θ=180°

and Fig. 32 for θ). If the linear interpolation and extrapolation of these two points are permitted, the value of τ_{SMP}/σ_{SMP} and then the failure criterion based on the SMP can be calculated even when the principal stress plane does not coincide with the bedding plane. Fig. 32 shows the proposed failure criterion for anisotropic soil analyzed by the above-mentioned method, the Mohr-Coulomb criterion for $(\sigma_1/\sigma_3)_f=4$ and 5, and the results of the true triaxial tests [11] on horizontally deposited sand samples ($\delta=0°$). In Fig. 32, δ denotes the angle between the principal stress plane and the bedding plane, and the bedding plane is rotated around the σ_Y axis (see Fig. 31). It can be seen from this figure that the failure criterion for $\delta=0°$ corresponds with $(\sigma_1/\sigma_3)_f=5$ at $\theta=0°$ and $(\sigma_1/\sigma_3)_f=4$ at $\theta=180°$. Fig. 33 represents the relationship between $(\sigma_1/\sigma_3)_f$ and δ calculated from this proposed failure criterion. The analytical results shown in Fig. 33 agree well with the tendency of some reported experimental values [16].

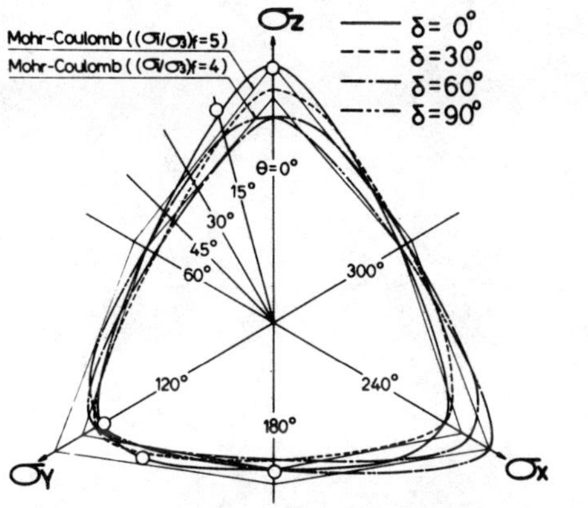

Fig. 32 Proposed failure criterion for anisotropic soil and measured values at $\delta=0°$ on octahedral plane

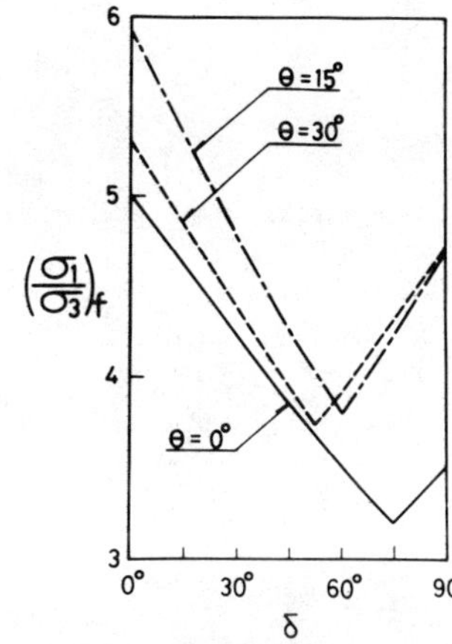

Fig. 33 Relationship between principal stress ratio at failure and δ

5.2 Analysis of Deformation for Anisotropic Soil

The stress-strain behaviour of anisotropic soil is analyzed on the basis of the theory of "Compounded Mobilized Planes"(CMP). The stress-strain relationships for isotropic soil based on the concept of the CMP are shown in Eqs. 40-45. In the stress-strain parameters (λ, μ, μ' and γ_0) of Eqs. 40-45, the effect of soil structure is most reflected by γ_0. Therefore, the parameter γ_0 is varied as γ_{0ij} (i,j=1, 2, 3; i<j) between the respective two principal stress directions to express the stress-strain behaviour of anisotropic soil. Eqs. 43 and 44 are thus changed for anisotropic soil as follows [11, 12]:

$$f'\left(\frac{\sigma_i}{\sigma_j}\right) = \frac{\gamma_{0ij}}{\mu'-\mu} \cdot \left(\frac{\mu - X_{ij}}{\lambda} + \frac{1}{2}\sqrt{\frac{\sigma_i}{\sigma_j}} \right) \cdot \exp\left(\frac{X_{ij}-\mu}{\mu'-\mu} \right) \tag{72}$$

$$g'\left(\frac{\sigma_i}{\sigma_j}\right) = \frac{\gamma_{0ij}}{\mu'-\mu} \cdot \left(\frac{\mu - X_{ij}}{\lambda} - \frac{1}{2}\sqrt{\frac{\sigma_j}{\sigma_i}} \right) \cdot \exp\left(\frac{X_{ij}-\mu}{\mu'-\mu} \right) \tag{73}$$

In the case of isotropic soil, γ_{0ij} are all equal between the respective two principal stress directions ($\gamma_{012}=\gamma_{023}=\gamma_{013}$). The principal strain increments for anisotropic soil can be calculated by supposing that as

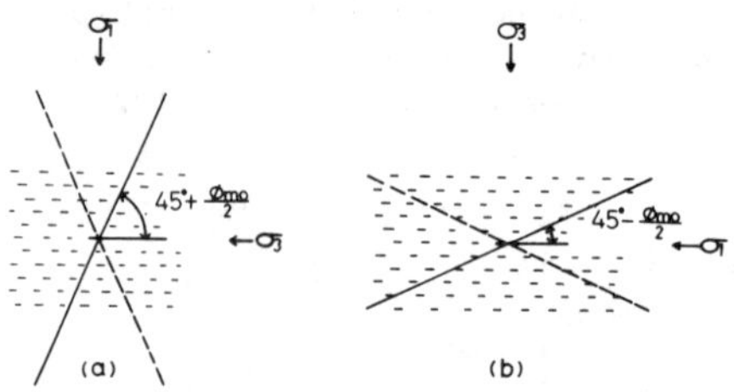

Fig. 34 Two-demensional angle between horizontal bedding plane and a mobilized plane

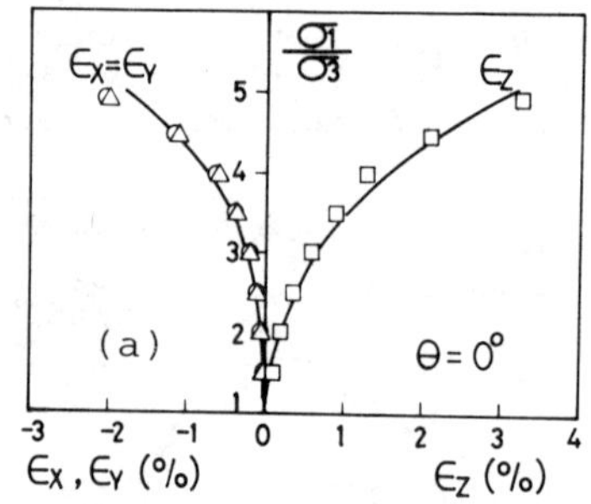

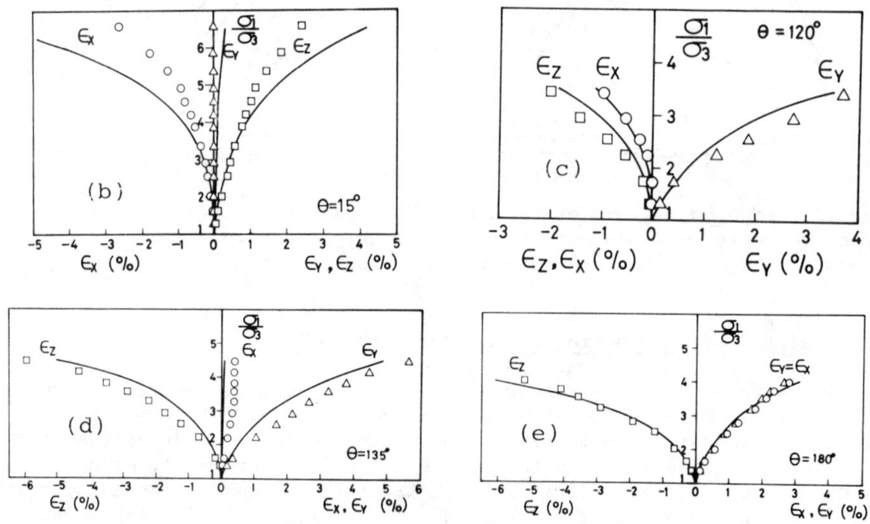

Fig. 35 (a)-(b) Principal stress ratio vs. principal strains relationships measured and analyzed under three principal stresses (σ_m=196kN/m^2)

the angle ζ' between the bedding plane and each of three mobilized planes becomes large, the strain increment caused by its mobilized plane becomes small and so γ_{0ij} takes a small value.

In the case that a principal stress plane coincides with the horizontal bedding plane as shown in Fig. 34, it may be assumed that $\gamma_{0ij}=\gamma_{0V}$ when the angle ζ' between the bedding plane and the mobilized plane is $(45°+\phi_{moij}/2)$, $\gamma_{0ij}=\gamma_{0H}$ when $\zeta'=(45°-\phi_{moij}/2)$, and $\gamma_{0ij}=\gamma_{0I}$ on the bedding plane $(\gamma_{0V}<\gamma_{0H}, \gamma_{0I}\fallingdotseq(\gamma_{0V}+\gamma_{0H})/2)$. All the above parameters (λ, μ, μ', γ_{0V} and γ_{0H}) can be determined by conventional triaxial compression and extension tests on the horizontally deposited soil samples. The stress-strain parameters used for calculation are $\lambda=1.2$, $\mu=0.23$, $\mu'=0.43$, $\gamma_{0V}=0.10\%$, $\gamma_{0H}=0.40\%$ and $\gamma_{0I}=0.25\%$. The calculated stress-strain relationships agree well with the experimental values obtained by the true triaxial tests on the artificially deposited sand sample as shown in Fig. 35. For example, Fig. 35 (c) shows the principal stress ratio vs. principal strains relationship under the stress condition of $\sigma_y>\sigma_z=\sigma_x$ ($\theta=120°$; see Fig. 32). The calculated values explain well the tendency of the experimental values that two principal strains under the same two principal stresses are different for anisotropy of soil structure. The strain normal to the bedding plane is inclined to be less compressible and that parallel to the bedding plane to be more compressible [11].

REFERENCES

1. Matsuoka, H., "A Microscopic Study on Shear Mechanism of Granular Materials", Soils and Foundations, Vol.14, No.1, pp. 29-43 (1974a).

2. Matsuoka, H. and Takeda, K., "A Stress-Strain Relationship for Granular Materials Derived from Microscopic Shear Mechanism", Soils and Foundations, Vol.20, No.3, pp. 45-58 (1980).

3. Matsuoka, H., "A Stress-Strain Model for Granular Materials Considering Mechanism of Fabric Change", Proc., U.S.-Japan Seminar on New Models and Constitutive Relations in the Mechanics of Granular Materials, Ithaca, (1982).

4. Geka, H. and Matsuoka, H., A" Stress-Strain Model for Granular Materials Derived from Microscopic Shear Mechanism", Proc., 36th Annual Meeting of JSCE, pp. 13-15 (in Japanese) (1981).

5. Matsuoka, H., "Stress-Strain Relationships of Sand Based on the Mobilized Plane", Soils and Foundations, Vol.14, No.2, pp. 47-61 (1974b).

6. Matsuoka, H., Geka, H. and Yasui, O., "Cyclic Shear Behavior of Granular Materials Considering Mechanism of Fabric Change", Proc., 17th Japan National Conf. SMFE, pp. 1673-1676 (in Japanese) (1982).

7. Matsuoka, H. and Nakai, T., "Stress-Deformation and Strength Characteristics of Soil under Three Different Principal Stresses", Proc., JSCE, No.232, pp. 59-70 (1974).

8. Matsuoka, H., "On the Significance of the Spatial Mobilized Plane", Soils and Foundations, Vol. 16, No. 1, pp. 91-100 (1976).

9. Matsuoka, H. and Nakai, T., "Stress-Strain Relationship of Soil Based on the SMP", Proc. of Specialty Session 9, 9th Int. Conf. on Soil Mechanics and Foundation Eng., pp. 153-162 (1977).

10. Murayama, S., "A Theoretical Consideration on a Behaviour of Sand", Proc. of IUTAM Symposium of Rheology and Soil Mechanics, Grenoble, pp. 146-159 (1964).

11. Matsuoka, H. and Ishizaki, H., "Deformation and Strength of Anisotropic Soil", Proc. of 10th Int. Conf. on Soil Mechanics and Foundation Eng., Vol. 1, pp. 699-702 (1981).

12. Matsuoka, H., "Constitutive Equation and FE Analysis for Anisotropic Soil", Proc. of 4th Int. Conf. on Numerical Methods in Geomechanics, Edmonton, pp. 223-233 (1982).

13. Matsuoka, H. and Nakai, T., "A Generalized Frictional Law for Soil Shear Deformation", Proc., U.S.-Japan Seminar on Continuum Mechanics and Statistical Approaches in the Mechanics of Granular Materials, Sendai, pp. 138-154 (1978).

14. Nakai, T. and Matsuoka, H., "Deformation of Soils in Three-Dimensional Stresses", IUTAM-Symposium on Deformation and Failure of Granular Materials, Delft, pp. 275-285 (1982).

15. Matsuoka, H. and Nakai, T., "A New Failure Criterion for Soils in Three-Dimensional Stresses", IUTAM-Symposium on Deformation and Failure of Granular Materials, Delft, pp. 253-263 (1982).

16. Oda, M., Koishikawa, I. and Higuchi, T., "Experimental Study of Anisotropic Shear Strength of Sand by Plane Strain Test", Soils and Foundations, Vol. 18, No. 1, pp. 25-38 (1978).

DEFORMATION AND STRENGTH OF GRANULAR MATERIALS BASED ON THE EXTENDED THEORY OF "SPATIAL MOBILIZED PLANE" (SMP*)

T. Nakai and H. Matsuoka

Department of Civil Engineering, Nagoya Institute of Technology, Nagoya, Japan

ABSTRACT

From the view point that soil is one of the granular materials whose mechanical behaviors are governed by the frictional law, a plane called "Spatial Mobilized Plane (SMP)" has been proposed, and a stress-strain model for soil under shear has been obtained based on the concept of the SMP. In this report, by introducing new amounts of strain increments on the SMP ($d\varepsilon^*_{SMP}$ and $d\gamma^*_{SMP}$) which denote the normal and parallel components of the principal strain increment vector to the SMP, a new stress-strain model is presented that explains the shear behavior of soil in three-dimensional stresses uniquely. Considering that the total strain increments of soil are given by the summation of the plastic component due to shear, the plastic component due to consolidation and the elastic component, a constitutive equation for soil which is applicable to the finite element analysis of soil foundation and earth structure is derived. Moreover, analytical results of the stress-strain behaviors under cyclic loading and liquefaction phenomena are presented.

1. STRESS-STRAIN RELATIONSHIP BASED ON THE EXTENDED THEORY OF "SPATIAL MOBILIZED PLANE" (SMP*) [1,2]

A stress-strain model for soil under shear has already been proposed based on the idea that there exist unique relations among the shear-normal stress ratio on the SMP (τ_{SMP}/σ_{SMP}) and the normal and shear strain increments on the SMP ($d\varepsilon_{SMP}$ and $d\gamma_{SMP}$). This model can explain various experimental results of shear tests [3,4]. However, strictly investigated, there are some differences with a trend between experimental results and values calculated using this model, which is described in the another report [5]. Therefore, considering the derivation process of this model, a new stress-strain model is proposed by introducing new amounts of strain increments based on the SMP.

The new amounts of strain increments based on the SMP is dicussed here. Fig.1 shows a cubical soil element and the Spatial Mobilized Plane (SMP) in the three-dimensional space, where I, II and III axes represent the directions to which three principal stresses σ_1, σ_2 and σ_3 are applied, respectively. Here, the SMP has been considered to be the plane where soil particles are most mobilized on the average, and the shear-normal stress ratio on the SMP (τ_{SMP}/σ_{SMP}) seems to govern the sliding of soil particles in the three-dimensional space. According to the former stress-strain model, the basic equations hold uniquely among

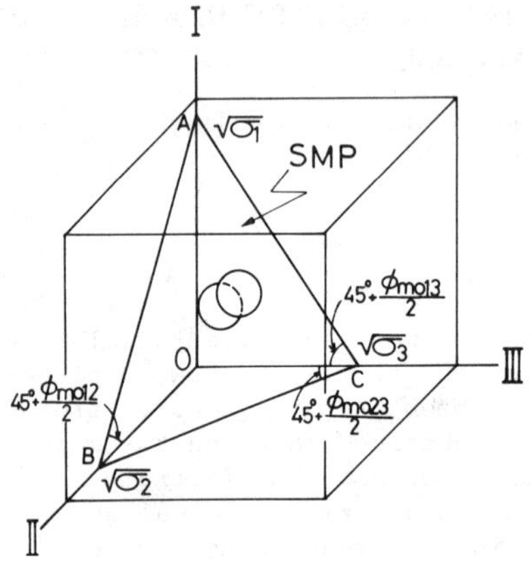

Fig. 1 A soil element and spatial mobilized plane (ABC) in three dimensional space

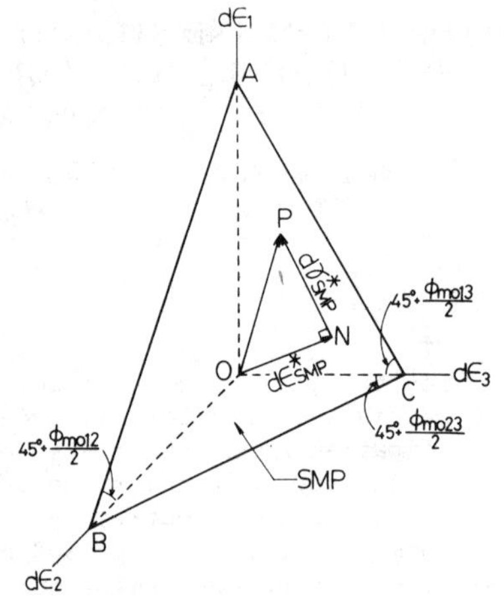

Fig. 2 Amounts of strain increments ($d\varepsilon^*_{SMP}$ and $d\gamma^*_{SMP}$) in principal strain increment space

τ_{SMP}/σ_{SMP} and the nomal and shear strain increment on the SMP ($d\varepsilon_{SMP}$ and $d\gamma_{SMP}$), because the average sliding direction of soil particles has been considered to coincide with the direction of the strain increment vector on the SMP. However, it is not of a general character that the behaviors of soil particles in the three-dimensional space are related to nothing but $d\varepsilon_{SMP}$ and $d\gamma_{SMP}$ which are the strain increment components on one plane in the space. It is rather reasonable to consider that the average sliding of soil particles and the principal strain increment vector are indentical in direction in the case that there exist randomly the interparticle contacts in the soil element. Therefore, new amounts of strain increments based on the SMP ($d\varepsilon^*_{SMP}$ and $d\gamma^*_{SMP}$), which denote the normal and parallel components of the principal strain increment vector to the SMP, are introduced, and a new stress-strain model under shear is derived using these new amounts ($d\varepsilon^*_{SMP}$ and $d\gamma^*_{SMP}$) instead of $d\varepsilon_{SMP}$ and $d\gamma_{SMP}$.

The formulas of the new stress-strain model are described here. The normal stress on the SMP (σ_{SMP}), the shear stress (τ_{SMP}) and the shear-normal stress ratio ($X\equiv\tau_{SMP}/\sigma_{SMP}$) are expressed as follows:

$$\sigma_{SMP} = \sigma_1 a_1^2+\sigma_2 a_2^2+\sigma_3 a_3^2 = 3\,\frac{J_3}{J_2} \tag{1}$$

$$\tau_{SMP} = \sqrt{(\sigma_1-\sigma_2)^2 a_1^2 a_2^2+(\sigma_2-\sigma_3)^2 a_2^2 a_3^2+(\sigma_3-\sigma_1)^2 a_3^2 a_1^2} = \frac{\sqrt{J_1 J_2 J_3-9J_3^2}}{J_2} \tag{2}$$

$$X \equiv \frac{\tau_{SMP}}{\sigma_{SMP}} = \sqrt{\frac{J_1 J_2 - 9J_3}{9J_3}} \tag{3}$$

Where (a_1, a_2, a_3) represent the direction cosines of the normal of the SMP, and are expressed by the following equation using the first, second and third effctive stress invariants (J_1, J_2 and J_3).

$$a_i = \sqrt{\frac{J_3}{\sigma_i J_2}} \qquad (i=1,2,3) \tag{4}$$

Fig.2 indicates the principal strain increment space, in which the principal strain increment vector $\overrightarrow{d\varepsilon_i} = (d\varepsilon_1, d\varepsilon_2, d\varepsilon_3)$ is defined as a vector $\overrightarrow{OP}$. On the assumption that the direction of the principal stresses and that of the principal strain increments are indentical, the new amounts of strain increments ($d\varepsilon^*_{SMP}$ and $d\gamma^*_{SMP}$), which are represented by the normal component $\overrightarrow{ON}$ and the parallel component $\overrightarrow{NP}$ of the vector $\overrightarrow{OP}$ to the SMP respectively, can be expressed by the following equations using the three principal strain increments ($d\varepsilon_1$, $d\varepsilon_2$ and $d\varepsilon_3$).

$$d\varepsilon^*_{SMP} = d\varepsilon_1 a_1 + d\varepsilon_2 a_2 + d\varepsilon_3 a_3 \tag{5}$$

$$d\gamma^*_{SMP} = \sqrt{(d\varepsilon_1 a_2 - d\varepsilon_2 a_1)^2 + (d\varepsilon_2 a_3 - d\varepsilon_3 a_2)^2 + (d\varepsilon_3 a_1 - d\varepsilon_1 a_3)^2} \tag{6}$$

Now, if the basic equations on the mobilized plane derived by Matsuoka [6, 7] are assumed to hold among τ_{SMP}/σ_{SMP}, $d\varepsilon^*_{SMP}$ and $d\gamma^*_{SMP}$ in the same manner as among τ_{SMP}/σ_{SMP}, $d\varepsilon_{SMP}$ and $d\gamma_{SMP}$ in the former model, the following equations can be obtained.

$$\frac{\tau_{SMP}}{\sigma_{SMP}} = \lambda^* \cdot \left(-\frac{d\varepsilon^*_{SMP}}{d\gamma^*_{SMP}}\right) + \mu^* \tag{7}$$

$$\frac{\tau_{SMP}}{\sigma_{SMP}} = \lambda^* \cdot \left(-\frac{\varepsilon^*_{SMP}}{\gamma^*_{SMP}}\right) + \mu'^* \tag{8}$$

$$(\varepsilon^*_{SMP} = \int d\varepsilon^*_{SMP} \text{ and } \gamma^*_{SMP} = \int d\gamma^*_{SMP})$$

where λ^*, μ^* and μ'^* are the soil parameters. By solving the differential equation obtained from Eqs.7 and 8, $d\varepsilon^*_{SMP}$ and $d\gamma^*_{SMP}$ are given as functions of the stress ratio.

$$d\gamma^*_{SMP} = \frac{\gamma^*_0}{\mu'^* - \mu^*} \cdot \exp\left(\frac{X - \mu^*}{\mu'^* - \mu^*}\right) \cdot dX \tag{9}$$

$$d\varepsilon^*_{SMP} = \frac{\mu^* - X}{\lambda^*} \cdot d\gamma^*_{SMP} \tag{10}$$

$$(X \equiv \tau_{SMP}/\sigma_{SMP})$$

The soil parameters (λ^*, μ^*, μ'^* and γ^*_0) in Eqs.9 and 10 have the same properties as those (λ, μ, μ' and γ_0) in the former stress-strain model.

Each of the parameters λ^*, μ^* and μ'^* is considered to be nearly constant for a given sample. On the other hand, γ_0^* is affected by the initial soil structure and the confining stress, and is empirically expressed by the following formula using the mean principal stress σ_m in the case that the initial structures of sample are similar.

$$\gamma_0^* = \gamma_{0i}^* + C_d \cdot \log_{10} \frac{\sigma_m}{\sigma_{mi}} \qquad (11)$$

where σ_{mi} is the basic value of σ_m, γ_{0i}^* is the value of γ_0^* at $\sigma_m = \sigma_{mi}$ and C_d^* is a coefficient determined by the sample and its initial structure.

The direction cosines of $d\varepsilon_{SMP}^*$ are equal to a_i (i=1,2,3) given by Eq.4 on the assumption that the direction of the principal stresses and that of the principal strain increments are indentical. If it is assumed that the parallel component of the average sliding of soil particles to the SMP coincides with τ_{SMP} in direction, the direction cosines of $d\gamma_{SMP}^*$ are expressed by the following direction cosines of τ_{SMP}

$$b_i = \frac{\sigma_i - \sigma_{SMP}}{\tau_{SMP}} \cdot a_i \qquad (i=1,2,3) \qquad (12)$$

Accordingly, the principal strain increments $d\varepsilon_i$ (i=1,2,3) can be calculated from $d\varepsilon_{SMP}^*$ and $d\gamma_{SMP}^*$ using the following formula.

$$d\varepsilon_i = a_i \cdot d\varepsilon_{SMP}^* + b_i \cdot d\gamma_{SMP}^* \qquad (i=1,2,3) \qquad (13)$$

The saturated Toyoura sand (the mean diameter D_{50}=0.2mm, the uniformity coefficient U_c=1.3,the specific gravity G_s=2.65, the maximum void ratio e_{max}=0.95 and the minimum void ratio e_{min}=o.58) is used in this investigation. Specimens are prepared by pouring the saturated sand into a mold in several layers and compacting each layer with a rod so that the specimens have a desired initial void ratio (e_0=0.68). It is checked using the isotropic consolidation tests that the specimens prepared in this manner have an isotropic initial structure. All the tests (triaxial compression, triaxial extension and true triaxial tests) are performed under constant mean principal stresses, because the deformation of soil produced by the change of stress ratio under a constant mean principal stress is usually considered to be the deformation due to shear.

Fig.3 shows the results of the triaxial compression and triaxial extension tests with respect to the relation between τ_{SMP}/σ_{SMP} and $d\varepsilon_{SMP}^*/d\gamma_{SMP}^*$ in Eq.9 (mean principal stresses σ_m=196kN/m² and 392kN/m²). Fig.4 indicates the results of the true triaxial tests (σ_m=196kN/m²). In Fig.4, θ represents the angle between the radial stress path and σ_1-axis on the octahedral plane (θ=0° and 60° correspond to the triaxial compression and triaxial extension condition respectively). It is seen from these figures that all the test results are arranged on a unique linear line in spite of the magnitude of mean principal stresses and the

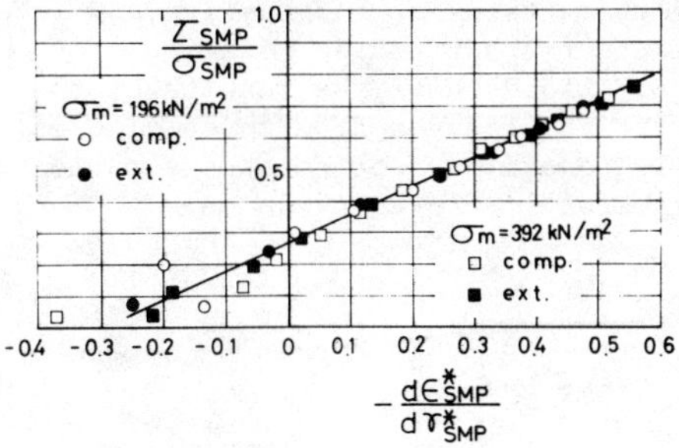

Fig. 3 Relation between τ_{SMP}/σ_{SMP} and $d\varepsilon^*_{SMP}/d\gamma^*_{SMP}$ obtained by triaxial compression and extension tests on sand

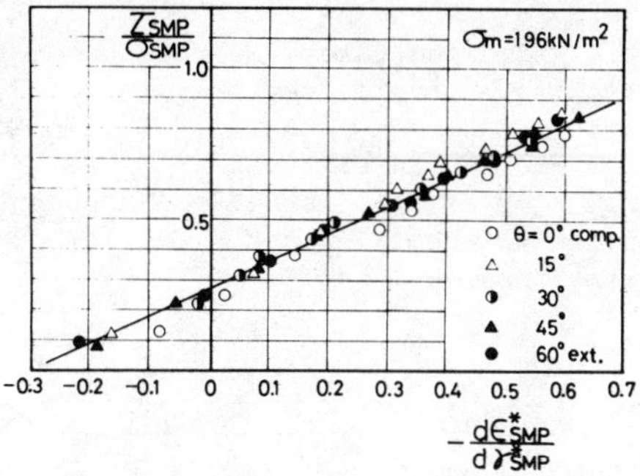

Fig. 4 Relation between τ_{SMP}/σ_{SMP} and $d\varepsilon^*_{SMP}/d\gamma^*_{SMP}$ obtained by true triaxial tests on sand

variety of stress paths on the octahedral plane, and Eq.7 has a good estimate. The soil parameters λ^* and μ^* can be determined as $\lambda^*=0.9$ and $\mu^*=0.27$ from the values of linear gradient and ordinate intercept of the line in these diagram. By the way, Fig.27 in the another report [5] indicates the relation between τ_{SMP}/σ_{SMP} and $d\varepsilon_{SMP}/d\gamma_{SMP}$ obtained by the same tests as in Fig.3. This is an arrangement based on the previous stress-strain relationship [3,4]. It is found that the tests results are arranged on the same straight line more closely in Fig.3 than in Fig.27 in the another report. Fig.5 shows the results of the same tests as in Fig.3, arranged with respect to the relation between the shear-normal stress ratio on the octahedral plane (τ_{oct}/σ_{oct}) and the normal-shear strain increment ratio on the octahedral plane ($d\varepsilon_{oct}/d\gamma_{oct}$). Roscoe et al. obtained the following relationship between stress ratio and strain increment ratio using the equation of energy dissipation [8,9].

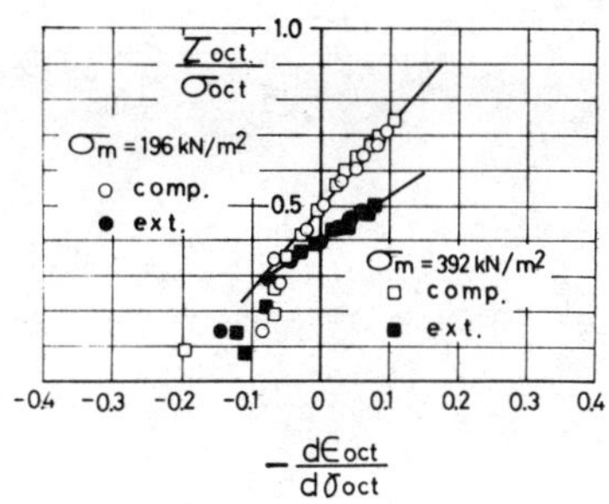

Fig. 5 Relation between τ_{oct}/σ_{oct} and $d\varepsilon_{oct}/d\gamma_{oct}$ obtained by same tests as Fig.3

$$\frac{\tau_{oct}}{\sigma_{oct}} = \frac{\sqrt{2}}{3}M - 2\frac{d\varepsilon_{oct}}{d\gamma_{oct}} \qquad (14)$$

However, it is not seen in Fig.5 that the results of triaxial compression and triaxial extension tests are arranged on a unique line.

Fig.6 indicates the results of the same triaxial compression and triaxial extension tests as in Fig.3, with respect to the relation between τ_{SMP}/σ_{SMP} and $\varepsilon^*_{SMP}/\gamma^*_{SMP}$ in Eq.8. Fig.7 shows the results of the same true triaxial tests as in Fig.4. It is observed from these figures that Eq.8 as well as Eq.7 holds uniquely regardless of the stress paths on the octahedral plane. The soil parameter μ'^* in Eq.8 can be determined as $\mu'^*=0.41$.

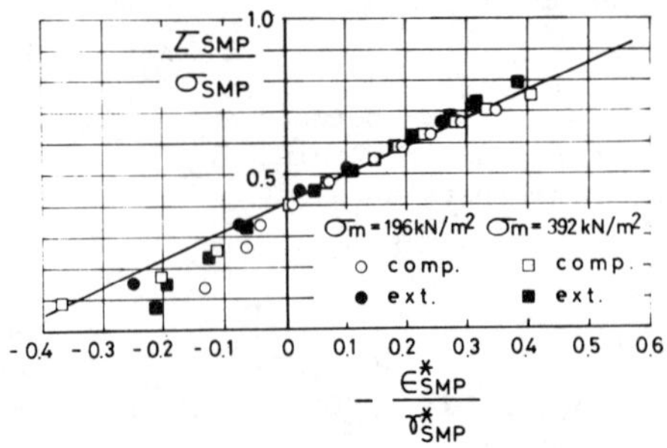

Fig. 6 Relation between τ_{SMP}/σ_{SMP} and $\varepsilon^*_{SMP}/\gamma^*_{SMP}$ obtained by same triaxial compression and extension tests as Fig.3

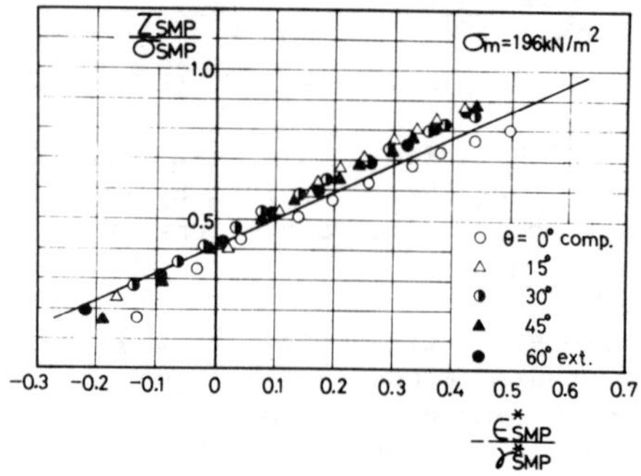

Fig. 7 Relation between τ_{SMP}/σ_{SMP} and $\varepsilon^*_{SMP}/\gamma^*_{SMP}$ obtained by same true triaxial tests as Fig.4

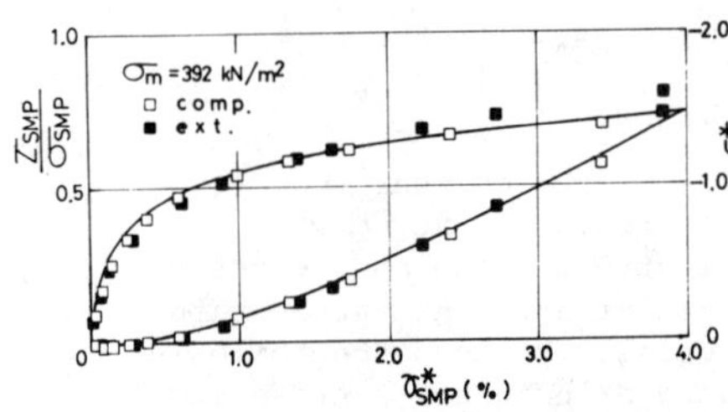

Fig. 8 Relation amang τ_{SMP}/σ_{SMP}, γ^*_{SMP} and ε^*_{SMP} obtained by triaxial compression and extension tests on sand

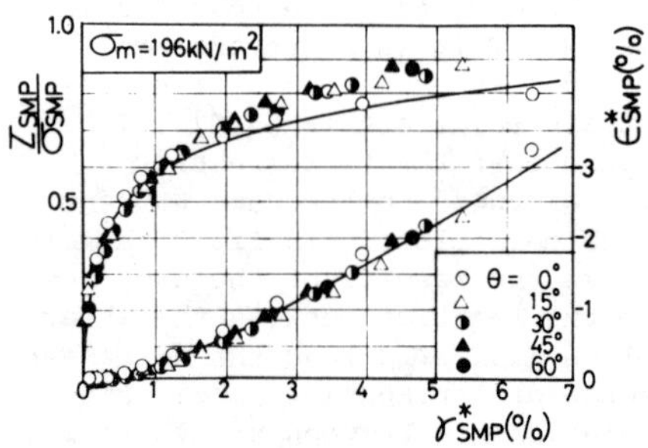

Fig. 9 Relation amang τ_{SMP}/σ_{SMP}, γ^*_{SMP} and ε^*_{SMP} obtained by true triaxial tests on sand

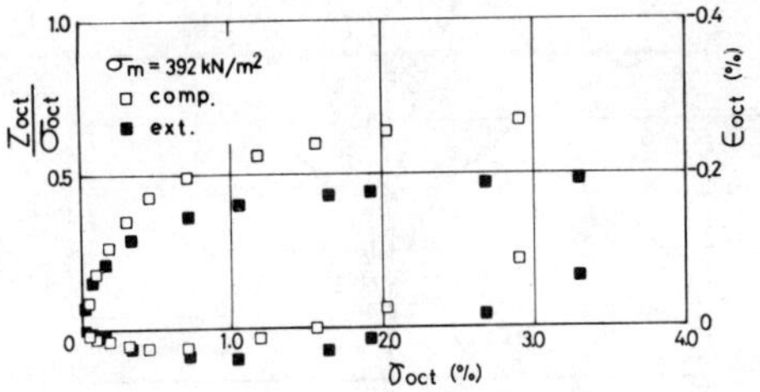

Fig. 10 Relation among τ_{oct}/σ_{oct}, γ_{oct} and ε_{oct} obtained by same tests as Fig.8

Figs.8 and 9 show the relation among τ_{SMP}/σ_{SMP}, γ^*_{SMP} and ε^*_{SMP} obtained from the triaxial compression and extension tests and the true triaxial tests, respectively. The plots represent the observed values and the solid curves the calculated values using Eqs.9 and 10. The stress-strain parameters for Toyoura sand are determined from the arrangements in Figs.3, 6 and 8 as follows; $\lambda^*=0.9$, $\mu^*=0.27$, $\mu'^*=0.41$, $\gamma^*_{0i}=0.10\%$, $C^*_d=0.066\%$ and $\sigma_{mi}=98\text{kN/m}^2$. It is seen in Figs.8 and 9 that regardless of the stress paths the observed values are plotted nearly on the same curves calculated using these soil parameters. Fig.10 shows the results of the same tests as shown in Fig.8 in terms of the relation among τ_{oct}/σ_{oct}, γ_{oct} and ε_{oct}. The test data arranged on the octahedral plane vary from the triaxial compression to the triaxial extension conditions with a trend in the same way as shown in Fig.5, while the results are uniquely analyzed in the case of Fig.8.

Figs.11 (a)-(e) show the results of the true triaxial tests ($\theta=0°$, $15°$, $30°$, $45°$ and $60°$) and the calculated curves with respect to the relation among the principal stress ratio (σ_1/σ_3) and the principal strains (ε_1, ε_2 and ε_3). It can be estimated from these figures that the plane strain condition exists between $\theta=15°$ and $\theta=30°$, because the intermediate principal strain ε_2 is expansive at $\theta=15°$ and is compressive at $\theta=30°$. This tendency also corresponds to that of other experimental results [10].

Figs.12 and 13 show the directions of projection of the strain increment vectors on the octahedral plane observed from the true triaxial tests ($\theta=15°$, $30°$ and $45°$) and calculated by the proposed stress-strain model respectively. The calculated vectors in Fig.13 explain well the tests results shown in Fig.12 that the directions of $d\gamma_{oct}$ deviate from those of τ_{oct} with a definite trend as the stress ratio increases. This tendency of the test results, however, was not explained by the former stress-strain model [3,4]. Although the solid curves in these figures represent the criterion based on the SMP which is described in the another report [5], both the observed and calculated vectors are not perpendicular to this criterion.

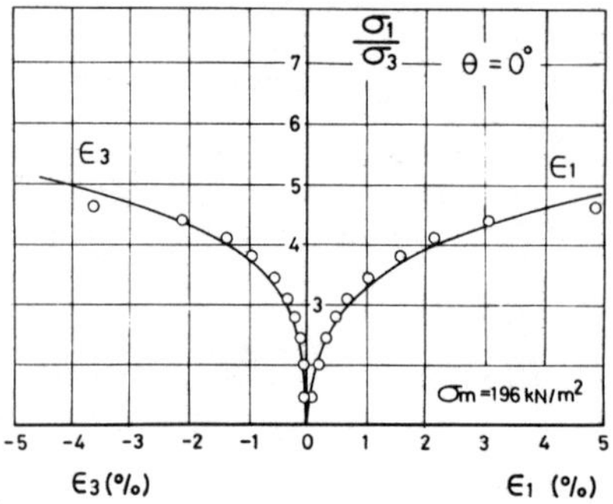

(a)

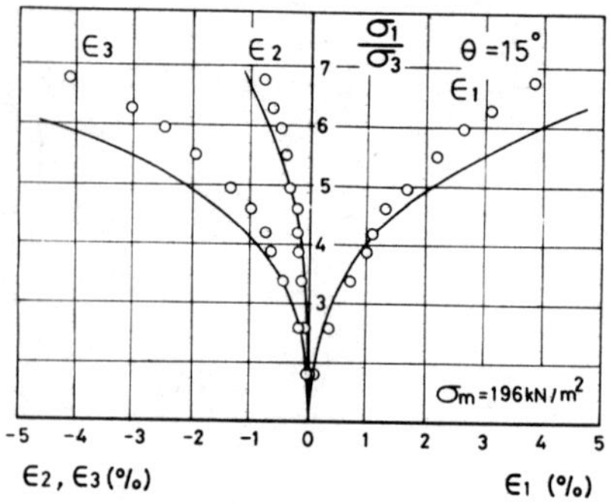

(b)

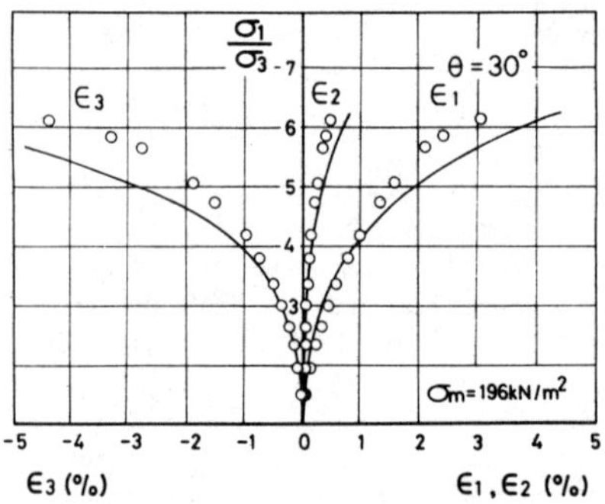

(c)

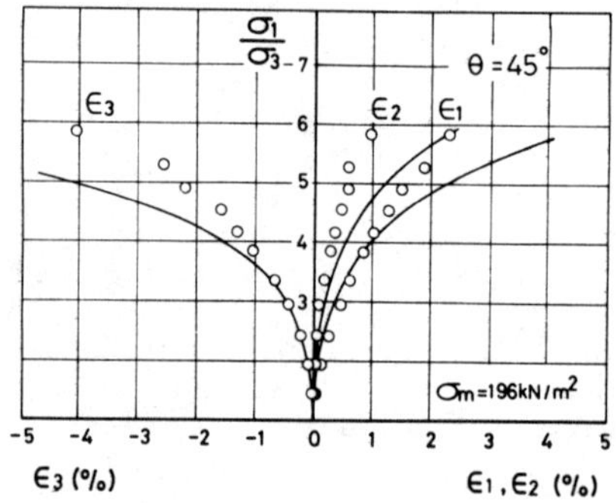

(d)

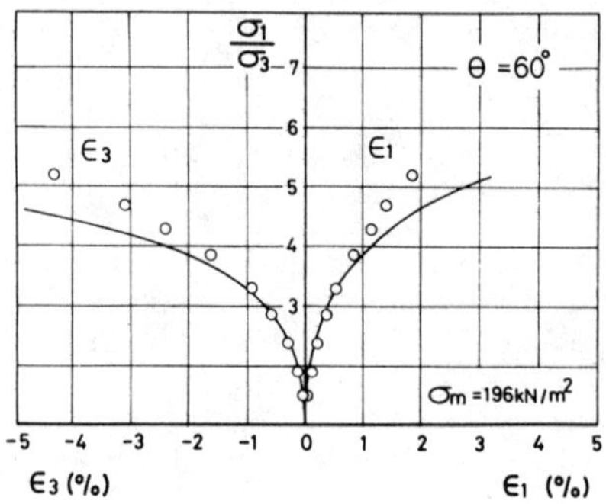

(e)

Figs. 11(a)-(e) Principal stress ratio vs. principal strains in true triaxial tests (θ= 0°, 15°, 30°, 45°, 60°) on sand

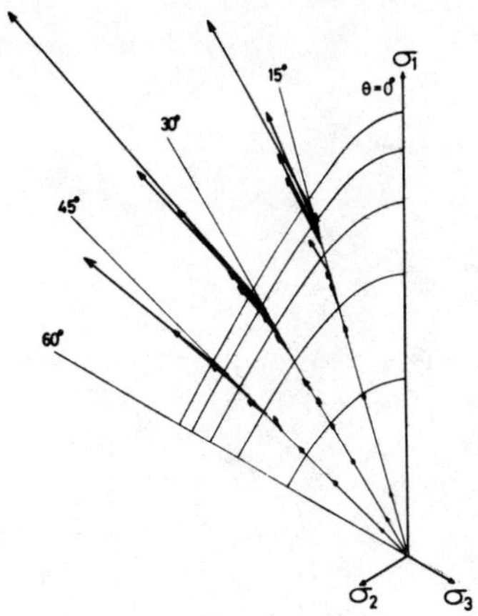

Fig. 12 Shear strain increment vectors on octahedral plane obtained by true triaxial tests on sand

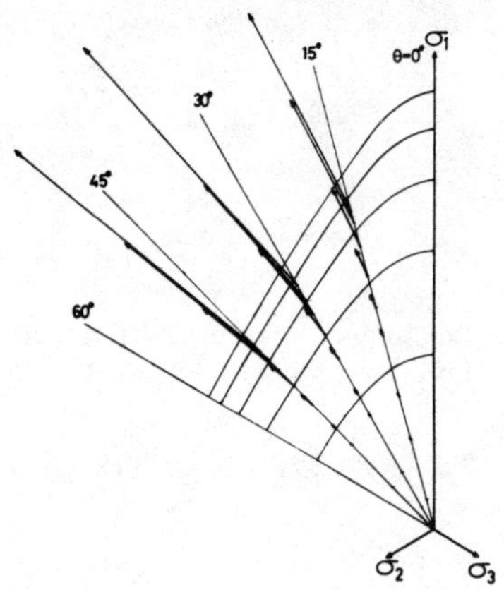

Fig. 13 Shear strain incrment vectors on octahedral plane calculated by proposed relationship

2. CONSTITUTIVE EQUATION FOR SOILS BASED ON THE EXTENDED THEORY OF "SPATIAL MOBILIZED PLANE" (SMP*) [11 , 12]

The stress-strain relationship described in the previous section is concerned with the soil behavior under shear. In this section, the outline of the constitutive equation for soil based on the concept of SMP* will be presented, which is derived by considering the consolidation component and the elastic component of the strains in addition to the above mentioned shear component. The total strain increments of soil $\{d\varepsilon\}$ are considered to be expressed as the summation of the plastic strain increments due to shear $\{d\varepsilon^s\}$ (the strain increments caused by the increase in stress ratio), the plastic strain increments due to consolidation $\{d\varepsilon^c\}$ (the strain increment caused by the increase in mean principal stress) and the elastic strain increments $\{d\varepsilon^e\}$.

$$\{d\varepsilon\}=\{d\varepsilon^s\}+\{d\varepsilon^c\}+\{d\varepsilon^e\} \tag{15}$$

Throughout this section, the superscript s demotes the plastic component due to shear, c the plastic component due to consolidation and e the elastic component.

Plastic strain increment due to shear $\{d\varepsilon^s\}$

The plastic strain increment due to shear is given by using the stress-strain relationship described in the previous section. From Eq.13, the principal strain increments due to shear $\{d\varepsilon_i^s\}$ are expressed by the following equations.

$$d\varepsilon_i^S = a_i \cdot d\varepsilon_{SMP}^{*S} + b_i \cdot d\gamma_{SMP}^{*S}$$
$$= (a_i \cdot E_1^* + b_i \cdot G_1^*) \cdot dX$$
$$\equiv E_i^S \cdot dX \qquad (i=1,2,3) \tag{16}$$

or

$$\{d\varepsilon_i^S\} = \{E_i^S\} \cdot dX \tag{17}$$

where a_i and b_i are the direction cosines of σ_{SMP} and τ_{SMP}, respectively G_1^* and E_1^* are expressed by the following equations from Eqs.9 and 10.

$$G_1^* = \{\gamma_0/(\mu'^* - \mu^*)\} \cdot \exp\{(X-\mu^*)/(\mu'^* - \mu^*)\} \tag{18}$$

$$E_1^* = (\mu^* - X)/\lambda^* \tag{19}$$

Performing the total differentiation of Eq.3, dX is represented by using the general stress increments $\{d\sigma\}$.

$$dX = \{\partial X/\partial \sigma\}^T \cdot \{d\sigma\} \equiv \{A\}^T \cdot \{d\sigma\} \tag{20}$$

On the assumption that the direction of the plastic principal strain increments and that of the principal stresses are identical, the matrix $[T]$ which transforms the principal strain increments $\{d\varepsilon_i^S\}$ into the general strain increment $\{d\varepsilon^S\}$ is given in a formula of stresses. Therefore, the general strain increments due to shear $\{d\varepsilon^S\}$ can be expressed by using the general stress increments $\{d\sigma\}$ as follows:

$$\{d\varepsilon^S\} = [T] \cdot \{d\varepsilon_i^S\} = [T] \cdot \{E_i^S\} \cdot \{A\}^T \cdot \{d\sigma\} \tag{21}$$

Plastic strain increments due to consolidation $\{d\varepsilon^C\}$

Upon the consideration about anisotropic consolidation tests, the plastic principal strain increments due to consolidation $d\varepsilon_i^C$ are assumed to be divisible into the components under isotropic consolidation $d\varepsilon_{i(iso)}^C$ and the components due to dilatancy under anisotropic consolidation $d\varepsilon_{i(dil)}^C$. $d\varepsilon_{i(iso)}^C$ are determined from the e-$\log_{10}\sigma_m$ relationship under isotropic consolidation. $d\varepsilon_{i(dil)}^C$ are expressed by the following amounts of strain increments ($d\varepsilon_{SMP}^{*C}$ and $d\gamma_{SMP}^{*C}$) in the same manner as under shear.

$$d\gamma_{SMP}^{*C} = 0.434K_c \cdot [\exp\{(X-\mu^*)/(\mu'^* - \mu^*)\} - \exp\{-\mu^*/(\mu'^* - \mu^*)\}] \cdot (d\sigma_m/\sigma_m)$$
$$\equiv G_2^* \cdot d\sigma_m \tag{22}$$

$$d\varepsilon_{SMP}^{*C} = \{(\mu^* - X)/\lambda^*\} \cdot d\gamma_{SMP}^{*C} \equiv E_2^* \cdot d\sigma_m \tag{23}$$

Accordingly, the principal strain increments $d\varepsilon_i^c$ are given as follows:

$$d\varepsilon_i^c = d\varepsilon_{i(iso)}^c + d\varepsilon_{i(dil)}^c$$

$$= \frac{0.434}{3} \cdot \frac{C_c - C_s}{(1+e_0)} \cdot \frac{d\sigma_m}{\sigma_m} + a_i \cdot d\varepsilon_{SMP}^{*c} + b_i \cdot d\gamma_{SMP}^{*c}$$

$$= \left(\frac{0.434}{3} \cdot \frac{C_c - C_s}{(1+e_0)\sigma_m} + a_i \cdot E_2^* + b_i \cdot G_2^* \right) \cdot d\sigma_m$$

$$\equiv E_i^c \cdot d\sigma_m \qquad (i=1,2,3) \tag{24}$$

or

$$\{d\varepsilon_i^c\} = \{E_i^c\} \cdot d\sigma_m \tag{25}$$

where C_c is the compression index, C_s the swelling index and e_0 the initial void ratio. $d\sigma_m$ is represented as follows by using the general stress increments $\{d\sigma\}$:

$$d\sigma_m = [1/3,\ 1/3,\ 1/3,\ 0,\ 0,\ 0] \cdot \{d\sigma\} \equiv \{B\}^T \cdot \{d\sigma\} \tag{26}$$

Therefore, the general strain increments due to consolidation $\{d\varepsilon^c\}$ can be given in the same manner as $\{d\varepsilon^s\}$ in Eq.21.

$$\{d\varepsilon^c\} = [T] \cdot \{d\varepsilon_i^c\} = [T] \cdot \{E_i^c\} \cdot \{B\}^T \cdot \{d\sigma\} \tag{27}$$

Elastic strain increments $\{d\varepsilon^e\}$

The elastic strain increments are given by the incremental stress-strain relationship for an isotropic elastic material.

$$\{d\varepsilon^e\} = [D_e]^{-1} \cdot \{d\sigma\} \tag{28}$$

In Eq.28, the tangential bulk modulus K_e is determined by the swelling index C_s, and Poisson's ratio ν_e is obtained from the measured unloading stress-strain relationship in a shear test.

By substituting Eqs.21, 27 and 28 into Eq.15, a stress-strain relationship in the general coordinate system can be derived. $\{d\varepsilon^s\}$ is considered to be 0 in the case of $dX \leq 0$, and $\{d\varepsilon^c\}$ is considered to be 0 in the case of $d\sigma_m \leq 0$. By the way, the coefficient K_c in Eq.22 is determined by using the K_0 value and the soil parameters on condition that this relationship satisfies the K_0-consolidation condition.

The validity of the constitutive equation is discussed here by analyzing various kinds of soil element tests. The soil parameters of the medium dense Toyoura sand (e_0=0.68) used in the finite element analyses are shown in Table I, which can be determined from constant mean principal stress tests and consolidation tests under triaxial compression. Fig.14 compares the computed stress-strain curves with the measured values by triaxial compression and triaxial extension tests under a constant mean principal stress. Fig.15 shows the comparison of computed values with measured values of triaxial compression and triaxial extension tests under constant principal stress ratios ($R\equiv\sigma_1/\sigma_3$) in terms of the volumetric strain ε_v-$\log_{10}\sigma_m$ relationship. It is seen from Fig.15 that the computed results explain well the measured dilatancy behaviors of soil under anisotropic consolidation. Figs.16 and 17 represent the computed results under plane strain condition with respect to the σ_1/σ_3-ε_1-ε_v relationship and the b-σ_1/σ_3 relationship respectively. It is clear from Fig.17 that the analytical results of b are different depending on the stress paths at low stress level, but converge to a certain value between 0.3 and 0.4 as the stress level increases. Such analytical results correspond with the experimental results [10].

Table I Soil parameters for Toyoura sand

λ^*		0.9
μ^*		0.27
μ'^*		0.41
γ_0^*	γ_{0i}^*	0.10 %
	C_d^*	0.066 %
	σ_{mi}	98 kN/m²
$C_c/(1+e_0)$		0.928 %
$C_s/(1+e_0)$		0.578 %
K_0		0.45
ν_e		0.3
ϕ (comp.)		40°

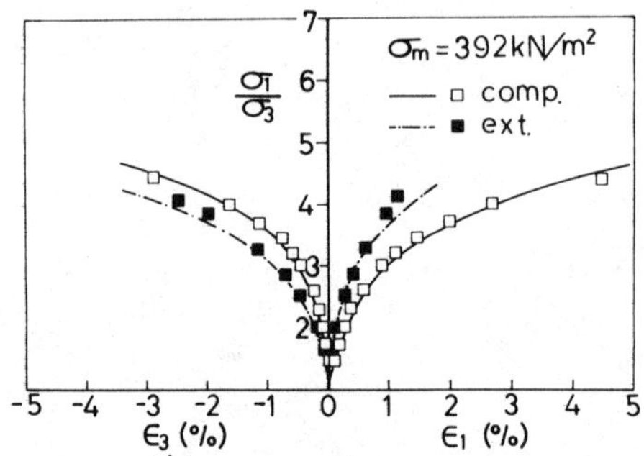

Fig. 14 Principal stress ratio vs. principal strains by shear tests

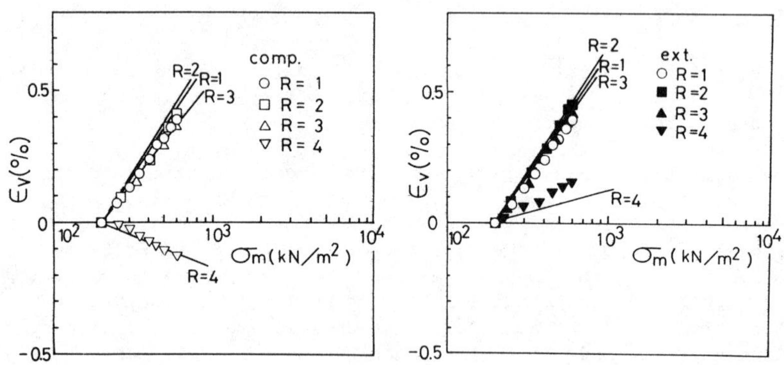

(a) (b)

Fig. 15 Volumetric strain vs. mean principal stress by consolidation tests

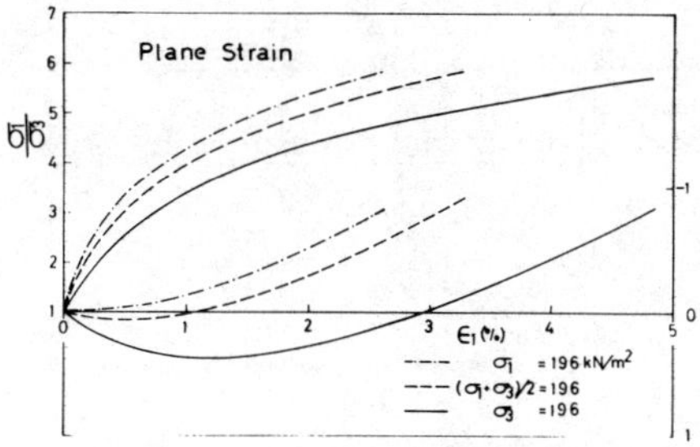

Fig. 16 Computed stress-strain curves under plane strain condition

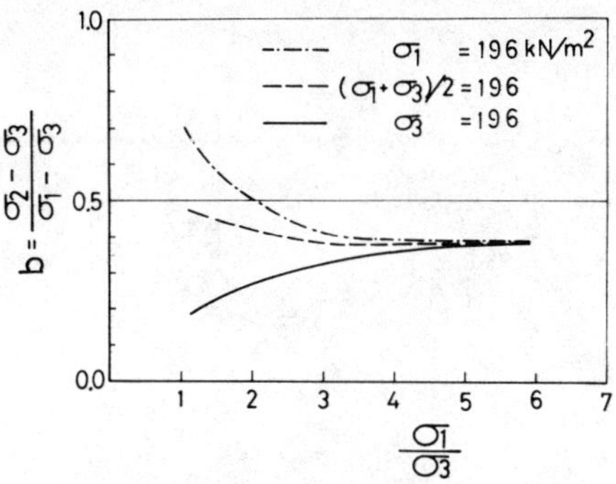

Fig. 17 Computed variations of b-value under plane strain condition

3. ANALYSIS OF STRESS-STRAIN BEHAVIORS UNDER CYCLIC LOADING [5,13]

Under cyclic triaxial compression ($\sigma_1>\sigma_2=\sigma_3$) and triaxial extension ($\sigma_1=\sigma_2>\sigma_3$) tests, the direction of τ_{SMP} is reversed in the same section of the SMP as shown in Fig.18, so the stress-strain relationships obtained from the two-dimensional analysis by Matsuoka, which are described in the another report [5], are considered to be also applicable to the SMP*. Fig.19 shows the results of triaxial compression and triaxial extension tests on the Toyoura sand described in the previous section. It can be seen from this figure that the shear strength is nearly constant ($\overline{\theta}_f$=const.) and γ^*_{SMP} from the start of shear up to failure is also nearly constant ($(\gamma^*_{SMP})_f$=const.) regardless of shear stress history, which are assumptions for this analysis as mentioned in the another report [5]. Fig.20 shows the τ_{SMP}/σ_{SMP} vs. $-d\epsilon^*_{SMP}/d\gamma^*_{SMP}$ relationship arranged from cyclic triaxial compression and extension test results on the Toyoura sand. In this test the principal stress ratio is applied up to 4.0 at the compression side for the 1st loading, and then it is applied up to the extension side for the 2nd loading. Finally it is applied up to failure at the compression side for the 3rd loading. It is seen from this figure that λ^* and μ^* are neary constant under cyclic loading (λ^*=0.9 and μ^*=0.27). As the principal stress ratio at failure R_f is 4.7, $\overline{\theta}_f$=0.60 is obtained from Eqs.5 and 54 in the another report [5]. Fig.21 shows the τ_{SMP}/σ_{SMP} vs. $-\epsilon^*_{SMP}/\gamma^*_{SMP}$ relationship arranged from the same test results as in Fig.19. From this figure μ'^*_{1st} is seen to be 0.41, and so κ=0.23 is calculated from Eq.27 in the another report for $\overline{\theta}_f$=0.60. The values of μ'^* for the 2nd and 3rd loading are thus calculated from Eqs.23, 26 and 54 in the another report, and the upper two solid lines in Fig.21 can be drawn for λ^*=0.9. Fig.22 represents the comparison between the cyclic triaxial compression and extension test results (the maximum principal ratio R_{max}=4.0) on the Toyoura sand [14], and the calculated values. In Fig.22 the measured stress-strain curve under the 3rd loading is more gentle than that under

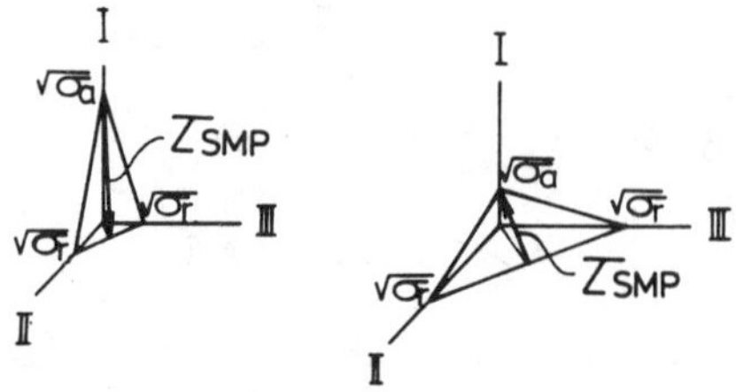

Fig. 18 Direction of τ_{SMP} under triaxial compression and triaxial extension conditions

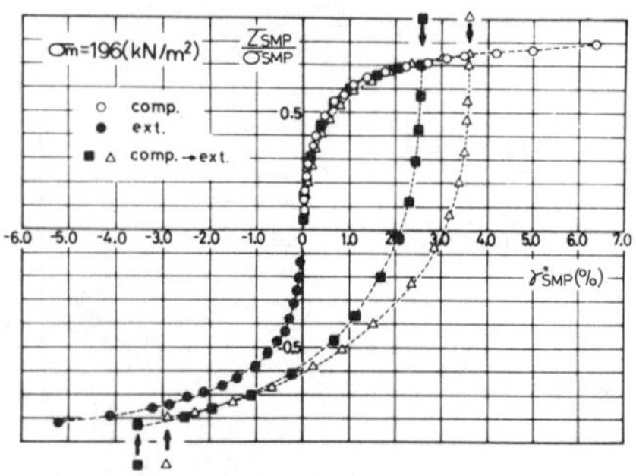

Fig. 19 τ_{SMP}/σ_{SMP} vs. γ^*_{SMP} relationships obtained from cyclic triaxial compression and extension tests on sand

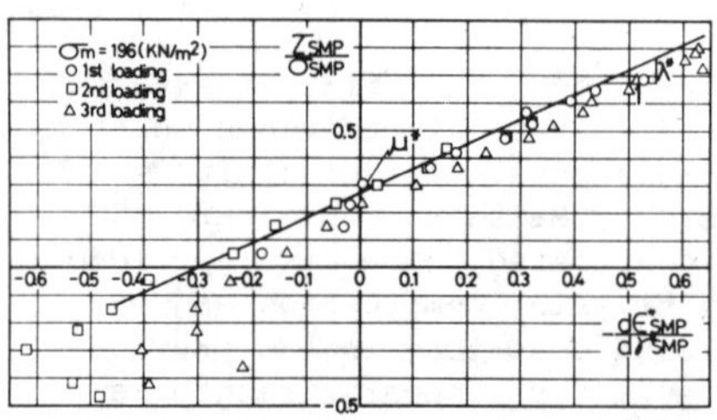

Fig. 20 τ_{SMP}/σ_{SMP} vs. $-d\varepsilon^*_{SMP}/d\gamma^*_{SMP}$ relationships obtained from cyclic triaxial compression and extension test on sand

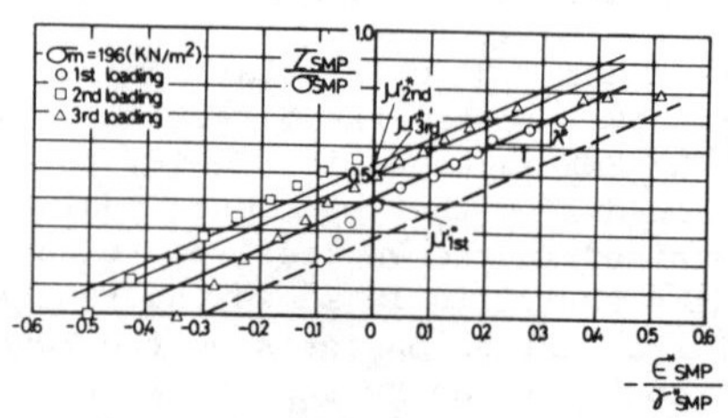

Fig. 21 τ_{SMP}/σ_{SMP} vs. $-\varepsilon^*_{SMP}/\gamma^*_{SMP}$ relationships obtained from cyclic triaxial compression and extension test on sand

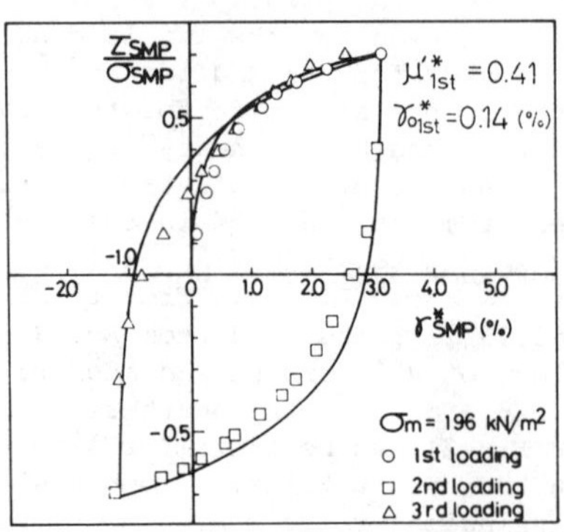

Fig. 22 Comparison between cyclic triaxial compression and extension test results and calculated values

the 1st loading, and the calculated curve also shows such tendency of the measured values.

Combining Eqs.23 and 28 in the another report to estimate the parameters μ' and γ_0 with the constitutive equation for soil based on the SMP*[11], the analyses of the undrained cyclic shear behaviors under triaxial compression and extension stress conditions are performed [15]. The undrained condition corresponds to the condition that the volumetric strain increment is always zero for saturated samples. The stress-strain parameters used are those for the relatively dense Toyoura sand; $\lambda^*=0.9$, $\mu^*=0.27$, $\mu'^*_{1st}=0.41$, $\gamma^*_{01st}=0.10\%$, $C_c/(1+e_0)=0.928\%$, $C_s/(1+e_0)= 0.578\%$, the principal stress ratio at failure $R_f=4.7$ (the

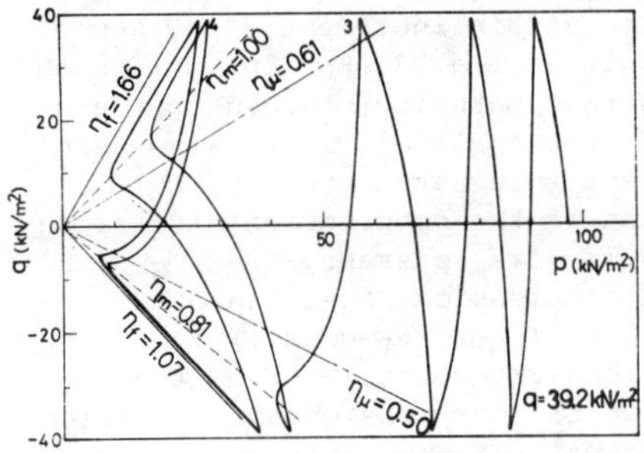

Fig. 23 Calculated effective stress path in undrained cyclic triaxial compression and extension test ($q=39.2kN/m^2$)

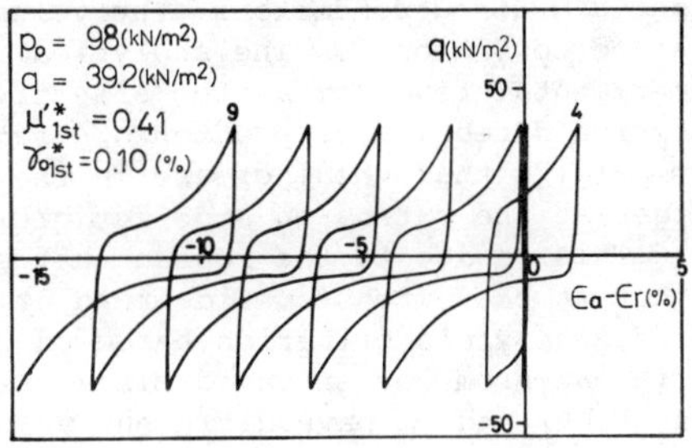

Fig. 24 Calculated stress-strain relationship in undrained cyclic triaxial compression and extension test ($q=39.2kN/m^2$)

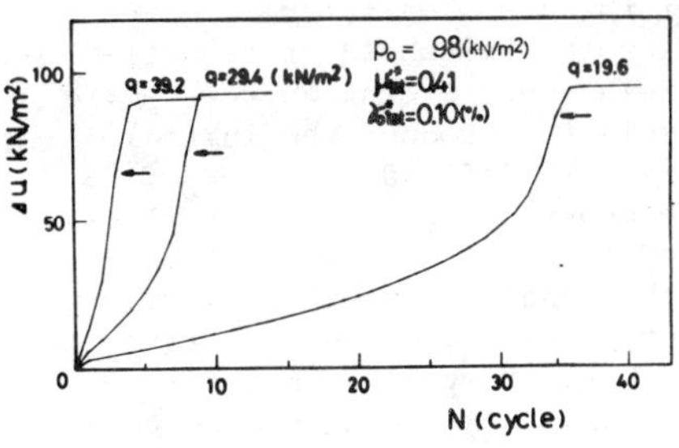

Fig. 25 Calculated cumulative excess pore water pressure vs. number of cycles relationship

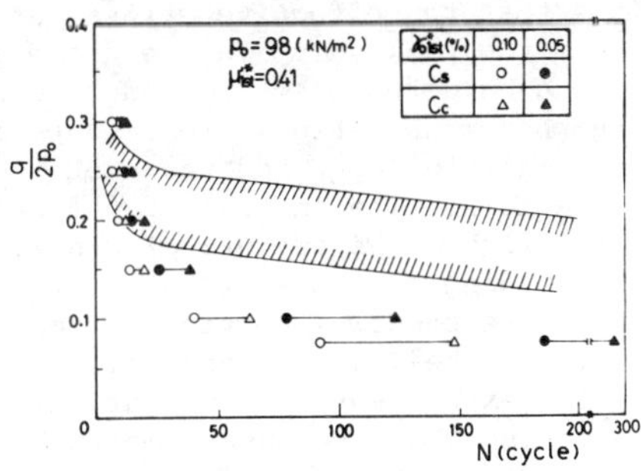

Fig. 26 Relationship between cyclic stress ratio and number of cycles to produce initial liquefaction

internal friction angle $\phi_d=40^\circ$). C_c is the compression index, C_s the swelling index and e_0 the initial void ratio. C_c is adopted when the effective mean principal stress p $(=(\sigma_1+\sigma_2+\sigma_3)/3)$ is increased and C_s is adopted when p_o is decreased. The initial effective mean principal stress p_o is 98 kN/m .

Fig.23 shows the analytical result of the effective stress path for the maximum principal stress difference $q=39.2$ kN/m^2 in undrained cyclic triaxial compression and extension test $(q=\sigma_a-\sigma_r$, σ_a:axial principal stress, σ_r:radial principal stress). In this figure η_f, η_m and η_μ represent the values of q/p at failure, at the "phase transformation" [16] and at $\tau_{SMP}/\sigma_{SMP}=\mu^*$ corresponding to the coefficient of interparticle friction respectively. The value of η_m can be determined from the condition of the effective mean principal stress increment dp=0 under undrained conditions (the volumetric strain increment $d\varepsilon_v=0$ under drained conditions). The analytical result in Fig.23 explains well such experimental tendency as the effective stress path is returned to the positive direction of p after q/p exceeds the η_m-line. It is also seen from Fig.23 that the increase in the excess pore water pressure is larger at the extension side (at $q<0$) than at the compression side (at $q>0$). This is due to the reason that even if q is constant, τ_{SMP}/σ_{SMP} is higher at the extension side than at the compression side. In other words, the yield criterion based on the SMP (τ_{SMP}/σ_{SMP}=const.) is different from the extended Mises criterion (q/p=const.). From the same reason, η_f and η_m have different values at the compression and extension sides. Fig.24 shows the calculated principal stress difference q vs. principal strain difference $(\varepsilon_a-\varepsilon_r)$ relationship (ε_a:axial principal strain, ε_r:radial principal strain). After the effective stress path exceeds the η_m-line, the principal strain difference increases suddenly to its negative direction. Fig.25 shows the calculated cumulative excess pore water pressure Δu vs. number of cycles N relationship. The values of q in this figure mean the maximum principal stress differences under cyclic loading. The arrow marks in Fig.25 correspond to the points that the effective stress paths exceed the η_μ-line. As Δu reaches the peak value in one cycle after the arrow mark, the η_μ-line may be considered to be an index to predict the threshold of liquefaction phenomena. Fig.26 represents the relationship between the cyclic stress ratio $q/(2p_o)$ and the number of cycles to produce the initial liquefaction. The initial liquefaction is here defined as $|\varepsilon_a-\varepsilon_r|\geqq 15\%$. In this figure the analytical results are shown by four kinds of plots corresponding to the four cases of $\gamma^*_{01st}=0.10\%$ and $\gamma^*_{01st}=0.05\%$ using C_s or C_c. The shadowed portion represents the range of experimental results on the Toyoura sand [17]. The analytical values do not agree with the experimental values at a large number of cycles. However, considering that the wave numbers by real earthquakes are no more than 20, Fig.26 may be considered to be good enough for the practical use. It is interesting to have analyzed the problem of liquefaction by the parameters obtained from drained monotonous triaxial tests without performing undrained vibrational triaxial tests.

REFERENCES

1. Nakai, T. and Matsuoka, H., "A Unified Law for Soil Shear Behavior under Three Dimensional Stress Condition," Proc., Japan Society of Civil Engineer, No.303, pp.65-77 (in Japanese) (1980).

2. Nakai, T. and Matsuoka, H., "Deformation of Soils in Three-Dimensional Stresses" Proc., IUTAM-Symposium on Deformation and Failure of Granular Materials, Delft, pp.275-285 (1982).

3. Matsuoka, H. and Nakai, T., "Stress-Deformation and Strength Characteristics of Soil under Three Different Principal Stresses, "Proc., Japan Society of Civil Engineer, No.232, pp.59-70 (1974).

4. Matsuoka, H. and Nakai, T., "Stress-Strain Relationship of Soil Based on the SMP," Proc., Specialty Session 9, 9th Int. Conf. on Soil Mech. and Found. Eng., pp.153-162 (1977).

5. Matsuoka, H., "Deformation and Strength of Granular Materials Based on the Theory of "Compounded Mobilized Plane" (CMP) and "Spatial Mobilized Plane" (SMP)," Advances in the Mechanics and the Flow of Granular Materials, (1982).

6. Matsuoka, H., "A Microscopic Study on Shear Mechanism of Granular Materials," Soils and Foundations, Vol.14, No.1, pp.29-43 (1974a).

7. Matsuoka, H., "Stress-Strain Relationship of Sands Based on the Mobilized Plane, "Soils and Foundations, Vol.14, No.2, pp.47-61 (1974b).

8. Roscoe, K. H., Schofield, A. N. and Thurairairajah, A., "Yielding of Clays in State Wetter than Critical," Geotechnique, Vol.13, No.3, pp.211-240 (1963).

9. Schofield, A. N. and Wroth, C. P., Critical State Soil Mechanics, Mcgraw-Hill, London (1968).

10. Lade, P. V. and Duncan, J. M., "Cubical Triaxial Tests on Cohesionless Soil," Proc., ASCE, Vol.99, No.SM10, pp.793-812 (1973).

11. Nakai, T. and Matsuoka, H., "A Unified Law for Soil Deformation Behavior under Various Stress Paths," Proc., Japan Society of Civil Engineer, No.306, pp.23-34 (in Japanese) (1981).

12. Nakai, T., "Analyses of Soil-Footing and Soil-Wall Interaction" Proc., 10th Int. Conf. on Soil Mech. and Found. Eng., Vol.2, pp.205-208 (1981).

13. Matsuoka, H., "A Stress-Strain Model for Granular Materials Considering Mechanism of Fabric Change", Proc., U.S.-Japan Seminar on New Model and Constitutive Relations in the Mechanics of Granular Materials, Ithaca, (1982).

14. Kanaya, K. and Suzuki, M., "Deformation Characteristics of a Sand under Axi-Symmetric Stress Condition", Graduation Thesis at Nagoya Institute of Technology (in Japanese) (1979).

15. Matsuoka, H., Geka, H. and Yasui, O., "Cyclic Shear Behavior of Granular Materials Considering Mechanism of Fabric Change", Proc., 17th Japan National Conf. SMFE, pp.1673-1676 (in Japanese) (1982).

16. Ishihara, K., Tatsuoka. F. and Yasuda, S., "Undrained Deformation and Liquefaction of Sand under Cyclic Stresses", Soils and Foundations, Vol.15, No.1, pp.29-44 (1975).

17. Ishihara, K., "Fundamentals of Soil Dynamics", Kajima-shuppankai (in Japanese) (1976).

ON STATIC AND DYNAMIC RESPONSE OF GEOMATERIALS

V. N. Nikolaevski

Institute of Physics of the Earth,
USSR Academy of Sciences
Moscow, USSR

SUMMARY

Discussed are the simple elasto-plastic dilatancy model of limit non-associated flows of geomaterials, some results of foundation settlements and spherical - symmetrical explosion motion, strain and permeability distribution, generated by explosion in high porous media, dynamic compressibility and the limit slip velocity, the estimation of corresponding delay time and pseudoviscosity, the low frequency repacking wave, created by ultrasonic pulse in sands, and acoustic emission due to solid friction slip.

1. INTRODUCTION

The experimental discovery of dilatancy of water-saturated sands as well as its own name belongs to O. Reynolds (1885). The dilatancy effect means that shear of a granular material leads to its essential volume changes. While nonelastic volume strains of metals become evident only after many stress cycles or under extremely high pressures, dilatancy of geomaterials is their characteristic feature and appears at very first loading. The pure kinematical nature of dilatancy can be illustrated by distortion of set of particles in contact, organizing a hinge system. The hinge distortion leads to essential change of void space. Such a kinematical interpretation is not so clear for crack voids in rocks. In the latter case the relation between volume, pressure and shear increments seems to be more natural. However, the obtained dilatant proportionality of shear and volume increments and following a nonassociated flow rule have shown extremely good agreement with granular material response as well as with brittle rock one. Such a coincidence has been observed if solid friction resistance at slip surfaces has the same order as the strength of intact material. It is a typical situation for granular masses. The depth interval for the Earth crust rocks with corresponding pressures and temperatures (Nikolaevsky, 1979) is seismogenic, that is, dilatancy effects play governing role for earthquake sources.

For model considered below the triaxial experiment data for sands were used. Therefore the model has to be in accordance with quasistatic problems. As usually, it is possible to use the resulting rheological laws for dynamic problems. However, I shall discuss also some dynamic effects which are out of the traditional elasto-plastic approach. They are weak wave propagation in plastic domains with elastic velocities, dynamic compressibility, existance of own internal frequencies of granulated media, etc.

2. ELASTO*PLASTIC DILATANCY MODEL

The basic model of dilatant plasticity is formulated (Nikolaevsky, 1971; Nikolaevsky, Syrnikov, Shefter, 1975) as a simple generalization of the Prandtl-Reuss plasticity. That is, the kinematical condition of plastic incompressibility $de^p_{ij}\ \delta_{ij} = 0$ is changed for the dilatancy condition

$$\Phi_e(de^p_{ij}) \equiv de^p_{ij}\ \delta_{ij} - \Lambda(\sigma_{ij},\chi)\,|d\gamma^p| = 0$$

$$d\gamma^p = (1/\sqrt{2})\{(de^p_{11} - de^p_{22})^2 + 6\ de^p_{12}\ de^p_{12} + \ldots\}^{\frac{1}{2}} \tag{I.1}$$

Here Λ is dilatancy rate, σ_{ij} - stresses, χ - hardening parameter de^p_{ij} - plastic strain increments, $d\gamma^p$ - plastic shear increment.

For geomaterials the Coulomb-Mohr limit condition is natural

$$\Phi_\sigma(\sigma_{ij}) \equiv \tau - \alpha(\chi)p - Y = 0$$

$$\tau = (1/2\sqrt{2})\{(\sigma_{11} - \sigma_{22})^2 + 6\sigma_{12}\ \sigma_{12} + \ldots\}^{\frac{1}{2}} \tag{I.2}$$

Here τ is shear stress, p - pressure, α - internal friction coefficient, Y - cohesion.

Two conditions (I.1) and (I.2) correspond to additional unknown functions ζ, λ in the nonassociated flow rule

$$de^p_{ij} = Q^p_{ijke}\ \sigma_{ke} = (\sigma_{ij} + p\delta_{ij})d\lambda - (H + p)\delta_{ij}\ d\zeta \tag{I.3}$$

The form is valid for linear isotropic case. Here Y = αH.

For elastic part of strains the traditional elastic law is valid

$$e^e_{ij} = e_{ij} - e^p_{ij} = Q^e_{ijke}\ \sigma_{ke} \tag{I.4}$$

and Q_{ijke} is elastic modulus tensor.

As the hardening parameter the ratio $\chi = e^p/e^p_*$ can be utilized. Here e^p_* is a critical value of $e^p = e^p_{ij}\ \delta_{ij}$, for which $\Lambda\ (e^p_*) = 0$.

Approximately it means $\chi = m/m_*$ where m_* is the critical porosity, a function of pressure.

The functions $\alpha(\Lambda)$ and $\alpha(\chi)$ for river sands are given in Figs. 1 and 2. The solid line in Fig. 1 corresponds to the play of solid friction forces at a microlevel (Nikolaevsky, Syrnikov, 1970), according to which the following form for the internal friction angle is valid

$$\phi = 2\phi_o + \nu \tag{1.5}$$

Here ν is the Hansen dilatancy angle and ϕ_o is the angle of true solid friction at grains contacts.

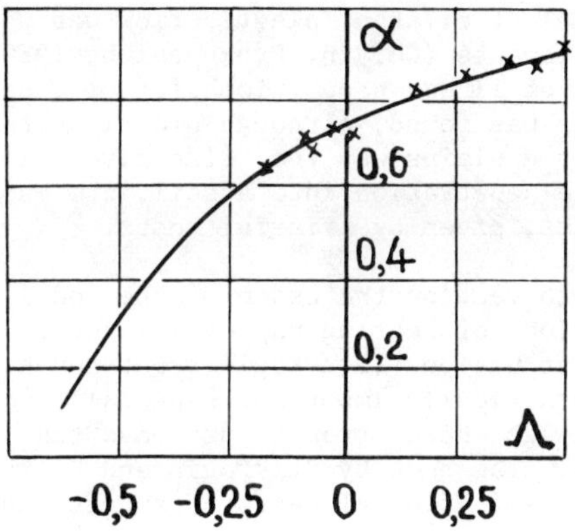

Fig. 1. Internal Friction coefficient α as a function of dilatancy rate Λ for river sands.

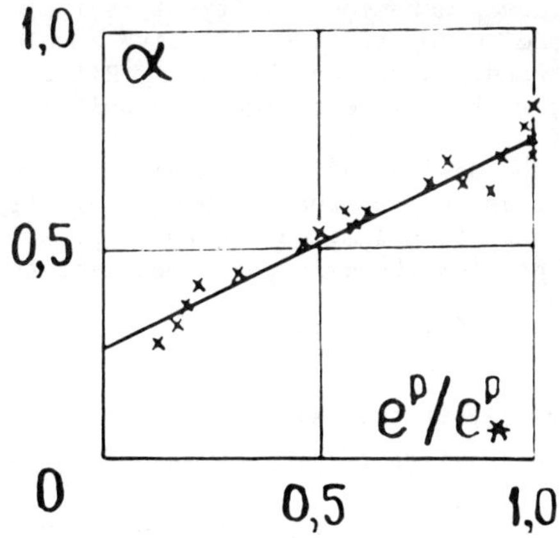

Fig. 2. Internal friction coefficient α as a function of hardening parameter χ for compaction of loose rive sands

The suggestion that $\chi = \gamma^p$ is also known, but this parameter is not so definite as a porosity. The parameter χ for sandstones is more complicated to analyze.

One can see that the model of plastic flow, associated with the Coulomb-Mohr condition is a particular case ($\alpha = \Lambda$) which gives too high dilatancy rate and leads to dissipation of mechanical work proportional only to a cohesion Y (and independent on internal friction). Nowdays the associated flow rule is used for "cap"-form of yield surfaces, which were suggested instead of the Coulomb-Mohr condition. Maybe it has a physical meaning for sublimit plasticity, connected with pore deformations. Really, for soft soils sublimit plastic strains have the same order as elastic ones, stress being lower than the limit yield. The transition to limit plasticity is smooth as usually. The boundary is determined by resolving of nonequality $de^p_{ij} >> de^e_{ij}$ on the base of experimental data.

The given model of dilatant plastic flow was used for calculation of foundation settlements (Goldin, Prokopovich, 1980; Bugrov, 1980). The important fact of linear proportionality of a settlement to moderate high pressures was found, although plastic zones appeared under the foundation. It explains why the "effective" linear modulus, measured by the die penetration into a soil, are essentially different than elastic modulus, given by triaxial tests.

It is worthy to mention the usage of the model (with condition - const) for estimations of bearing capacities and displacements of wall of underground constructions (Stavrogin, Protosenya, 1979). Such calculations are practically important especially for postfailure regime of ambient rock masses where strength decrease take place due to dilatant loosening. The book by Stavrogin and Protosenya (1979) contains also detail of triaxial tests of sandstones and other rocks.

Sometimes one can hear that usage of the associated flow rule is quite admissible because difference of calculations of settlements has the order of 15%, see, for example, Bugrov (1980). However, this value is essential, and moreover nonassociativity of real plastic flows of granular materials lead to important qualitative conclusions which are connected with shear band generation (Rudnicki, Rice, 1975; Nikolaevsky, Rice, 1979), that is, with failure of rock and soil masses.

The question of the Coulomb slip surfaces is important for limit design of slopes. Let us suppose that mathematical slip surface in reality is a narrow band of nonlimited growth of plastic shear strain. In the case of plane problem the dilatancy condition (I.1) has the form

$$\partial v_2/\partial \chi_2 = \sin \nu \sqrt{(\partial v_2/\partial \chi_2)^2 + (\partial v_1/\partial \chi_2)^2} \qquad (I.6)$$

because $\partial v_i/\partial \chi_2 >> \partial v_i \; \partial \chi_1$ if the axis χ_1 is along the band, χ_2 - orthogonal to it. The equality (I.5) also means $dv_2/dv_1 = \mathrm{tag}\ \nu$.

The integrand of the equation (I.5) is the following one

$$[v_2] = [v_1] \text{ tg } \nu \tag{3}$$

if ν = const. That is, generally speaking, the jump of tangential displacement velocities is interconnected with the jump of normal ones. (Such conclusion was originally given by Shield (1953) but for the associated rule model.) So, the slip surface can be modeled by the band of finite thickness. The geomaterial structure inside the band is developing to critical porosity state, according to figures 1, 2 ($\nu \to 0$). The decrease of friction resistance of the band follows mainly in the finite zone (Palmer, Rice, 1973) near the tip of spreading shear band. In particular boundary problems generation of shear bands is connected with nonuniqness of solution and in case of rigid-plastic simplification can be found by intersection of one-family characteristics.

The local orientation of shear bands was investigated by Rice and Rudnicki (1975) as a strain bifurcation in the inclined band with arbitrary outside elasto-plastic fields. They utilized the deformation model which is equivalent to the constitutive laws (I.1) - (I.4) for proportional loadings. This question as well as difficulties of limit transition to rigid-plastic regime were the theme of additional discussion.

The space interval between shear bands is governed by boundary conditions because the elasto-plastic model does not have its own length scale. The shear in sands has shown proportionality of the space interval to external length scale (of experimental box). The same interval for the crack set at the surface of granular material layer is proportional to the layer thickness if the plane shear of the layer bottom is given (Merzer, Freund, 1976). We have to note that the internal physical length scale (the grain or block diameter) of geomaterial can also influence but this renders a much more complicated mathematical model with internal particle distortion.

3. DILATANCY UNDER EXPLOSION MOTIONS

The complete dilatant plastic model under consideration was developed for underground explosion calculations (Nikolaevsky, 1967; Rodionov et al., 1971). In particular, it was shown that the integrand of the velocity field

$$v(r_1 t) = C(t)/r^n \tag{3.1}$$

for motion with spherical symmetry was a consequence of the dilatancy condition (I.1) if Λ = const. The value $n < 2$ corresponds to geomaterial loosening behind the diverging explosion wave front due to generation of cracks and pores.

The second kinematical integrand (for density distribution) was also found (Dunin, Sirotkin, 1977) and it gives the possibility to make the explosion calculation method originated by Chadwick, Cox and Hopkins

(1964), more realistic. By computer calculations it was found (Bovt, Nikolaevsky, 1981) that a characteristic break of curve I of maximum displacement velocity and of curve 2 of maximum pressure in explosion wave (Fig. 3) coincides with external boundary r_τ of plastic dilatant

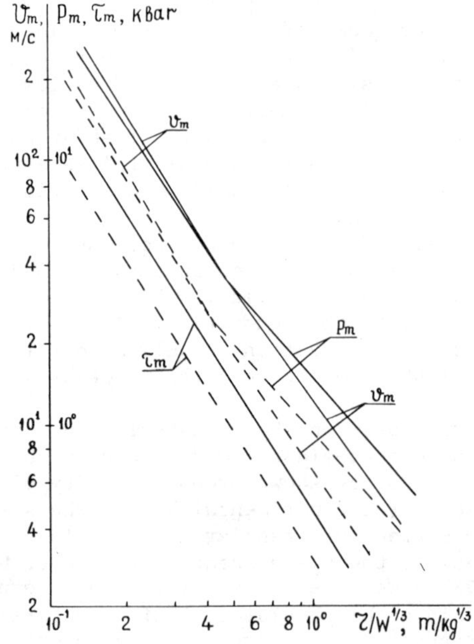

Fig. 3. Dependance of maximum parameters of explosion wave on distance.

zone. Dilatancy of rocks, corresponding to crack void generation ($\Lambda > 0$) can be also estimated because of its essential permeability $k >> k_o$. Really, the break of the curve σ_i^m and external radius of permeable zone after explosions in granites coincide: $\bar{r}_\tau \approx \bar{r}_k \approx I\ m/kg^{1/3}$, r - radial distance from the explosion center, W - explosion yield (Bovt, Nikolaevsky, 1981), $\bar{r} = r/w^{1/3}$.

The wave profiles for a time sequency (Figure 4) show two local maxima. The first is interpreted as a wave front and the second one as a maximum of residual stresses around explosion cavity. Here p_o is a lithostatic pressure. The broken lines correspond to the case $\Lambda = 0$. If $\Lambda > 0$, near the cavity the specific volume V is much more than the initial V_o. (Explosions in intact rocks give dilatant loosening.) Correspondingly the material volume, which is dynamically pressed into elastic external zone, also is increasing. The latter gives the growth of effective seismic source for real contained underground explosion (Rodionov et al., 1971).

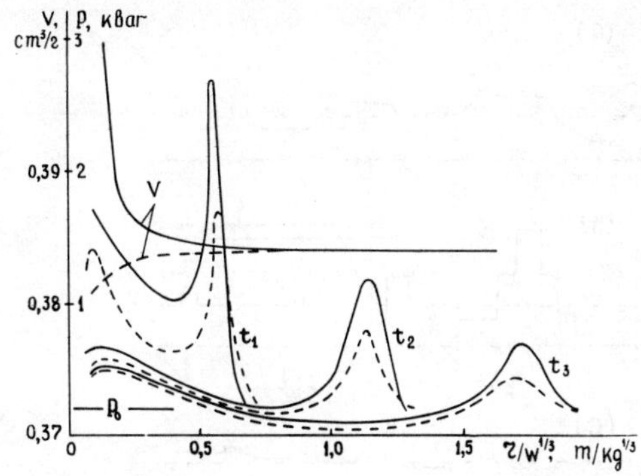

Fig. 4. Explosion wave pressures for three moments of time.

On the contrary the explosion action in high porous geomaterials is mainly compaction up to the critical level after which the reverse dilatant loosening begins. It means the change of sign Λ. In the laboratory model experiments (Bovt, Nikolaevsky, 1981) the spherical ring zone of compaction was found.

In the figure 5 the postexplosion density ρ, the sound velocity in radial c_{11} and in hoop c_1 directions are given for cemented sands with initial porosity m = 25%. The first two curves (5a, 5b) are in a good correlation but curve 5c corresponds better to relative change of permeability (5d). The explosion wave creates the evident sound velocity anisotropy, but the permeability, which was estimated by the air blow-through tube model wells, appears to be less than initial one (k_o = 150 mdarcy) everywhere. The reason is that the dilatancy loosening effect in the zone $0.16 \leq \bar{r} \leq 0.32$ m/kg$^{1/3}$) is surpressed by additional fragmentation of sand grains. Fractures, which give high postexplosion permeability in rocks are not essential for high porous geomaterials. For these two competitive processes - fracturization and compaction the initial porosity $m_o^{\sim}$ ~ 15% plays the bound role (it was shown by computer calculations).

The external boundary of dilatant plastic flow which is estimated by the break of maximum stress $\sigma_r^m(\bar{r})$ experimental curve has the coordinate $\bar{r}_\tau \approx 0.48$ (m/kg$^{1/3}$) and it is a good agreement with external radius of real dilatant compaction.

However, explosion in natural sands of the same initial porosity gives the break of the curve σ_r^m at $\bar{r}_\tau = 3$ (m/kg$^{1/3}$). This discrepancy can be resolved, if one introduces the nondimensional radius $R = r(\rho c^2/w^{1/3})$ which accounts the difference of geomaterial acoustical rigidity (for the

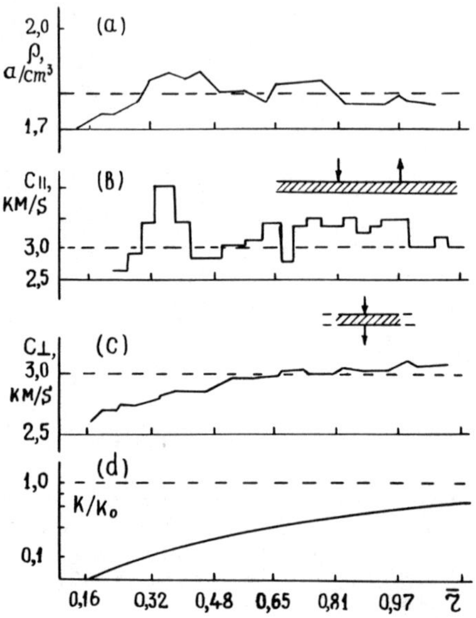

Fig. 5. Density, sound velocities and permeability near explosion cavity in a dry high porous medium.

sand ρc^2 = 7.86 MPa). Indeed, the coordinates have a coinciding value: R = 12.4 ($mMPa^{1/3}/kg^{1/3}$) in both cases.

The outer boundary of permeability decrease has the coordinate $\bar{r}_k$ = 2 or R = 50 for cemented sand. On the otherhand the explosions in soft soils (see, Bovt, Nikolaevsky, 1981) gave sublimit plastic strains up to the radius $\bar{r}$ = 6 or to R = 50 because ρc^2 = 5.94 MPa. So, the permeability changes of high porous geomaterials are connected with a small residual deformation of a porous space.

The explosion in a saturated porous media gives nonmonotonous changes of permeability: decrease at $0.45 \leq \bar{r} \leq 1.0$ ($m/kg^{1/3}$) and increase at $\bar{r} \leq 0.45$ (Bovt, Nikolaevsky, 1981).

Unfortunately, in considered above experiments there were no simultaneous registrations of stresses and displacement velocities inside explosion waves. The problem is the following one. According to traditional concepts in porous materials the maximum pressure of the wave has to coincide with maximum compaction. However, one can see the delay of volume compression strains. It is important to know the nature of this effect. Theoretically the delay may be connected with a dilatancy developing with shear behind diverging wave front or with soil viscosity.

Moreover, the elasto-plastic dilatant model is nonholonomic one, that is, the compressibility law $\rho = \rho(p)$ depends on the loading path. Indeed, the systematic essential difference of compressibility of powders in plane and spherical shock waves are known (see, Bovt, Nikolaevsky, 1981). In the first case (sublimit plastic strains) compaction is smaller than in the second one (dilatant strains), but both are smaller than in the corresponding static ones. The difference of dynamic and static dilatant curves $e^p\ (p, \gamma^p)$ at standard triaxial tests of rocks has the order of 10 - 20%. The corresponding deviation of yield surfaces (of a limestone with m = 5%) are not essential. For granites (with porosity 1%) dynamical and statical curves $\sigma_1 = \sigma_1(e_1)$ coincide but for porous geomaterials (with porosity ~5%) they are essentially different: dynamical compaction is 70% of a static one.

4. LIMIT SLIP VELOCITY AND DYNAMIC COMPRESSIBILITY

Some measurements show that hoop stresses of explosion waves in porous dry materials are greater than static yield limit permits. Besides at the explosion in a sand with vacuum pores (or in confined sand) there is a constant velocity of nonelastic strain front (or plastic wave) which follows the elastic precursor. It is the effect of limit slip velocity, the existance of which follows the Broberg (1978) solution of dynamic relative slip problem for two contact blocks under action of solid friction. The confinement of sand (or its vacuumization) leads to more uniform distribution of solid friction microforces than in naturally loose sand.

At spherical - symmetry of explosion motion in confined sand the velocity of elastic precursor has the value c = 810 m/s while the maximum of compression wave has practically constant velocity D = 451.6 m/s (the confining pressure is 2.0 MPa). The compression modulus, corresponding to mentioned value c (which is measured in static tests under the 4th loading cycle up to 2.0 MPa) was equal to K = 5.1 kbar, because the shear modulus G = 3.86 kbar. Therefore the Poisson coefficient $\mu = 0.198$ and the shear wave velocity $c_t = c(I - 2\mu)/(2 - 2\mu) = 0.614$ $c = 497$ m/s. The Rayleigh wave velocity has a value $c_R = 0.91\ c_t$ if $\mu = 0.198$. It means

$$c_R = 452.5 \text{ m/s} = D \tag{4.1}$$

So, one can think that maximum of explosion wave coincides with slip front (inside radial shear bands or cracks) in confined sands.

At analogous explosion in sands with vacuum pores there is also essential distance interval where explosion wave maximum has constant velocity D = 200 m/s (the sound velocity was twice more: D = 0.5 c). The absence of data for c_t does not permit to check the equality (4.1) in this case.

In sands with density 1.8 g/cm^3 under vacuum conditions plane shock waves have two fronts - of precursor and of "steady" structure (Stoll Ebeido, 1965). The elastic precursor velocity c is equal to 260, 350 or 400 m/s in dependence of vacuum levels. The velocity D of the second front (of nonelastic deformations), corresponding to c=400 m/s and to minimum amplitude of 1.4 bar is equal to 230 m/s - 0.576 c. Again the absence of data for shear wave velocities prohibited to argue that D is equal to physically limited value.

Let us look at another aspect of dynamical deformation. The experiments by Stoll and Ebeido (1965) have shown that the velocity D grows linearly with the amplitude σ_* of the impact:

$$D = 0.6c(0.6 + 0.285\ \sigma_*) \tag{4.2}$$

where $1.4 \leq \sigma_* \leq 6$ (bars). For estimation of D-value I can utilize also balance equations at the discontinuity, that is,

$$\rho(u - D) = -\rho_o D; \quad \rho_o Du = \sigma_* - \sigma_*^0 \tag{4.3}$$

if the function $e = (\rho - \rho_o)/\rho_o = f(\sigma_*)$ for uniaxila compression is given. Such curves are known from static tests. They are plotted in figure 6 as broken lines for two levels of vacuumization. However they determine lower values than (4.2). One can make the reverse calculations, that is, find effective dynamic compressibility $e_{dyn}(\sigma_*)$. Two corresponding solid curves are also given in the figure 6. So, the dynamic compressibility of vacuum sand appears to be less than static one. The analogous effect was mentioned also in experiments with sands under natural conditions.

The explanation of dynamic compressibility is necessary to consider the dynamics of an individual pore. The analysis of such a problem corresponds to sublimit plasticity and it was made earlier under assumption on quasistatic strains around the pore. It is equivalent to phenomenological description of the compressibility law $e = e(p, \partial e)$, where values ∂e are different for loading and unloading paths.

If one introduces inertial forces corresponding to spherical-symmetrical convergence motion of material with velocity v_r to the center of collapsing pore, then dynamic compressibility appears to be more (Carroll, Holt, 1973) than static one. Moreover, the introduction of additional viscous damping does not change this wrong result. Really, experimental dynamic compressibility of pore materials (and even of gas suspension in a liquid) is less than a static one. In this connection we have suggested to add a mean "turbulent"

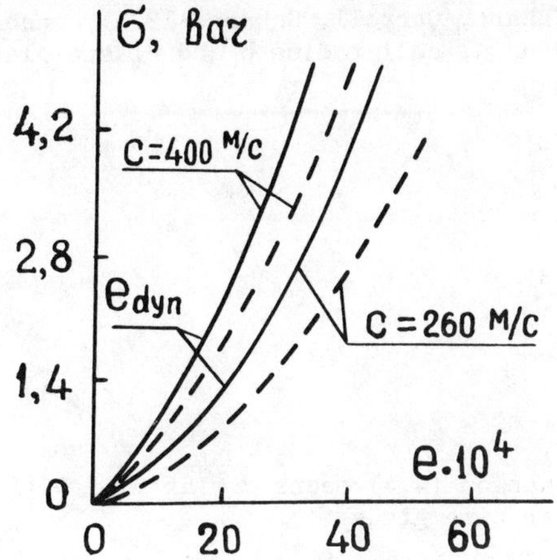

Fig. 6. Dynamic and static compressibility of sand under different level of vacuum in pore space.

pressure $P_t = \langle -\rho v_r v_r \rangle$ into the macrobalance impulse equation. It will enable us to introduce a correction factor of a proper sign

Let us suppose that the effect of limit slip velocity is acting during pore collapse. Real granular material with internal solid friction and voids can be simulated (see Figure 7) by the set of cells with solid walls from a plastically incompressible medium, governed by the Mohr-Coulomb yield condition.

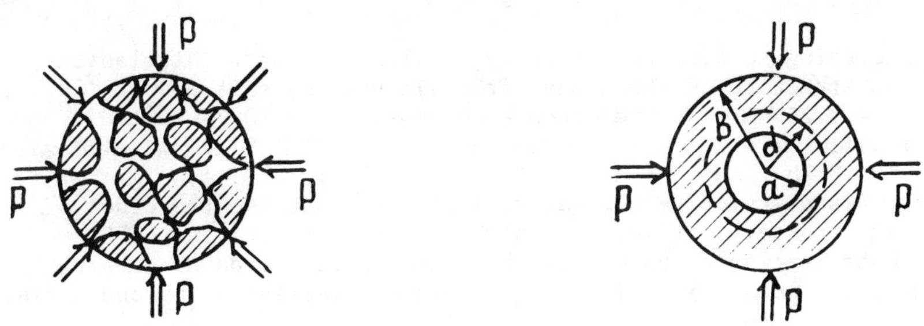

Fig. 7. Real and mathematically simulated picture of a pore collapse.

Quasistatic solution (Bhatt, Carroll, Schatz, 1975), enables us to find the connection of pore radius a, cell radius b and elasto-plastic boundary radius d with pressure p:

$$p = p_c(d) + (4/3)(d^{-3} - b^{-3}),$$

$$p_c(d) = (Y/M)\{(d/a)^{2M} - I\} \qquad (4.4)$$

$$\frac{d}{a} = \{\frac{2G}{Y}\frac{\beta_o - \beta}{\beta - 1}\}^{1/(2M+3)}$$

$$M = \frac{2}{7}(2 - \alpha)$$

$$B = Ya^3/(2b), \quad \beta = I/(I-m) .$$

The latter of equation (4.4) means d = af(m). Differentiation of two equations (4.4) over time gives

$$\dot{p} = 2Y\,(d/a)^{2M - I}\,(\dot{d}/a) - (\dot{d}a/a^2) = -\,(4GB/d^4)\dot{d} \qquad (4.5)$$

If one suggests that $\dot{d} = v_L$, where v_L - the limit slip velocity, then the compressibility law follows:

$$\dot{p} = \dot{F}(m) + (K(m)/\theta), \quad \theta = L(p)/v_L$$

$$L(p) = d^4/b^3, \quad K(m) = (4/3)GB(a/b)^3$$

$$F(m) = 2Y \int f^{2M - I}\,\dot{f}(m)\,dm \qquad (4.6)$$

In other words, the approximate account of limit slip velocity leads to the Sokolovsky - Malvern model of volume strain of granular materials. The estimation of relaxation time θ may be done as follows:

$$v_L = 10^3 m/s, \quad L = 10^{-3} m, \quad \theta = 10^{-6} s$$

According to B.D. Christoforov et al. (see, Bovt, Nikolaevsky, 1981) the structure of shock wave front in pressed salt powder (ρ_o = 1.87 g/cm^3, c = 3.25 km/s) corresponds to viscosity $\eta = 10^4 - 10^5$ poise and the more the pressure, the less is viscosity. This material has value $\rho_o c^2$ = 197.5 kbar and $\theta = \eta/(\rho_o c^2) = 0.5\ (10^{-7} : 10^{-6})$ s. We can recall that the viscosity of intact material of salt rock has the order of $10^{14} : 10^{15}$ poise, that is, much higher than viscosity in shock experiments mentioned above. It shows the special mecahnics of viscosity (or pseudoviscosity) of dynamic compression of porous media.

Shock pseudoviscosity η of plane waves in vacuumized sands, considered above, was equal to 300,000 poise (this value corresponds to measured relaxation time $\theta = 10^{-4}$ s). In accordance with pseudoviscosity concept the amplitude of elastic precursor have to decay with distance. Such effects were observed for limestones and for dolomites.

5. NONELASTIC STRAINS AND ACOUSTIC EMISSION

There are some effects, connected with limit velocity of elementary mechanics of nonelasticity (of slip or of brittle fracture), for which the considered viscous approximation is not enough. Between them there is wear shock wave propagation in plastically deformed zones with elastic wave velocity, corresponding to unloading branch of $e(p,\partial e)$ instead of the velocity, determined by tangent to loading branch, see Stoll, Ebeido (1966). Therefore, seismic waves in the earth crust do not "feel" its plastic states.

Dynamical response of a geomaterial consists of waves of different frequencies, between which there are frequencies, determined by internal structure of the medium. For recognition of the latters and experiments of our postgraduate student, Nora Vilchinska (1982) are exemplary. She has observed the ultrasound (20-150 kHz) pulse propagation in beach sands. The ultrasound pulse disappears after very short distances: 10 cm in a dry sand and 70 cm in a wet one. However, this pulse generates the packet of low frequency oscillations with maximum amplitude approximately at 25 Hz. What is a nature of such frequencies? The answer was found when she had checked the hypothesis that it was the acoustic emission manifestation. Really, the spectrum of acoustic emission at quasistatic loading before limit plastic yield and the spectrum of the already discussed low frequency wave practically coincide.

So, small slips, nonavoidable even at small sand deformations, generate the acoustic emission pulses. The corresponding frequencies can be estimated as $f = v/u$, where v is a displacement velocity and u is displacement itself. Another argument is that the front of this low frequency oscillation propagates exactly with the velocity of repacking wave (Nikolaevsky et al., 1970) which in dry sand is the wave, seismically observed, and in saturated sand is the wave of the second type, according to the Frenkel-Biot theory. In fully saturated sands only ultrasound pulse velocity (of the first type wave) was measured, because the second wave was very weak. In practically dry sand only second wave velocity was measured. But in the intermediate cases both wave velocity could be measured. The frequencies of the second wave depended slightly on saturation and sand granulometry.

The possible explanation of ultrasound generation of slips and therefore of connected acoustic emission may be the following one. The effective internal friction angle $\phi = 2\phi_0 + \nu$ depends on true contact solid friction angle ϕ_0 but the latter decreases under ultrasound action (Archangelsky, 1969). The dilatancy angle ν depends only on relative position of grains, and so the microsituation becomes unstable, that is, the tendency to nonelastic repacking and to emission will appear.

It is known that a vibrating plate placed on the beach sand surface will sink (Gomes, Graves, 1962) and absolute maximum of its settlement velocity corresponds exactly to 25 Hz. So, this frequency has a resonance character.

It is known also that 25 Hz is a predominant frequency of explosions in sands, and the explosion of group charges with the delay equal to predominant frequency leads to resonance amplification (Zayakin, 1981). Another source of resonance response of sand masses and the amplification factor is connected with its stratification.

6. REFERENCES

[1] Archangelsky, M.E., 1969, Decreasing of solid dynamical friction by ultrasound oscillations. Trans. Acad. Sci. USSR, Mech. Solids (MTT), No. 2.

[2] Bhatt, J., Carroll, M., Schatz J., 1975, A spherical model calculation of volumetric response of porous rocks. Trans. ASME, E42, No. 2, 363 - 368.

[3] Bovt, A.N., Nikolaevsky, V.N., 1981, Dilatancy and mechanics of underground explosion. In: Itogi nauki i tehniki, Mech. solids, v. 14, M, VINITI, 129 - 169.

[4] Broberg, K., 1978, On transient sliding motion, Geophys. J. Roy. Astr. Soc., v. 52, No. 3, 397 - 432.

[5] Bugrov, A.K., 1980, Calculations of elasto-plastic soil structures. Archiwum hydrotechniki, v. 27, No. 3, 403 - 414.

[6] Carroll, M., Holt, A., 1973, Steady waves in ductile porous solids. J. Appl. Phys., v. 44, No. 10, 511 - 518.

[7] Chadwick, P., Cox, A., Hopkins, H., 1964, Mechanics of deep underground explosions. Phil. Trans. Roy. Soc. London, A256, No. 1070, 235 - 300.

[8] Dunin, S.Z., Sirotkin, V.K., 1977, The expanding of gas cavity in a brittle rock with dilatancy account. Appl. Mech. Techn. Phys. (PMTF), No. 4, 106 - 109.

[9] Goldin, L., Prokopovich, V., 1980, Bearing capacity of foundations and nonassociated flow rule of soils. Proc. VNIIG, Leningrad, v. 137, 3 - 7.

[10] Gomes, L., Graves, L., 1962, Stabilization of beach sand by vibration. Highway Res. Board, Bull. No. 325, Wash. D.C.

[11] Merzer, A., Freund, R., 1976, Equal spacing of strike-slip faults. Geophys, J. Roy. Astr. Soc., v. 45, No. 1, 177 - 188.

[12] Nikolaevsky, V.N., 1967, On volume and shear strain relations and on shock waves in soils. Proc. Acad. Sci. USSR (DAN), v. 177, No. 3, 542 - 545.

[13] Nikolaevsky, V.N., 1971, Governing equations of plastic flow of granulated medium, Sov. Appl. Math. Mech. (PMM), v. 35, no. 6.

[14] Nikolaevsky, V.N., 1979, The Mohorovicic boundary as a limit depth of brittle - dilatant states of rocks. Proc. Acad. Sci. USSR (DAN), v. 249, No. 4, 817 - 821.

[15] Nikolaevsky, V.N., Basniev, K.S., Gorbunov, A.T., Zotov, G.A., 1970, Mechanics of saturated porous media. Moscos, Nedra, 335 p.

[16] Nikolaevsky, V.N., Rice, J.R., 1979, Current topics in the nonelastic deformation of geological materials, High pressure sci. & Technol. v. 2, NY-L, Plenum, 455 - 464.

[17] Nikolaevsky, V., Syrnikov, N., 1970, On plane limit flow of a granular dilating medium. Trans. USSR Acad. Sci., Mech. Solids (MTT), No. 2, 159 - 166.

[18] Nikolaevsky, V., Syrnikov, N., Shefter, G., 1975, Dynamics of elasto-plastic dilating media, Advances in mechanics of solids (Galerkin volume), Moscos, Nauka, 397 - 413.

[19] Palmer, A., Rice, J., 1973. The growth of slip surfaces in the progressive failure of overconsolidated clay, Proce. Roy. Soc. London, A332, 527 - 548.

[20] Reynolds, O., 1885, On the dilatancy of media composed of rigid particles in contact. Philos. Mag., ser. 5, No. 127, 469 - 481.

[21] Rodionov, V.N., Adushkin, V.V., Kostuchenko, V.N., Nikolaevsky, V.N., Romashov, A.N., Zvetkov, V.M., 1971, Mechanical effect of underground explosion, Moscow, Nedra, 222 p.

[22] Rudnicki, J., Rice, J., 1975, Conditions for the localization of deformation in pressure-sensitive dilatant materials, J. Mech. Phys. Solids, v. 23, No. 6, 371 - 394.

[23] Shield, R.T., 1953, Mixed boundary value problems in soil mechanics, Quart. Appl. Math., vol. II, No. 1, 61 - 75.

[24] Stavrogin, A.N., Protosenya, A.G., 1979, Rock plasticity, Moscos, Nedra, 301 p.

[25] Stoll, R.D., Ebeido, J.A., 1965, Shock waves in granular soil, J. Soil Mech. Found. Div., Proc. Am. Soc. Civ. Eng., SM4, 107 - 125.

[26] Stoll, R.D., Ebeido, J.A., 1966, Unloading effects in the dynamic response of granular soil, Exptl. Mech., v. 6, No. 10, 502 - 507.

[27] Vilchinska, N.A., 1982, Repacking wave and acoustic emission in sands Proc. Acad. Sci. USSR (DAN), v. 262, No. 3, 568 - 572.

[28] Zayakin, V.N., 1981, Study of seismic action of short-delayed explosion Armenian Acad. Sci. Publications, 82 p.

ON THE ROTATION OF PRINCIPAL STRESS AXES IN THE SIMPLE SHEAR TEST AND ITS UTILIZATIONS

H. Ochiai[1], T. Yamanouchi[2] and Y. Tanabashi[3]

[1,3]Department of Civil Engineering, Nagasaki University,
Nagasaki, Japan
[2]Department of Civil Engineering (Suiko), Kyushu University,
Fukuoka, Japan

ABSTRACT

The principal stress axes rotate during the increase of shear stress in a simple shear test. The following simple relation between the stress ratio τ_{yx}/σ'_y on the horizontal plane and the inclination angle ψ of the maximum principal stress axis to the vertical direction has been proposed by Oda and Konishi[14]

$$\tau_{yx}/\sigma'_y = \kappa \cdot \tan\psi$$

It is shown that this relation makes it possible to formulate various interesting relations if the constant κ can be expressed with the other soil parameters. The determinations of the principal stresses during the simple shear test and of the constant κ are presented. The availabilities of these relations and their applications in other problems are discussed.

1. ROTATION OF PRINCIPAL STRESS AXES DURING SIMPLE SHEAR TEST

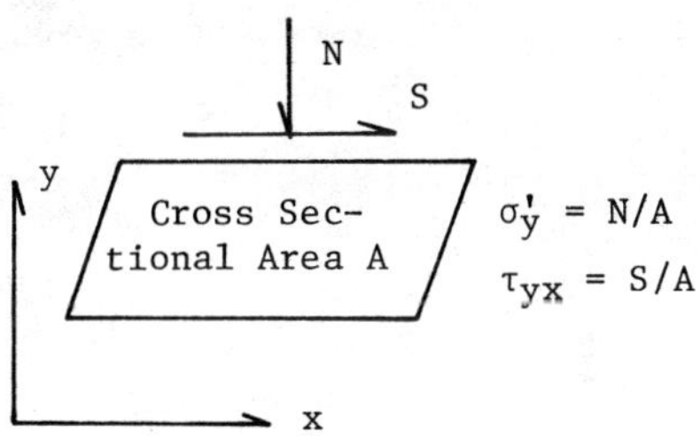

Fig.1 Stress measurements in simple shear test

Principal stress axes gradually rotate during monotonous increase of shear stress applied parallel to the horizontal plane of a simple shear apparatus. Cole[5] performed simple shear tests on Leighton Buzzard Sand and determined the inclination angle ψ of the maximum principal stress axis to the vertical direction on the basis of the constructions of Mohr's circle of stresses. By plotting the average stress ratio τ_{yx}/σ'_y on the central third of specimen against ψ , he found that there is a unique relationship between τ_{yx}/σ'_y and ψ and this relationship is independent of test path. He postulated that $\psi = F(\tau_{yx}/\sigma'_y)$ and gave the following empirical equation linking these two parameters:

$$\psi = 1.82\,(\tau_{yx}/\sigma'_y) - 0.75\,(\tau_{yx}/\sigma'_y)^2 \quad \text{(rad.)} \tag{1}$$

His test results (from his Fig.5.26) and equation (1) are shown in Fig.2.

Oda and Konishi[13,14] carried out the two-dimensional simple shear test on an assembly of cylinders (made of photoelastic material) packed at random, and showed that the principal stress axes gradually rotate during the progressive shearing in agreement with the gradual rotation of the prefered direction of N_i (N_i is the direction normal to the tangential plane at the contact). From the theoretical study on the microscopic contact force in a granular assembly, they derived the relation between the stress ratio τ_{yx}/σ'_y acting on the apparent shear plane (parallel to the horizontal plane) and the inclination angle ψ of the maximum principal stress axis to the vertical direction in the simple shear test as follows:

$$\tau_{yx}/\sigma'_y = \frac{(1+\tan^2\alpha)K + (\tan^2\alpha-1)\tan\psi}{\tan^2\psi + \tan^2\alpha + (\tan^2\alpha-1)K\tan\psi}$$
$$\alpha = \pi/4 + \phi_\mu/2 \tag{2}$$

where ϕ_μ is an interparticle friction angle and K is a parameter determined only by the spatial arrangement of solid particle in the granular assembly. They also tried to simplify equation (2) on the basis of the consideration that there must be an interrelation between $\tan\psi$ and K, because both of them are determined only by the spatial arrangement of solid particle for a given material, and they proposed the following

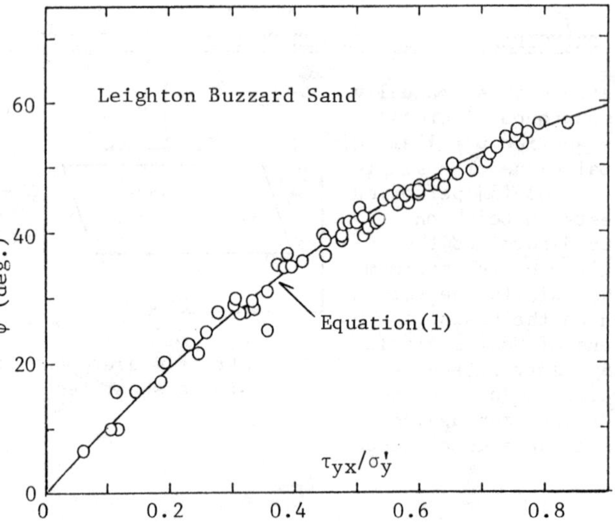

Fig.2 Relationship between ψ and τ_{yx}/σ'_y in simple shear test

linear relationship between τ_{yx}/σ'_y and $\tan\psi$:

$$\tau_{yx}/\sigma'_y = \kappa \cdot \tan\psi \tag{3}$$

in which κ is a material constant which may be determined by the inter-particle friction angle ϕ_μ .

Cole's test results in Fig.2 are plotted again in Fig.3 as τ_{yx}/σ'_y against $\tan\psi$. It can be seen that there is a good linear relationship between τ_{yx}/σ'_y and $\tan\psi$ as expected from equation (3). Hence, equation (3) is very useful to know the stress state acting on the specimen of simple shear test, if the value of κ can be estimated for a given soil.

2. PRINCIPAL STRESSES IN SIMPLE SHEAR TEST

Ochiai[10,11] has shown that the principal stresses σ'_1 and σ'_3 acting on the specimen in simple shear appararus can be formulated from equation (3) and consideration of the geometry of Mohr's circle of stress. The Mohr's circle of stress representing the mutual relation among σ'_1 , σ'_3 , σ'_y , τ_{yx} and ψ is shown in Fig.4. Using the stresses σ'_y and τ_{yx} measured on the horizontal plane and the corresponding principal stresses σ'_1 and σ'_3 , we obtain the pole $(\sigma'_1-\kappa\sigma'_y\ ,\ \tau_{yx})$ of the Mohr's circle of stress. Moreover, the three points $(\sigma'_y\ ,\ \tau_{yx})$,$(\sigma'_1\ ,\ 0)$ and $(\sigma'_1-\kappa\sigma'_y\ ,\ \tau_{yx})$ must be on the circle. Consequently, the principal

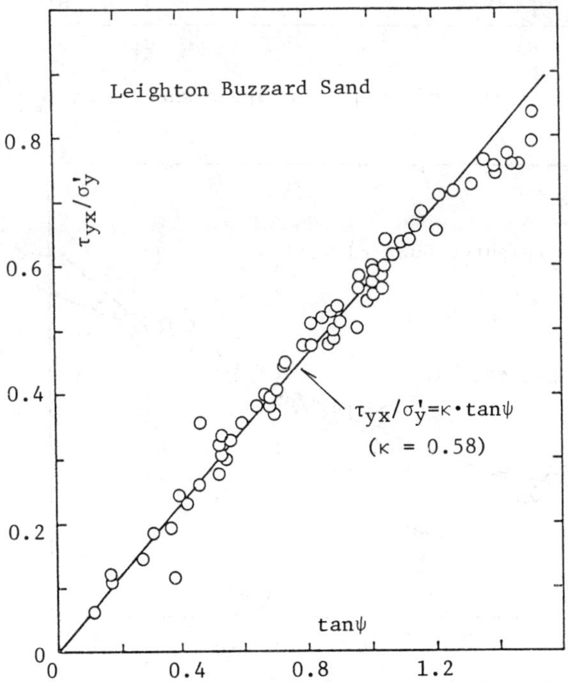

Fig.2 Linear relatioship between τ_{yx}/σ'_y and $\tan\psi$ in simple shear test

stresses σ_1' and σ_3' are formulated as function of σ_y', τ_{yx} and κ as follows:

$$\sigma_1' = (\kappa\sigma_y'^2 + \tau_{yx}^2) / \kappa\sigma_y' \tag{4}$$

$$\sigma_3' = (1 - \kappa)\sigma_y' \tag{5}$$

Equation (5) suggests that σ_3' should be constant in a simple shear test conducted with constant normal stress, irrespective of the magnitude of applied shear stress. Konishi[8] has experimentally found that σ_3' is nearly constant during the simple shear test on the two-dimensional assembly of photoelastic rods. Budhu[3] has also confirmed this interesting characteristics in the simple shear test on a sand.

From equations (4) and (5), the mean normal stress $s=(\sigma_1'+\sigma_3')/2$ and the shear stress $t=(\sigma_1'-\sigma_3')/2$ are formulated as follows:

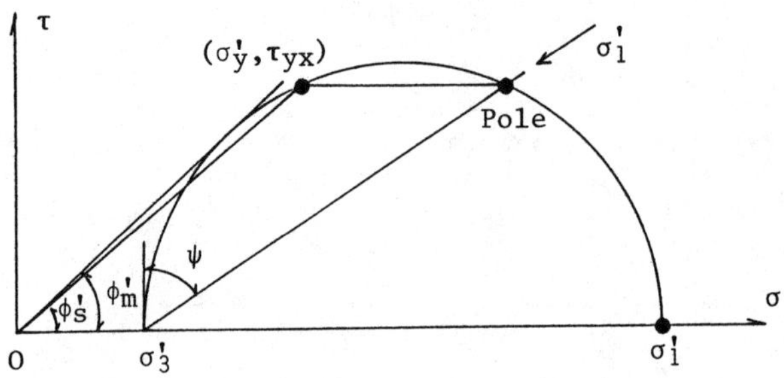

Fig.4 Mohr's circle of stress in simple shear test

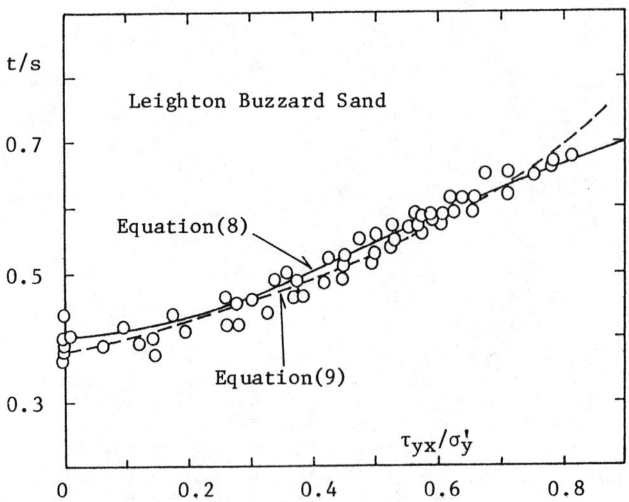

Fig.5 Relationship between the stress ratios t/s and τ_{yx}/σ_y' in simple shear test

$$s = \frac{1}{2}(\sigma_1'+\sigma_3') = \frac{1}{2}\;\frac{\kappa(2-\kappa)\sigma_y'^2+\tau_{yx}^2}{\kappa\sigma_y'} \tag{6}$$

$$t = \frac{1}{2}(\sigma_1'-\sigma_3') = \frac{1}{2}\;\frac{\kappa^2\sigma_y'^2+\tau_{yx}^2}{\kappa\sigma_y'} \tag{7}$$

The stress ratio $t/s=(\sigma_1'-\sigma_3')/(\sigma_1'+\sigma_3')$, therefore, is given by

$$t/s = (\sigma_1'-\sigma_3')/(\sigma_1'+\sigma_3') = \frac{\kappa^2 + (\tau_{yx}/\sigma_y')^2}{\kappa(2-\kappa) + (\tau_{yx}/\sigma_y')^2} \tag{8}$$

Cole[5] postulated that $t/s = G(\tau_{yx}/\sigma_y')$ and gave the following empirical equation for small values of τ_{yx}/σ_y' :

$$t/s = 1/\{2.64 - 1.50(\tau_{yx}/\sigma_y')\} \tag{9}$$

Cole's six drained tests at various initial void ratios and normal stresses (from his Fig.5.28) and equation (9) are shown in Fig.5. The relationship given by equation (8) is also represented in this figure by the solid line as $\kappa = 0.58$ for Leighton Buzzard Sand , and it can be seen that equation (8) is in good agreement with the test results.

Returning to Fig.4, the maximum obliquity of stress ϕ_m' and the friction angle ϕ_s' mobilized on the horizontal plane are represented as follows :

$$\sin\phi_m' = (\sigma_1'-\sigma_3')/(\sigma_1'+\sigma_3') \tag{10}$$

$$\tan\phi_s' = \tau_{yx}/\sigma_y' \tag{11}$$

Equation (8) can be, therefore, rewriten by ϕ_m' and ϕ_s' as

$$\sin\phi_m' = \frac{\kappa^2 + \tan^2\phi_s'}{\kappa(2-\kappa) + \tan^2\phi_s'} \tag{12}$$

3. DETERMINATION OF THE CONSTANT κ

The constant κ in equation (3) can be expressed in three simple forms with ϕ_{cv}' , ϕ_μ or K_0 , as described by Ochiai[10,12].

Roscoe et al[16] showed that at large shear strains the principal axes of stress and strain increment almost coincide for medium to loose sand, though there is some divergence for dense sand. It may be, therefore, assumed from practical purposes that at the critical state the principal axes of stress and strain increment coincide. At the critical state strain increment of y-direction is zero in simple shear test, so that the inclination angle ξ of the maximum principal strain increment axis to the vertical direction is equal to $\pi/4$ (cf. Fig.6). Hence, $\xi = \psi = \pi/4$ at the critical state. Substituting equations (3) and (11) into equation (12) we get

$$\sin\phi_m' = \frac{\kappa(1+\tan^2\psi)}{2-\kappa+\kappa\tan\psi} \tag{13}$$

and using the relations that $\phi'_m = \phi'_{cv}$ and $\psi = \pi/4$ at the critical state, we get the first expression of κ as follows:

$$\kappa = \sin\phi'_{cv} \qquad (14)$$

The constant κ can be also formulated experimentally by the angle of interparticle friction ϕ_μ. At the start of a simple shear test $\tau_{yx} = 0$ and equation (12) becomes

$$\sin\phi'_{m,o} = \kappa/(2-\kappa) \qquad (15)$$

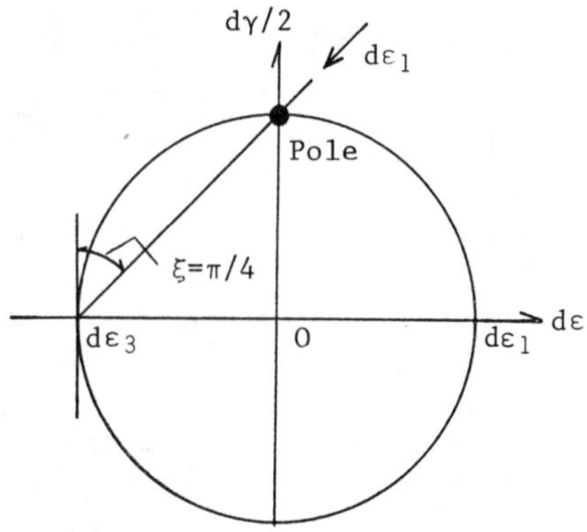

Fig. 6 Mohr's circle of strain increment in simple shear test at the critical state

According to the experimental study on photoelastic rods by Oda and Konishi[13], there are some critical contacts satisfying the relation that the mobilized angle at the contact is equal to ϕ_μ even in the granular assembly which was expressed uniaxially without any application of shear force in a simple shear test. This finding means that $\phi'_{m,o} = \phi_\mu$, and substituting this ralation into equation (15) we get the second expression of κ as

$$\kappa = 2\sin\phi_\mu/(1 + \sin\phi_\mu) \qquad (16)$$

Consider the start of a simple shear test that $\tau_{yx} = 0$, again. At this state the specimen is at rest as known as K_o-consolidation and equations (4) and (5) produce a relationship

$$\sigma'_3/\sigma'_1 = K_o = 1 - \kappa \qquad (17)$$

Thus, we get the third expression of κ as follows:

$$\kappa = 1 - K_o \qquad (18)$$

According to Cole[5], $\phi'_{cv} = 35°$ and $\phi_\mu = 24°$ for Leighton Buzzard Sand. Using these values, equations (14) and (16) give that $\kappa = 0.57$ and $\kappa = 0.58$, respectively. He also showed that the value of K_o ranges from 0.33 to 0.47 for dense samples and from 0.42 to 0.49 for loose samples (cf. Fig.8). With these values, equation (18) gives that κ = 0.67 - 0.53 (dense) and κ = 0.58 - 0.51 (loose), and the mean value of κ is 0.59. The average value of κ obtained from equations (14), (16) and (18) is about 0.58 for Leighton Buzzard Sand, and this value is consistently used in Figs.3 and 5.

4. ON THE PLANE OF MAXIMUM OBLIQUITY IN SIMPLE SHEAR TEST

The commonest assumtion to determine the principal stresses in simple shear test is that the horizontal plane of the apparatus is a plane of maximum stress obliquity. This assumption means that in Fig.4 $\phi'_m = \phi'_s$. Substituting this relation into equation (12) and rearranging, we get

$$\sin\phi'_s = \kappa/(1 - \kappa) \qquad (19)$$

It must be satisfied that $0 \leqq \sin\phi'_s \leqq 1$ in simple shear tests, and from equation (19)

$$0 \leqq \kappa \leqq 0.5 \tag{20}$$

That is, for the material with $\kappa > 0.5$ the above assumption is not valid in all stage of shearing. If κ is calculated from equation (16), then $\kappa > 0.5$ corresponds to $\phi_\mu > 19.5°$. It may be, therefore, concluded that the above assumption is not true because the values of ϕ_μ of usual natural soil particle are higher than 20°.

5. RELATIONSHIP BETWEEN ϕ'_{cv} AND ϕ_μ

Both of ϕ'_{cv} and ϕ_μ are basic important soil parameters. Several equations have been proposed for predicting the value of the Mohr-Coulomb angle of friction at the critical state (ϕ'_{cv}) from the value of the interparticle friction angle (ϕ_μ). Caquot[4] derived the following expression for plane strain conditions:

$$\tan\phi'_{cv} = (\pi/2)\tan\phi_\mu \tag{21}$$

Bishop[1] by an approximate analysis suggested that

a) for triaxial compression where $\sigma'_2 = \sigma'_3 < \sigma'_1$

$$\sin\phi'_{cv} = 15\tan\phi_\mu/(10 + 3\tan\phi_\mu) \tag{22}$$

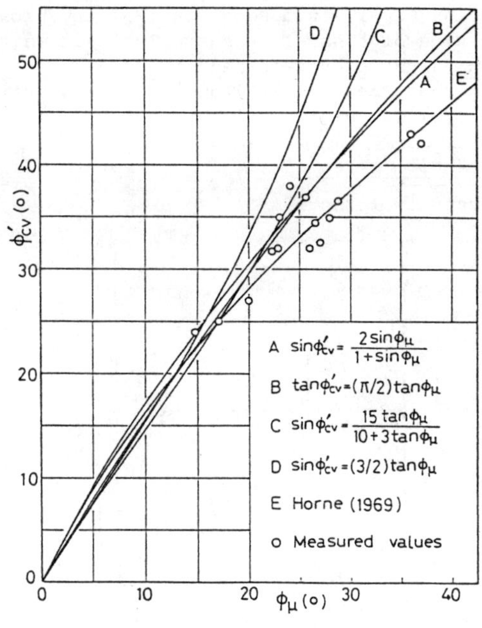

Fig.7 Relationship between ϕ'_{cv} and ϕ_μ

b) for plane strain conditions taking $\sigma_2' = (\sigma_1' + \sigma_3')/2$

$$\sin\phi_{cv}' = (3/2)\tan\phi_\mu \quad (23)$$

Horne[6] derived a theoretical relationship between ϕ_{cv}' and ϕ_μ for triaxial compression, where $\sigma_2' = \sigma_3' < \sigma_1'$. Horne's relationship gives better predictions of ϕ_{cv}' than the other ones for a very wide range of ϕ_μ , but his relationship is very complicated in form.

A simple relationship between ϕ_{cv}' and ϕ_μ can be obtained from equations (14) and (16) as follows:

$$\sin\phi_{cv}' = 2\sin\phi_\mu/(1 + \sin\phi_\mu) \quad (24)$$

Equation (24) is shown in Fig.7 with the other proposed relationships and the experimental results reported in the past, and it can be seen that equation (24) provides nearly the same value of ϕ_{cv}' as Caquot's one.

6. COEFFICIENT OF EARTH PRESSURE AT REST

The following expression of K_o can be obtained from equations (14), (16) and (18) :

$$K_o = 1 - \sin\phi_{cv}' = \tan^2(\pi/4 - \phi_\mu/2) \quad (25)$$

It should be noted that this equation would predict equal values of K_o for dense and loose sand. Although the independence of K_o from initial density is not born out by experiment, equation (25) may provide approximately an average K_o-value of a sand, as shown in Fig.8 which shows the experimental results on Leighton Buzzard Sand by Cole[5]. If we assume from practical purposes that $\tan\phi_\mu = (1/F)\tan\phi'$, then substituting this relation into equation (25) we get

$$K_o = \frac{\sqrt{F^2 + \tan^2\phi'} - \tan^2\phi'}{\sqrt{F^2 + \tan^2\phi'} + \tan^2\phi'} \quad (26)$$

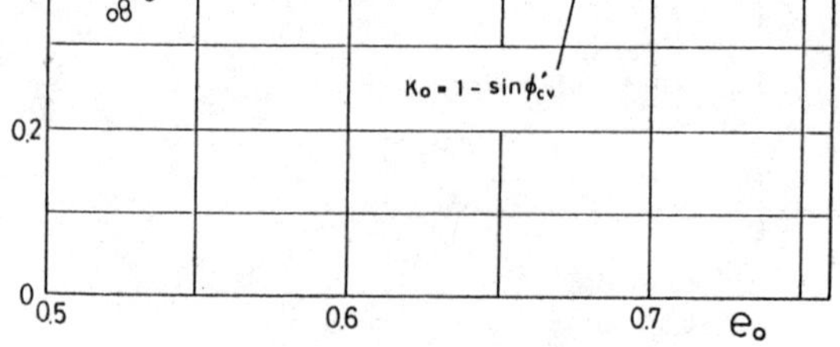

Fig.8 Measured values of K_o of a sand

and taking $1/F = 2/\pi$, we get an empirical expression of K_o as follows :

$$K_o = \frac{\sqrt{(\pi/2)^2 + \tan^2\phi'} - \tan^2\phi'}{\sqrt{(\pi/2)^2 + \tan^2\phi'} + \tan^2\phi'} \qquad (27)$$

This equation provides almost the same value of K_o as Jaky's formula ($K_o = 1 - \sin\phi'$) [7], as shown in Fig.9.

7. UNDRAINED STRENGTH RATIO OF NORMALLY CONSOLIDATED CLAY MEASURED IN SIMPLE SHEAR TEST

The undrained strength ratio s_u/p is an important index expressing the increasing rate of the undrained strength due to consolidation, in which s_u is defined as

$$s_u = (\sigma_1 - \sigma_3)_{max}/2 = \tau_{max} \qquad (28)$$

When the ratio s_u/p is determined by means of the laboratory shear tests, the tests should simulate the in situ modes of failure because most clays exhibit anisotropic strength behavior [9]. The simple shear test has some advantages over other tests in this regard, and the shearing resistance measured under the simple shear test conditions may provide a very useful measure of shear strength for stability analyses of field loading conditions. It should be noted, however, that the shear stress τ_{yx} measured in the simple shear test is not the maximum shear stress τ_{max} and so that the ratio $(\tau_{yx})_{max}/p$ is not equal to the undrained strength ratio s_u/p .

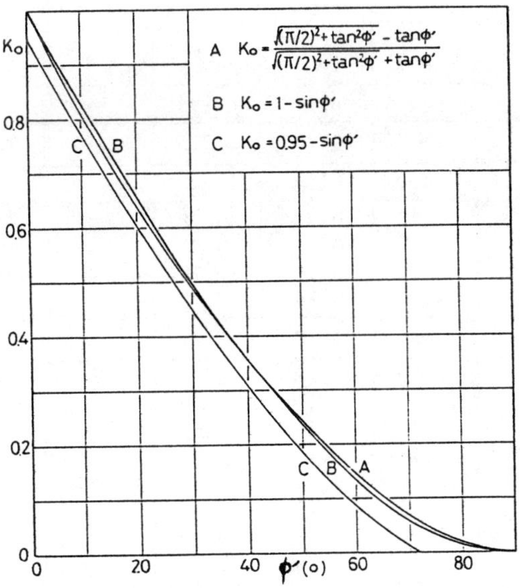

Fig.9 Relationship between K_o and ϕ'

In undrained simple shear test, the normal effective stress σ'_y on the horizontal plane is related to the initial normal stress (consolidation pressure) p as follows:

$$\sigma'_y = \beta p \tag{29}$$

in which β is a quantity related to the pore water pressure u during the test and is expressed as

$$\beta = (p - u)/p \tag{30}$$

Using these relations and $\kappa = 1-K_o$, equations (4) and (5) can be rewriten in the following forms :

$$\sigma'_1 = p\,\frac{\beta^2(1-K_o) + (\tau_{yx}/p)^2}{\beta(1-K_o)} \tag{31}$$

$$\sigma'_3 = K_o\beta p \tag{32}$$

The undrained strength ratio $(s_u/p)_s$ of normally consolidated clay measured in the simple shear test is, therefore, expressed as

$$(s_u/p)_s = \frac{(\sigma'_1-\sigma'_3)_{max}}{2p} = \frac{\beta^2(1-K_o)^2 + \{(\tau_{yx})_{max}/p\}^2}{2\beta(1-K_o)} \tag{33}$$

Table 1 The experimental results of three clays and the calculated values of $(s_u/p)_s$ and ϕ'_m

Measured values						Calculated values	
p (t/m²)	$(\tau_{yx})_{max}$ (t/m²)	u (t/m²)	$\frac{(\tau_{yx})_{max}}{p}$	$\frac{\sigma'_y}{p} = \beta$	K_o	$(\frac{s_u}{p})_s$	ϕ'_m (deg.)
Boston Blue Clay [9]							
-	-	-	0.200	0.575	0.50	0.213	25.2
Kaolin Clay [15]							
5.0	1.20	1.10	0.24	0.78	0.51	0.27	23.8
10.0	2.50	2.30	0.25	0.77		0.27	24.1
15.0	3.70	3.30	0.25	0.78		0.27	23.8
20.0	4.80	4.20	0.24	0.79		0.27	23.8
Manglerud Quick Clay [2]							
6.0	1.10	2.28	0.18	0.62	0.51	0.21	23.2
6.0	0.98	1.86	0.16	0.69		0.21	22.0
6.0	1.16	2.16	0.19	0.64		0.22	23.4
6.7	1.05	3.28	0.16	0.51		0.18	23.8
6.8	1.14	2.11	0.21	0.69		0.23	23.5
6.8	1.16	2.99	0.17	0.56		0.19	23.5
6.9	1.50	2.90	0.22	0.58		0.23	26.1
7.5	1.49	3.00	0.20	0.60		0.21	24.2
7.7	1.67	3.00	0.22	0.61		0.23	25.1
8.7	1.65	3.74	0.19	0.57		0.20	24.3

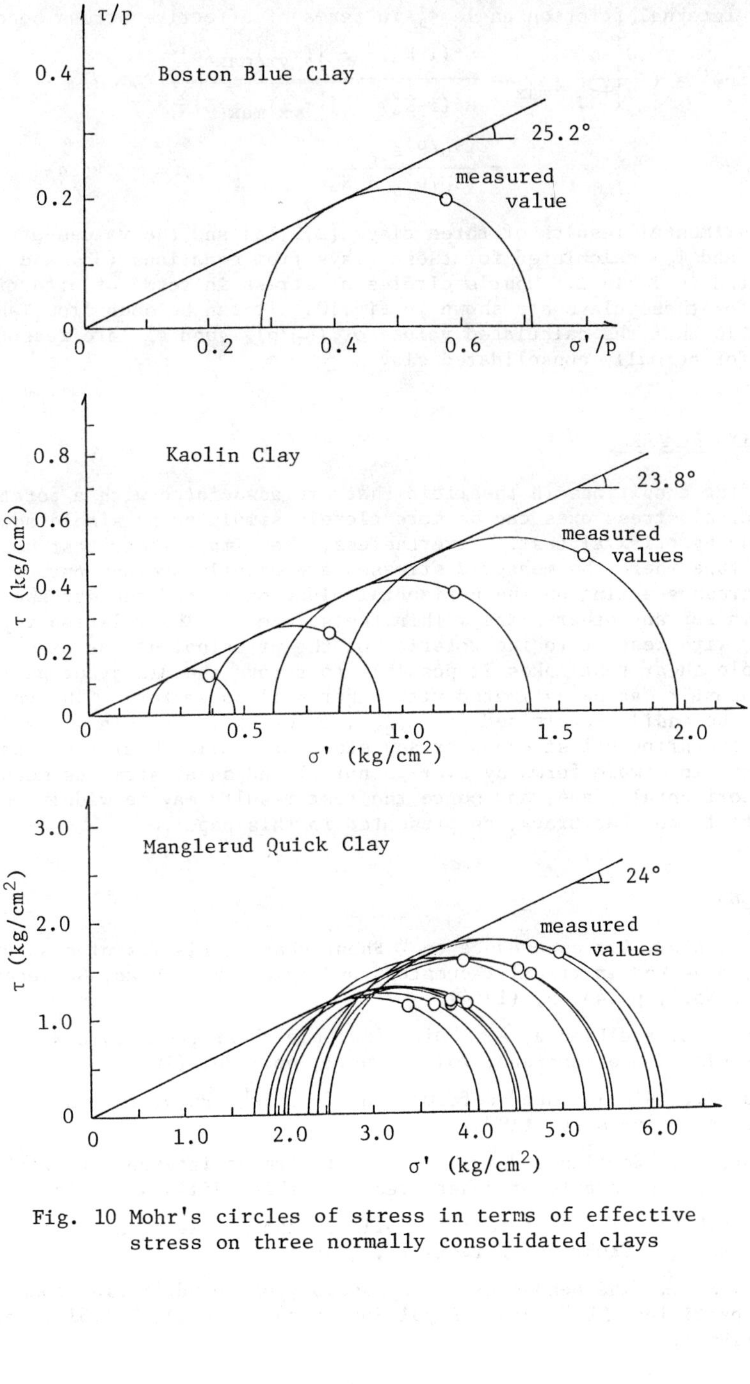

Fig. 10 Mohr's circles of stress in terms of effective stress on three normally consolidated clays

And the internal friction angle ϕ'_m in terms of effective stress becomes

$$\sin\phi'_m = \left(\frac{\sigma'_1-\sigma'_3}{\sigma'_1+\sigma'_3}\right)_{max} = \frac{\beta^2(1-K_o)^2 + \{(\tau_{yx})_{max}/p\}^2}{\beta^2(1-K_o^2) + \{(\tau_{yx})_{max}/p\}^2}$$

$$= \frac{(s_u/p)_s}{(s_u/p)_s + K_o\beta} \qquad (34)$$

The experimental results of three clays [2,9,15] and the values of $(s_u/p)_s$ and ϕ'_m calculated for these clays from equations (33) and (34) are listed in Table 1. Mohr's circles of stress in terms of effective stress for these clays are shown in Fig.10. It can be seen from Table 1 and Fig.10 that the calculated values of $(s_u/p)_s$ and ϕ'_m are reasonable values for normally consolidated clay.

CONCLUDING REMARKS

Loading conditions in the field that are associated with a rotation of principal stress axes can be more closely simulated by simple shear test than by triaxial test. Nevertheless, the simple shear test has a disadvantage where the measured stresses are usually average normal and shear stresses acting on the horizontal plane only, and the stresses are not known for any other plane within the specimen. The relation $\tau_{yx}/\sigma'_y = \kappa\cdot\tan\psi$ with respect to the rotation of the principal stress axes in the simple shear test makes it possible to remove the disadvantage if the constant κ can be estimated with other soil parameters. The value of κ can be easily determined from ϕ'_{cv} , ϕ_μ or K_o . By using this relation, the principal stresses at any stage of simple shear test can be formulated in simple forms by average normal and shear stresses measured on the horizontal plane, and hence the test results may be widely used with satisfactory accuracy, as presented in this paper.

REFERENCES

1. Bishop, A.W., Correspondence on " Shear characteristics of a saturated silt, measured in triaxial compression" by A.D.M. Penman, Geotechnique, Vol.4, No.1, pp.43-45, (1954).

2. Bjerrum, L. and Landva, A., "Direct simple shear tests on a Norwegian quick clay", Geotechnique, Vol.16, No.1, pp.1-20, (1966).

3. Budhu, M., "Simple shear deformation of soils", Ph.D. thesis, Cambridge University, (1979).

4. Caqout, A., "Equilibre des massifs à frottement interne. Stabilité des ferres pulvérents et cohérentes", Gauthier-Villars, Paris, (1934).

5. Cole, E.R.L., "The behaviour of soils in the simple shear apparatus", Ph.D. thesis, Cambridge University, (1967).

6. Horne, M.R., "The behaviour of an assembly of round, rigid, cohesionless particles, III ", Proc. Royal Soc. London, Ser.A, Vol.310, pp.21-34, (1969).

7. Jaky, J., "Pressure in silos", Proc. 2nd Int. Conf. on Soil Mechanics Foundation Engineering, Vol.1, pp.103-107, (1948).

8. Konishi, J., "Stress state in the simple shear test on the two-dimensional assembly of rods", Proc. 30th Annual Meeting of the Japanese Society of Civil Engineers, Vol.3, pp.9-10, (1975).

9. Ladd, C.C., "Discussion, Main Session 4", Proc. 8th Int. Conf. on Soil mechanics and Foundation Engineering, Vol.4-2, pp.109-115, (1979).

10. Ochiai, H., "The behavior of sands in direct shear tests", Jour. of the Japanese Society of Soil Mechanics and Foundation Engineering, Vol.15, No.4, pp.93-100, (1975).

11. Ochiai, H., "The relation between the shear strength of sands in direct shear and plane strain tests", Jour. of the Japanese Society of Soil Mechanics and Foundation Engineering, Vol.15, No.4, pp.113-118, (1975).

12. Ochiai, H., "The coefficient of earth pressure at rest of sands", Jour. of the Japanese Society of Soil Mechanics and Foundation Engineering, Vol.16, No.2, pp.105-111, (1976).

13. Oda, M. and Konishi, J., "Microscopic deformation mechanism of granular material in simple shear", Soils and Foundations, Vol.14, No.4, pp.25-38, (1974).

14. Oda, M. and Konishi, J., "Rotation of principal stresses in granular material during simple shear", Soils and Foundations, Vol.14, No.4, pp.39-53, (1974).

15. Ohara, S. and Matsuda, H., "Dynamic shear strength of saturated clay", Jour. of the Japanese Society of Civil Engoneers, No.274, pp.69-78, (1978).

16. Roscoe, K.H., Bassett, R.H. and Cole, E.R.L., "Principal axes observed during simple shear of a sand", Proc. Geotechnical Conf., Oslo, Vol.1, pp.231-237, (1967).

EFFECTS OF FABRIC CHANGE ON DEFORMATION CHARACTERISTICS OF SAND IN CYCLIC SHEARING

Yoshio Tobita

Department of Civil Engineering, Tohoku University
Sendai, 980, Japan

ABSTRACT

The deformation characteristics of saturated sand in the cyclic shearing are influenced to a great extent by the fabrics created in the preceeding shear loading as well as consolidation. The characteristics obsereved in the triaxial cyclic shearing test and in the simple shearing test are discussed based on the microscopic behavior of granular materials. The microscopic behavior of granular material obsereved in the bi-axial compression test with photoelastic techniques are examined. Based on the observation, two fundamental microscopic properties are proposed; the induced anisotropy and the dead particles. The effects of these fabric change on the deformation characteristics of saturated sand are discussed. Brief comments on the continuum mechanical description of the behavior of granular materials are also presented.

1. INTRODUCTION

Liquefaction behavior of a saturated sand deposit in an earthquake excitation is an important problem for the earthquake resistant design of the soil deposit and super structures. A great number of experimental works have been done for the determination of the number of cycles required to the initial liquefaction, which has been related to the magnitude of an earthquake.(1) Attention has also been paid to the stress strain relationship of a saturated sand in a liquefaction test as well as the cyclic shearing test in a drained condition. Recently some usefull relationships for the liquefaction behavior in the cyclic loadings have been developed for the sake of numerical analysis. It is understood, however, that the relationships have been developed up to date can't account for the whole behavior obsereved in the liquefaction test, e.g.; the effects of initial anisotropy and stress history on the cyclic behavior of saturated sand.

It has been understood that the liquefaction potential of saturated sand is not only a function of the averaged density of the sand specimen but also a function of the fabric (2),(3),(4),(5). The study on the effects of prestressing on the deformation characteristics of a saturated sand specimen is important for the cases where any shear stress exist; slope stability etc. The liquefaction behavior of saturated sand in these cases may be quite different from that in the horizontally layered sand deposits, for which the ordinary experimental works are performed to determine the liquefaction potential.

In this paper the deformation characteristics of the saturated sand which was subjected to the various stress histories, which were obsereved in the triaxial testing apparatus and in the simple shear testing apparatus, are discussed; some of which have already discussed by Nemat Nasser and Tobita(6). These deformation characteristics are important for the better understanding of the cyclic behavior of saturated sand.

We first discuss the fundamental microscopic behavior of granular materials obsereved in the bi-axial compression test with photoelastic techniques performend by Konishi,Oda and Nemat Nasser(7). After discussing the microscopic behavior, we postulate two fundamental microscopic properties in the shearing deformation of granular materials: the induced anisotropy and the dead particles(coordination number). Based on the fundamental properties, we discuss the complicated defromtion characteristics of the saturated sand specimen in the cyclic shearing tests.

It has been found that these two fundamental properties of microscopic behavior of granular materials can be taken into consideration by a fabric tensor; a second rank tensor which characterizes a fabric state of granular materials like sand. The introduction of the fabric tensor lead to a continuum mechanical description of granular material behavior taking the internal change of the fabric of granular material into consideration. Brief comments on the continumm mechanical description of granular material like sand are also presented.

2.EXPERIMENTAL PROCEDURES

2-1 Biaxial Compression Test with Photoelastic Techniques

Rod-shaped particles with oval section are made of polyurethane rubber which is a photoelastically sensitive material. Table 1(a) shows the properties of particles used in this study. Initial dimension of the sample surrounded by the frame is about 330^{mm} in height and 207^{mm}in width. The apparatus can be tilted to a desired angle to get a various bedding angle: the angle between the normal direction of bedding plane and the direction of the vertical loading. Each sample consists of equal numbers of particles of the three sizes.

The sample is confined laterally by a constant dead load of 0.45 kgf and is compressed vertically by giving a displacement step by step to the top frame. Vertical load is measured by a load transducer. Vertical and lateral displacements of the frame are measured by a dial gauge and LVDT respectively. Photoelastic pictures are taken at appropriate stages.

The main aim of this experimental works is to consider the effect of inherent anisotropy and induced anisotropy(fabric change) on the deformation and strength characteristics of granular materials(7).

2-2. Experimental Procedures of Simple Shear Test and Triaxial Test

Monterey No.0 Sand was used in the simple shear test, which was performed at Northwestern University by Nemat Nasser and Tobita(6). The physical properties of the sand are listed in Table 1. The diameter and height of the samples were, on the average, 7.10cm and 3.10cm respectively. The sand mixed with 1.5 % water by weight was poured into the mould by a tea sppon and then tamped with a specially designed rod guided through a transparent plastic plate to get a desired height. Sand grains or the waterproof sand paper was gluded to the surfaces of the top cap and the bottom pedestal in order to prevent slippage between these plates and the sand specimen. The height of each layer was 1.03 cm and the number of tamping was adjusted to get the desired height.

Shinchi sand sampled at Shinchi town, Fukushima Prefecture, was used in the triaxial test, which was performed at Tohoku University. The physical properties of the sand are listed in Table 1. The diameter and height of the samples were, on the average, 5.00 cm and 12.50 cm respectively. The samples were prepared by spooning air-dried grains into the mould filled with deaired water and by tapping the sides of the mould by use of a wooden hammer to get a desired density. For one layer five spoonful sand was poured into and tapped.

A high degree of saturation was achieved by circulating carbon-dioxicide(CO_2) and deaired water through the sample and applying a back pressure of 100 (kN/m^2) for both tests. The degree of saturation was considered to be sufficient when B-value exceeded 0.95. All tests in both tests series were performed on the isotropically consolidated samples with the mean effective stress of 100 (kN/m^2). The loading and measuring systems for both tests are shown in Fig.1.

3. EXPERIMENTAL RESULTS

We first discuss the microscopic properties of deformation of granular assembly composed of rod shaped photoelastic particles. Some experimental results of saturated sand in the cyclic shearing test by use of the simple shear testing apparatus and the triaxial one are discussed based on the microscopic behavior of granular material.

Before discussing the experimental results, we briefly discuss the general properties of a solid material as a preliminary discussion.

3-1. Hardening and Softening effects of Granular Material in Deformation

It is well known that the strength of a solid material is determined by the mutual effects between a hardening effect and a weakenig effect. For clayey soils Vialov et.al introduced the concept of a defect density composed of voids and microscopic cracks as a weakening effect. The hardening effect of clayey soils was attributed to the rearrangement of particles and the mechanism of healing defects.

For granular material like sand, Arthur et.al(12) developed a semi microscopically founded theory for the strength of sand; where the rearrangement of grain particles and the generation of new contact between grains were considered as a hardenig effect and the local failure (slippage) in a granular assembly was considered as a weakening effect.

In this study, two fundamental microscopic mechanisms in the shear deformation of granular material are inroduced based on the bi-axial compression test with photoelastic techniques. The micromechanical behavior of granular material are induced anisotropy: the rearrangement of particles with an oriented fabric, and the generation and disappearance of dead particles: in other terms, coordination number (an average number of cantact a particle). These two micro-mechanical properties are closely related to the hardening effect and the weakening effect of granular materials. The effects of fabric change of sand in the course of shearing on the deformation characteristics of sand in the succeeding loadings are discussed in the following sections.

3-2. Two Fundamental Mechanisms in Deformation of Granular Materials

3-2-1. Induced Anisotropy

The degree of anisotropy of granular materials can be expressed in terms of the normal direction of the contact planes between grain particles. The vector magnitude V.M. is defined by the following equation (9):

$$V.M. = \sqrt{\sum(n_i \sin 2\theta_i)^2 + \sum(n_i \cos 2\theta_i)^2} \times \frac{100}{N} \quad (\%) \tag{1}$$

where N: the number of all contact

θ_i: an angle measured form the direction of maximum principal stress

With the increase of V.M. from 0 % (complete isotropic) to 100 %, the intensity of the concentraition of normal directions of contact planes becomes high.

Fig.2 shows a typical stress- strain-dilatancy relationship of the granular assembly in the bi-axial compression test, which simulates the deformation characteristics of sand in the triaxial testing apparatus very well. The vector magnitude V.M. in this test is shown in Fig.3. The vector magnitude increases up to the maximum density point then decreases gradually with the increase of axial strain. In the analysis, three types of contacts were considered, namely; active contact, dead contact and contact between dead particles and active particles. The vector magnitude of dead contact is not so

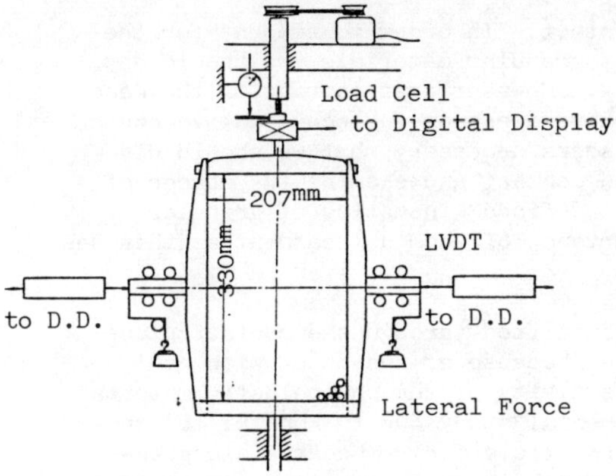

Fig.1-(a) The bi-axial compression test.

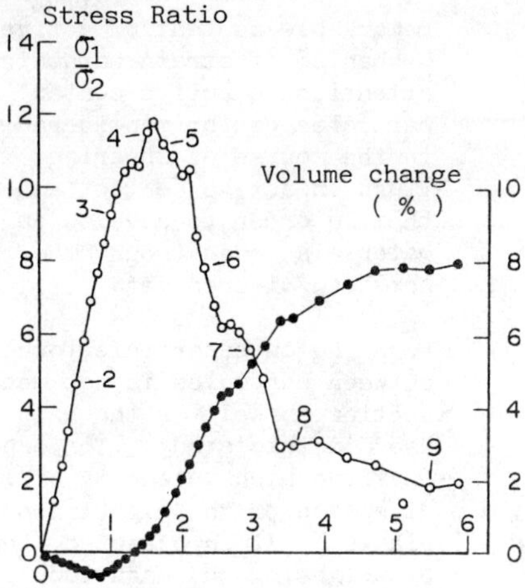

Fig.2 Stress strain relation in the bi-axialcompression test.

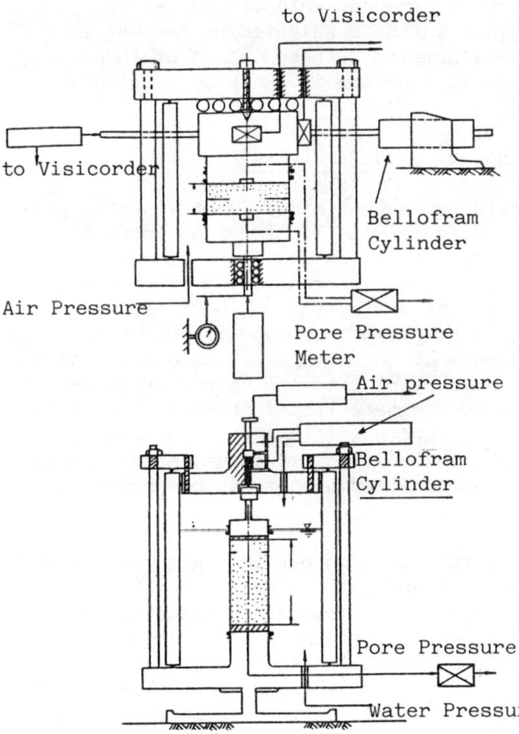

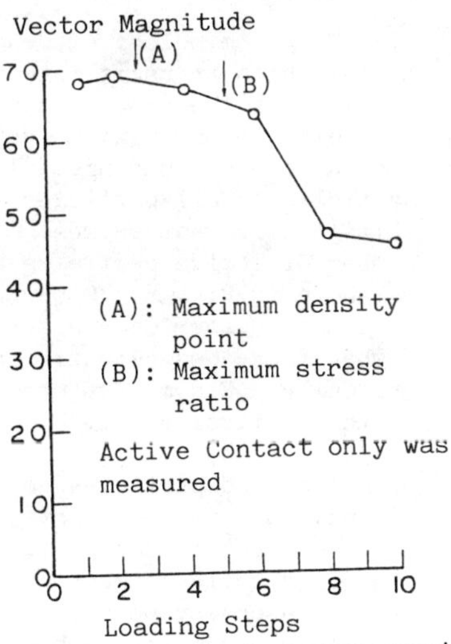

Fig.3. The change of vector magnitude in the course of shearing

Fig.1-(b) The simple shearing apparatus
Fig.1-(c) The triaxial testing device

remarkable as that of active contact. In order to account for the mechanism of strain hardening of granular materials, we should pay attention to active contact only. However,dead contact of the dead particles can be considered to have a chance to become active contact in the course of shearing. It seems necessary that we should distinguish active contact and dead contact and examine the change of them in order to develop an well defined kinematics of granular materials, even though the importance of such a treatment as this has not been cleared yet.

The interparticle force transmitted through the contact plane between particles is not measured because of the shape with oval section as well as the high sensitivity of the photoelastic material used in this study. The experimental study due to Konishi (9) showed that the high intensity of interparticle force appeares along the direction of the high intensity of the normal directions of cantact planes. It is expected, therefore, that the intensity of anisotropy by weighting the magnitude of interparticle force will increase up to the peak stress ratio. The concept of induced anisotropy used in this study can be considered as that which is weighted by the magnitude of interparticle force. The schematic representation of induced anisotropy in the course of monotonic shearing is shown by a solid line in Fig.6.

3-2-2 Generation and disappearance of Dead Particles

The terminology of dead particles in this study identifies the particles which transmit interparticle force with little or no intensity. Fig.4 shows the experimental results on the disappearance of the dead particles to the maximum density point and the generation of them in the succeeding loadings. In a critical state, the ratio of the dead particles to all particles seems to be constant, irrespective of the initial different values. An example of the change of the coordination number of active particles and dead particles in the course of shearing is shown in Fig. 5. Other data show the same trend: the active particles have more number of contact than the dead particles, the coordination number increases slightly up to the maximum density point then decreases gradually. Minimum value of the coordination number can be obsereved in the critical state.

The dead particles and the coordination number in a granular assembly may have the same physical meaning and may be interchangeable, since the increase of dead particles is always accompanied by the decrease of the coordiantion number and vice versa. Up to the present study, there is no reason why we must take the dead particles instead of the coordination number. It can be said, however, that the dead particles implies the existence of particles which have no role in the stability of granular assembly. In other words, the dead particles is more comprehensive than the coordination number.

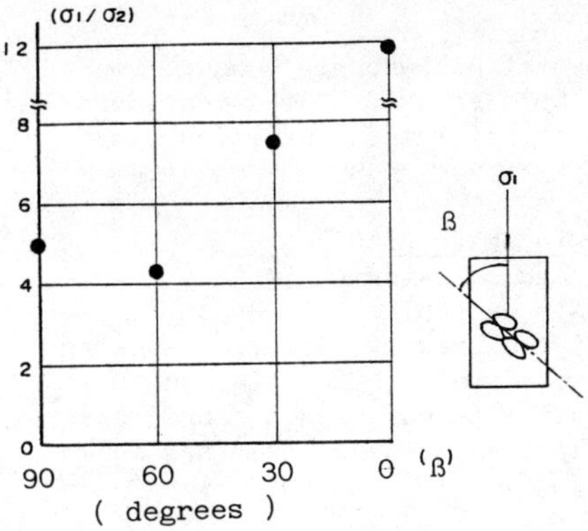

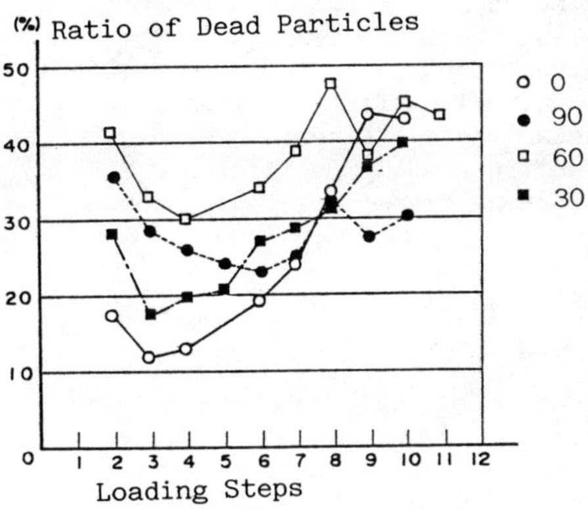

Fig.4. The change of the ratio of dead particles in the course of shearing

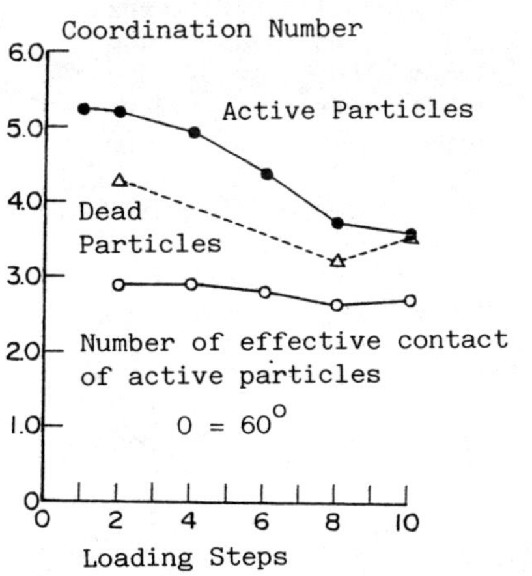

Fig.5 The change of coordination number in the bi-axial compression test

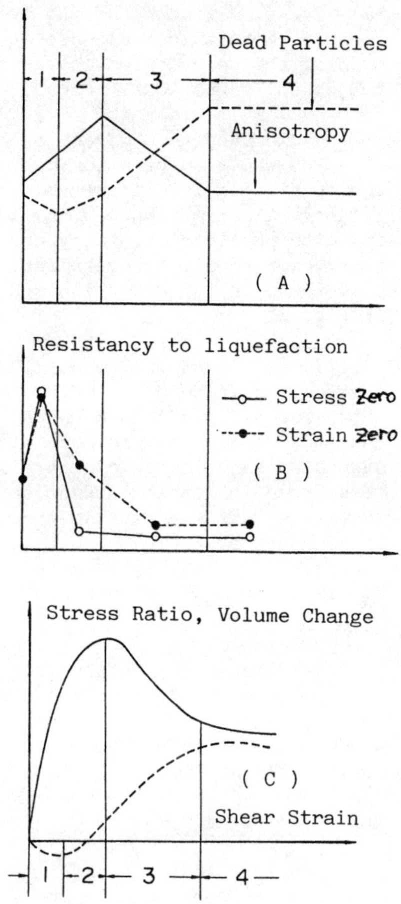

Fig.6(A): Microscopic change of granular materials in the course of shearing.
Fig.6(B): The effect of previous stress history on the resistancy to liquefaction
Fig.6(C): Stress-Strain-Volume change of sand in a monotonic shearing.

The significance of the two fundamental microscopic behavior of granular materials mentioned above will be discussed in the consideration of the deformation characteristics of saturated sand which was subjected to previous history.

3-3 Loading Regime of Granular Material in Shear Loading

The whole deformation process of granular material in shear loading is divided into four regime: 1; strain hardening with negative dilatancy, 2; strain hardening with positive dilatancy, 3; strain softening with positive dilatancy and 4; critical state. The relations between macroscopic behavior and microscopic internal change fabric are summerized in Table 2. The schematic representation of them are shown in Fig. 6.

In what follows, we use the loading regime in order to identify the loading state of granular materials. The subdivision described above was already used by Nemat Nasser and Tobita (6) in order to discuss the effects of previous history on the liquefaction potential of saturated sand in the cyclic simple shear tests. It should be also remembered that the distinction between the loading regime 1 and 2 implies almost the same as the line of phase transformation proposed by Ishihara et.al(10)

4. DISCUSSIONS ON THE DEFORMATION CHARACTERISTICS OF SAND IN CYCLIC SHEARING

We discuss the deformation characterisitcs of saturated sand in the cyclic shearing test based on the two fundamental microscopic mechanisms of granular materials proposed in the previous section. We deal with this problem by adopting the loading regime proposed in the last section.

4-1. Loading Regime 1.

Consider that a cyclic shearing test with a small strain(stress) amplitude is performed on the sample of saturated sand in undrained condition; the pore water in the sample will not allow to flow in and out. The particle rearrangement due to the small stressing lead to a more stable fabric than that formed during the sample preparation and the suceeding consolidation if it is in drained condition. In undrained condition, the negative dilatancy due to particle arrangement results in the pore pressure increase, as a result of this, the effective confining pressure necessarily decreases, which lead to the increase of the effective stress ratio under the stress controlled cyclic test with a constant stress amplitude. Since dilatancy can be believed to be mobilized by the applied effective stress ratio rather than the shear stress itself, more negative dilatancy generates in the succeeding loadings; which results in the progressive pore pressure increase. Finally the effective stress path of the sample in undrained condition reaches to the line of phase transformation. The sample which reaches to this line shows completely different behavior from that in loading regime 1; the pore pressure increase in unloading, the large shear strain develops when a dynamic shear stress becomes zero, which can be compared to the flow of a viscous fluid.

Consider that the liquefaction test is stopped before the effective stress path reaches to the line of phase transformation and then the pore pressure in the sample is allowed to dissipate, in this case the sand sample shows the lower rate of the pore pressure increase, namely the high resistancy to the succeeding liquefaction test (Ishihara (2) Finn et.al.,(3)). This is believed to be a result of the fact that the negative dilatancy in the succeeding shearing becomes small, because the deffects involved in the virgin sample have already disappeared in the small pre-shearing. The hardenig effect due to the small shearing is also found even for the pre-liquefied sample.(see Table 3)

In a granular assembly, the defects is believed to be a local volume fraction which has a loose prcking compared with the ovarall packing. The disappearance of the defects, therefore, lead to the increase of the coordination number as a neccesity, it is also believed that the newly formed fabric will be effective to withstand the external force.

4-2 Loading Regime 2

A sand sample which was previously sheared up to the loading regime 2, the sample may be sheared in drained condtion or in undrained condition, shows almost elastic behavior in the same direction of loading and the lower rigidity and the large densification in the oppasite direction of loading. The sample has a directional dependency of loading. (see Fig.7,Fig.8 and Fig.9 and their captions).

According to Oda (11), the normal directions of cantact between grains concentrate toward the maximum principal stress in the course of shearing in order to withstand the external loadings. This finding is supported by the two-dimesional compression and/or simple shear test. (see Konishi (9), Oda & Konishi(12)) The fabric of the sand sample which sheared up to loading regime 2 has a preferred orientation, which has more number of contact in the maximum principal stress and less number of cantact in the minimum principal stress direction, such a state of fabric as this shows a kind of elastic behaivor in the same direction of loading, because the resistant fabric was already created in the previous loading. On the other hand, the created fabric in the previous loading needs a great number of grain movements in order to withstand the loading in the opposite direction, which results in the large densification and the lower rigidity. Remember that the coordination number decreases in the loading regime 2, which also accenturates the densification process.

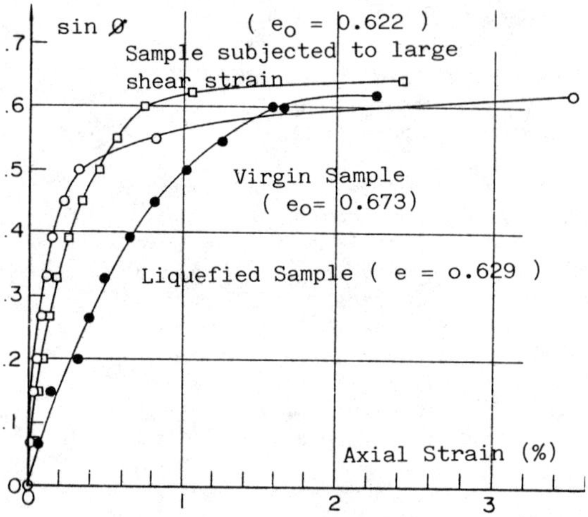

Fig.7 : The lower rigidity of the sample subjected to large shearing in the opposite direction of loading

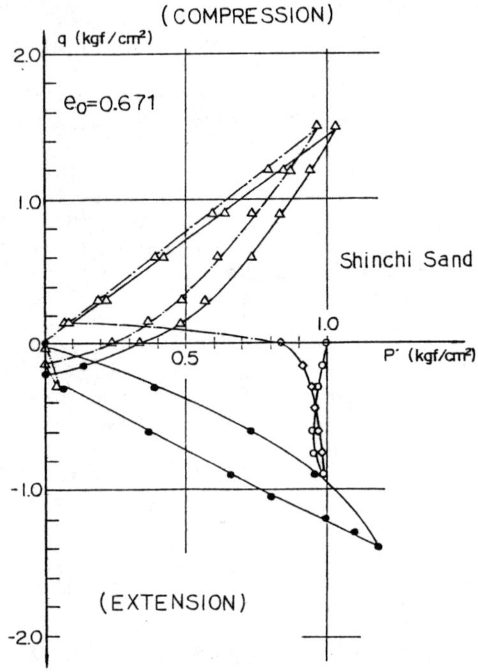

Fig.8: The sudden increase of pore pressure in the sand specimen which was subjected to large shear stress in the direction of extension

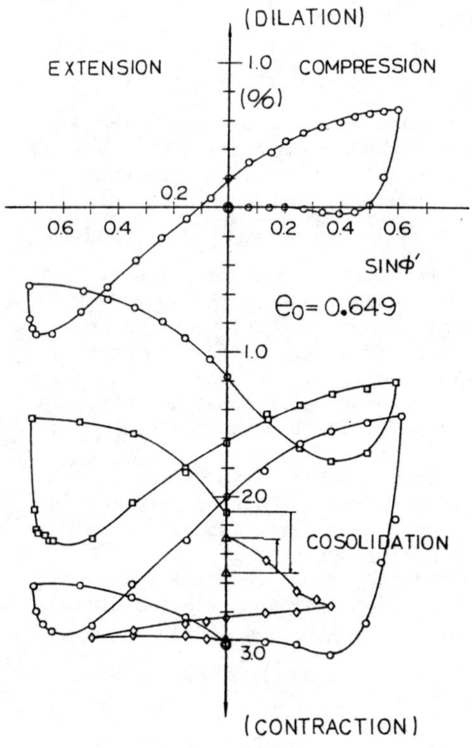

Fig.9: Volumetric change of sand specimen in a large cyclic shear test. ←

These behaviors of a previously large sheared sample:lower rigidity and large densification in the opposite direction of loading, are key points in the consideration of the re-liquefaction problem as well as the physical process of liquefaction itself. The conclusion on whether a liquefied sand deposit in situ shows a high liquefaction potential at a future big earthquake or not has not been cleared yet. However, we can say that the sand deposit with the highly oriented fabric due to the earthquake excitation and the succeeding densification will show the higher liquefaction potential.

The argument discussed above is not the case with the strain controlled test, where unloading means the zero shear strain state. At zero shear strain the highly oriented fabric is already destroyed in the loading from zero shear stress to the stress state at which zero shear strain can be attained. Consequently, the remarkable directional dependency can't be observed (see Fig.3, where the liquefaction potential of the sample which was previously sheared up to loading regime 2 and then unloaded to the zero shear strain state is not so high compared to the virgin sample.(see the experimental results in Nemat Nasser and Tobita (6))

Unfortunately we can't draw any cinclusion on the problem of reliquefaction in situ, since the stress state and the boundary condition of the testing apparatus are not the same, e.g.; in situ we can't expect any shear stress for the sand deposit when it is requefied. However this state can't be reproduced in the ordinary testing apparatus.

The densification mechanism of the sample, which was previously subjected to a direction of loading beyond loading regime 2, in the other direction of loading has been already discussed by Nemat Nasser & Tobita(6) and Matsuoka (14). In their treatments, the change of dilatancy angle along a slip plane was considered. However, such treatment as these don't seem to be sufficient for the description of the stress deformation behavior of liquefied sand.

The other key point of the liquefied sand sample is the large densification in unloading (see Fig.8), which lead to the complete loss of the confining pressure at zero shear stress state and the succeeding flow. The interpretation of the densification in unloading will need a consideration of the decrease of the coordination number as well as the induced fabric of the liquefied sand sample. The satisfactory explanation of this problem should be postponed in the future study.

The large densification of the liquefied sample under reconsolidation is also an interesting and important feature from practical point of view as well as theoretical point of view, since the large settlement of sand deposit lead to the failure of super-structures. The microscopic mechanism of this behavior has not been made clear yet. This behavior may be related to the relationship between the coordination number and the void ratio in the granular assembly; which is sometimes considered as a unique one irrespective of the preparation mehtod, the large densification of the liquefied sample may raise a question on this unique relationship.

A comprehensive study from microscopic point of view is neccesary in order to understand the mechanism of the behaviors mentioned above: the large densification of the sample subjected to a large shearing and the large densification of the liquefied sample in reconsolidation. Especially the importance of dead particles and/or the coordination number in the granular assembly should be clearified in the future study in order to make clear the several important features of the deformation characteristics of the sand sample subjected to a large shearing and/or a liquefaction history.

Fig.10 is a result of the continuous cyclic shearing of saturated sand with a relatively high shear stress. The effect of an opposite previous loading on the undrained deformation behavior of the densified sample due to reconsolidation is clearly understood. The sand specimen shows a liquefaction behavior at each shearing stage irrespective of the fact that the sample becomes denser and denser after each shearing. It is understood, therefore, that the weakening effect due to the oriented fabric will override the hardening effect due to the densification. It should be noted that the developed shear strain becomes smaller and smaller according as the cyclic number increases. It can be said that the densification of the liquefied sample during reconsolidation lead to small displacement of the sample, probably due to the high potential of positive dilatancy of the densified sample.

4-3 Loading Regime 3

According to Finn et.al (3), the sample which was previously subjected to a fairly large shear strain shows a high liquefaction potential even in the strain controlled test. Since the residual state is zero shear strain, we can't account for the high liquefaction potentail by the oriented fabric only. Other interpretation is necessary to explain this behavior. The increase of the dead particles in the granular assembly during the previous large shearing: loading regime 3, may take part in the interpretation of the high liquefaction potential, since the dead particles must get new contact in order to become active and withstand the succeeding loadings, which lead to the large densification as a necessity. However, the interpretation mentioned above can't be a sufficient one but a possible one.

4-4 Summary

From the discussions above, it can be understood that the deformation characteristics of a saturated sand in the cyclic shearing have not been explained in a sufficient manner from microscopic point of view. The problems associated with liquefaction behavior are inter-

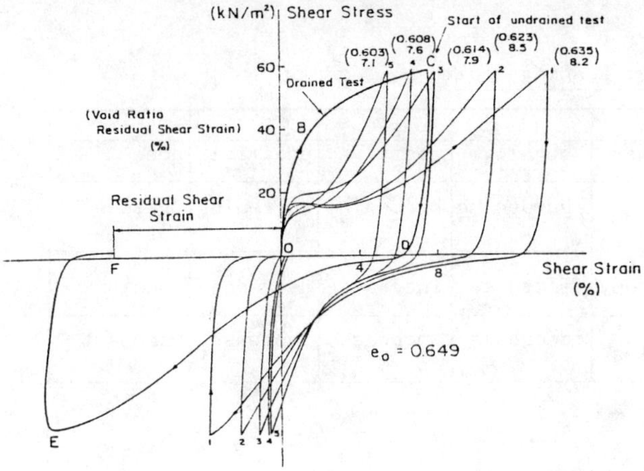

Fig.10: Continuous liquefaction test with a large stress amplitude: the effect due to induced anisotropy seems to be more important than that due to the increase of density (see the detail, Nemat Nasser and Tobita(6))

TABLE 1. Physical Properties of Materials

(a) Photoelastic Material

Friction Angle	Cohesion	Cross Section	r_1/r_2
26°	none	oval 2	1.4

(b) Monterey No.0 Sand

Specific Gravity G_s	Maximum Void Ratio e_{max}	Minimum Void Ratio e_{min}	Average diameter of Grains D_{50}
2.65	0.83	0.53	not measured

(c) Shinchi Sand

G_s	e_{max}	e_{min}	D_{50}
2.738	0.854	0.531	0.35^{mm}

TABLE 2. Loading Regime

	1	2	3	4
dilatancy	negative	positive	positive	none
state	hardening	hardening	softening	critical state
anisotropy	increase	increase	decrease	constant
dead particles	decrease	increase	increase	constant

TABLE 3. Liquefaction tests with various strain history

Shinchi Sand, Triaxial test

Loading steps	1	2	3	4	5	6	7
Stress state	σ_3=1.0 σ_d=0.5 (kgf/cm2)	"	σ_d=0.2	σ_d=0.5	σ_d=0.3	σ_d=0.5	σ_d=0.6
Cyclic number to liquefaction	N_L= 483	1	n= 200	N_L= 3	n= 1391	n= 900	N_L= 1063
max. shear strain	5.90 (%)	7.77	——	4.55	——	——	2.19
Void ratio	0.644	0.623	0.585	0.584	0.558	0.550	0.549
Volume reduction	1.29 (%)	2.29	0.07	1.65	0.52	0.09	0.82
note			$\frac{U}{\sigma_0}$=0.1		$\frac{U}{\sigma_0}$=0.8	$\frac{U}{\sigma_0}$=0.12	
(Dr = %)	(65.0)	(71.5)	(83.3)	(83.6)	(91.6)	(94.1)	(94.4)

esting and important for the constitution of granular mechanics It is of course understood that the better understandings of these problem lead to the better prediction of the behavior of sand deposit under earthquake shaking. In order to get a sufficent answer to the problems discussed so far, we must attack to a mountain of problem which have not been discussed up to date, which will be discussed in the next section.

5. CONTINUUM MECHANICAL DESCRIPTION OF GRANULAR MATERIALS

So far, we have discussed the deformation characteristics of saturated sand in the cyclic shearing apparatus, which are closely related to liquefaction behavior , from microscopic movements of grain particles obsereved in the bi-axial compression test. In order to discuss this problem in a more efficient way, a fabric tensor was introduced in the previous paper (13). The fabric tensor was derived based on the spatial distribution of the contact planes in the granular assembly.

The problems discussed in this paper can be treated based on the fabric tensor. The introduction of the fabric tensor is aiming at the description of granular materials within the framework of continuum mechanics. It seems to be possible that we develop a comprehensive constitutive equation which can account for stress path dependent behavior and the coupling effect between shear and consolidation, in which the fabric tensor and its evolution in the course of loadings should be described explicitly in the mathmatical formulation.

Before treating such a constitutive equation, we should solve the fundamental problems on the behavior of granular materials. They can be summerized as follows:

(1) A stress tensor for granular materials; how to determine a stress tensor for granular materials?

(2) Kinematics of granular materials; how to incorporate the sliding and the rolling of grain particles into the continuum mechanical description?

(3) An evolutionary equation; How does the fabric tensor of granular materials change when a set of loadings is applied?

(4) A stiffness tensor: how to relate the change of the fabric tensor (second rank tensor) to the stiffness tensor(fourth rank tensor) of granular materials?

(5) Heterogeneity; how to treat the heterogeneity involved in the granular assembly in the formulation of constitutive equation?

The problems listed above are important for the better understanding of the deformation characteristics of saturated sand in the cyclic shearing apparatus. Actually we can't draw any conclusion on the problem discussed in this paper before we can get a sufficient answer to the problems listed above.

6. CONCLUDING REMARKS

The deformation characteristics of saturated sand in the cyclic shearing apparatus are discussed from microscopic point of view, which are important for the better understanding of liquefaction behavior.

Based on the bi-axial compression test performed by Konishi, Oda and Nemat Nasser (7), two fundamental microscopic mechanism of deformation of granular materials are proposed: the induced anisotropy and the dead particles(or the coordination number). The important characteristics of the deformation of saturated sand specimen which was subjected to a previous stress history have been discussed based on the proposed mechanisms. From the study reported in this paper, the effect of the induced anisotropy and the change of dead particles (or coordination number,i.e., the internal change in the granular assembly, on the sucseeding behavior in the cyclic loadings are found to be significant.

It is also understood, however, that we have a lot of problem associated with liquefaction behavior which should be solved in the future study. Any conclusions on the problem discussed in this paper should be postponed until we get a better understanding of the fabric state of granular materials and the change of it in the course of loadings and we can get a relationship between the macroscopic quantities e.g.; a stiffness matrix,ovarall stress tensor, deformation rate tensor etc. and the microscopic state and its change.

7. ACKNOWLEDGEMENTS

This work was initiated at Northwestern University under the supervision of Prof. Nemat Nasser, The author wishes to express his sincere gratitude to him for his valuable suggestions and discussions. He also appreciate Prof.E. Yangisawa of Tohoku University for his continuous guidance. Prof. Konishi and Prof. Oda were kind enough to permit him to analyse their original data on the bi-axial compression test. Their kindness and helpful discussions are also appreciated.

8. REFFERENCES

(1) Seed H.B., P.P. Martin and J.M. Lysmer(1976) " Pore water pressure changes during soil liquefaction " ASCE, Vol.102, GT4, 323

(2) Ishihara K. (1978) " Stress history effects on the behavior of sand" Proceedings of U.S.-Japan Seminar on Continuum Mechanical and Statistical Approaches in the Mechanics of Granular Materials, Gakujutsu Bunken Fukyukai, Tokyo

(3) Finn W.D.L., P.L.Bransby and D.J.Pickering(1970):" Effect of strain history on liquefaction of sand " ASCE, Vol.96, SM6, 1917

(4) Ladd, R.S. (1977):" Specimen preparation and cyclic stabilty of sands " ASCE, Vol.103, GT6, 535

(5) Mulilis,J.P., H.B.Seed, C.K.Chan, J.K.Mitchell and K.Aralanandan (1977):" Effects of sample preparation on sand liquefaction " ASCE, Vol.103, GT2,91

(6) Nemat Nasser S. and Y. Tobita (1982):" Influence of fabric on liquefaction and densification potentail of cohesionless sand " Mechanics of Materials, Vol.1, No.1, 43

(7) Konishi, J., M.Oda and S.Nemat Nasser (1982):" Induced Anisotropy in Assemblies of oval cross sectional rods in bi-axial compression"

(8) Vialov, S.S.,YU.K.Zaretsky,P.V.Maximyak and N.K.Pekarskaya (1973): " Kinetics of structural deformation and failure of clays " Proc. of 8th ICSMFE, Moscow, Vol.2-1, 459

(9) Konishi J. (1978):" Microscopic studies on the deformation and strength characteristics of granular material like sand " Doctorial thesis of Kyoto University (in Japanese)

(10) Ishihara, K.,F. Tatsuoka and S. Yasuda (1975):" Undrained deformation and liquefaction of sand under cyclic stresses " Soils and Foundations, Vol.15, No.1, 29

(11) Oda, M. (1972):" The mechanism of fabric changes during compressional deformation of sand " Soils and Foundations, Vol.12, No.2, 1

(12) Oda, M. and J.Konishi (1974):" Microscopic deformation mechanism of granular material in simple shear " Soils and Foundations, Vol.14, No.4, 24

(13) Tobita Y. (1982):" A continuum mechanical description of granular material and its application to liquefaction problem " U.S.-Japan Seminar on New Models and Constitutive Relations in the Mechanics of Granular Materials, ITHACA, (in print)

(14) Matsuoka, H. and K. Takeda (1980):" A stress strain relationship for granular materials derived from microscopic shear mechanism " Soils and Foundations, Vol.20, No.3, 45

RESPONSE OF SAND IN CYCLIC TORSIONAL LOADING INFLUENCED BY ROTATION OF PRINCIPAL STRESS DIRECTIONS

I. Towhata and K. Ishihara

Department of Civil Engineering, University of Tokyo
Tokyo, Japan

ABSTRACT

In order to examine effects of continuous rotation of principal stress directions on the cyclic response of sand, three types of undrained cyclic loading tests were conducted on hollow cylindrical specimens of loose sand using a triaxial torsion shear test apparatus in which torsional mode and triaxial mode of shear stress can be applied to the test specimen. The first type of test was the conventional cyclic torsion shear test and the second type was the cyclic triaxial shear test executed only on the triaxial compression side. The third type of test consisted of simultaneous execution of the above two modes of loading in which continuous rotation of principal stress directions is induced during the cyclic loading. The test results disclosed the fact that the effect of rotating principal stress axes acts towards fostering pore water pressure buildup in the test specimen, as compared to the cyclic torsional tests in which no continuous rotation of principal stress directions is involved.

1. INTRODUCTION

Laboratory cyclic loading tests by means of the triaxial, simple shear and torsional shear test apparatus have provided most of our understanding on the response of saturated sands under cyclic loading conditions. In these modes of cyclic load application, the axes of principal stresses are changed suddenly by 90 degrees at the instant of load reversal. In many of the mode of load application such as those encountered in seismic enviroments, the manner of change in the principal stress directions as above may be deemed as duplicating the manner of actual stress changes occurring in the soil deposits in the field. However, several other cases of cyclic loading enviroments exist where the mode of load application is associated with the continuous rotation of the principal stress directions. It was shown by Ishihara and Towhata (1983a) that the change in shear stress induced in the seafloor deposit by waves travelling overhead involves a continuous rotation of the principal stress axes. Also, the shear stresses induced by traffic loads in the subgrade of road pavements or railways are associated with the rotation of principal stress directions.

The effects of principal stress axis rotation on the response of clean sand have been meticulously investigated by Arthur et al (1979,1980, 1981) by means of a specially designed flexible boundary shear test apparatus. Some of their test results revealed the important features of the influence of rotating principal stress directions for understanding the deformation and strength characteristics of sand. In an effort to simulate the complete rotation of principal stress directions occurring in a seabed deposit due to travelling waves, Ishihara and Towhata (1983a) conducted a series of cyclic undrained tests on saturated specimens of loose sand using a triaxial torsion shear test apparatus. The results of their tests revealed that the continuous rotation of principal stress directions significantly reduces the cyclic resistance of sand, as compared to the conventional types of cyclic loading tests involving no continuous rotation of principal stress axes. Ishihara and Towhata (1983b) also investigated the effect of continuous rotation of principal stress axes executed only on the side of triaxial compression by means of the triaxial torsion shear test apparatus.

The study described in the following pages is a continuation of the previous investigation as cited above and mainly deal with the influence of a particular type of cyclic principal stress rotation on the deformation response of sand.

2. TRIAXIAL TORSION SHEAR APPARATUS

The test apparatus used was a triaxial torsion shear apparatus which permits torsional shear stress to be applied to a specimen independent of vertical and horizontal stresses. A cross section of the test device is shown in Fig. 1. A hollow cylindrical test specimen measuring 10 cm in outer diameter, 6 cm in inner diameter and 10.4 cm in length and encased in rubber membranes is placed in the triaxial chamber as shown in the figure. The test apparatus was designed so that the pressure inside the hollow cylindrical specimen could be varied, if necessary, independently of the pressure outside the specimen. In the present study, however, the chamber pressures both inside and outside were held at identical levels. Therefore, the three components of stress, i.e. vertical stress, lateral stress and

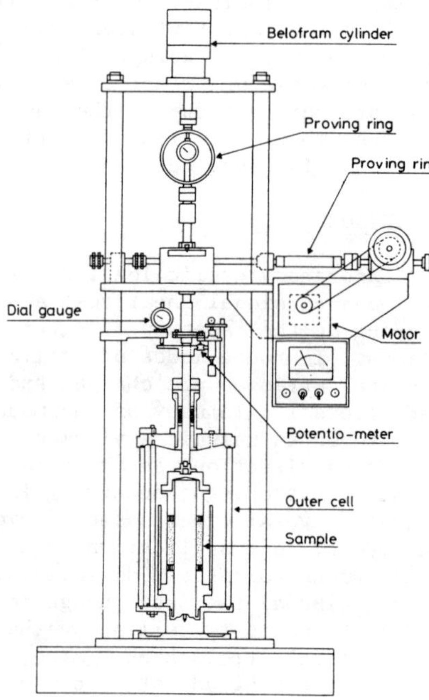

Fig.1 Triaxial torsion shear test apparatus

torsional stress, were varied to produce cyclic excursion of shear stresses in the test specimen.

The vertical load was applied through a vertical rod connected to a Belofram cylinder which was operated by air pressure. A proving ring was used to monitor the vertical load and a dial gauge to measure the vertical displacement of the test specimen. The lateral stress was applied by a change in the chamber pressure. The torsional shear stress was applied by a rotational force transmitted to the vertical rod above the triaxial cell. The rotational force produced by a motor was monitored by a proving ring and the rotational angle measured by means of a potentiometer. Pore water pressures were monitored by an electric transducer installed at the base of the triaxial cell.

3. TEST MATERIAL AND PROCEDURES

The sand used in the present study was a type of Japanese standard sand called Toyoura sand. The mean particle size, D_{50}, and uniformity coefficient, U_c, of this sand are 0.17 mm and 2.0, respectively. Subrounded to subangular in grain shape, the sand has maximum, e_{max}, and minimum, e_{min}, void ratios of 0.98 and 0.60, respectively. The specific gravity, G_s, of this sand equals 2.65.

Air-dried sand mixed with carbon dioxide gas, CO_2, was poured into the annular space between the outer and inner molds by means of a vinyl pipe. A 10 to 13 cm height of fall was employed to spread the sand to a relative density, D_r, of 40 to 50 %. After forming the specimen, it was saturated under a back pressure of 29.4 kN/m^2. With the aid of the CO_2 gas, full saturation was achieved with a B-value exceeding 0.99. The specimen was consolidated under a confining stress of 294 kN/m^2. The cyclic shear stress application was made statically with a shear strain rate of 0.14 percent per minute. All the tests reported in this paper were conducted under undrained conditions with pore water pressure measurements.

4. LOADING SCHEME

The cyclic loading scheme involving changes in both the torsional shear stress and the axial stress difference was programmed so that the direction of the principal stress could be continuously rotated. Conventional types of cyclic torsion shear and cyclic triaxial shear tests without continuous rotation of the principal stress were also performed to provide a basis of comparison with the tests involving the continuous rotation of the principal stress.

Let the shear stress be denoted by τ_{vh}, and the vertical and horizontal effective normal stresses by σ_v and σ_h, respectively, as illustrated in Fig. 2 and 3. Then, the major principal stress, σ_1, and the minor principal stress, σ_3, are given by,

$$\left.\begin{aligned} \sigma_1 &= \frac{\sigma_v + \sigma_h}{2} + \sqrt{\left(\frac{\sigma_v - \sigma_h}{2}\right)^2 + \tau^2_{vh}} \\ \sigma_3 &= \frac{\sigma_v + \sigma_h}{2} - \sqrt{\left(\frac{\sigma_v - \sigma_h}{2}\right)^2 + \tau^2_{vh}} \end{aligned}\right\} \quad (1)$$

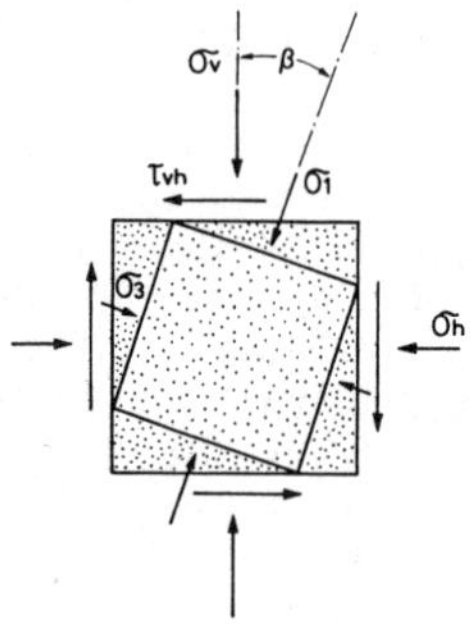

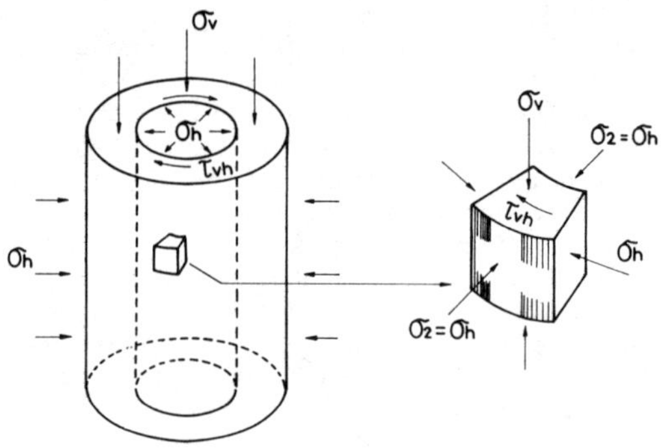

Fig.2 Definition of angle of major principal stress directions

Fig.3 State of Stresses in the hollow cylindrical sample

The angle between the major principal stress direction and the vertical, β, are given by (Fig. 3),

$$\tan 2\beta = \frac{2\tau_{vh}}{\sigma_v - \sigma_h} \tag{2}$$

Therefore, the deviator stress, defined as the difference between the major and minor principal stresses, is given by,

$$\frac{\sigma_1 - \sigma_3}{2} = \sqrt{\left(\frac{\sigma_v - \sigma_h}{2}\right)^2 + \tau^2_{vh}} \tag{3}$$

Corresponding to the stress components, it is possible to define the principal strain components as follows:

$$\left.\begin{aligned} \varepsilon_1 &= \frac{\varepsilon_v - \varepsilon_h}{2} + \frac{1}{2}\sqrt{(\varepsilon_v - \varepsilon_h)^2 + \tau^2_{vh}} \\ \varepsilon_3 &= \frac{\varepsilon_v - \varepsilon_h}{2} - \frac{1}{2}\sqrt{(\varepsilon_v - \varepsilon_h)^2 + \gamma^2_{vh}} \end{aligned}\right\} \tag{4}$$

$$\tan 2\beta = \frac{\gamma_{vh}}{\varepsilon_v - \varepsilon_h} \tag{5}$$

where ε_v and ε_h denote vertical and horizontal strains, respectively, and γ_{vh} is the shear strain in the torsional mode of deformation. ε_1 and ε_3 are, respectively, the major and minor principal strains.

Three types of cyclic loading tests were conducted by changing the

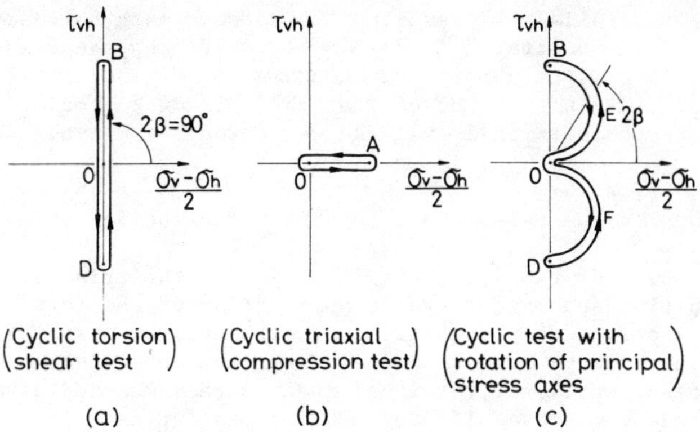

Fig.4 Loading schemes adopted in the tests

two components of shear stress either singly or in combination. The loading schemes of these tests are illustrated in the stress space shown in Fig. 4, in which the shear stress and the stress difference are represented in a rectangular coordinate system.

1. Cyclic Torsion Shear Test

In this type of test, only the shear stress, τ_{vh}, is applied cyclically to the specimen. The test specimen was isotropically consolidated and then subjected to the cyclic torsional shear stress under undrained conditions, while holding the vertical and horizontal stresses unchanged. This type of test is identical to the conventional cyclic torsion shear tests such as those conducted by Ishihara and Yasuda (1975), and Ishibashi and Sherif (1974). In this type of test, the axis of the major principal stress is oriented in a direction 45 degrees from the vertical towards the left in the course of increasing the shear stress, τ_{vh}, towards the right. However, when the shear stress application is directed to the right, the axis of the major principal stress is oriented at 45 degrees from the vertical towards the left as illustrated in Fig. 4(a). Therefore, an abrupt 90 degree rotation in the axis of the major principal stress occurs each time the direction of the shear stress is reversed during the cyclic loading.

2. Cyclic Triaxial Compression Test

Only the stress difference, $(\sigma_v - \sigma_h)/2$, is cyclically applied to the test specimen, while the other component of shear stress, τ_{vh}, is kept equal to zero. The test specimen was first consolidated isotropically and then the vertical stress was increased and decreased under undrained conditions, thereby applying the cyclic stress difference only on the side of triaxial compression. Tbe stress path followed in this type of test is demonstrated in Fig. 4(b), where the shear stress component, τ_{vh}, is plotted versus the component of stress difference, $(\sigma_v - \sigma_h)/2$. It is to be noted that the axis of the major principal stress is directed vertically all the way through the cyclic loading test.

3. Cyclic Torsion Shear Test with Rotation of Principal Stress Axes

In this type of test, the cyclic stress on the triaxial compression side was superimposed on the cyclic application of the torsional shear stress. An isotropically consolidated test specimen was first subjected to the torsional shear stress on one direction, while increasing the vertical stress, until the torsional shear stress reached a value half that of a predetermined amplitude. During the further increase in the torsional shear stress, the vertical stress was reduced so as to become equal to zero when the torsional shear stress reached its maximum value. The stress path during this phase of loading is indicated by points 0 to E to B in Fig. 4(c) in which the shear stress, τ_{vh}, is plotted versus the stress difference, $(\sigma_v - \sigma_h)/2$, in the rectangular coordinate system. Then, the torsional shear stress was reduced, while increasing and decreasing the vertical stress, to the state of initial zero stress. The stress change in this unloading phase was executed following the same stress path as above in a backward direction. Next, the torsional shear stress was cycled in the opposite direction, while changing the vertical stress in the same manner as above. The stress path during this phase of the load-reload cycle is indicated by points 0 to F to D and back to F to 0 in Fig. 4(c). The one complete loading cycle as above was repeated as many times as necessary to produce a large amount of shear strain or a state of initial liquefaction in the test specimens where the developed pore water pressure became equal to the initial confining pressure. It is to be noted that in the course of the above cycle of load excursion, the direction of the principal stresses rotates continuously between angles of + 45 and - 45 degrees to the vertical, as illustrated in Fig. 5. It is also to be noted, however, that the magnitude of the deviator stress as defined by Eq. (3) is changing while the axes of the principal stresses are rotated during the cyclic loading test. It should be noted that, because the intermediate stress, σ_2, remains unchanged ($\sigma_h = \sigma_2$), the magnitude of the major and minor principal stresses changes relative to the intermediate principal stress during the cyclic loading. However, since the cyclic changes in the deviator stress components, $\sigma_1 - \sigma_2$, and $\sigma_2 - \sigma_3$, take place on one side of loading under the conditions of $\sigma_1 - \sigma_2 \geqq 0$ and $\sigma_2 - \sigma_3 \geqq 0$, the plastic component of deformation caused by these complementary stress alteration can generally be deemed as small. Therefore, their effects on the pore water pressure buildup and residual strains will be neglected in the following presentation.

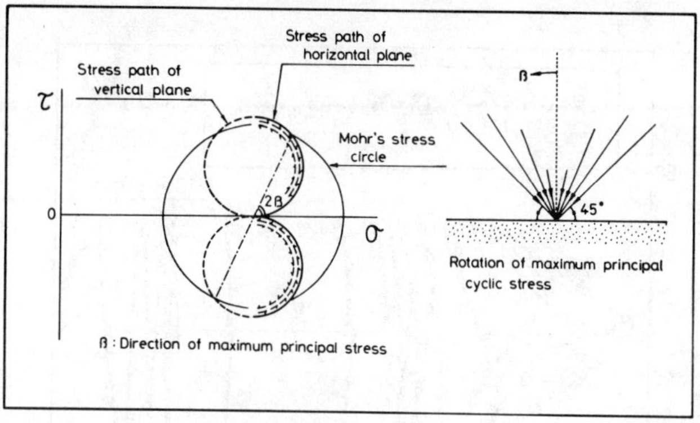

Fig.5 Illustration for two semi-circle type rotations of principal stress directions

5. TEST RESULTS

1. Cyclic Torsion Shear Test

One result of this type of test is presented in Fig. 6. In this particular example, a specimen was isotropically consolidated to a void ratio, e, equal to 0.784 under a confining pressure of σ_0' equal to 294 kN/m^2. The torsional shear stress with a single amplitude of τ_{vh} equal to 55.4 kN/m^2 was applied cyclically under undrained conditions until the specimen was sufficiently strained to bring about a state of initial liquefaction. It may be noted in Fig. 6(a) that the cyclic loading steadily moves the effective stress state of the sample towards the failure envelope and when the developed pore water pressure reaches a certain value, the stress path starts to show a turnaround and moves up towards the failure envelope. The straight line connecting points of the turnaround in the stress path referred to as the line of phase transformation by Ishihara et al (1975). When the sample is in a state of stress just on the line of phase transformation at the last stage of the cyclic loading, the effective confining stress is minimum but it increases as the stress path moves up towards the failure line. In conformity to such an increase in effective confining stress, the stiffness of the sample increases. Such an increase in stiffness can definitely be observed in the stress-strain plot shown in Fig. 6(b). In the last cycle, the specimen deformed to a shear strain of 3.8 percent and upon unloading, the initial liquefaction is seen to have been arrested.

The results of other tests in this test series are presented in Figs. 7 through 9 in the order of increasing amplitude of shear stress employed in each test. The amplitudes of shear stress used in this test series were chosen so as to correspond to each of the shear amplitudes employed in the aforementioned series of the cyclic triaxial compression test.

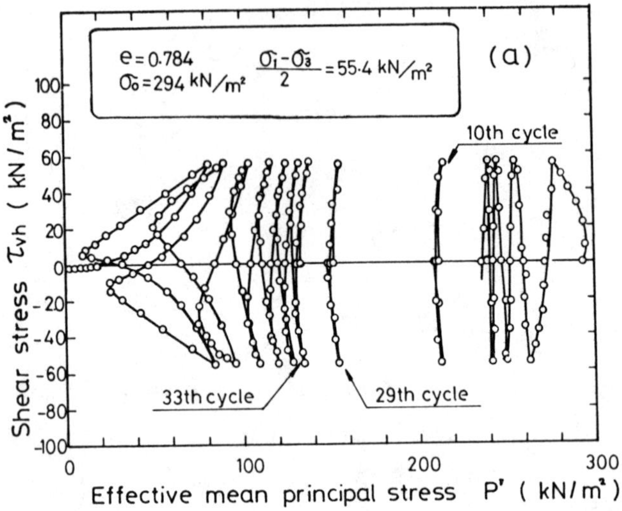

Fig.6(a) Effective stress path in cyclic torsion shear test (Undrained test)

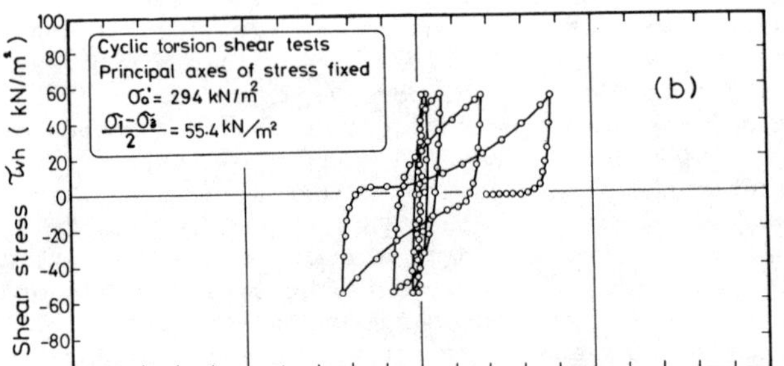

Fig.6(b) Stress versus strain relation in cyclic torsion shear test (Undrained test)

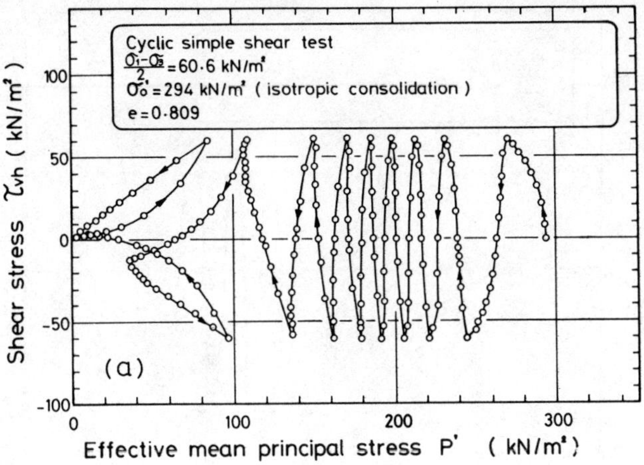

Fig.7(a) Effective stress path in cyclic torsion shear test (Undrained test)

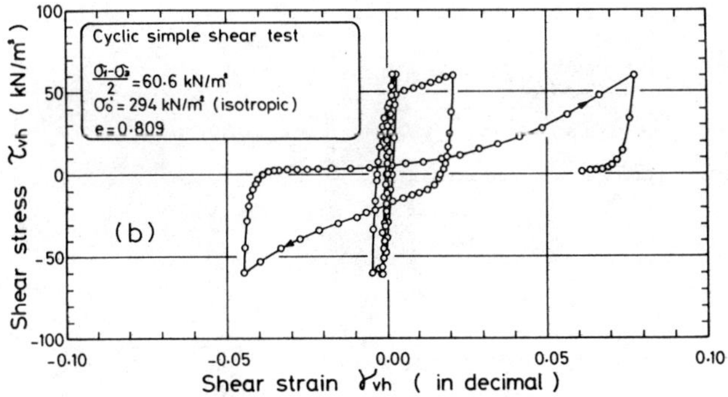

Fig.7(b) Stress versus strain relation in cyclic torsion shear test (Undrained test)

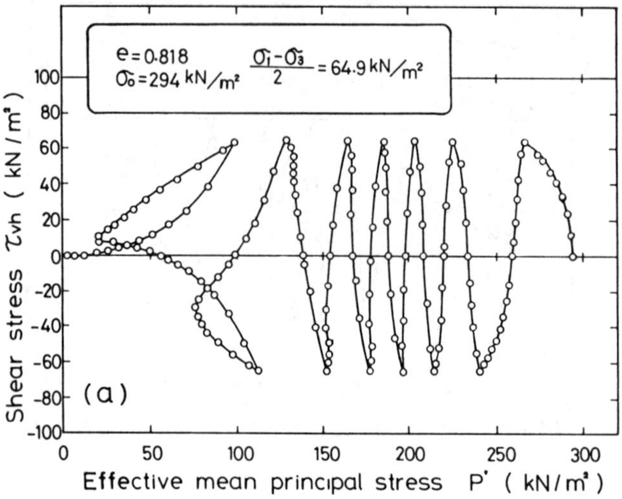

Fig.8(a) Effective stress path in cyclic torsion shear test (Undrained test)

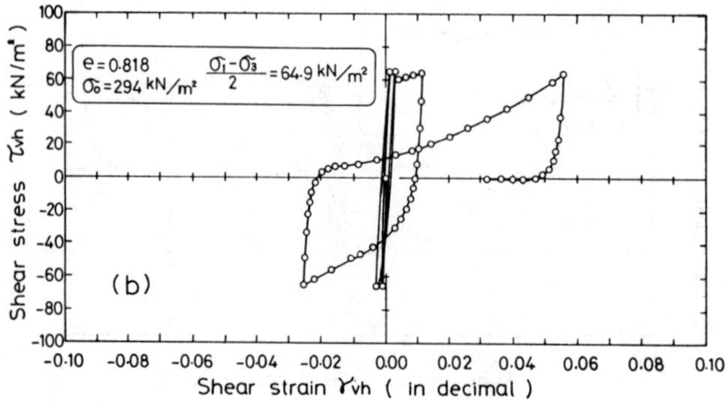

Fig.8(b) Stress versus strain relation in cyclic torsion shear test (Undrained test)

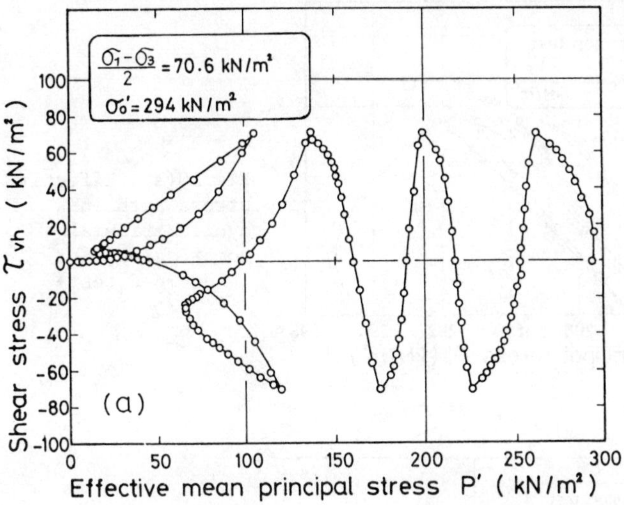

Fig.9(a) Effective stress path in cyclic torsion shear test (Undrained test)

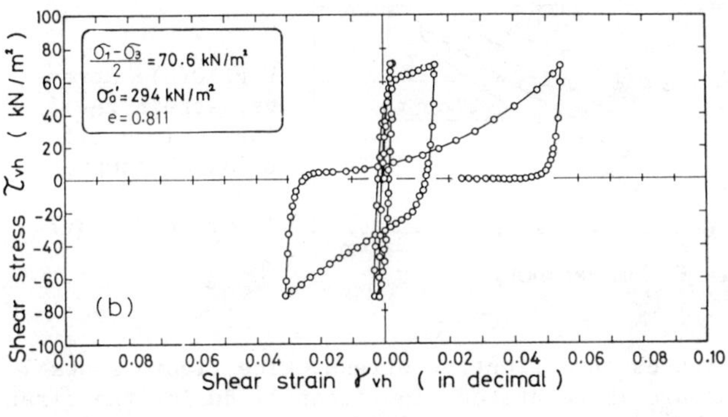

Fig.9(b) Stress vs. strain relation in cyclic torsion test (Undrained test)

2. Cyclic Triaxial Compression Test

A test specimen was first consolidated isotropically under a confining pressure, σ_0', of 294 kN/m^2 to a loose state of packing with a void ratio, e, of 0.823. Then, the vertical stress was applied cyclically on the side of triaxial compression with an amplitude, σ_v, equal to 55.4 kN/m^2 under undrained conditions, while keeping the horizontal stress unchanged. In this test condition, the amplitude of cyclic deviator stress was $(\sigma_1 - \sigma_3)/2$ = 27.7 kN/m^2 as defined by Eq. (3). The results of this test are presented in Fig. 10. On the basis of the measured pore water pressures, the effective mean principal stress, $p' = (\sigma_v' + 2\sigma_h')/3$, was obtained and used to construct the effective stress path during the cyclic loading as shown in Fig. 10(a), where the prime

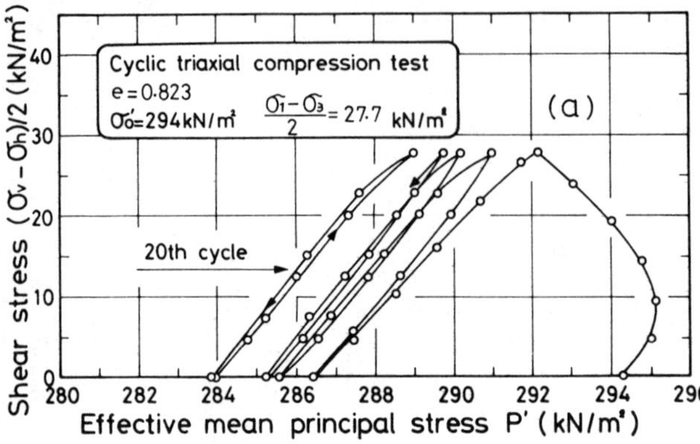

Fig.10(a) Effective stress path in cyclic triaxial compression test (Undrained test)

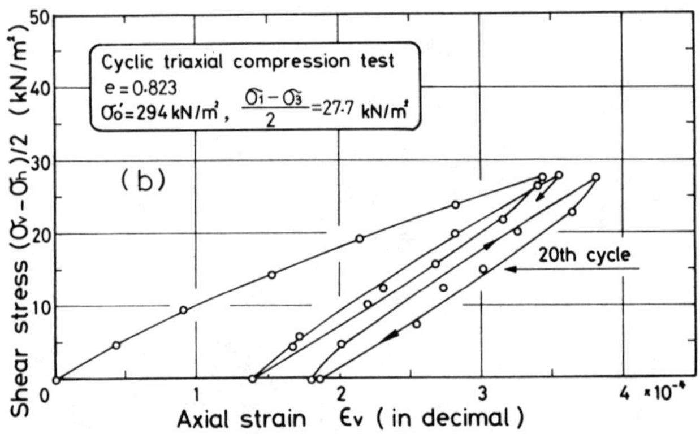

Fig.10(b) Stress vs. strain relation in cyclic triaxial compression test (Undrained test)

above the notations indicates the effective stress. The figure shows that the pore water pressure is generated significantly during the first cycle but the rate of pore water pressure buildup per cycle becomes smaller thereafter as the number of cycles proceeds, until the pore water pressure approaches a terminal value asymptotically. It may be seen that this terminal state of the stress path is determined in such a way that the cyclic undrained stress path stabilizes along the failure envelope. It can be seen in Fig. 10(a) that all the stress paths at a later stage of cyclic loading have nearly equal angles of slope to the horizontal axis. This implies that the cyclic loading produces predominantly recoverable deformation. It should also be noted that at the initial stage of the first loading, the stress path is almost parallel in direction to the stress paths observed at the later stage of cyclic loading. This would appear to indicate that within the limit of small shear stress, the pore water pressure response is recoverable and that a truly residual type of pore water pressure can be produced when the applied shear stress exceeds this limit. The stress-strain relation obtained in the same test is presented in Fig. 10(b), where it also be observed that the stress-strain plot tends to stabilize as the number of cycles increases.

The results of other tests in this series are presented in Figs. 11 through 13.

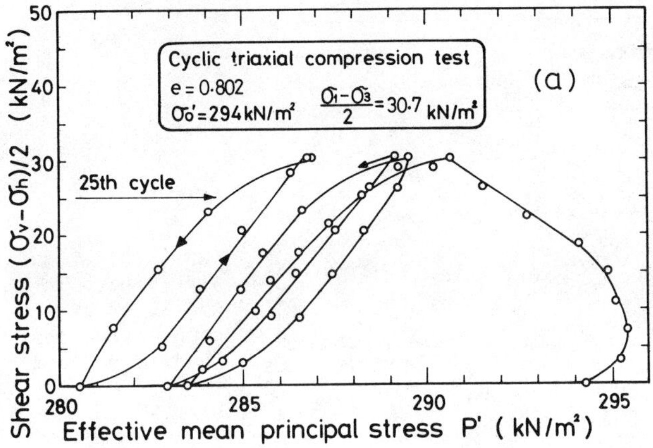

Fig.11(a) Effective stress path in cyclic triaxial compression test (Undrained test)

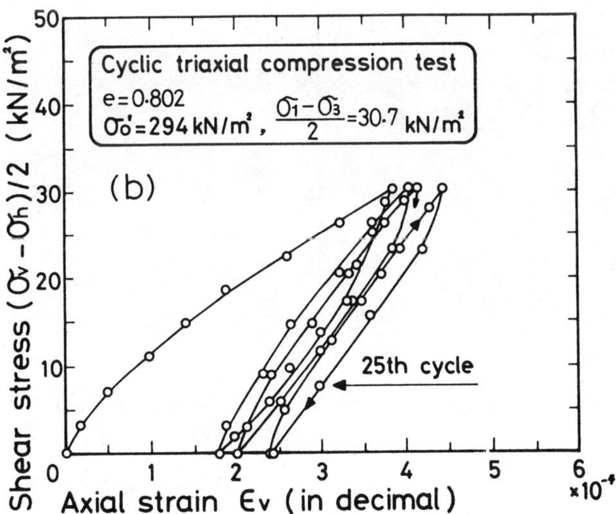

Fig.11(b) Stress vs. strain relation in cyclic triaxial compression test (Undrained test)

In these tests, the amplitudes of deviator stress were increased successively to (σ_1 - σ_3)/2 = 30.7 kN/m^2, 32.7 kN/m^2, and further up to 35.3 kN/m^2. The test specimens were compacted equally loose to a void ratio, e, ranging between 0.800 and 0.834. All the results of these tests show the characteristic deformation response which is similar to the behavior explained above.

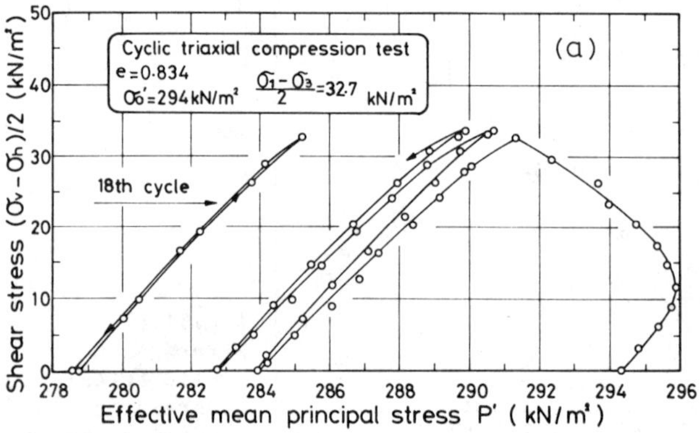

Fig.12(a) Effective stress path in cyclic triaxial compression test (Undrained test)

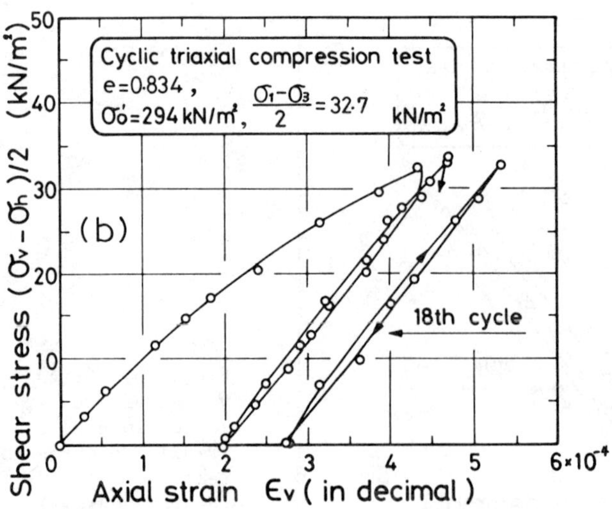

Fig.12(b) Stress vs. strain relation in cyclic triaxial compression test (Undrained test)

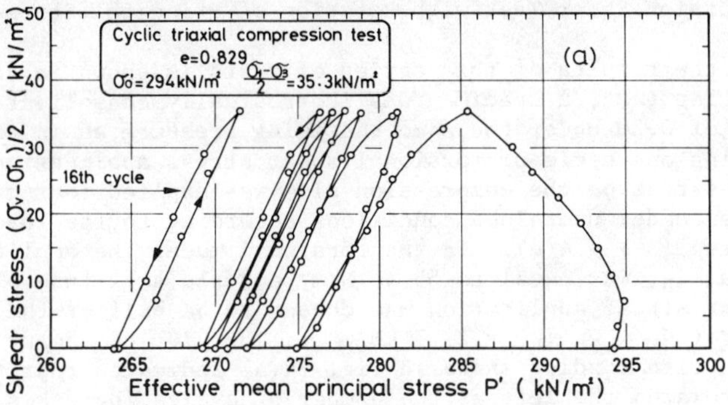

Fig.13(a) Effective stress path in cyclic triaxial compression test (Undrained test)

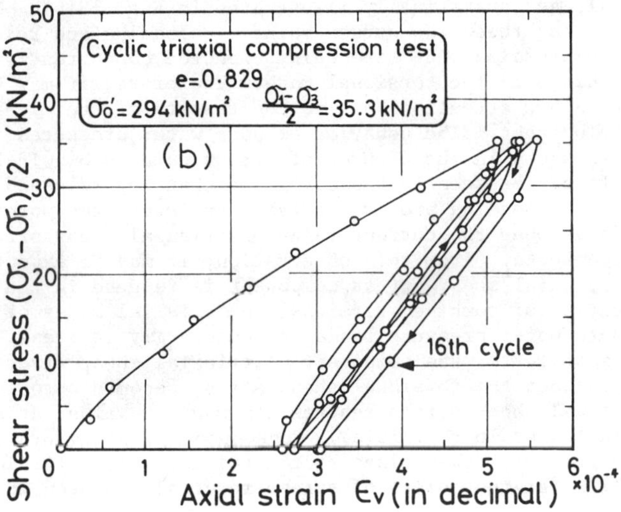

Fig.13(b) Stress versus strain relation in cyclic triaxial compression test (Undrained test)

3. Cyclic Torsion Shear Test with Rotation of Principal Stress Axes

One of the results of this series of tests is shown in Fig. 14. In this particular test, a specimen was isotropically consolidated to a void ratio of 0.80 under the same confining pressure as in the other tests. During one cycle of torsional shear stress application, the triaxial shear stress on the compression side was applied four times to the test specimen under undrained conditions according to the loading scheme as illustrated in Fig.4(c). In the torsional mode, the amplitude of shear stress, τ_{vh}, was equal to 55.4 kN/m^2 and the amplitude of triaxial mode of shear stress application was chosen to be half of the above value, i.e., $(\sigma_v - \sigma_h)/2 = 27.7$ kN/m^2 . The plot of the stress path during the cyclic loading shown in Fig. 14(a) indicates that the stress path moves towards the left as the number of cycles increases. As in the case of the cyclic torsional shear test, the development of pore water pressures is significantly large particularly in the first cycle and in the last few cycles where the applied shear stress takes the specimen to a state of stress beyond the line of phase transformation in the stress space. In the picture of the stress path, one of the features which is different from the stress path of the cyclic torsion shear test is that tiny kinks exist at point of zero shear stress and also at points of stress reversal, as schematically illustrated in Fig. 18(b). It should be recalled that at these two stress points unloading and reloading in the triaxial compression mode are being executed concurrently with the shear stress change in the torsional mode. Therefore, it may be of interest to turn our attention back to the results of the cyclic triaxial compression test. The behavior of pore water pressures in this type of test is shown in the picures of stress paths shown in Figs.10 through 13. It can be seen in these figures that regardless of whether the unloading or reloading process is being executed, the pore water pressure is large when the current value of triaxial shear stress is small, and vice versa. From this observation, it may be expected that whenever the triaxial shear stress component is reduced in the course of cyclic application of combined torsional and triaxial modes of shear stress, the pore water pressure tends to temporarily increase. In the present series of tests, the component of triaxial shear stress is brought to zero when the torsional shear stress becomes zero and also when the torsional shear stress reaches its maximum value, as schematically illustrated in Fig. 18(a). Consequently, a temporary increase in pore water pressure takes place at the points of zero torsional shear stress as well as at the points of stress reversal, as seen in the stress path in Figs. 14(a).

The stress versus strain behavior of the test specimen during the present test is shown in Fig. 14(b). It is of interest to note that the stiffness of the specimen does not temporarily increase in the last portion of the cyclic loading, although the effective confining pressure is increasing with the increase in shear stress as can be seen in Fig. 14(a). This behavior, illustrated schematically by a curve A-B in Fig. 18(c), is in contrast to what was observed in the case of the aforementioned cyclic torsional shear test (curve A-C in Fig. 18(c)).

The results of other tests in this test series are presented in Figs. 15 through 17 in the order of increasing amplitude of shear test employed in each test. The amplitudes of shear stress in this series were chosen so as to correspond to each of the shear stress amplitudes employed in the two aforementioned series of cyclic loading tests.

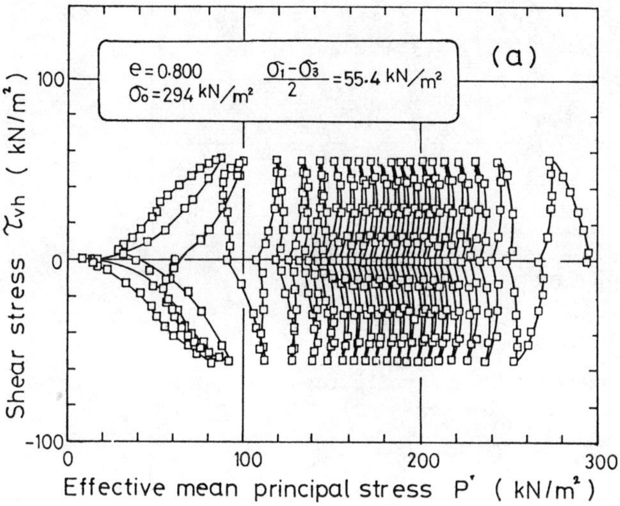

Fig.14(a) Effective stress path in cyclic torsion test with principal stress rotation (Undrained test)

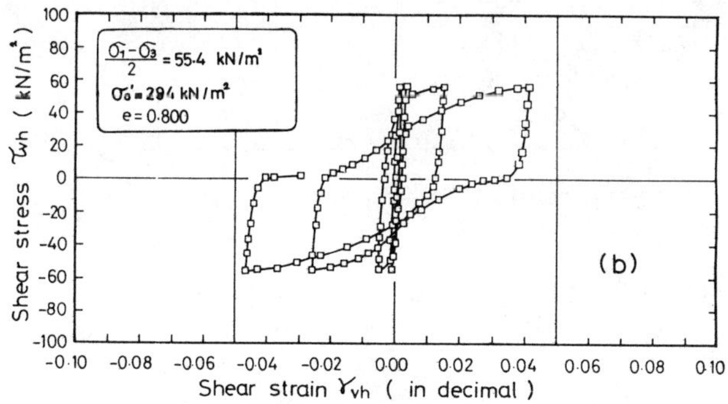

Fig.14(b) Stress vs. strain relation in cyclic torsion test with principal stress rotation (Undrained test)

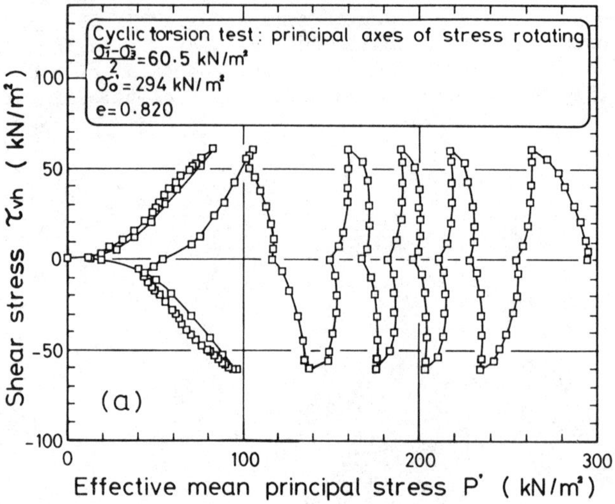

Fig.15(a) Effective stress path in cyclic torsion test with principal stress rotation (Undrained test)

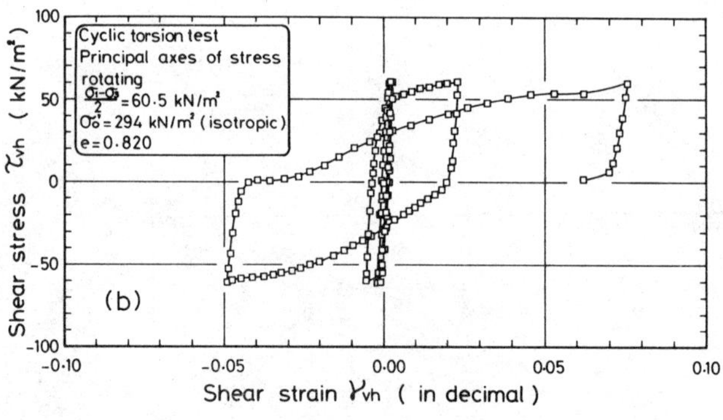

Fig.15(b) Stress versus strain relation in cyclic torsion test with principal stress rotation (Undrained test)

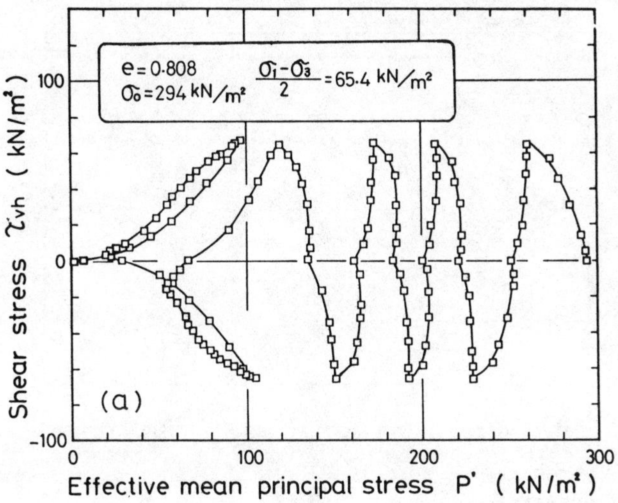

Fig.16(a) Effective stress path in cyclic torsion test with principal stress rotation (Undrained test)

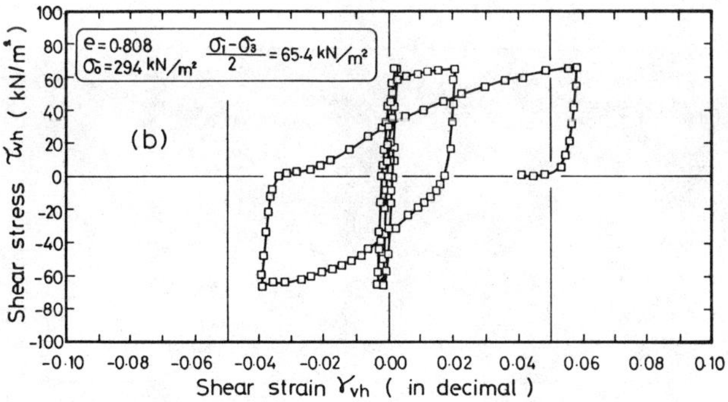

Fig.16(b) Stress versus strain relation in cyclic torsion test with principal stress rotation (Undrained test)

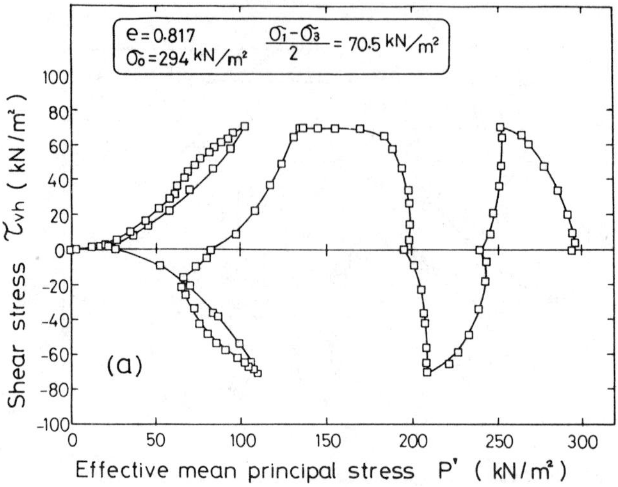

Fig.17(a) Effective stress path in cyclic torsion test with principal stress rotation (Undrained test)

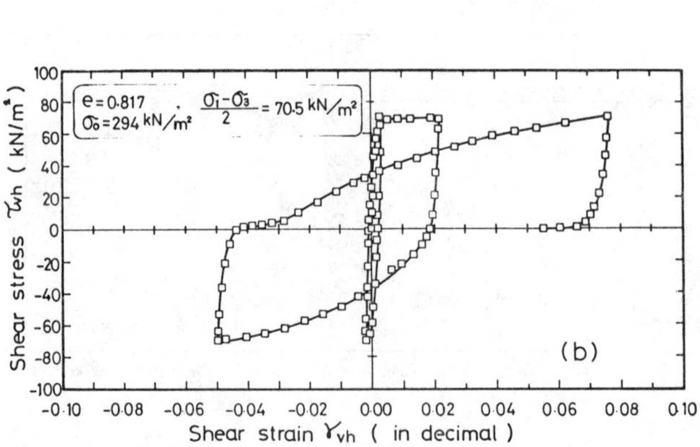

Fig.17(b) Stress versus strain relation in cyclic torsion test with principal stress rotation (Undrained test)

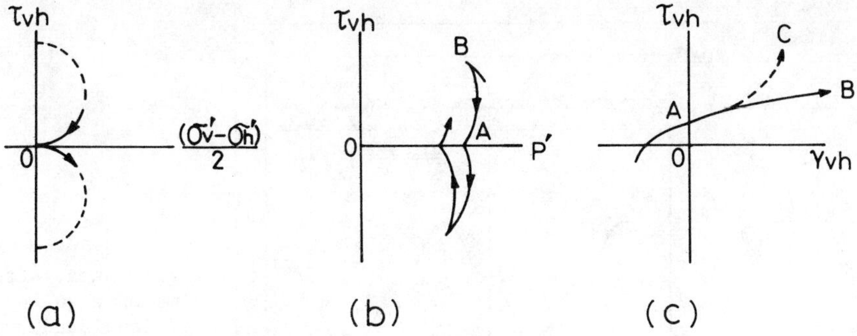

Fig.18 Characteristic patterns of stress and strain changes

6. COMPARISONS BETWEEN TWO TESTS WITH AND WITHOUT ROTATION OF PRINCIPAL STRESS AXES

The stress path traced during the cyclic test with rotation of principal stress axes may be regarded as a superposition of two stress paths; the one followed in the cyclic torsion shear test and the other traced in the cyclic triaxial compression test, as envisioned from the three pictures of stress path shown in Fig.4. The time history of stress application executed in the cyclic test with rotation of principal stress can be decomposed into sinusoidal change in torsional shear stress and non-reversing sinusoidal change in the stress difference applied at a frequency four times as many. Each component of the decomposed shear stress time histories is demonstrated in Fig. 19. Therefore, it may be of interest to observe how pore water pressure can be generated in the combined mode of stress application in comparison to the pore water pressure generation in each mode of shear stress application. In Fig. 20, the pore water pressure generated during a cyclic test with rotation of principal stress axes is plotted with rectangles versus the number of cycles for the case of the amplitude of shear stress, $(\sigma_1 - \sigma_3)/2 = 55.4$ kN/m^2. Also plotted in Fig. 20 with circles is the pore water pressure buildup during a torsional shear test performed with the same shear stress amplitude. It can be seen that the cyclic test with rotation of principal stress axes always produces a greater amount of pore water pressure as compared to the cyclic torsion shear test performed without continuous rotation of principal stress axes. The number of cycles resulting in the attainment of initial liquefaction in the former test is shown to be 20 as against the corresponding number of 39 in the latter.

The pore water pressure generated during the cyclic triaxial compression test performed with a corresponding amplitude of 55.4/2 = 22.7 kN/m^2 can be read off from the plot of the stress path shown in Fig. 6. The value of pore water pressures thus obtained at each cycle of loading were added to the pore water pressures for corresponding cycle of

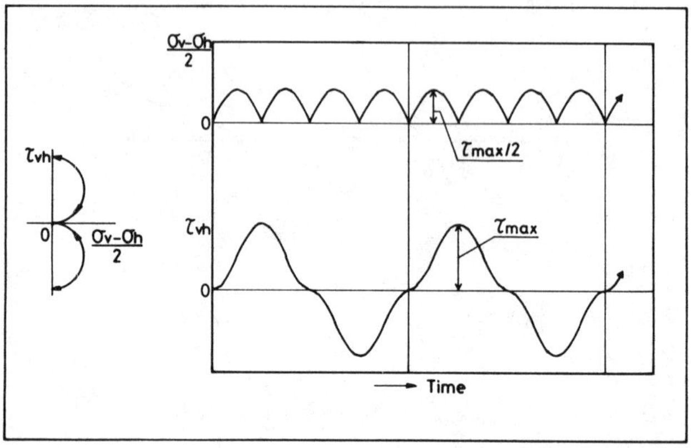

Fig.19 Shear stress time changes in torsional and triaxial compression modes

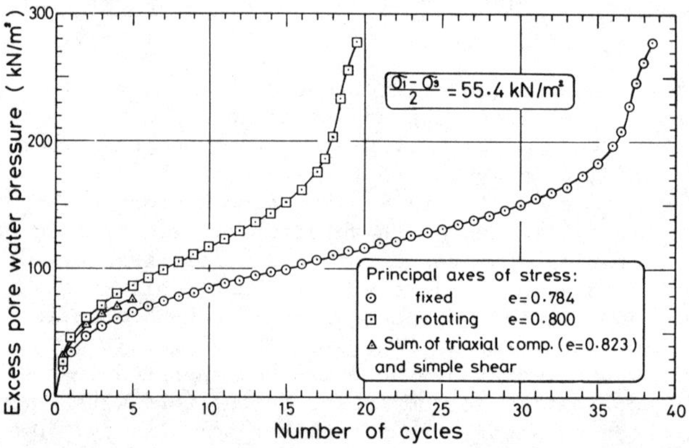

Fig.20 Comparison of pore water pressure generations

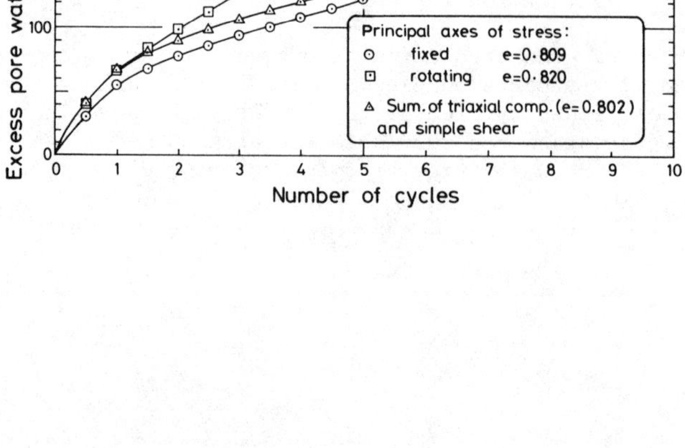

Fig.21 Comparison of pore water pressure generations

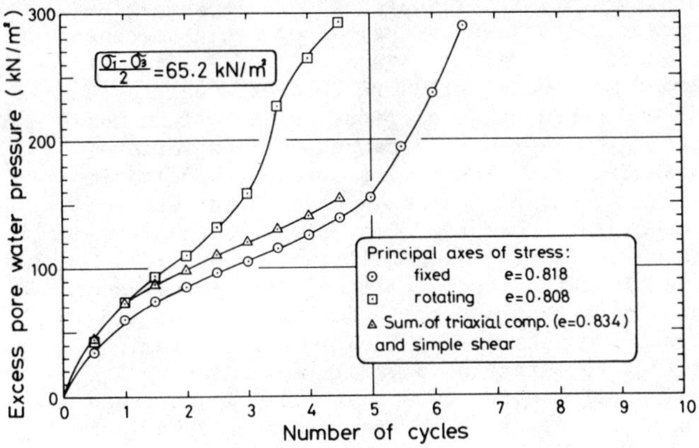

Fig.22 Comparisons of pore water pressure generations

the cyclic torsional shear test performed with the amplitude of $(\sigma_1 - \sigma_3)/2 = 55.4$ kN/m^2. The pore water pressure thus superposed is also shown in Fig. 20 with triangles. It can be seen that the superposed pore water pressure is smaller than the pore water pressure generated during the combined application of cyclic triaxial compression and torsional shear stress.

Similar data processing was also carried out for other sets of test results obtained with identical shear stress amplitude. For example, Fig. 21 compares the pore water pressure buildup between the cyclic torsional shear test and the cyclic test with rotation of principal stress axes both performed with the amplitude of $(\sigma_1 - \sigma_3)/2 = 60.6$ kN/m^2. Also plotted with triangular marks in Fig. 21 is the sum of the pore water pressures obtained from the cyclic torsion shear test and from the cyclic triaxial compression test. Fig. 21 also indicates that the pore water pressure measured in the application of combined torsion and triaxial compression is greater than the summation of pore water pressures obtained from the separate application of cyclic torsion and triaxial compression. The results of similar test series conducted with increased deviator stress amplitudes are presented in Fig. 22 and 23. As is the case with the aforementioned tests, it is observed that the combined application of cyclic torsion and triaxial compression involving the rotation of principal stress directions produces a larger amount of pore water pressure than can be produced in the case of separate application of these two modes of cyclic loads. Therefore, it may be concluded that, if two components of cyclic loads are applied so as to induce the rotation of principal stress directions, its effect acts towards increasing the plastic deformation accompanied by a larger amount of pore water pressure buildup as compared to the case where two components of cyclic stresses are applied seperately without inducing the rotation of principal stress directions.

In order to examine the other aspect of the effect of principal stress rotation, the results of tests presented above are summarized in terms of the cyclic stress ratio versus the number of cycles necessary to cause failure in the specimen. Firstly, incidence of initial liquefaction where the developed pore water pressure becomes equal to the initial confining stress was taken up as a criterion of failure under cyclic loading conditions. The test data arranged on this failure criterion are presented in Fig. 24. The cyclic stress ratio is simply defined as the amplitude of torsional mode of applied shear stress divided by the initial confining pressure. Fig. 24 shows that the cyclic stress ratio required to cause initial liquefaction is about 10 % smaller if the rotation is induced in the principal stress directions, as compared to the corresponding cyclic stress ratio for the case of the conventional type torsional shear stress application. As a failure criterion in cyclic loading, the state of a sample deforming to 5 % double-amplitude deviator strain in the torsional mode may be taken up. The test data arranged on this failure criterion are presented in Fig. 26. It may be seen in the figure that, when the rotation is induced in principal stress directions, the cyclic stress ratio causing the 5 % double-amplitude deviator shear strain is also approximately 10 % smaller than the corresponding cyclic stress ratio for the test without continuous rotation of principal stress directions.

7. CONCLUSIONS

A first series of cyclic undrained torsional tests was conducted on hollow cylindrical specimens of loose sand using a triaxial torsion shear test apparatus. Using the same test apparatus, a second series of cyclic triaxial shear tests was carried out in which the axial stress was cycled only on the side of triaxial compression. A third series of cyclic tests consisted of simultaneous application of these two types of cyclic loads, i.e., during one-half cycle of the torsional mode of shear stress application, a complete load-unload cycle was executed on the side of triaxial compression. In this type of cyclic loading scheme, continuous rotation of principal stress directions was induced in the test specimen, in contrast to the intermittent change by 90 degrees occurring at the instant of stress reversal in the first and second types of cyclic loading schemes. Hence, comparisons of the response of sand in the third type of tests with that of the first and second types of tests appear to indicate the effects of the rotation of principal stress directions. The arrangements of the test data from such a point of view revealed that the rotation of the principal stress directions significantly influences the deformation response of sand under cyclic loading conditions. It was shown that the pore water pressure buildup during the cyclic test involving the rotation of principal stress axes was significantly larger as compared to the sum of the pore water pressures obtained separately in each of the aforementioned first and second types of cyclic test without continuous rotation of principal stress axes.

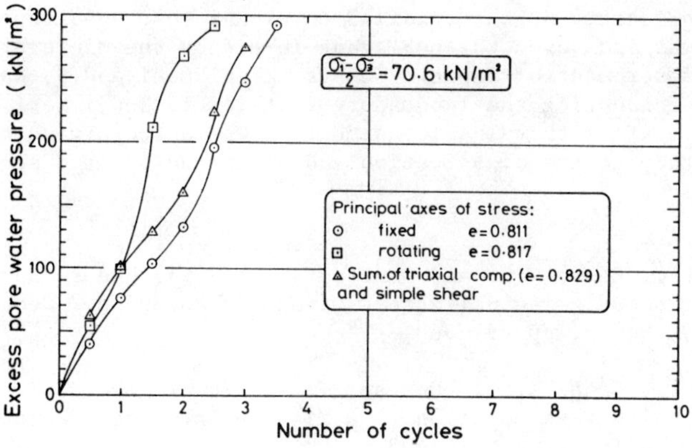

Fig.23 Comparison of pore water pressure generation

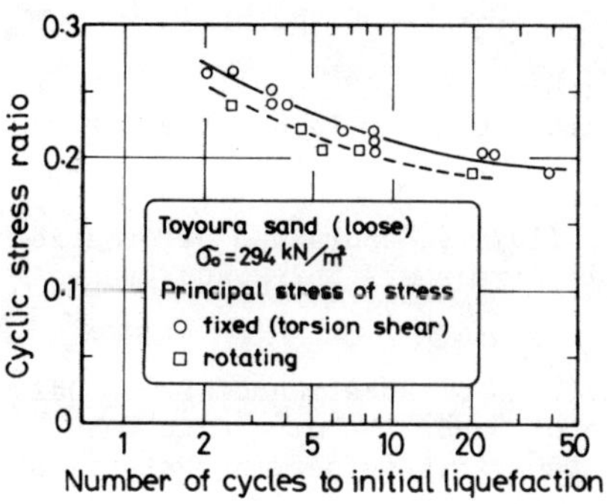

Fig.24 Cyclic stress ratio versus number of cycles to cause initial liquefaction

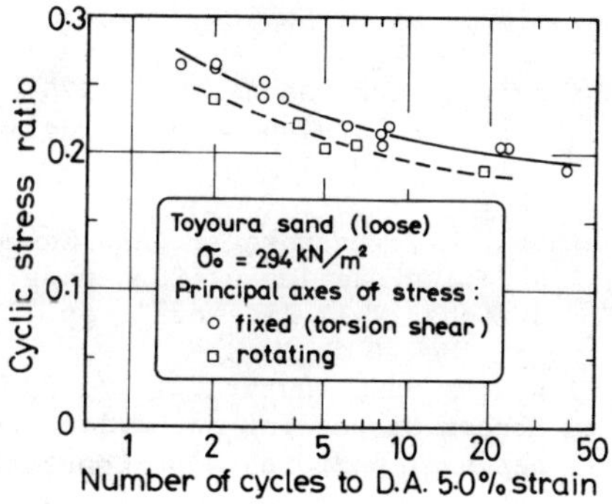

Fig.25 Cyclic stress ratio versus number of cycles to cause 5% double amplitude deviator strain

8. ACKNOWLEDGEMENTS

The manufacture of the triaxial torsional test apparatus used in this study was sponsored through grant-in-aid by the Ministery of Education, Government of Japan. Messrs. Y. Kikuchi and S. Nakazato assisted in conducting the laboratory tests. Dr. Kenji Mori kindly reviewed the draft with valuable comments. The authors wish to express their gratitude to the organization and individuals cited above.

REFERENCES

Arthur, J.R.F., Chua, K.S. and Dunstan, T. (1979), "Dense Sand Weakened by Continuous Principal Stress Direction Rotation," Geotechnique, Vol. 29, No. 1, pp. 91-96.

Arthur, J.R.F., Chua, K.S., Dunstan, T. and Rodoriguez, J.I. (1980), "Principal Stress Rotation : A Missing Parameter," Proc. ASCE, Vol. 106, GT. 4, pp. 419-433.

Arthur, J.R.F., Bekenstein, S., Germaine, J.T. and Ladd, C.C. (1981), "Stress Path Tests with Controlled Rotation of Principal Stress Directions," Laboratory Shear Strength of Soil, ASTM, STP 740, pp. 516-540.

Ishibashi, I. and Sheriff, M.A. (1974), "Soil Liquefaction by Torsional Simple Shear Device," Proc. ASCE, Vol. 100, GT. 8, pp. 871-888.

Ishihara, K. and Yasuda, S. (1975), "Sand Liquefaction in Hollow Cylinder Torsion under Irregular Excitation," Soils and Foundations, Vol. 15, No. 1, pp. 45-59.

Ishihara, K., Tatsuoka, F. and Yasuda, S. (1975), "Undrained Deformation and Liquefaction of Sand under Cyclic Stresses," Soils and Foundations, Vol. 15, No. 1, pp. 29-44.

Ishihara, K. and Towhata, I. (1983a), "Effects of Rotation of Principal Stress Directions on Cyclic Response of Sand," Submitted to International Conference on Constitutive Laws for Engineering Materials, University of Arizona, U.S.A.

Ishihara, K. and Towhata, I. (1983b), "Cyclic Behavior of Sand During Rotation of Principal Stress Axes," Submitted to the U.S. Japan Seminar on New Models and Constitutive Relations in the Mechanics of Granular Materials to be published by Elsevier Scientific Publishing Company.

THREE-DIMENSIONAL DEFORMATION MODEL FOR SAND UNDER DRAINED CONDITIONS

Y. Yamada and K. Ishihara

Department of Civil Engineering, University of Tokyo
Tokyo, Japan

ABSTRACT

A three-dimensional deformation model for sand has been developed using experimental data obtained from true triaxial tests. The proposed model is based on the postulate that any principal strains in three-dimensional deformation of sand can be represented as a summation of strain components which are supposed to develop in two-dimensional slip planes, The model is able to take into account the inherent anisotropy formed by deposition of sand as well as the anisotropoc yielding in the three-dimensional stress space. The strains predicted by the model well simulated measured strains of the sand using the true triaxial device.

1. INTRDUCTION

Investigation of the behavior of sand under three-dimensional stress conditions have revealed that two kinds of anisotropy play important roles in stress-strain relationships of sand. The first type is inherent anisotropy formed by sand deposition. Studies made by Oda [1] and Arthur and Menzies [2] showed that sand specimens with tilted deposition angle exhibited different stress-strain and strength characteristics in triaxial compression tests. Yamada and Ishihara [3, 4] performed a series of true triaxial tests on loose sand specimens and showed that the deformation of sand was affected not only by the stress state represented by relative magnitude of three principal stresses but also by the orientation of the three principal stresses relative to the direction of sand deposition. Using an apparatus capable of controlling normal and shear stresses in two mutually perpendicular directions simultaneously, Arthur et al. [5] also showed the inherently anisotropic nature of deformation behavior of sand specimens caused by deposition under gravity.

The other is anisotropy manifested during the yieldeng of sand. According to the assumption of isotropic yielding, the yield surface expands uniformly from the hydrostatic axis ($\sigma_1 = \sigma_2 = \sigma_3$) in three dimensional stress space. Therefore, when the sand is sheared in compression mode in the triaxial test, yield locus expands not only in the triaxial compression side but also in the triaxial extension side. However, cyclic triaxial tests performed by Tatsuoka and Ishihara [6] indicated that the sand yields independently between triaxial compression and extension. Using the true triaxial device, Yamada and Ishihara [7] performed multi-directional cyclic loading tests and confirmed the anisotropic nature of yielding that can occur in the sand under three-dimensional stress conditions.

In modeling deformation behavior of samd under cyclic loading conditions, therfore, it is necessary to take into account the two types of anisotropy as described above. Based on the postulate proposed by Matsuoka [8], that the deformation of sand can be represented as a linear summation of strains produced in the imaginary two-dimensional stress systems, Tatsuoka [9] developed an elasto-plastic model in which the effects of these two anisotropies can be taken into consideration. Yamada and Ishihara [10] proposed an empirical model in coformity to the above assumption and examined its adequecy by test data obtained using the true triaxial apparatus. This paper is intended to demonstrate the anisotropic yielding in the three-dimensional stress space incorporated into the model in comparison with some additional data of multi-directional loading tests.

2. OUTLINE OF MODEL

The proposed model is based on the postulate that any of the principal strains in three-dimensional deformation of sand can be represented as a summation of strain components that are supposed to develope in the imaginary slip planes. Let i, j and k denote the directions of the three effective principal stresses, σ_i', σ_j' and σ_k'. In the case of the stress conditions, $\sigma_i'>\sigma_j'>\sigma_k'$, three imaginary slip planes are envisaged for three pairs of two-dimensional stress systems $\sigma_i'>\sigma_j'$, $\sigma_i'>\sigma_k'$ and $\sigma_j'>\sigma_k'$, as schematically shown in Fig. 1. Therfore, the three principal strains, ε_i, ε_j and ε_k are decomposed each into two components as follows,

$$\left.\begin{aligned} \varepsilon_i &= f_{ij} + f_{ik} \\ \varepsilon_j &= g_{ij} + f_{jk} \\ \varepsilon_k &= g_{ik} + g_{jk} \end{aligned}\right\} \qquad (1)$$

where f_{ij} means compressive (positive) strain in i-direction and g_{ij} means expansive (negative) strain in j-direction, which are caused by the two-dimensional slippings in $\sigma_i'>\sigma_j'$ stress system. Similarly f_{ik} and g_{ik} are compressive and expansive strains, respectively, for $\sigma_i'>\sigma_k'$ stress system, and f_{jk} and g_{jk} are compressive and expansive strains, respectively, for $\sigma_j'>\sigma_k'$ stress system.

Both Matsuoka [8] and Tatsuoka [9] assumed that the positive and negative strains generated in each two-dimensional

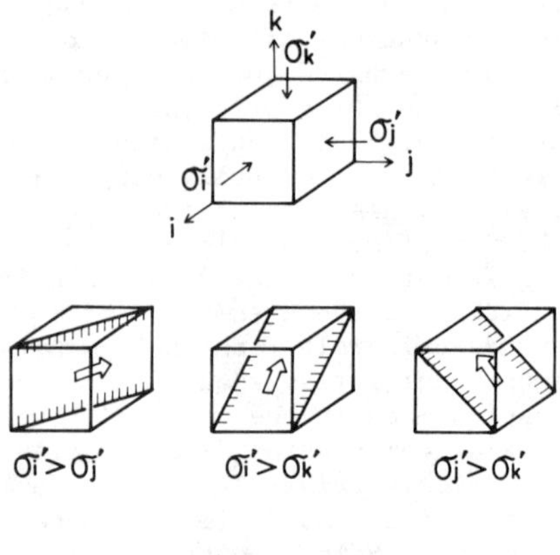

Fig. 1 Three imaginary two-dimensional slippings considered in $\sigma_i'>\sigma_j'>\sigma_k'$ stress condition

stress system are given as a function of stress ratio. For $\sigma_i'>\sigma_j'>\sigma_k'$ stress conditions, therefore, Eq. (1) can be written in the following form

$$\left.\begin{aligned} \varepsilon_i &= f_{ij}(\sigma_i'/\sigma_j') + f_{ik}(\sigma_i'/\sigma_k') \\ \varepsilon_j &= g_{ij}(\sigma_i'/\sigma_j') + f_{jk}(\sigma_j'/\sigma_k') \\ \varepsilon_k &= g_{ik}(\sigma_i'/\sigma_k') + g_{jk}(\sigma_j'/\sigma_k') \end{aligned}\right\} \quad (2)$$

In this paper, however, a more general representation as follows is examined for $\sigma_i'>\sigma_j'>\sigma_k'$ stress conditions,

$$\left.\begin{aligned} f_{ij} &= f_{ij}(\sigma_i', \sigma_j', p') = f_{ij}(r_{ij}) \\ g_{ij} &= g_{ij}(\sigma_i', \sigma_j', p') = g_{ij}(r_{ij}) \\ f_{ik} &= f_{ik}(\sigma_i', \sigma_k', p') = f_{ik}(r_{ik}) \\ g_{ik} &= g_{ik}(\sigma_i', \sigma_k', p') = g_{ik}(r_{ik}) \\ f_{jk} &= f_{jk}(\sigma_j', \sigma_k', p') = f_{jk}(r_{jk}) \\ g_{jk} &= g_{jk}(\sigma_j', \sigma_k', p') = g_{jk}(r_{jk}) \end{aligned}\right\} \quad (3)$$

where p' means effective mean principal stress, and $r_{ij}=(\sigma_i'+p')/(\sigma_j'+p')$ is a new variable. For an isotropic material, it is sufficient to define only one set of f and g which is common to all of the two-dimensional stress systems. However, for sand specimens having anisotropic characteristics between vertical and horizontal directions which is caused by deposition of sand, six functions must be specified, i.e. f_{hv}, f_{hh}, f_{vh}, g_{hv}, g_{hh} and g_{vh}, where suffixes "h" and "v" mean horizontal and vertical directions, respectively. It is assumed that the yield conditions are independent for each of the two-dimensional deformation along the three slip planes shown in Fig. 1. However, because two directions in each slip plane are assumed to cause yielding independently, there are six stress systems associated with the yielding, i.e. $\sigma_i'>\sigma_j'$, $\sigma_j'>\sigma_i'$, $\sigma_j'>\sigma_k'$, $\sigma_k'>\sigma_j'$, $\sigma_k'>\sigma_i'$ and $\sigma_i'>\sigma_k'$. In this paper the parameter, r_{ij}, itself is used as a yield function, that is r_{ij} = const. represents a yiels locus for the i' j' two-dimensional stress system.

During unloading and reloading, only elastic deformations are supposed to occur in order to simplify the composition of the model. These elastic deformations during unloading and reloading are also given as a function of parameter, r.

3. TRUE TRIAXIAL TESTS

In this investigation experimental data obtained from drained true triaxial tests on loose sand specimens reported by Yamada and Ishihara [3, 4] are used to construct and evaluate the model. In these tests saturated sand was poured vertically into a cubical mold filled with deaired water to form a test specimen.

In order to specify the direction of sand depositoin a coordinate system shown in Fig. 2 is adopted, where σ_z' axis is chosen vertically

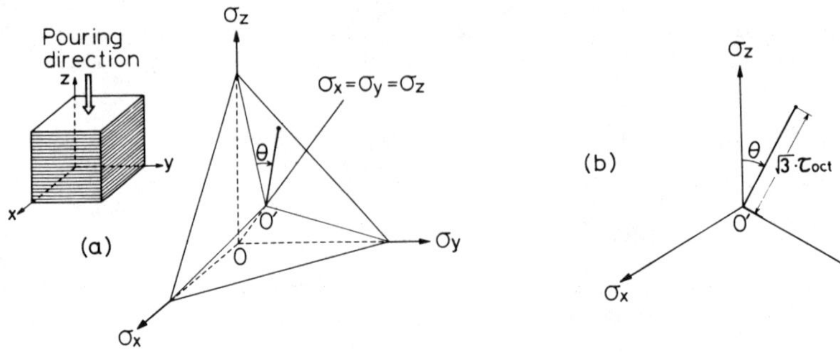

Fig. 2 Representation of stress conditions in the octahedral plane

coincident with the direction of pouring. The three-dimensional stress condition can also be represented by effective mean principal stress, p', octahedral shear stress, τ_{oct}, and an angular parameter, θ, which are defined as follows,

$$\left.\begin{aligned} p' &= 1/3\ (\sigma_x' + \sigma_y' + \sigma_z') \\ \tau_{oct} &= 1/3\sqrt{(\sigma_x' - \sigma_y')^2 + (\sigma_y' - \sigma_z')^2 + (\sigma_z' - \sigma_x')^2} \\ \tan\theta &= \sqrt{3}\ (\sigma_y' - \sigma_x')/(2\sigma_z' - \sigma_x' - \sigma_y') \end{aligned}\right\} \quad (4)$$

In conformity with the definition for the stress state, volumetric strain, v, octahedral shear strain, γ_{oct}, and an angular parameter, ω, are defined by,

$$\left.\begin{aligned} v &= \varepsilon_x + \varepsilon_y + \varepsilon_z \\ \gamma_{oct} &= 2/3\sqrt{(\varepsilon_x - \varepsilon_y)^2 + (\varepsilon_y - \varepsilon_z)^2 + (\varepsilon_z - \varepsilon_x)^2} \\ \tan\omega &= \sqrt{3}\ (\varepsilon_y - \varepsilon_x)/(2\varepsilon_z - \varepsilon_x - \varepsilon_y) \end{aligned}\right\} \quad (5)$$

where ε_x, ε_y and ε_z are principal strains in x, y and z directions, respectively.

All the specimens were first consolidated isotropically to a confining pressure of p' = 98 kN/m², and were then subjected to shear stresses so as to keep p' constant. A series of monotonic loading tests was performed along radial shear stress paths in the right half side of the octahedral plane as illustrated in Fig. 3(a) (Yamada and Ishihara [3, 4]). The stress condition of these tests can be specified by a parameter, b, defined by,

$$b = (\sigma_2 - \sigma_3)/(\sigma_1 - \sigma_3) \quad (6)$$

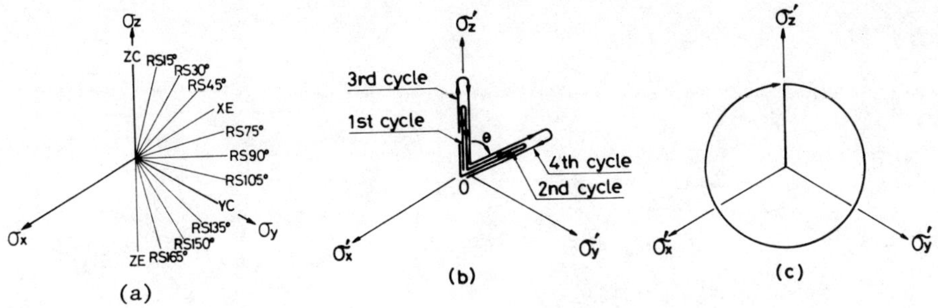

Fig. 3 Stress paths in the octahedral plane; (a) radial stress path, (b) multi-directional cyclic loading path, (c) circular stress path

which represents the relative magnitude of intermediate principal stress. The ZC and YC tests have a b-value of zero corresponding to the triaxial compression states, while the XE and ZE tests have a value of b = 1.0 corresponding to the triaxial extension states. Between the triaxial compression and extension stress states, three other classes of tests with b = 0.268 (RS15°, RS105° and RS135°), b = 0.5 (RS30°, RS90° and RS150°) and b = 0.732 (RS45°, RS75° and RS165°) were carried out. Each stress path having an identical b-value differs only in the directions in which the major, intermediate and/or minor principal stresses are applied.

A series of multi-directional cyclic loading tests was carried out following stress paths schematically shown in Fig. 3(b) (Yamada and Ishihara [7]). In these tests the first and third loading cycles were carried out along ZC stress path, and the second and fourth loading cycles were carried out along RSθ° stress path. The values of angle, θ, were chosen at an interval of 15 degrees. Other types of multi-directional loading tests were also performed: cyclic loading tests with the reverse loading order to that shown in Fig. 3(b) and tests with three or four different loading directions in the octahedral plane. Another type of tests with a circular stress path in the octahedral plane illustrated in Fig. 3(c) were performed to examine the yield condition in the octahedral plane.

4. STRESS-STRAIN RELATIONSHIP IN TWO-DIMENSIONAL STRESS SYSTEM

The stress-strain relationships obtained from four monotonic loading tests, ZC, XE, YC and ZE tests shown in Fig. 3(a), in which the triaxial compression or extension states of stress were applied were used to establish r-f, g curves for virgin loading conditions. In each of these tests, two imaginary two-dimensional slippings are considered as shown in Fig. 4. Therfore, three principal strains observed during these tests are represented by strain components listed in table 1.

As can be seen in Table 1, f_{vh} corresponds to half of the ε_z

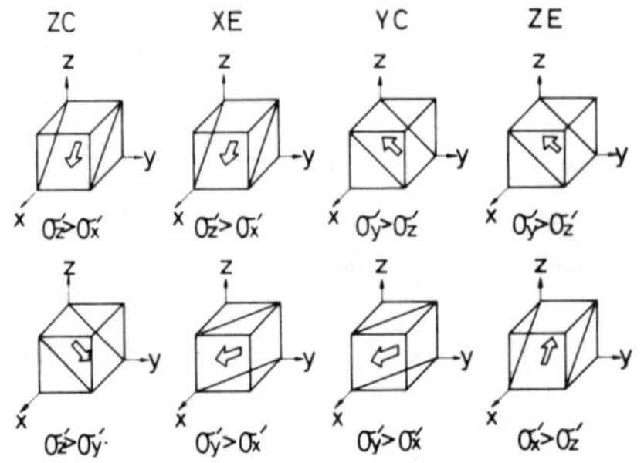

Fig. 4 Imaginary two-dimensional slippings taking place in triaxial stress conditions

stress path	stress state	ϵ_x	ϵ_y	ϵ_z
Z C	$\sigma_z' > \sigma_x' = \sigma_y'$	g_{vh}	g_{vh}	$2 \times f_{vh}$
X E	$\sigma_y' = \sigma_z' > \sigma_x'$	$g_{vh} + g_{hh}$	f_{hh}	f_{vh}
Y C	$\sigma_y' > \sigma_z' = \sigma_x'$	g_{hh}	$f_{hh} + f_{hv}$	g_{hv}
Z E	$\sigma_x' = \sigma_y' > \sigma_z'$	f_{hv}	f_{hv}	$2 \times g_{hv}$

Table 1 Strain components measured in triaxial stress conditions

observed in a ZC test as well as the total ε_z observed in a XE test. Similarly g_{hv} equals half of the ε_z measured in a ZE test as well as the total ε_z measured in a YC test. Using these relationships it is possible to determine the form of parameter, r, and functions, f_{vh} and g_{hv}. The form of the parameter, r, can also be examined by observing that ε_x values measured in ZC, YC and XE tests correspond to g_{vh}, g_{hh} and $g_{vh} + g_{hh}$, respectively, and that ε_y values measured in XE, ZE and YC tests correspond to f_{hh}, f_{hv} and $f_{hh} + f_{hv}$, respectively.

Two simple and common variables, σ_i'/σ_j' and $(\sigma_i' - \sigma_j')/p'$, whose constant lines in the octahedral plane are shown in Fig. 5 were first evaluated to see if these variables can be adopted as a proper parameter for r_{ij}. Adopting σ_i'/σ_j' as the parameter means that deformation behavior in two-dimensional stress system $\sigma_i' > \sigma_j'$ is governed by Mohr-Coulomb's criterion, while adopting $(\sigma_i' - \sigma_j')/p'$ assume the validity of extended Tresca's criterion. Results of the evaluation are partly shown in Fig. 6. Comparing Figs. 6(a) and 6(b), f_{vh} obtained from the ZC test is larger than that obtained from the XE test when plotted versus σ_i'/σ_j', however, the inverse trend is observed when plotted versus $(\sigma_i' - \sigma_j')/p'$. The similar trend was recognized in the

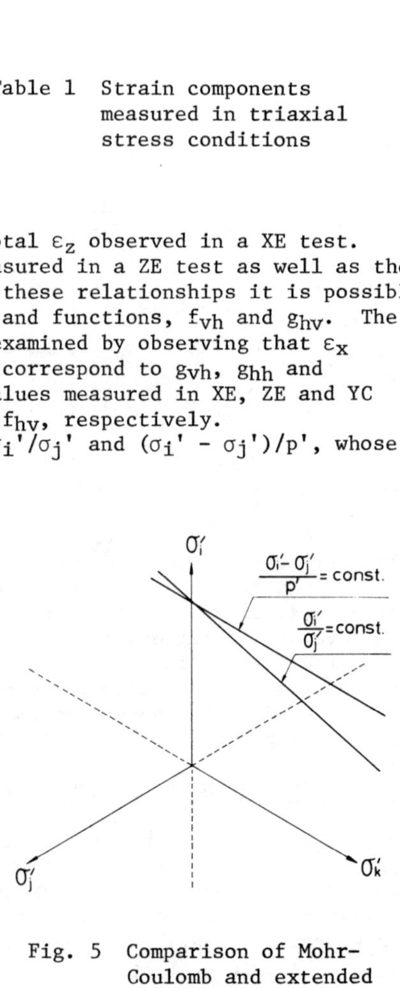

Fig. 5 Comparison of Mohr-Coulomb and extended Tresca's criteria

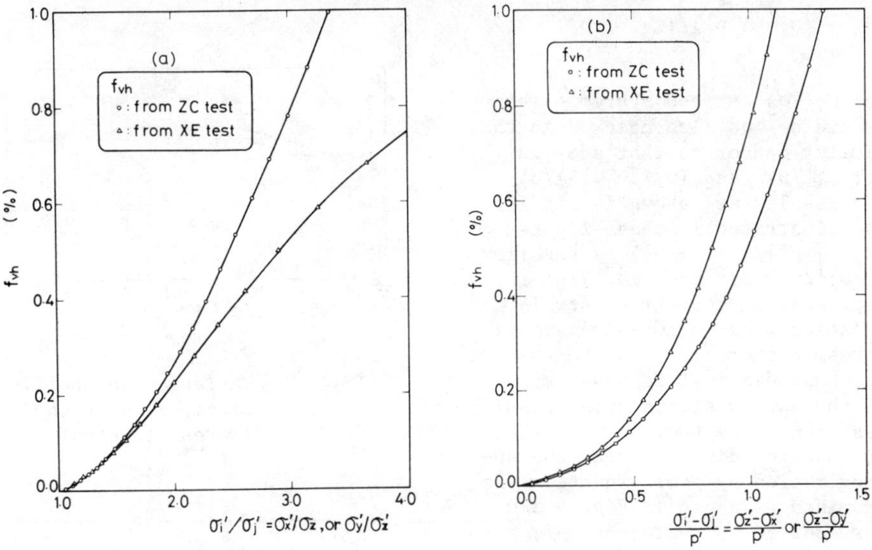

Fig. 6 Stress ratio vs. strain for two-dimensional stress system; (a) σ_i'/σ_j' - f_{vh}, (b) $(\sigma_i' - \sigma_j')/p'$ - f_{vh}

plots of f_{hh}, g_{hh} and g_{hv}, although the results are not shown here.

These findings indicate that either σ_i'/σ_j' or $(\sigma_i' - \sigma_j')/p'$ is not suitable for the parameter representing the stress-strain relationships in two-dimensional stress system. A better parameter should exist somewhere between σ_i'/σ_j' and $(\sigma_i' - \sigma_j')/p'$ in the octahedral plane shown in Fig. 5. Assume, therefore, that the parameter is expressed as $r_{ij} = (\sigma_i' + ap')/(\sigma_j' + ap')$, which is equivalent to σ_i'/σ_j' when $a = 0$ and to $(\sigma_i' - \sigma_j')/p'$ when $a = +\infty$. The coefficient, a, was determined using the values of f_{vh} obtained from ZC and XE tests, and g_{hv} obtained from YC and ZE tests.

The results are plotted in Fig. 7 in the form of variation in coefficient, a, with f_{vh} and g_{vh}. In the case of f_{vh}, for example, values of a were calculated such that the parameter $r_{ij} = (\sigma_i' + ap')/(\sigma_j' + ap')$ would have the same value for each value of f_{vh} in ZC and XE tests. Although there is considerable scatter at small strains, the coefficient, a, has values between 0.5 and 1.5 for strains larger than 0.1%. The coefficient a = 1.0 was adopted in this study for the following reasons: 1) at small strains, that is in the range of small r, r changes very little with the variation in the value of a; 2) a = 1 is a round number and nearly equals the average value. Therfore, r is defined by,

$r_{ij} = (\sigma_i' + p')/(\sigma_j' + p')$ (7)

for $\sigma_i' > \sigma_j'$ stress system. This parameter was also checked in the similar manner to that adopted for σ_i'/σ_j' and $(\sigma_i' - \sigma_j')/p'$ and results are shown in Fig. 8. As illustrated in these figures $(\sigma_i' + p')/(\sigma_j' + p')$ is superior to σ_i'/σ_j' or $(\sigma_i' - \sigma_j')/p'$ in representing the stress-strain relationships in two-dimensional stress system.

Mathematical representation of the stress-strain relationship was then attempted. As hyperbolic function did not work, the approximation by power function was examined as shown in Fig. 9 and was adopted to represent r-f, g curves.

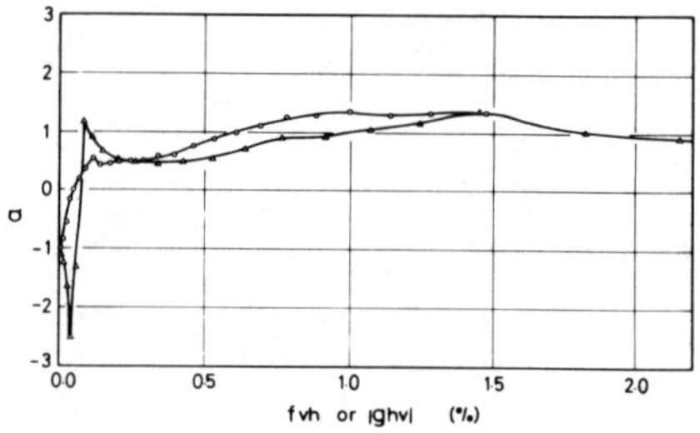

Fig. 7 Variation in coefficient, a, with the increse in strain

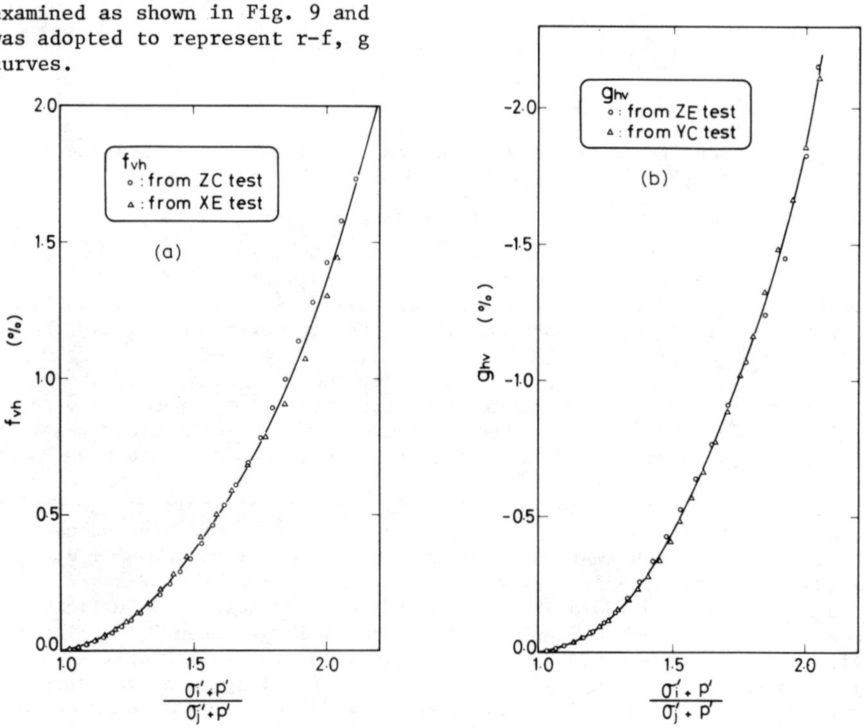

Fig. 8 Stress ratio, $(\sigma_i' + p')/(\sigma_j' + p')$, vs. strain for two-dimensional stress system; (a) f_{vh}, (b) g_{hv}

The stress-strain relationships during unloading and reloading were also examined using the results of cyclic loading tests. As a results of this examination, an unique linear elastic stress-strain relationship was selected for both unloading and reloading. This simplification necessarily results in deformation without any volume change during unloading and reloading.

The stress-strain relationship in two-dimensional stress system adopted in the present model is schematically illustrated in Fig. 10. As can be seen in this figure, the virgin loading curve represents a skelton curve and functions, f and g, are given by the following equations,

$$\left.\begin{array}{ll} \begin{matrix} f \\ g \end{matrix} = m\,(r-1)^n & r \geq r_0 \\ \begin{matrix} f \\ g \end{matrix} = m\,(r_0-1)^n \mp 0.18(r_0-r) & r < r_0 \end{array}\right\} \qquad (8)$$

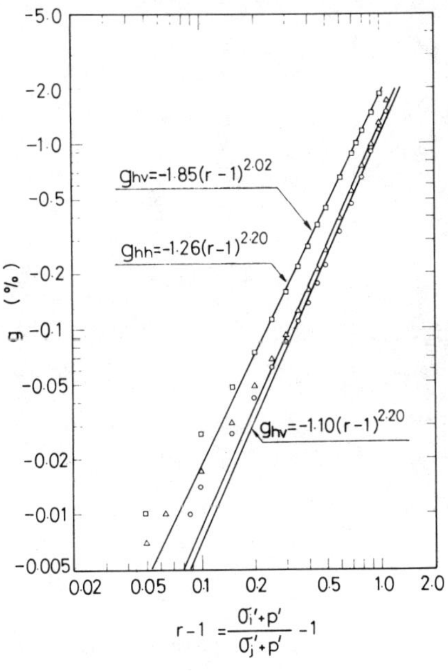

Fig. 9 Representation of stress-strain relationships by power function

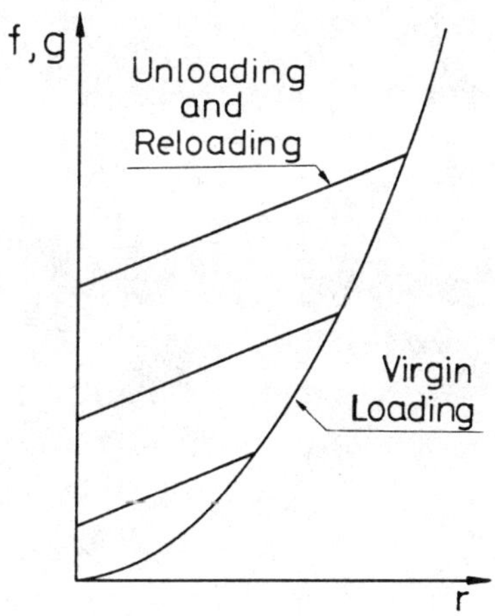

Fig. 10 Schematic illustration of stress-strain relationships for two-dimensional stress system

where r_0 means the past maximum value of r, and values of m and n are listed in Table 2. As recognized in Fig. 10 and Eq. (8), the parameter, r_{ij}, itself serves as the yield function in $\sigma_i'>\sigma_j'$ two-dimensional stress system. The yield conditions are independent for each two-dimensional stress system, $\sigma_i'>\sigma_j'$, $\sigma_j'>\sigma_i'$, $\sigma_j'>\sigma_k'$, $\sigma_k'>\sigma_j'$, $\sigma_k'>\sigma_i'$ and $\sigma_i'>\sigma_k'$, therfore the overall yield condition in the octahedral plane is necessarily anisotropic. Fig. 11 illustrates the development of yield loci in the octahedral plane for a cyclic loading

	m	n
f_{vh}	1.31	1.85
f_{hh}	1.60	1.85
f_{hv}	2.30	1.74
g_{vh}	-1.10	2.20
g_{hh}	-1.26	2.20
g_{hv}	-1.85	2.02

Table 2 Values of parameters used in the model

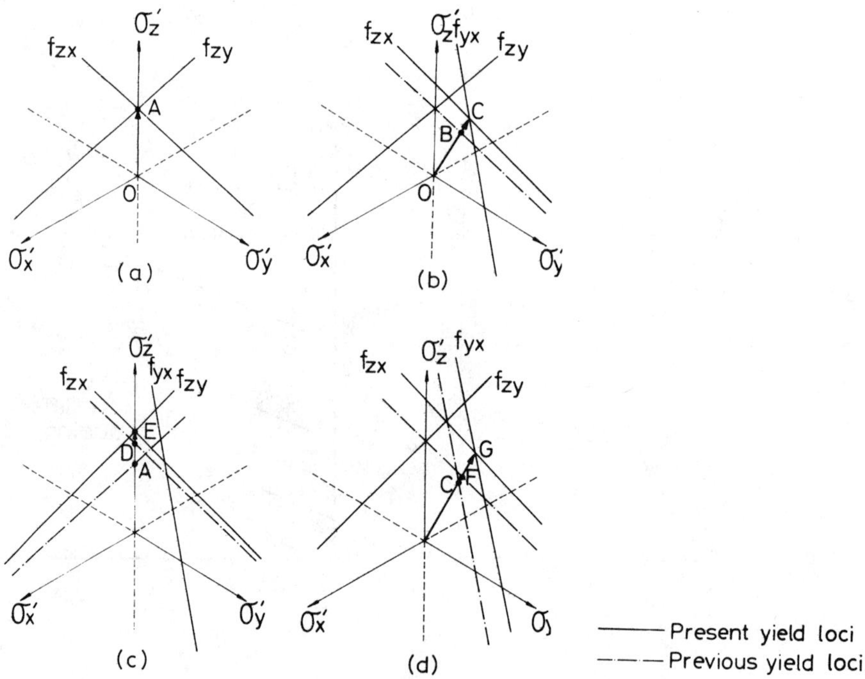

Fig. 11 Development of Yield loci during multi-directional cyclic loading test predicted by the model

test whose stress path is shown in Fig. 3(b). On the other hand, Fig. 12 demonatrates how the six yield loci are constructed during the loading test with a circular stress path shown in Fig. 3(c). In this figure the dotted areas correspond to the domain where the specimen has completely yielded for any conceivable two-dimensional stress system.

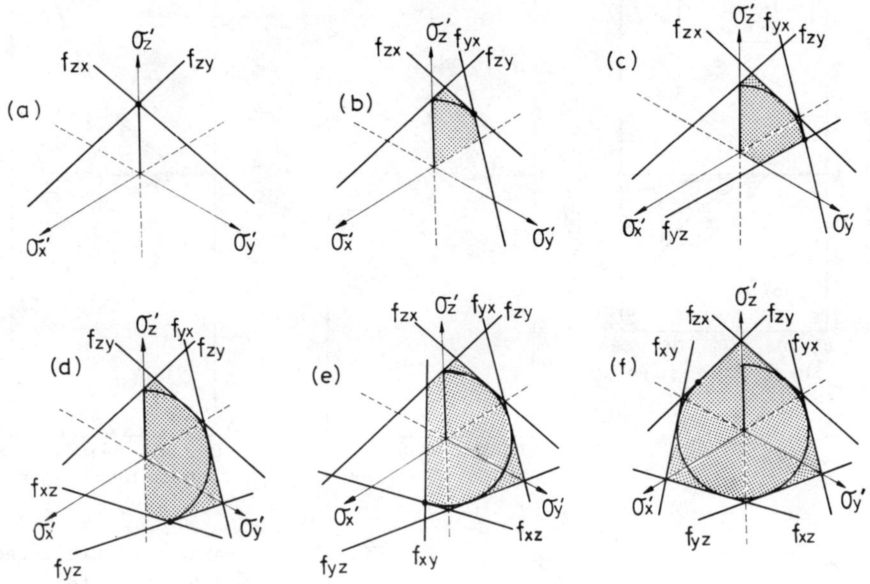

Fig. 12 Development of yield loci in a loading test with a circular stress path

6. COMPARISON BETWEEN PREDICTED AND MEASURED STRAINS

Fig. 13 demonstrates the comparison between predicted and measured principal strains for monotonic loadings with a b-value of 0.268, i.e. the tests performed along the RS15°, RS105° and RS135° stress paths shown in Fig. 3(a). In these figures solid lines correspond to the strains predicted by the model and individual points indicate the experimental data (Yamada and Ishihara [3]).

As illustrated in Fig. 13, the model predicted the variation in stress-strain behavior with the change in radial stress path. For the other stress conditions designated by b-values of 0, 0.5, 0.732 and 1.0 the model also simulated fairly well the measured stress-strain relationships (Yamada and Ishihara [10]). This means that the model can take into account the effects of the stress state expressed by b-value, as well as the effects of the inherent anisotropy of the specimen formed by sand deposition.

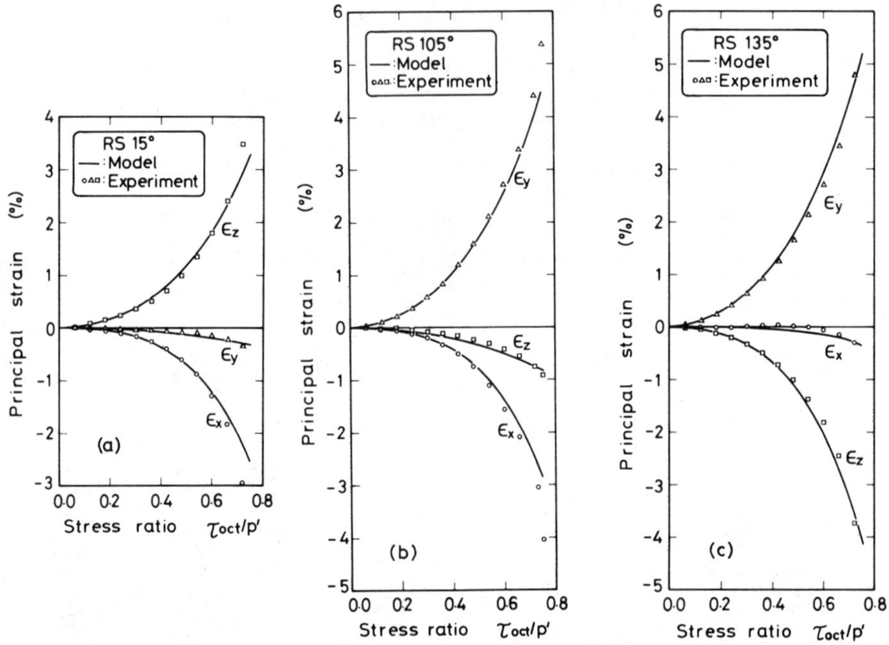

Fig. 13 Comparison of predicted and measured strains in radial shear tests
(a) RS15° test
(b) RS105° test
(c) RS135° test

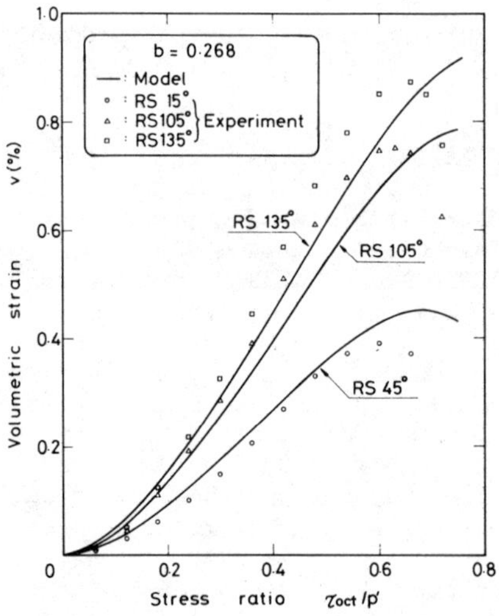

Fig. 14 Comparison of predicted and measured volumetric strains in radial shear tests; RS15°, RS105° and RS135° tests

In Fig. 14 the predicted and measured volumetric strains for tests with b = 0.268 are plotted versus stress ratio τ_{oct}/p'. Although the model simulates well the tendency that the volume change becomes larger as the stress path changes from RS15°, through RS105°, to RS135°, the predicted values are not coincident with the measured values. As the volumetric strain, which is given by the sum of three principal strains, has relatively small values, the difference between the predicted and measured volumetric strains appears more prominent.

Figs. 15(a) and 15(b) show the measured and predicted volumetric strains during ZC-XE and ZC-YC multi-directional loading tests, respectively, whose stress path is schematically illustrated in Fig. 3(b). These figures show that the model is able to represent the volumetric deformation characteristics during multi-directional loadings. However, the model differs from actual behavior in that it does not produce any volumetric strain during unloading and reloading and that it has clear yield points.

In Figs. 16 to 18 the strain paths projected on the octahedral plane are compared for several types of multi-directional loading tests. Figs. 16(a) and 16(b) show the measured and predicted strain paths for the ZC-XE and ZC-YC tests whose volumetric strain behavior are shown in Fig. 15. On the other hand Figs. 17(a) and 17(b) illustrate the comparison

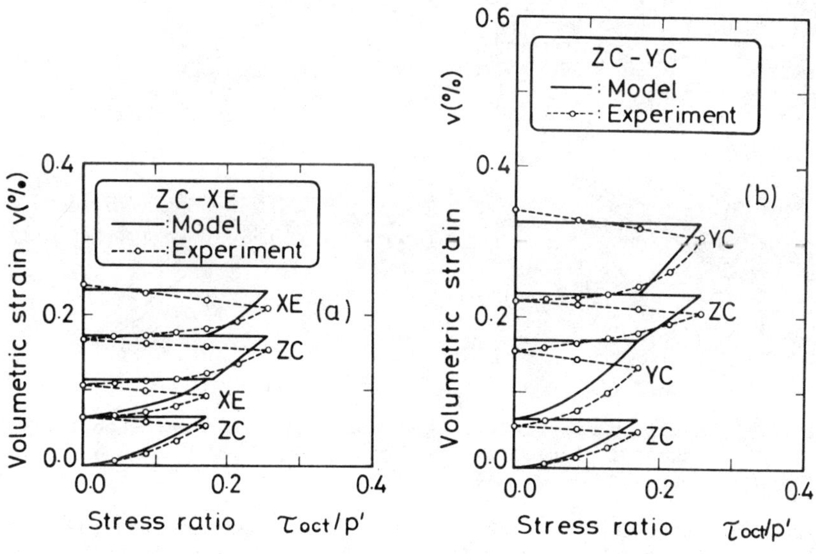

Fig. 15 Comparison of predicted and measured volumetric strains of multi-directional cyclic loading tests; (a) ZC-XE test, (b) ZC-YC test

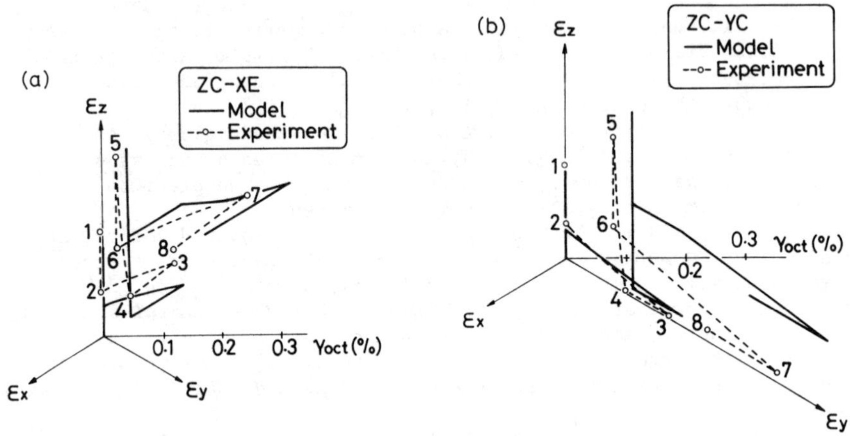

Fig.16 Comparison of predicted and measured strain paths on the octahedral plane; (a) ZC-XE test, (b) ZC-YC test

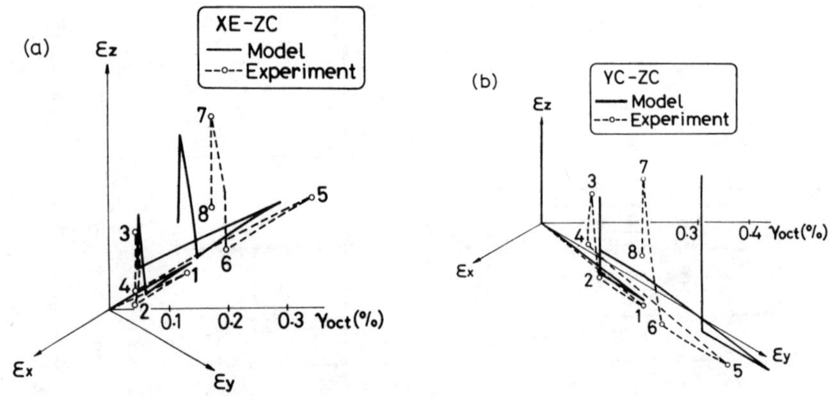

Fig. 17 Comparison of predicted and measured strain paths on the octahedral plane; (a) XE-ZC test, (b) YC-ZC test

of strain paths during the XE-ZC and YC-ZC tests performed in the reverse order of loading to those of ZC-XE and ZC-YC tests, respectively. In Fig. 18 the trajectories of shear strain on the octahedral plane for cyclic loading tests with three or four different directions of shear stress application are demonstrated. Although the model is able to

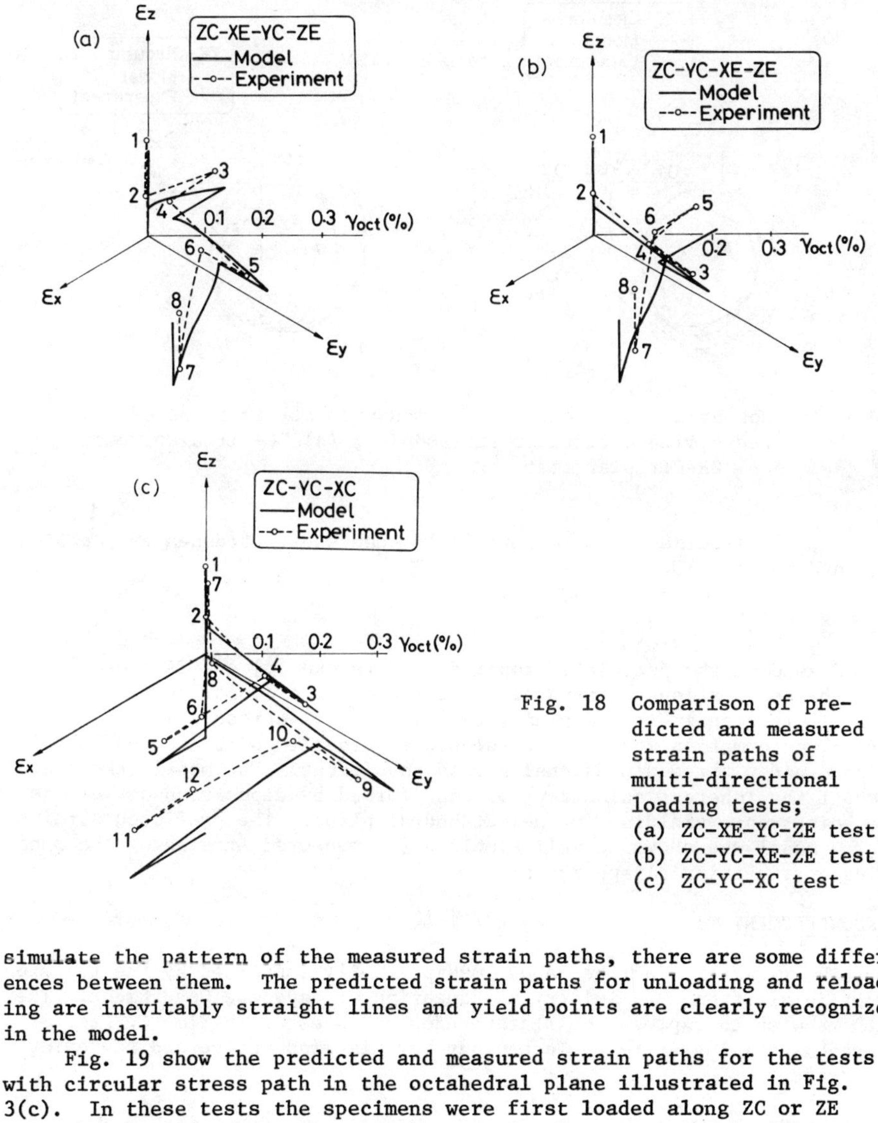

Fig. 18 Comparison of predicted and measured strain paths of multi-directional loading tests; (a) ZC-XE-YC-ZE test, (b) ZC-YC-XE-ZE test, (c) ZC-YC-XC test

simulate the pattern of the measured strain paths, there are some differences between them. The predicted strain paths for unloading and reloading are inevitably straight lines and yield points are clearly recognized in the model.

Fig. 19 show the predicted and measured strain paths for the tests with circular stress path in the octahedral plane illustrated in Fig. 3(c). In these tests the specimens were first loaded along ZC or ZE stress paths and, then, sheared along a circular stress path with a constant value of $\tau_{oct}/p' = 0.170$. In the first cycle the predicted strain paths represent the characteristics of the measured strain paths quite well. However, the predicted paths trace a loop composed of only elastic component in the subsequent cycles. On the other hand, the measured strain paths move downward being affected by the inherent anisotropy,

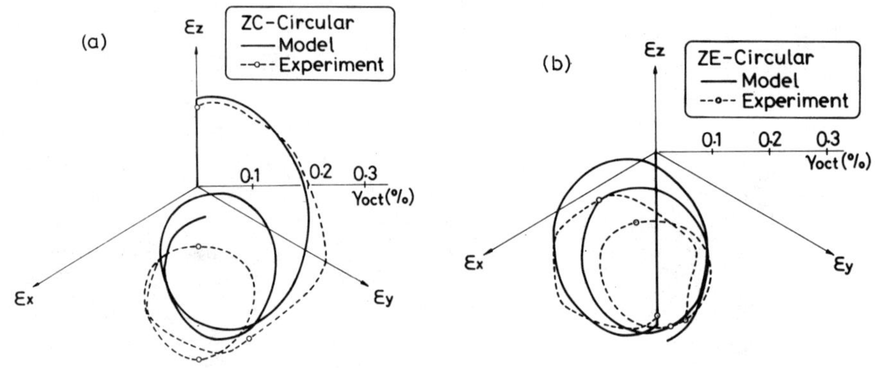

Fig. 19 Comparison of predicted and measures strain paths of loading tests with a circular stress path; (a) ZC-Circular test, (b) ZE-Circular test

that is, the specimen is more deformable when shear stresses are applied downward in Fig. 19.

7. CONCLUSION

Based on the postulate proposed by Matsuoka [8] and Tatsuoka [9] that the deformations of sand are represented as a linear summation of strains produced in the imaginary two-dimensional stress systems, an empirical deformation model was developed in order to simulate the behavior of sand under three-dimensional stress conditions. The model takes into account the inherent anisotropy of sand formed by deposition as well as the anisotropic yielding in the octahedral plane. The predicted strains by the model was shown to well simulate the measured strains of the sand using a true triaxial apparatus.

ACKNOWLEDGMENT

The research described above was partially supported by the research grant-in-aid from the Ministry of Education of Japanese Government. The authors wish to express their gratitude to Professor Charles C. Ladd of Massachusetts Institute of Technology for his kind advice and reviewing the draft.

REFERENCES

1. Oda, M., "Initial Fabrics and Their Relations to Mechanical Properties of Granular Material," Soils and Foundations, Vol. 12, No.1, pp. 17-36 (1972).

2. Arthur, J.R.F. and Menzies, B.K., "Inherent Anisotropy in a Sand," Geotechnique, Vol. 22, No. 1, pp.115-128 (1972).

3. Yamada, Y. and Ishihara, K., "Anisotropic Deformation Characteristics of Sand under Three Dimensional Stress Conditions," Soils and Foundations, Vol. 19, No. 2, pp.79-94 (1979).

4. Yamada, Y. and Ishihara, K., "Undrained Deformation Characteristics of Loose Sand under Three-Dimensional Stress Conditions," Soils and Foundations, Vol. 21, No. 1, pp. 97-107 (1981).

5. Arthur, J.R.F., Bekenstein, S., Germaine, J.T. and Ladd, C.C., "Stress Path Tests with Controlled Rotation of Principal Stress Direction," ASTM STP740, pp. 516-540 (1981).

6. Tatsuoka, F. and Ishihara, K., "Drained Deformation of Sand Under Cyclic Stress Reversing Directions," Soils and Foundations, Vol. 14, No. 3, pp.51-65 (1974).

7. Yamada, Y. and Ishihara, K., "Yielding of Loose Sand in Three-Dimensional stresses," Soils and Foundations, Vol. 22, No. 3, pp. 15-31 (1982).

8. Matsuoka, H., "Stress-Strain Relationship as Sand Based on the Mobilized Plane," Soils and Foundations, Vol. 14, No. 2, pp.47-61 (1974).

9. Tatsuoka, F., "Stress-Strain Behavior of an Idealized Anisotropic Granular Material," Soils and Foundations, Vol. 20, No. 3, pp.75-90 (1980)

10. Yamada, Y. and Ishihara, K. "Modeling of Drained deformation of Sand under Three-Dimensional Stress Conditions," Proc. IUTAM Symp., Deformation and Failure of Granular Materials, P.A. Vermeer and H.J. Luger (editors), A.A. Balkema, Rotterdam, Netherlands, pp.331-342 (1982).

INFLUENCE OF VOID RATIO AND STRESS CONDITION ON THE DYNAMIC SHEAR MODULUS OF GRANULAR MEDIA

E. Yanagisawa

Department of Civil Engineering, Tohoku University
Sendai, Japan

ABSTRACT

Distribution of contact point in granular materials, which are randomly packed in a specimen, is determined and the method for estimating the elastic constants of uniform spheres is developed by applying the mathematical solution of the elastic spheres in contact. The shear moduli of granular materials are experimentally determined by the resonant column tests and these results are compared with the theoretical values. Then the effect of octahedral normal stress and shear stress on the dynamic properties of sand is studied by applying stress in various ways. An empirical equation for evaluating the shear modulus of sand is derived from the test results.

1. INTRODUCTION

Many methods for evaluating the dynamic response of soils and foundation under seismic loading conditions are developed recently and it becomes possible to predict the seismic risks of structures. The accuracy of such procedures depends mainly on the information of the ground and the dynamic properties of soils and structures. However it is quite difficult in many cases to evaluate the vibrational characteristics of soil deposits because the soil conditions are often different in different places and moreover because soils have intrinsically nonlinearity in the stress strain relationship.

In cases of calculation of the seismic response of the ground, soil deposits are assumed as visco-elastic materials, and the response is generally determined by considering the horizontally layered system to the shear wave incidence. The response is considerably influenced by the shear modulus and the damping characteristics of the soil strata. As a first step toward evaluating the dynamic properties of soils, it is useful to consider the soil particles to be represented by small spheres. The purpose of this paper is to investigate the dynamic shear modulus of granular materials by evaluating the elastic constants of the regular packing of spheres, which are obtained from the mathematical solution of the elastic spheres in contact. Then the influence of void ratio and stress condition on the dynamic properties of soils are studied.

2. CONTACT POINTS IN GRANULAR MATERIALS

Uniform spheres which are arranged in uniform arrays have limited co-ordination numbers, namely 6, 8, 10 and 12 contact points per sphere corresponding to the types of arrays. Smith, Foote and Busang (1929) investigated the distribution of contact points between each sphere by means of uniform lead spheres, and proved that there exist intermediate structures with contact points in odd numbers and some spheres with more than 12 contact points. They assumed that the arrays of uniform spheres are consist of simple cubic array and close hexagonal array, and found the relationship between porosity and numbers of contact point. After Filep (1936) proposed a simple empirical equation between porosity and numbers of contact point, the distribution of contact points in granular materials were investigated and the relations between the co-ordination number and the void ratio or the porosity were obtained by many authors.

(a) Smith, Foote and Busang (1929)

$$N = 26.486 - \frac{10.726}{1-n} \quad (1)$$

(b) Filep (1936)

$$N = \frac{3.1}{n} \quad (2)$$

(c) Field (1963)

$$N = \frac{12}{1+e} \quad (3)$$

(d) Meissner (1964)

$$N = 2\ \exp(2.4n) \quad (4)$$

(e) Ridgway and Turback (1969)

$$N = 1.072 - 0.1193\ n + o.oo431\ n^{2} \quad (5)$$

(f) Yanagisawa (1978)

$$N = 3.183^{(2.469-e)} \quad (6)$$

Marsal (1973) performed experiments to determine the relationship among grain size, density and contact points of rock fill materials. Statistical distribution of contact points per particle of those materials was determined and a close relation between void ratio and the distribution of the contact point number was observed. Oda (1977) made a similar investigation on mixtures of glass balls with two different diameters, and found that the experimental results indicates smaller co-ordination number than expected from the equations (1) to (5).

In order to determine the distribution of contact points in gravel mass, gravel particles with three kinds of particle shape, round, oval and flat, respectively, of about 20 mm diameter were selected. After the particles were packed randomly into a cylindrical vessel, dilute suspension of slaked lime was poured into the vessel and the particles were flooded with the slaked lime suspension from bottom to top. Then the suspension was drained and the gravels were allowed to dry for 24 hours. After the sample was dried out, the numbers of contact points on 200 particles in the interior of specimen were recorded. Fig.1 shows the test results obtained by the experiment and theoretical values of uniform spheres in uniform arrays. The empirical formulae proposed by the other authors and the results obtained by Oda (1977) are also shown in the same figure.

As far as the gravels concerned, the relation between void ratio and average number of contact points can be approximated by Eq. 2. The significance of average contact point will be recognized in the following discussion.

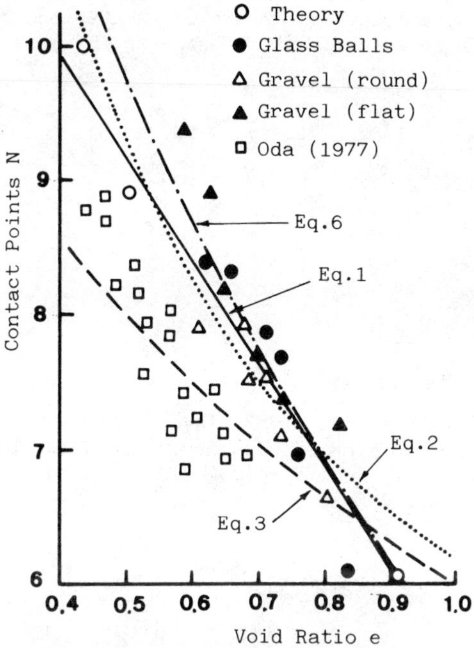

Fig.1 Relation between void ratio and contact points

3. ELASTIC CONSTANTS OF UNIFORM SPHERES IN CONTACT

Elastic constants of uniform spheres in uniform arrays can be calculated by applying the Hertz's theory of contact. The solution is obtained in the following manner:

(a) A representative unit volume is chosen.

(b) The stress acting on the unit volume is defined by the stress component of σ_{xx}, σ_{yy}, σ_{zz}, τ_{xy}, τ_{xz} and τ_{yz}.

(c) The strains of the unit volume ε_x, ε_y, ε_z, γ_{xy}, γ_{yz} and γ_{zx}

are defined by considering the relative displacement of neighboring spheres.

(d) The stress increments are superimposed on an initial hydrostatic state of stress.

(e) The stress acting on the surface of the unit volume is distributed on the contacts in proportion to the cross sectional area.

(f) No torsional moments are produced on the contact point.

(g) The twisting moments of each sphere are negligible.

The elastic constants of spheres in contact are defined by the following equations:

$$
\begin{aligned}
\sigma_{xx} &= C_{11}\varepsilon_x + C_{12}\varepsilon_y + C_{13}\varepsilon_z \\
\sigma_{yy} &= C_{21}\varepsilon_x + C_{22}\varepsilon_y + C_{23}\varepsilon_z \\
\sigma_{zz} &= C_{31}\varepsilon_x + C_{32}\varepsilon_y + C_{33}\varepsilon_z \\
\tau_{xy} &= C_{44}\gamma_{xy} \\
\tau_{xz} &= C_{55}\gamma_{xz} \\
\tau_{yz} &= C_{66}\gamma_{yz}
\end{aligned}
\qquad (7)
$$

Deresiewicz (1957) investigated the elastic constants of elastic spheres in symple cubic array and determined the expression

$$
\begin{aligned}
C_{11} &= C_{22} = C_{33} = \left[\frac{3\mu_s^2\sigma_0}{2(1-\nu_s)^2}\right]^{\frac{1}{3}} \\
C_{12} &= C_{13} = C_{21} = C_{23} = C_{31} = C_{32} = 0 \\
C_{44} &= C_{55} = C_{66} = \left[\frac{3\mu_s^2\sigma_0}{2(1-\nu_s)}\right]^{\frac{1}{3}} \frac{2(1-\nu_s)}{2-\nu_s}
\end{aligned}
\qquad (8)
$$

where μ_s; the shear modulus of the material

ν_s; Poisson's ratio of the material

σ_0; a hydrostatic pressure.

It should be noted that the elastic constants are in proportion to the one third power of σ_0 .

The elastic constants of face-centred and hexagonal close-packed arrays were studied by Duffy and Mindlin (1957), and Duffy (1958) respectively. For cubical tetrahedral and tetragonal sphenoidal arrays, Makhlouf and Stewart (1967) and Makhlouf (1963) made investigation on the elastic constants. For further details the reader should refer to the literature on this subject.

Takahashi and Sato (1949) derived the elastic constants of isotropic body as a function of elastic constants of anisotropic body, by equating the energy per unit volume of the isotropic body to the average energy per unit volume of the anisotropic parts.

$$C^0_{11} = C^0_{22} = C^0_{33} = \frac{3}{15}(C_{11}+C_{22}+C_{33}) + \frac{2}{15}(C_{23}+C_{31}+C_{12}) + \frac{4}{15}(C_{44}+C_{55}+C_{66})$$

$$C^0_{23} = C^0_{31} = C^0_{12} = \frac{1}{15}(C_{11}+C_{22}+C_{33}) + \frac{4}{15}(C_{23}+C_{31}+C_{12}) - \frac{2}{15}(C_{44}+C_{55}+C_{66}) \qquad (9)$$

$$C^0_{44} = C^0_{55} = C^0_{66} = \frac{1}{2}(C_{11}-C_{12})$$

where C^0_{11}, C^0_{12}, etc., are the elastic constants of the isotropic body
C_{11}, C_{12}, etc., are the elastic constants of the anisotropic array.

If a random packing of uniform spheres could be considered to consist of known parts of regular arrays, elastic constants of randomly packed spheres could be predicted by weighted average of the distributionof each regular array. Assuming that G* of randomly packed spheres are written in the form

$$G = \int G^*p(e)de \qquad (10)$$

where $G^* = C^0_{44} = C^0_{55} = C^0_{55} = \frac{1}{2}(C_{11}-C_{12})$

p(e) is the probability density function and e is the integral variable.

Assuming that void ratio is the only quantity which characterizes the state of soil, then the integral variable in Eq.10 can be regarded as void ratio. If the relation between contact points and void ratio holds for every packing of spheres, Eq.10 can be written as follows:

$$G = \int_0^\infty G^*(n)P(n)dn \qquad (11)$$

The shear modulus G*(n) of uniform arrays can be calculated by substituting the material constants of particles into the equation. The distribution function of contact points can be determined by the procedures which are shown in the previous section.

4. THE SHEAR MODULUS OF GRANULAR MEDIA

The shear modulus G* of uniform arrays can be calculated theoretically by using the material constants of glass balls. It should be noted that G* is theoretically in proportion to the one third power of . Table 1 shows the theoretical values of void ratio e and the shear modulus G* of uniform arrays with contact points of 6, 8, 10 and

Table 1. Theoretical Values for Uniform Arrays

Arrays	Symple Cubic	Cubical Tetrahedral	Tetragonal Sphenoidal	Face-centred	Close-packd hexagonal
e	0.910	0.654	0.433	0.350	0.350
contact points N	6	8	10	12	12
G for =1 (kg/cm^2)	5.94	7.18	9.88	12.03	12.03
G/N	0.99	0.90	0.99	1.00	1.00

12 for σ_o= 1 and 3 kg/cm^2. The material constants used for this calculation are ones of crown glass. As is evident from the table, the values of G* increase with increasing numbers of contact points N. It must be noted, however, that the ratio of the shear modulus to numbers of contact points G/N is almost equal except for the cubical tetrahedral array, and G/N takes approximately the values of unity as shown in Table 1.

$$\frac{G^*(N)}{N} \doteqdot 1$$

Strictly speaking, this relation is valid only for the limited numbers of N. From the statistical point of view, however, there exist intermediate values of N for random packing of spheres as shown in section 3. Assuming that the relation

$$\frac{G^*(N)}{N} = C \qquad (12)$$

where N: number of contact point in regular array
G*(N): the shear modulus of regular packing
C: a constant

holds for all values of N of regular packing, we can easily extend the above equation to random packing with mean number of contact point $\bar{N}$.

Provided that the distribution of contact points in random pack ing can be represented by the normal distribution, then p(n) in Eq.11 can be described as follows

$$p(n) = \int_0^\infty \exp\left\{-\frac{(N-\bar{N})}{2\sigma^2}\right\} dN \qquad (13)$$

where σ: standard deviation
$\bar{N}$: mean of contact points
N: contact point

From Eqs.12 and 13

$$G = \int_0^\infty G^*(N)p(N)dN$$

$$= \int_0^\infty CN \exp\left\{-\frac{(N-\bar{N})^2}{2\sigma^2}\right\} dN$$

$$= CN\int_0^\infty \exp\left\{-\frac{(N-\bar{N})^2}{2\sigma^2}\right\} dN - C\sigma^2\left[\exp\left\{-\frac{(N-\bar{N})^2}{2\sigma^2}\right\}\right] \qquad (14)$$

$$= C\,\bar{N} + \sigma^2 \exp\left(-\frac{\bar{N}^2}{2\sigma^2}\right)$$

It is easily recognized that the shear modulus of randomly packed spheres is proportional to the mean of contact points and the standard deviation of the contact point distribution. Assumed that the mean number of contact points of granular materials is represented in terms of void ratio as shown by Filep's equation (Eq.2),

$$\bar{N} = \frac{3.1}{n} = 3.1\left(1 + \frac{1}{e}\right) \qquad (15)$$

then the shear modulus of the granular materials can be calculated from Eq.14. Neglecting the effect of the deviation, Eq.14 shall be written in the form

$$G = 3.1C\left(1 + \frac{1}{e}\right) \qquad (16)$$

From Eq.8 we can calculate the shear modulus of uniform spheres with arbitrary contacts for unit confining pressure using Eq.16. Fig.2 shows the theoretical and experimental values of the shear modulus of uniform glass balls. The shear moduli of the glass balls are measured by the resonant column method. In the resonant column test a cylindrical sample of granular material is contained within a rubber membrane and a confining pressure is applied to maintain the shape of specimen during the test. Then the torsional oscillation is applied on the top of the specimen and resonant frequency is measured. The shear wave velocity or the shear modulus of the material can be calculated by the resonant frequency and the dimension of the specimen.

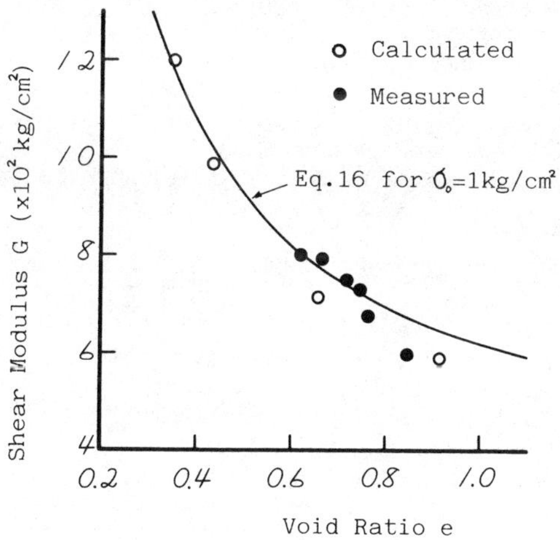

Fig.2 Values of G for packing of spheres

Fig.3 shows experimental results of the shear modulus of granular materials together with the theoretical values. The theoretical values calculated for corresponding void ratio to the test are 1.6 times as large as the experimental

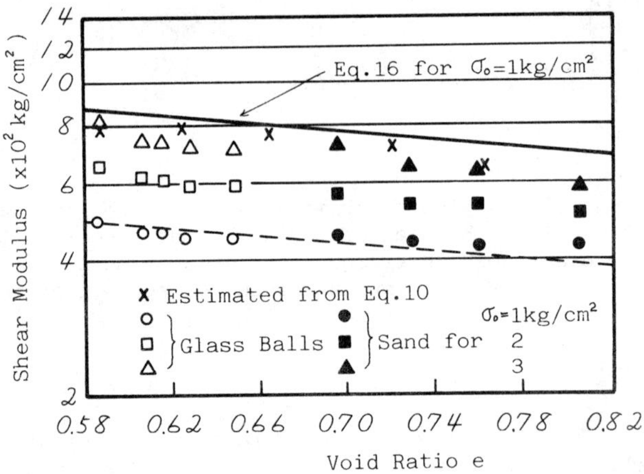

Fig.3 Effect of void ratio on shear modulus

values. As it is clear from the figure, the shear modulus is affected considerably by void ratio, and the tendency is said to be almost the same for both theoretical and experimental values.

If there is any difference in the diameter of the sphere, the elastic constants of the mass decrease considerably. Duffy and Mindlin investigated the variation of the velocity of the compression waves with the confining pressure for the bars composed of steel spheres. These steel spheres in one group had a high tolerance on the diameter of 1/8 + 0.00001 in. and those in another group had a tolerance of 1/8 + 0.00005 in. Even for these spheres there was a definite difference between the values predicted by theory and those obtained by experiments. If inhomogeneity of the material, irregularity in grain shape and inequality in grain size are taken into account, the estimation of the shear modulus by the procedure presented in this study is said to be rather accurate.

5. INFLUENCE OF STRESS CONDITION ON THE SHEAR MODULUS

The shear modulus is influenced significantly also by the confining pressure. The theory indicates that the shear modulus is in proportion to the one third power of the confining pressure σ_0. Fig.4(a) shows the experimental values of glass balls obtained by the resonant column tests for e=0.60 and predicted values of G by Eq.16 for the same void ratio. The experimental values increase in proportion to the 1/2.3 power of σ_0. A similar tendency is displayed in the case of sand as shown in Fig.4(b). The experimental results do not agree with the theoretical values in the strict sense.

Iwasaki and Tatsuoka (1977) investigated the effect of hydrostatic pressure on the shear modulus of sand, glass balls, steel balls and

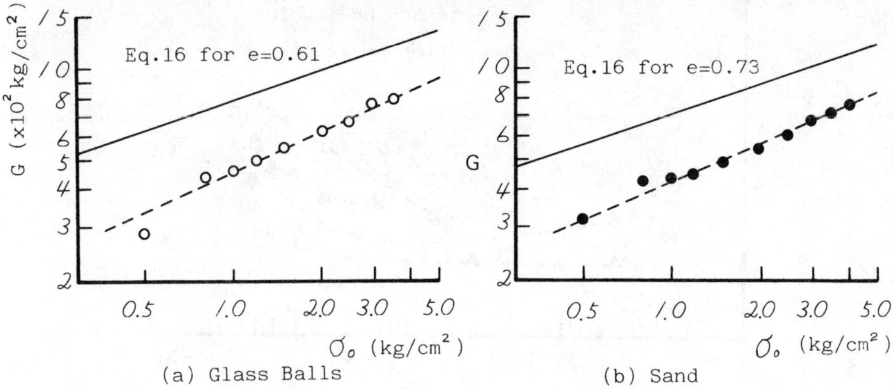

Fig.4 Effect of σ_0 on the shear modulus

lead shots, and showed that the shear moduli of these materials are approximately in proportion to 0.38th power of hydrostatic pressure. Hardin and Black (1966) derived an empirical equation which indicates the effect of void ratio and hydrostatic pressure.

$$G_o = 793\frac{(2.17-e)^2}{1+e}\sigma_0^{1/2} \tag{17}$$

It is not yet well known, whether the sandy soils, which have ever been compressed by some loading in the past, behave differently from the soils which have not. The specimen is first prestressed by confining pressure of 3 to 5 kg/cm², then the shear modulus and damping factor are measured under lower confining pressure. Fig.5 shows the results of tests which were performed under the condition of 4.0 kg/cm² pre-

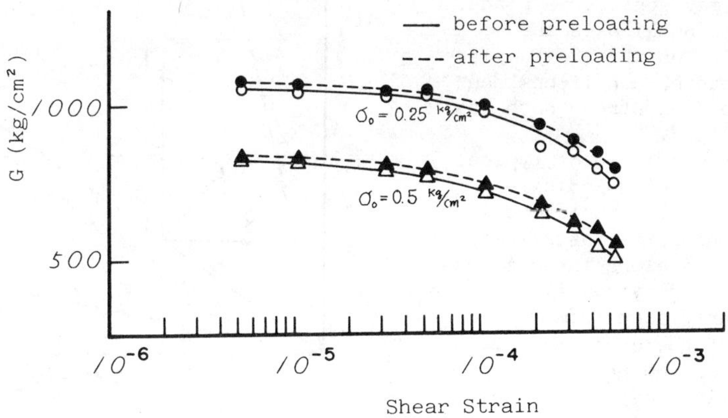

Fig.5 Effect of preloading on the shear modulus

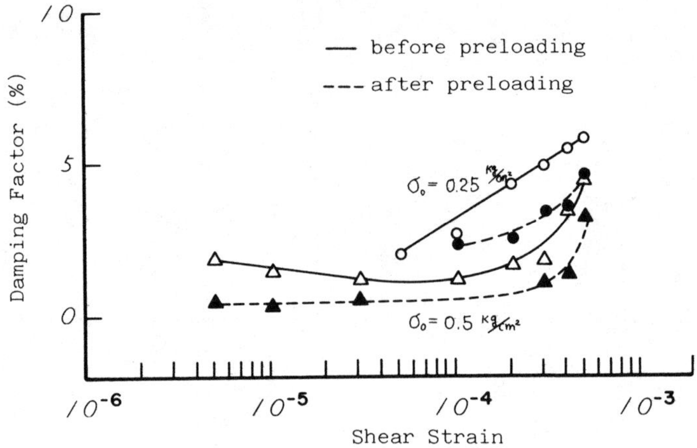

Fig.6 Effect of preloading on the Damping

loading. It is clearly seen that the shear modulus increases by a little amount after preloading. Preloading causes densification of loose parts in the specimen and averages the distribution of void. The damping factors are also affected by preloading as shown in Fig.6. Energy dissipation will likely decrease, since the specimen becomes homogeneous after preloading.

6. EFFECT OF OCTAHEDRAL SHEAR STRESS

By considering the three dimensional principal stress space, any stress state in triaxial test can be represented as a point in the space. Since σ_2 and σ_3 are always equal in the conventional tests, we may consider a plane, which contains σ_1 axis and $\sigma_2 = \sigma_3$ line on the $\sigma_2 - \sigma_3$ plane and any loading conditions can be described as stress paths as shown in Fig.7.

Fig.8 shows one of typical test results which indicate variations of the shear modulus of sand during the tests which are performed along three stress paths in the previous figure. It is clearly seen from this figure that the shear modulus increases as the principal stress ratio σ_1/σ_3 increases under constant confining pressure condition. In case of

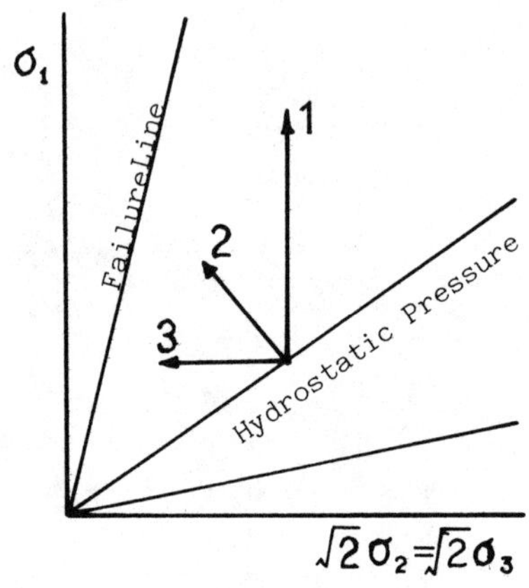

Fig.7 Stress Paths

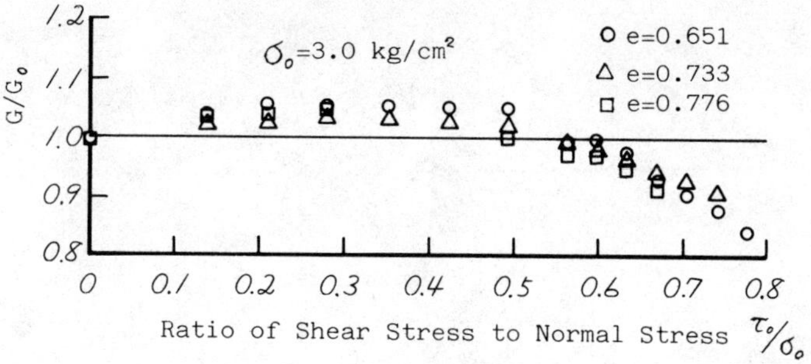

fig.9 Effect of octahedral shear stress

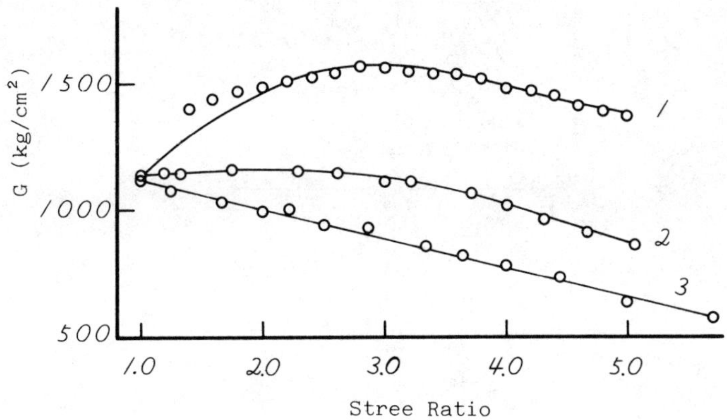

Fig.8 Influence of stress path on the shear modulus

constant axial stress test, the shear modulus decreases linearly as the stress ratio increases. If the octahedral normal stress σ_0 is held to be constant, the shear modulus varies as the octahedral shear stress τ_0 increses. The effect of τ_0 on the shear modulus is shown in Fig.9. The relation con be approximated by the equation

$$\frac{G}{G_0} = 1 + \frac{1}{2}\frac{\tau_0}{\sigma_0}(0.6-\frac{\tau_0}{\sigma_0}) \tag{18}$$

From the test results of different stress conditions, an isogram of the shear modulus can be drawn on the σ_1, $\sigma_2=\sigma_3$ plane, as shown in dotted lines in Fig.10.

As for damping factor D, changes in values of D for different stress paths are shown in Fig.11. It is clearly seen that the damping factor increases as the stress ratio increases, but stress path does not have significant influence on the damping factor.

The shear modulus of sand in small amplitude could be predicted

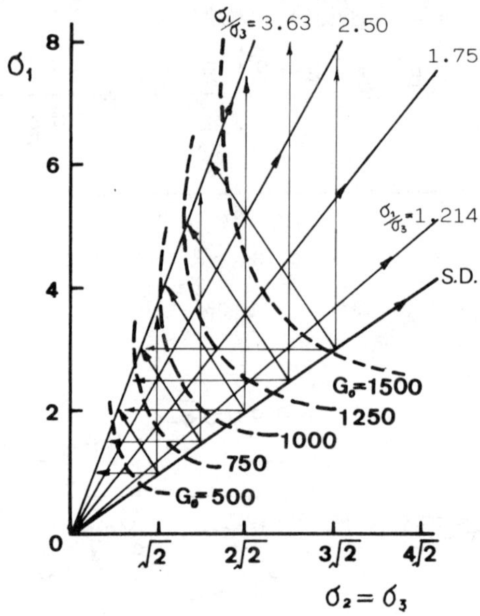

Fig.10 Isogram of the shear modulus

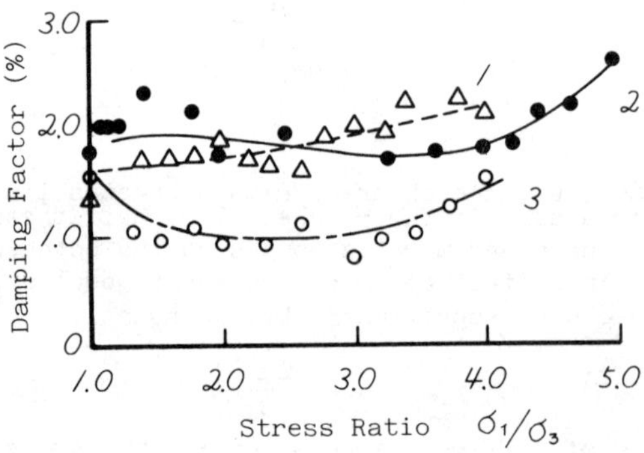

Fig.11 Influence of stress path on damping factor

by the expression

$$G_0 = 793 \frac{(2.17-e)^2}{1 + e} \sigma_0^{0.53} \left\{1 + \frac{1}{2}\frac{\tau_0}{\sigma_0}\left(0.6 - \frac{\tau_0}{\sigma_0}\right)\right\} \tag{19}$$

This equation could be used for dynamic problems in which considerable shear stress is statically acting in the soil mass. The effect of preloading on the dynamic properties of sands is negligible in comparison with the other factors which were described in this paper.

7. CONCLUSIONS

Considering the relation of void ratio and number of contact points, a theoretical method for estimating the elastic constants of randomly packed spheres are mentioned. The influence of stress condition and void ratio on the dynamic properties of granular material is discussed based on the laboratory tests and theoretical studies, and the following conclusions are made:

1) Using gravels and small glass balls which are packed randomly, the distribution of contact points is determined, and the average number of contact points is related to void ratio and it is approximated by an equation.

2) The shear modulus of randomly packed shperes can be estimated from the elastic constants of uniform spheres in uniform array by assuming normal distribution of contact points.

3) Experimental results obtained by resonant column test shows a similar tendency to those predicted by the theory.

4) The shear modulus of sand is approximately in proportion to a half power of octahedral normal stress σ_0. For Toyoura sand the index number of σ_0 proved to be 0.53.

5) Specimen prestressed at higher confining pressure before test causes as increase in the shear modulus and decrease in damping factor.

6) Effect of octahedral shear stress τ_0 is not so significant as octahedral normal stress, but there is still evident effect on the shear modulus.

7) The isogram of the shear modulus is drawn on the Rendulic plane. The effect of octahedral shear stress is approximated by hyperbolic parabolas.

8) An empirical equation for estimating the shear modulus of sand is obtained.

REFERENCES

1. Smith, W.O., Foote, P.D. and Busang, P.F., "Packing of Homogeneous Spheres," Phys. Rev., Vol.34, pp1271-1274 (1929)
2. Graton, L.C. and Fraser, H.J., "Systematic Packing of Spheres with Particular Relation to Porosity and Permeability," Journ. Geol., Vol.43, pp 785-909 (1935)
3. Joseph, R.C., "The Analysis of two-dimensional Orientation Data," Journ. Geol., Vol. 64, pp 117-120 (1968)
4. Field, W.G., "Towards the Statistical Definition of a Granular Mass," Proc. 4th Aust. and N.Z. Conf. on Soil Mech., pp 143-148 (1963)
5. Gray, W.A., "The Packing of Solid Particles," Chapman and Hall Ltd. (1968)
6. Oda, M., "Co-ordination Number and its Relation to Shear Strength

of Granular Material," Soils and Found., Vol. 17, No.2, pp 29-42 (1977)
7. Yanagisawa, E., "Relation between the Shear Modulus and Void Ratio in Granular Media," The Technology Reports of Tohoku Univ., Vol.43 No.1, pp 211-220 (1978)
8. Mindlin, R.D., "Conpliance of Elastic Bodies in Contact," Journ. Appl. Mech., Trans. ASME Vol.71 pp259-268 (1949)
9. Duffy, J. and Mindlin R.D., "Stress Strain Relations and Vibrations of a Granular Medium," Trans. ASME, Vol.58 No.A53, pp 88-94 (1959)
10. Duffy, J. "A Differential Stress Strain Relation for the Hexagonal Close-Packed Array of Elastic Spheres," Journ. Appl. Mech. Trans. ASME Vol.79, pp 1-9 (1957)
11. Takahashi, T. and Sato, Y., "On the Theory of Elastic Waves in Granular Substance I," Bull. Earthquake Res. Inst. Vol.27 pp 37-43 (1949)
12. Makhlouf, H.M. and Stewart, J.J., "Elastic Constants of Cubical and Tetragonal-Sphenoidal Arrays of Uniform Spheres," Proc. Intn. Symp. on Wave propagation and Dynamic Properties of Earth Materials, New Mexico, pp 825-836 (1967)
13 Marsal, R.J., "Mechanical Properties of Rockfill," Embankment Dam Engineering, pp 109-145 (1973)
14. Richart, F.E., Hall, J.R. and Woods, R.D., Vibration of Soils and Foundations, Prentice-Hall, Chap.6 (1970)
15. Hardin, B.O. and Richart, F.E., "Elastic Wave Velocities in Granular Soils," Journ. Soil Mech. & Found. Div., Proc. ASCE, Vol.89 No.SM1, pp33-41 (1963)
16. Drnevich, J.R., Hall, J.R. and Richart, F.E., "Effect of Amplitude of Vibration on the Shear Modulus of Sand," Proc. Intn. Conf. on Wave Propergation and Dynamic Properties of Earth Materials, New Mexico, pp189-199 (1967)
17. Hardin, B.O. and Black, W.L. "Vibration Modulus of Normally Consolidated Clay," Journ. Soil Mech. & Found. Div. Proc. ASCE, Vol. 94, pp 353-369 (1968)
18. Hardin, B.O. and Drnevich, V.P., "Shear Modulus and Damping in Soils," Journ. Soil Mech. & Found. Div. Proc. ASCE Vol.98 No.Sm6 pp 603-624 (1972)
19 Iwasaki, T. and Tatsuoka, F., "Effect of Grain Size and Grading on Dynamic Shear Moduli of Sand," Soils and Found. Vol.17 No.3 pp 19-35 (1977)
20. Hardin, B.O. and Black, W.L. "Sand Stiffness Under Various Triaxial Stresses," Journ. Soil Mech. & Found. Div. Proc. ASCE, Vol92 No. Sm2, pp 27-42 (1966)

Index to References

Match the numbers in the parentheses with the numbers following the author(s) name(s) in the authors index for the corresponding page number in the text.

Supplementary Index to References